MW01518902

FLUORESCENT ANALOGS OF BIOMOLECULAR BUILDING BLOCKS

FLUORESCENT ANALOGS OF BIOMOLECULAR BUILDING BLOCKS

Design and Applications

Edited by

PROF. DR. MARCUS WILHELMSSON
Department of Chemistry and Chemical Engineering
Chemistry and Biochemistry
Chalmers University of Technology
Kemivägen 10
SE-412 96 Göteborg
Sweden

PROF. DR. YITZHAK TOR
Department of Chemistry and Biochemistry
University of California, San Diego,
9500 Gilman Drive, 0358
La Jolla, CA 92093
USA

Library of Congress Cataloging-in-Publication Data has been applied for

ISBN - 9781118175866

Typeset in 10/12pt TimesLTStd by SPi Global, Chennai, India

10 9 8 7 6 5 4 3 2 1

CONTENTS

13 Site-Specific Fluorescent Labeling of Nucleic Acids by Genetic Alphabet Expansion Using Unnatural Base Pair Systems   297

Michiko Kimoto, Rie Yamashige, and Ichiro Hirao

14 Fluorescent C-Nucleosides and their Oligomeric Assemblies   320

Pete Crisalli and Eric T. Kool

17 Fluorescent Neurotransmitter Analogs   393

James N. Wilson

Index   409

LIST OF CONTRIBUTORS

Bo Albinsson, Chemistry and Chemical Engineering/Chemistry and Biochemistry, Chalmers University of Technology, Gothenburg, Sweden

Bruce A. Armitage, Department of Chemistry and Molecular Biosensor and Imaging Center, Carnegie Mellon University, Pittsburgh, PA, USA

Subhendu S. Bag, Department of Chemistry, Indian Institute of Technology, Guwahati, India

Nediljko Budisa, Department of Chemistry, Berlin Institute of Technology/TU Berlin, Biocatalysis Group, Berlin, Germany

Marek Cebecauer, Department of Biophysical Chemistry, J. Heyrovský Institute of Physical Chemistry of the Academy of Sciences of the Czech Republic, Prague 8, Czech Republic

Amitabha Chattopadhyay, Centre for Cellular and Molecular Biology, Council of Scientific and Industrial Research, Hyderabad, India

Arunima Chaudhuri, Centre for Cellular and Molecular Biology, Council of Scientific and Industrial Research, Hyderabad, India

Kirby Chicas, Department of Chemistry, The University of Western Ontario, London, Canada

Pete Crisalli, Department of Chemistry, Stanford University, Stanford, CA, USA

Anaëlle Dumas, Department of Chemistry, University of Zurich, Zurich, Switzerland

Patrick M. Durkin, Department of Chemistry, Berlin Institute of Technology/TU Berlin, Biocatalysis Group, Berlin, Germany

Ichiro Hirao, Institute of Bioengineering and Nanotechnology (IBN), Singapore, Singapore

Robert H.E. Hudson, Department of Chemistry, The University of Western Ontario, London, Canada

Gregor Jung, Biophysical Chemistry, Saarland University, Saarbruecken, Germany

Takashi Kanamori, Education Academy of Computational Life Sciences, Tokyo Institute of Technology, Yokohama, Japan

Michiko Kimoto, Institute of Bioengineering and Nanotechnology (IBN), Singapore, Singapore

Eric T. Kool, Department of Chemistry, Stanford University, Stanford, CA, USA

Nathan W. Luedtke, Department of Chemistry, University of Zurich, Zurich, Switzerland

Guillaume Mata, Department of Chemistry, University of Zurich, Zurich, Switzerland

Bengt Nordén, Chemistry and Chemical Engineering/Chemistry and Biochemistry, Chalmers University of Technology, Gothenburg, Sweden

Akihiro Ohkubo, Department of Life Science, Tokyo Institute of Technology, Yokohama, Japan

Radek Šachl, Department of Biophysical Chemistry, J. Heyrovský Institute of Physical Chemistry of the Academy of Sciences of the Czech Republic, Prague 8, Czech Republic

Isao Saito, Department of Materials Chemistry and Engineering, School of Engineering, Nihon University, Koriyama, Japan

Kohji Seio, Department of Life Science, Tokyo Institute of Technology, Yokohama, Japan

Mitsuo Sekine, Department of Life Science, Tokyo Institute of Technology, Yokohama, Japan

Sandeep Shrivastava, Centre for Cellular and Molecular Biology, Council of Scientific and Industrial Research, Hyderabad, India

Renatus W. Sinkeldam, Department of Chemistry and Biochemistry, University of California, San Diego, La Jolla, CA, USA

Yitzhak Tor, Department of Chemistry and Biochemistry, University of California, San Diego, La Jolla, CA, USA

L. Marcus Wilhelmsson, Chemistry and Chemical Engineering/Chemistry and Biochemistry, Chalmers University of Technology, Gothenburg, Sweden

James N. Wilson, Department of Chemistry, University of Miami, Coral Gables, FL, USA

Rie Yamashige, RIKEN Center for Life Science Technologies (CLST), Yokohama, Kanagawa, Japan

PREFACE

Fluorescence spectroscopy, an established and highly sensitive analytical technique, has been extensively used by the scientific community for many years. For decades, however, the majority of users have relied on a limited number of established fluorophores, either naturally occurring or of synthetic origin. This has dramatically changed in recent years.

Major technological advances in fluorescence-based instrumentation and techniques, including single-molecule spectroscopy, have triggered a renewed interest in the synthesis and development of new fluorescent probes and labels. Two major paths have been taken that are fundamentally related to the above-mentioned two (i.e., of biosynthetic or synthetic origin) but differ in their accommodation of the challenges presented by modern techniques and contemporary scientific questions. A particularly intriguing and emerging area of research, which is highlighted in this book, is the fabrication of minimally perturbing fluorescent analogs of otherwise nonemissive biological building blocks, including amino acids, lipids, and nucleosides.

To share with the reader the renaissance in this field of fluorescent biomolecules and their building blocks, we open with a general and concise tutorial of fluorescence spectroscopy. As readers would appreciate, it is practically impossible to capture all the nuances associated with the development of new fluorescent probes in such a book. To partially correct for this "deficiency," the second chapter provides a condensed overview of naturally occurring and synthetic fluorescent biomolecular building blocks, addressing the core issues and key advances in this field. Selected topics are then elaborated on in individual chapters.

While most laboratories utilize steady state and perhaps basic time-resolved techniques, a great deal of information can be obtained from more sophisticated experiments. Albinsson and Nordén discuss the theory and applications of polarized light spectroscopy-based techniques and their application for the study of biomolecules. Such experiments can be done in bulk solution as well as in microscopy and single-molecule modalities to provide information about the separation and orientation of chromophores.

Before moving on to discuss new synthetic chromophores in later chapters, we first cover fluorescent proteins as they have become the cornerstone of modern biophysics. Two main approaches are typically considered. One relies on the genetic expression of the classical green fluorescent protein and its variants, where the chromophore is generated from the spontaneous condensation of naturally occurring amino acids as discussed by Jung. A distinct approach, presented by Durkin and Budisa, relies on the incorporation of intrinsically fluorescent noncanonical amino acids by *in vitro* translation techniques, which exploit an expanded genetic code. Both techniques are extremely powerful and provide experimentalists with an enhanced toolbox of emissive proteins, but rely on rather sophisticated biochemical techniques for protein expression. A simplified approach is discussed by Armitage, where genetically encoded antibody fragments and fluorogenic dyes assemble noncovalently to form bright fluorescent complexes.

One element, distinguishing protein biochemists from the community interested in nucleic acids is that, unlike aromatic amino acids that are emissive, the canonical DNA and RNA nucleosides are all practically nonemissive. This has triggered rather extensive efforts aimed at the synthesis and implementation of fluorescent nucleoside analogs. Several approaches are covered here. Saito and Bag discuss diverse families of solvatochromic nucleosides produced by either covalently linking known chromophores to the native nucleosides or by conjugating additional aromatic rings to the native nucleobases. Chicas and Hudson specifically discuss fluorescent cytidine analogs, with emphasis on pyrrolo-C and its derivatives, both in the context of oligonucleotides and in PNAs. Sekine and coworkers elaborate on another family of pyrimidine analogs built around the pyrimidopyrimidoindole motif. While diverse applications have previously been reported, the authors focus here on the implementation of this responsive family of emissive C analogs within triple-stranded motifs. In contrast to the responsive families of fluorescent C analogs mentioned above, Wilhelmsson describes a family of minimally responsive chromophores, which makes them ideal for FRET studies. Well-matched FRET pairs, unique among nucleoside analogs, can then be used to accurately assess nucleobase–nucleobase distance and orientation, generating high-resolution 3-D structural information.

Although the birth of fluorescent nucleoside analogs as a field is frequently attributed to Stryer's 1969 disclosure of 2-aminopurine, an archetypical and extensively employed emissive nucleoside, the number of newly developed and useful purine analogs is substantially smaller compared to their pyrimidine counterparts. This is partially due to synthetic considerations but also likely reflects that modifying the purine core, unlike that of the pyrimidines, frequently hampers their WC and Hoogsteen pairing abilities as well their accommodation within higher structures. In this context, Luedtke describes useful 8-modified purine analogs, which are exploited for the study of G-quadruplexes without detrimental structural effects. Sinkeldam and Tor then discuss the design and implementation of minimally perturbing yet responsive fluorescent nucleoside analogs, frequently referred to as isomorphic surrogates. Structural and functional elements imparting sensitivity to environmental factors (such as polarity, viscosity, and pH) are introduced into the nucleosidic skeleton with the smallest possible size and functional perturbation.

While all analogs described were designed to form WC pairs and be paired with their native complementary nucleobases, Hirao and coworkers discuss unnatural base pair systems, where both partners selectively recognize one another and discriminate against the canonical nucleobases. While some of the analogs made are in fact emissive, such selective pairing practically expands the genetic code and facilitates the incorporation of other bright fluorescent labels with high efficiency and selectivity. Deviating even further from the canonical structure of the native nucleosides, Crisalli and Kool replace the native heterocyclic nucleobases with aromatic fluorophores, while maintaining the phosphate–sugar backbone. Due to their chromophore–chromophore interactions, such DNA-like oligomers, coined fluorosides, display unique photophysical features and provide a fertile motif for the combinatorial discovery of new sensors and labels.

Similarly to the biomolecular building blocks of proteins and nucleic acids, the majority of membrane components are nonemissive. Designing emissive analogs to study these unique assemblies imposes certain structural and functional issues. Chattopadhyay and colleagues review several popular membrane probes and highlight their potential for extracting information on the environment, organization, and dynamics of membranes. Cebecauer and Šachl then take a rather comprehensive look at diverse fluorescent probes that have been developed to assess lipid phases and their separation, membrane viscosity, and curvature as well as pH and potential. They conclude by discussing future directions and cell biology questions that may be addressed in future using lipophilic fluorescent probes.

We conclude this book with a rather unique chapter discussing small fluorophores that don't serve as components of higher molecular weight biomolecules or assemblies. Wilson discusses the design and utility of fluorescent neurotransmitter analogs as tools for exploring neurotransmission and its regulation. Such analogs can be used to investigate receptors, enzymes, and transporters that interact with native neurotransmitters.

As most readers appreciate, contemporary fluorescence spectroscopy, with all its experimental variations, touches numerous and very diverse fields. Yet, with all the technological advances, in its most fundamental level, this amazing spectroscopy relies on the availability of suitably designed fluorescent probes. The creative and elegant approaches presented here highlight how judiciously designed and implemented fluorescence probes could significantly promote advances in biophysics, biochemistry, and structural biology. What is perhaps less obvious is that the design and implementation of such probes remains an empirical exercise. Our ability to predict the intricate photophysical features of designer probes and their response to diverse environmental effects is still rather primitive and, for the most part, qualitative. It is likely (and it is certainly our hope) that computational approaches developed in coming years will refine the experimentalists' approach, which frequently relies on trial and error. Nevertheless, as evidenced by two Nobel prizes awarded in recent years

(R. Y. Tsien, M. Chalfie, and O. Shimomura in 2008 and W. E. Moerner, S. W. Hell, and E. Betzig in 2014), fluorescence spectroscopy continues to pave the road forward in critical scientific disciplines. We hope that this book inspires the next generation of young scientists to dive into this fascinating field and spend their creative years ensuring that the future of this field remains bright and colorful!

Assembling such a collection of quality chapters, as any editor knows, takes far longer than originally expected and planned. It requires the ultimate cooperation of authors, reviewers, and publishers. We thank them all. We feel the end product is clearly worth the effort and wait.

Marcus Wilhelmsson, Chalmers University of Technology, Gothenburg, Sweden
Yitzhak Tor, University of California, San Diego, La Jolla, CA, USA

1

FLUORESCENCE SPECTROSCOPY

RENATUS W. SINKELDAM

Department of Chemistry and Biochemistry, University of California, San Diego, CA, USA

L. MARCUS WILHELMSSON

Chemistry and Chemical Engineering/Chemistry and Biochemistry, Chalmers University of Technology, Gothenburg, Sweden

YITZHAK TOR

Department of Chemistry and Biochemistry, University of California, San Diego, CA, USA

1.1 FUNDAMENTALS OF FLUORESCENCE SPECTROSCOPY

Fluorescence spectroscopy is unique in its combination of sensitivity with experimental versatility. While all optical spectroscopy techniques benefit from the very short timescale of the photon absorption and emission sequence (Fig. 1.1), an additional and major advantage of fluorescence spectroscopy is the energy difference in the wavelength of excitation and emission. Unlike UV–vis or infrared spectroscopy, where the minimal loss of incident light intensity due to sample absorption is measured, fluorescence spectroscopy yields an energetically distinct signal, frequently remote from, and therefore free of interference by the excitation wavelength (Fig. 1.1).

In short, light of an appropriate energy, the excitation wavelength, elevates a chromophore to the Franck–Condon state, normally a higher vibrational level of S_1, S_2 or higher (S_n) within 10^{-15} s. This extremely fast process is followed by internal conversion (ic) and vibrational relaxation (vr) within 10^{-12}–10^{-10} s to the lowest vibronic and potentially emissive S_1 state. Due to these processes, there is an energy difference

Fluorescent Analogs of Biomolecular Building Blocks: Design and Applications, First Edition.
Edited by Marcus Wilhelmsson and Yitzhak Tor.
© 2016 John Wiley & Sons, Inc. Published 2016 by John Wiley & Sons, Inc.

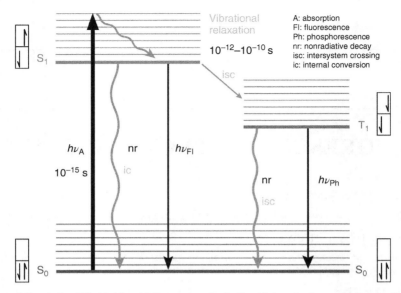

Figure 1.1   A simplified Jablonski diagram not including higher singlet excited states than S_1.

between the photons required for excitation and the photons emitted. This difference ($\nu_{abs} - \nu_{em}$), typically expressed in cm^{-1}, is called the Stokes shift and is an intrinsic property of a fluorophore in a given set of conditions. The excited state lifetime (τ), with typical values of 0.5–20 ns for organic fluorophores, is the result of the sum of all nonradiative (k_{nr}) and radiative (Γ) decay rates reflecting the processes returning the fluorophore to its ground state (Eq. 1.1).

$$\tau = \frac{1}{\Gamma + k_{nr}} \qquad (1.1)$$

Several factors impact the potential utility of any fluorophore. The efficiency of the excitation process is dependent on the chromophore's molar absorptivity (ε), which itself is proportional to the cross section (σ). The efficiency of the emission process, the fluorescence quantum yield (Φ), reflects the fraction of emitted photons with respect to the absorbed ones. Expressed in rate constants, the quantum yield (Φ) is determined by the radiative rate constant (Γ) over the sum of the radiative (Γ) and all nonradiative (k_{nr}) rates (Eq. 1.2):

$$\Phi = \frac{\Gamma}{\Gamma + k_{nr}} \qquad (1.2)$$

The combined efficiency of the excitation and emission is expressed by the brightness ($\varepsilon \times \Phi$), the product of molar absorptivity (ε), and fluorescence quantum yield (Φ). Hence, poorly emissive fluorophores can still enjoy sufficient brightness if their low quantum yield is compensated by a high molar absorptivity. Or, *vice versa*, highly emissive fluorophores possessing high quantum yields can still suffer from low brightness due to a low molar absorptivity.

Note that a spin forbidden additional pathway, named intersystem crossing (isc), populates the much longer lived triplet (T_1) state. The generally slow radiative decay from T_1 to the ground state is known as phosphorescence and not further discussed here (Fig. 1.1).

The advent of relatively affordable, robust, yet sophisticated, benchtop fluorimeters in conjunction with the vast and growing number of commercially available fluorescent probes have contributed to the accessibility and popularity of fluorescence spectroscopy. It has become one of the most important analytical techniques for the *in vitro* study of biomolecules and *in vivo* cellular imaging, providing spatial and temporal information.[1,2] The "*in situ*" study of intricate and large biomolecules in their complex environment is further facilitated by exclusive excitation of fluorescent probes to minimize background emission. Provided noninterfering probes are used, the inherently nonperturbing fluorescence measurement delivers valuable insights into biomolecules in their native environments.

The fundamentals of excitation and emission, as depicted in the simplified Jablonski diagram, form the foundation of any fluorescence technique (Fig. 1.1). The versatility ranges from exotic one-of-a-kind studies requiring very sophisticated instrumentation to straightforward, but yet very informative, techniques available on most modern benchtop fluorimeters. The majority of techniques commonly used in the study of biomolecules fall in the latter category and are briefly discussed in the following section.[1,3–5] For more specialized fluorescence and microscopy techniques, the reader is recommended to turn to other chapters in this book or to journal articles focused on a certain technique.

1.2   COMMON FLUORESCENCE SPECTROSCOPY TECHNIQUES

1.2.1   Steady-State Fluorescence Spectroscopy

The quintessential fluorescence-based technique is steady-state fluorescence spectroscopy. The emission spectrum of a fluorophore is recorded upon excitation with a constant photon flux light source (e.g., a xenon arc lamp), typically at its absorption maximum or where it can be selectively excited if other chromophores are present. The fluorescence spectrum obtained provides the fluorophore's emission signature, its wavelength-dependent emission intensity, and emission maximum (see example in Fig. 1.2). Instrument settings and detector sensitivity aside, the emission intensity is dependent on the fluorescence quantum yield of the fluorophore and is proportional to its concentration provided sufficiently dilute samples are used (absorbance <0.05). The emission maximum is an inherent property of the fluorophore but could

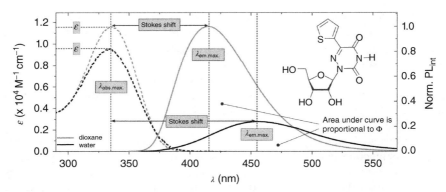

Figure 1.2 Absorption (dashed lines) and fluorescence (solid lines) spectra of 5-(thiophen-2-yl)-6-aza-uridine in water (black) and dioxane (gray). Annotations illustrate the most important parameters that can be obtained. The difference in Stokes shift in water and dioxane reveals the environmental polarity sensitivity of this isomorphic fluorescent nucleoside. *Note:* Stokes shifts are typically reported in energy units, commonly cm^{-1}.

be highly dependent on its immediate environment and subject to diverse effects (e.g., solvent polarity, viscosity, pH). The emission intensity measured in steady state can be used to estimate the fluorophore's quantum yield (Φ) using reference fluorophores with known quantum yields emitting at similar wavelengths as the fluorophore under investigation and the same instrument settings.[1]

Fluorophores possessing a different dipole moment in their excited state compared to the ground state frequently reveal sensitivity to environmental polarity. This behavior, termed solvatochromism, results from a solvent's ability to accommodate and thereby lower the fluorophore's excited state energy by solvent molecule rearrangement. By definition, a fluorophore is said to show positive solvatochromism if the emission maximum undergoes a bathochromic (to longer wavelength) shift upon increasing solvent polarity and negative solvatochromism if the emission maximum undergoes a hypsochromic (to shorter wavelength) shift. While potentially complex and subjected to artifacts, fluorogenic probes possessing such traits have been used to examine local polarity in biomolecules, including DNA,[6-10] proteins,[11-15] and membranes.[16-19]

Traditionally, polarity has been expressed using dielectric constants (ε), a parameter reflecting bulk property, and its derived orientational polarizability (Δf).[20,21] Newer microenvironmental polarity parameters (e.g., Reichardt's $E_T(30)$ scale), utilizing zwitterionic solvatochromic chromophores with a polarity-sensitive ground state and hence absorption maximum, enable polarity measurements on the molecular level.[22] This is especially relevant for probing biomolecular cavities, environments that deviate significantly in polarity from the aqueous bulk. In comparison to the dielectric constant and orientational polarizability, the $E_T(30)$ scale typically better describes changes in spectral phenomena, like Stokes shift, as a response to changing solvent polarity.[23]

1.2.2 Time-Resolved Fluorescence Spectroscopy

Despite the increased complexity and the sophisticated optics and electronics required, the additional layer of information obtained from time-resolved fluorescence experiments makes it complementary to steady-state spectroscopy. The informational content in a steady-state fluorescence spectrum is limited to an averaged emission profile of the entire population of excited fluorophores. Distinguishing between individual fluorophores in a heterogeneous sample and/or the same kind of fluorophore experiencing different local environments is therefore not possible. In such cases, time-resolved measurements are frequently invaluable. In its simplest form, a time-resolved fluorescence measurement gives a monoexponential decay curve from which the concentration-independent fluorescence lifetime can be calculated. This is an important parameter since it reflects the time available for a chromophore to diffuse or interact with its environment in its excited state. Hence, time-resolved fluorescence spectroscopy has the potential to provide insight into the excited state dynamics of a chromophore by comparing its lifetime under different experimental conditions to its natural lifetime (τ_n). The latter is the fluorescence lifetime (Eq. 1.1) in the absence of nonradiative processes (Eq. 1.3). Albeit complex, the radiative decay (Γ) rate can be calculated from the absorption spectrum, the molar absorptivity, and the emission spectrum of the chromophore.[1]

$$\tau_n = \frac{1}{\Gamma} \tag{1.3}$$

In most biophysical studies, where the binding, structure, and folding of biomolecules are studied, fluorescent probes could simultaneously exist in different environments. Each environment, bound/unbound, exposed to/shielded from solvent, likely has a unique influence on the fluorophore's excited state and is reflected by changes in emission maximum, quantum yield, and fluorescence lifetime. In contrast to steady-state fluorescence spectroscopy, time-resolved fluorescence analysis can facilitate the simultaneous analysis of multiple emissive states with overlapping spectral bands, each with its own fluorescence decay, by deconvolution of a sample's multiexponential decay curve. For example, the folding of an enzyme containing two emissive tryptophan residues might position each in a different local environment, a situation likely undistinguishable with steady-state fluorescence spectroscopy. A time-resolved fluorescence measurement, however, will likely give a biexponential intensity decay with a different contribution for each tryptophan residue. Changes in the relative contributions upon interaction of the enzyme with its substrate may reveal which tryptophan residue is most affected by the binding event, thereby revealing its proximity to the binding site.

Quenching experiments also greatly benefit from time-resolved fluorescence measurements by distinguishing between static (ground-state complex formation) and collisional (diffusion) quenching. In the former, the fluorescence lifetime is unaffected, whereas collisional quenching does affect the lifetime. Similarly, analysis of time-resolved fluorescence spectra, when applied to resonance energy transfer (RET) studies (*vide infra*), reveals whether all, or a subset of donors, engage in the RET process. See the following additional discussion.

1.2.3   Fluorescence Anisotropy

In most common solution phase fluorescence-based experiments, a fluorophore is excited with unpolarized light and the emission is measured without polarization. When a fluorophore is excited with polarized light, the emission remains polarized if the chromophore's Brownian motion, or tumbling, in the excited state prior to emissive decay to the ground state is slower than the excited state lifetime. A small molecule fluorophore in a nonviscous environment of ambient temperature typically has a tumbling rate faster than its fluorescence lifetime. Hence, if excited with polarized light under such conditions, the resulting emission will be completely depolarized and isotropic. If the fluorophore, however, is attached to a large (bio)molecule (e.g., a protein), or exposed to a highly viscous medium, the fluorophore's tumbling rate will slow down. The emission retains, at least in part, the polarized excitation if the tumbling rate is slower than the fluorescence lifetime. The extent of fluorescence polarization (P) is then calculated using Equation 1.4. Herein, $I_\parallel$ and $I_\perp$ stand for parallel and perpendicular polarized emission intensity, respectively.

$$P = \frac{I_\parallel - I_\perp}{I_\parallel + I_\perp} \tag{1.4}$$

Polarization (P) is interchangeable with anisotropy (r) (Eq. 1.5) since both are expressions of the same phenomenon.

$$r = \frac{I_\parallel - I_\perp}{I_\parallel + 2I_\perp} \tag{1.5}$$

To measure fluorescence anisotropy, excitation and emission polarizers have to be installed in a standard steady-state fluorescence spectrometry setup. In a tandem fluorescence experiment, a fluorophore is excited with vertically polarized light and the intensity of its vertically polarized emission is recorded. This is followed by a second vertically polarized excitation, but now the intensity of the horizontally polarized emission is recorded. To take the instrumental properties into account, one has to also measure horizontal excitation polarization combined with horizontal and vertical emission polarization, respectively (G-factor).[1] The fluorescence lifetime of the fluorophore plays a crucial role in the sensitivity of the fluorescence anisotropy experiment. In a biomolecular binding study, the fluorescence lifetime of the fluorophore needs to be sufficiently long to give a close-to-zero anisotropy when unbound. As a result, a significant drop in the tumbling rate due to binding to a much larger biomolecule (e.g., a protein) yields a maximum retention of polarization.

In addition to the aforementioned biomolecular binding studies, fluorescence anisotropy has found use in protein dynamics,[24,25] as well as in studying protein–protein[26] and protein–nucleic acid interactions.[27–29] Fluorescence anisotropy is also used in membrane fluidity and microviscosity studies,[30–32] and to determine aqueous bulk-membrane partition coefficients of fluorescent probes.[33] The fundamentals and various applications of fluorescence anisotropy including time-resolved fluorescence anisotropy have been the topic of selected recent reviews.[34,35]

1.2.4   Resonance Energy Transfer and Quenching

Steady-state fluorescence spectroscopy, time-resolved fluorescence spectroscopy, and fluorescence anisotropy, as described above, are typically, although not necessarily, concerned with monitoring a single fluorescent probe. Fluorescence techniques that exploit interactions between chromophores (such as a fluorophore and a quencher or a fluorophore and another distinct fluorophore) are extremely powerful and have been widely used in the study of biomolecules.

Valuable molecular information can be obtained from two commonly studied quenching mechanisms: dynamic and static. The former, also called collisional, quenching, is described by a linear relationship between the quenching effect and the quencher concentration as defined by the Stern–Volmer equation (Eq. 1.6),[36] and its modification, the Lehrer equation.[37] In Equation 1.6, $[Q]$ is the quencher concentration and F_0 and F are the fluorescence intensities in the absence and presence of quencher, respectively. The bimolecular quenching constant, fluorescence lifetime of the fluorophore in the absence of quencher, and the Stern–Volmer quenching constant are denoted by k_q, τ_0, and K_D, respectively.

$$\frac{F_0}{F} = 1 + k_q\tau_0[Q] = 1 + K_D[Q] \tag{1.6}$$

Hence, the Stern–Volmer quenching constant is given by Equation 1.7.

$$K_D = k_q\tau_0 \tag{1.7}$$

Deviation from linearity implies the contribution of static quenching due to formation of a ground-state complex between the fluorophore and quencher, as stated by the Perrin model.[38] The utility of fluorescence quenching experiments is illustrated in selected recent reviews for multiple fields including the study of RNA folding, dynamics, and hydridization[39–41] protein folding, structure and dynamics,[42,43] protein–membrane interactions,[44–46] and membrane microdomains.[47–50] Nevertheless, extracting molecular information from fluorescence quenching studies can be challenging since apparent quenching can also result from unrelated technical issues, such as the sample's turbidity or high optical density.

The disadvantage associated with quenching experiments can be largely overcome by exploiting RET, a nonradiative process between two molecular entities typically referred to by donor and acceptor. The RET process is facilitated by either a Dexter or Förster mechanism. The Dexter mechanism requires orbital overlap and hence close proximity to the donor and the acceptor. In contrast, the Förster process, based on a dipole–dipole coupling between a donor and an acceptor, operates over larger distances and is viable when there is significant spectral overlap of donor emission with acceptor absorption. The size of biomolecules (30–60 Å)[1] is in the same range as the Förster critical distance of many D/A pairs, the distance at which the energy transfer efficiency is 50%. This makes Förster resonance energy transfer (FRET) broadly applicable for biomolecular studies. As in quenching studies, the emission of the donor fluorophore is quenched, but a sensitized acceptor emission at a longer wavelength is frequently observed.

Prior to any FRET experiments, the Förster distance (R_0) in Å, for the donor–acceptor pair used, must be calculated using values for donor quantum yield (Φ_D), spectral overlap of donor emission and acceptor absorption $(J(\lambda))$, relative orientation of donor and acceptor (κ^2), and the refractive index of the medium (n) (Eq. 1.8).

$$R_0 = 0.211(\kappa^2 n^{-4}\Phi_D J(\lambda))^{1/6} \tag{1.8}$$

The rate of the energy transfer process, $k_{ET}(r)$, can now be calculated based on the distance between the donor and the acceptor (r), the fluorescence lifetime of the donor in absence of the acceptor (τ_D), and the Förster distance (R_0) (Eq. 1.9).

$$k_{ET}(r) = \frac{1}{\tau_D}\left(\frac{R_0}{r}\right)^6 \tag{1.9}$$

The efficiency of the RET process can be expressed as the Förster distance (R_0) over the sum of the Förster distance (R_0) and the donor–acceptor distance (r) (Eq. 1.10).

$$E = \frac{R_0^6}{R_0^6 + r^6} \tag{1.10}$$

In general, $\kappa^2 = 2/3$, reflecting random interchromophore orientation, is used when calculating R_0. It must be noted, however, that in certain cases this assumption could be a crude oversimplification leading to false interpretations.[51] Moreover, by making the approximation that $\kappa^2 = 2/3$, the opportunity is lost to get orientational information about the system under study.[52] With the Förster distance (R_0), determined for the chosen FRET pair (Eq. 1.8), the FRET efficiency is strongly dependent on the distance between the donor and acceptor (Eq. 1.10). This enables calculation of the distance (r) between donor and acceptor sites on a macrobiomolecule or its complexes, provided the distance does not change during the excited state lifetime.[1] Therefore, the FRET phenomenon has been termed a "spectral ruler."[53,54]

FRET is not limited to distance measurements in biomolecules but can also be used in binding, folding, and hybridization studies. Because of its wide applicability, FRET measurements have been used, for example, in membrane research to study microdomain formation[47] and transmembrane peptides in surface-supported bilayers.[55] Applications of FRET in nucleic acid research have been widely described in selected reviews on structure, folding, hybridization, and dynamics of RNA[41,56–58] and the sequence-dependent structure, stability, and dynamics of nucleosomes.[59] FRET measurements have also been exploited to investigate protein folding, protein–protein interactions, and cellular signaling events in live cells.[42,60–62]

1.2.5 Fluorescence Microscopy and Single Molecule Spectroscopy

Advancement in instrumentation and increased availability of bright (and sometimes organelle specific dyes) fluorophores have led to increased sensitivity of fluorescence-based spectroscopy techniques. Developments in cellular visualization include total internal reflection (TIRF), confocal, and two- or multiphoton fluorescence spectroscopy.[63–68] Another fairly recent development is fluorescence lifetime imaging microscopy (FLIM), where a fluorescent probe is used to stain a biological sample (e.g., a cell).[50,69,70] Image contrast is based on differences in fluorescence lifetime as a result of probe distribution over multiple unique locations (e.g., cellular components).

Further advancements in single photon excitation in the late 90s of the last century have led to single molecule spectroscopy, the ability to follow the emission of just one molecule at a time facilitated by optical "tweezers" or trapping.[71–74] The magnitude of this achievement is easily appreciated by the realization that a typical 1 mL, 1 μM fluorescent probe sample contains (1×10^{-9} mol fluorescent probe * 6.02×10^{23} (Avogadro's number)) $\sim 6 \times 10^{14}$ fluorescent molecules! As outlined in the previous section, the averaged emission profile of this unfathomable number of fluorescent probes is informative and sufficient for numerous studies. The ability to follow complex biological processes at the single molecule level, however, is greatly beneficial. An example of such a complex process is the conformational changes a ribosome undergoes during the translation of messenger RNA into proteins.[75] The development of single molecule spectroscopy has benefitted virtually all areas of biomolecular research as described in selected reviews.[42,76–83] The advent of single molecule spectroscopy was quickly exploited to enable single pair FRET studies.[72] Such studies proved instrumental in the areas of nucleic acid (DNA/RNA) structure, folding, and dynamics,[84–87] DNA–protein interactions,[88,89] and nucleosome conformations.[90,91] It must be noted here that the conjunction of the discovery[92] and development[93] of the highly emissive green fluorescent protein (GFP), with the advancement of single molecule spectroscopy forged one of the most useful tools in modern biology.[94]

1.2.6 Fluorescence-Based *in vivo* Imaging

Arguably, the pinnacle of fluorescence spectroscopy applications in the life sciences is fluorescence-based *in vivo* imaging. This is the most recent addition to invaluable existing imaging techniques including X-ray, positron emission tomography (PET), ultrasound, and magnetic resonance imaging (MRI). Besides potential cost reduction, development of fluorescence-based imaging techniques brings the advantage of improved resolution and contrast. An additional benefit of a fluorescence-based approach, which is lacking in established imaging techniques, is the potential of fluorescent probes to respond in real time to specific physiological changes.[95]

The majority of fluorescent probes absorbs and emits in the ultraviolet and visible domain of the electromagnetic spectrum. To efficiently penetrate through living tissue, avoiding absorption by water, lipids, as well as[96] oxy- and deoxyhemoglobin, light of near-infrared (NIR) wavelengths (700–1000 nm) is used.[97–99] Hence, ideal fluorescent probes for *in vivo* imaging combine a low-energy excitation wavelength with a large Stokes shift. Diverse examples of probes suitable for, but not limited to, *in vivo* use exist[100,101] and include modified nucleosides[102–104] and amino acids[105,106] in addition to dendrimers, nanoparticles, and quantum dots.[106,107] Most promising is the development of (near)infrared fluorescent proteins (IFPs) with a recent example characterized by an excitation maximum of 684 nm ($\varepsilon > 90,000\,M^{-1}\,cm^{-1}$), emission maximum of 708 nm and a quantum yield of 0.07.[108] Fortunately, the long wavelength excitation required to excite NIR probes is deemed safe, making whole body fluorescence tomography an exciting prospect.[109] Alternatively, to minimize absorption by the surrounding tissue, suitable short wavelength absorbing fluorophores can be subjected to two- or multiphoton excitation using long-wavelength laser excitation.[110–113]

Besides probe development, technological improvements have also contributed to the advancement of fluorescence-based imaging techniques. For instance, differences in fluorescence lifetimes enabled isolation of probe emission from emission of the surrounding tissue.[114,115] The emergent field of NIR fluorescent probes and their *in vivo* imaging applications has been the topic of several reviews.[95,99,109,116–118]

1.3 SUMMARY AND PERSPECTIVE

In the preceding sections, the most commonly used fluorescence spectroscopy techniques are discussed. Their importance and applicability in research areas involving biomolecular building blocks is illustrated with selected examples. Due to the vast scope of fluorescence techniques available, this chapter cannot be comprehensive. Important developments not mentioned here include, for example, fluorescence correlation spectroscopy (FCS). This technique is based on fluctuations of fluorescently labeled compounds (e.g., biomolecular building blocks) in very small volumes. FCS is most useful in the study of dynamic molecular processes in living cells (e.g., diffusion, ligand–protein, protein–protein, and protein–DNA interactions).[119–122]

The next chapter discusses the fundamental features of the native fluorophores found in biomolecules, followed by a concise overview outlining the development of fluorescent analogs of fluorescent building blocks. Each of the following chapters discusses a specific use of such fluorescent analogs in conjunction with fluorescence spectroscopy. Together, these chapters illustrate not only the diversity in fluorescence techniques used but also the plethora of research areas that greatly benefit from it. Undoubtedly, the desire to explore new research areas has pushed the technological development of fluorescence instrumentation. These advancements, *vice versa*, enabled exploration of new scientific frontiers. Perhaps we find ourselves at the mere beginning with many exciting fluorescence-based discoveries ahead of us.

REFERENCES

1. Lakowicz, J. R. Principles of Fluorescence Spectroscopy. 3rd ed.; Springer: New York, **2006**.
2. Bacia, K.; Schwille, P. *Methods* **2003**, *29*, 74.
3. Lakowicz, J. R. *J. Biochem. Biophys. Methods* **1980**, *2*, 91.

4. Valeur, B. Molecular Fluorescence, Principles and Applications. Wiley-VCH: Weinheim, **2002**.

5. Turro, N. J. Modern Molecular Photochemistry of Organic Molecules. University Science Books: New York, **2009**.

6. Jin, R.; Breslauer, K. J. *Proc. Natl. Acad. Sci. U. S. A.* **1988**, *85*, 8939.

7. Jadhav, V. R.; Barawkar, D. A.; Ganesh, K. N. *J. Phys. Chem. B* **1999**, *103*, 7383.

8. Kimura, T.; Kawai, K.; Majima, T. *Org. Lett.* **2005**, *7*, 5829.

9. Okamoto, A.; Tainaka, K.; Saito, I. *Bioconjug. Chem.* **2005**, *16*, 1105.

10. Sinkeldam, R. W.; Greco, N. J.; Tor, Y. *ChemBioChem* **2008**, *9*, 706.

11. Schutz, C. N.; Warshel, A. *Proteins* **2001**, *44*, 400.

12. Cohen, B. E.; McAnaney, T. B.; Park, E. S.; Jan, Y. N.; Boxer, S. G.; Jan, L. Y. *Science* **2002**, *296*, 1700.

13. Sundd, M.; Robertson, A. D. *Nat. Struct. Biol.* **2002**, *9*, 500.

14. Vazquez, M. E.; Rothman, D. M.; Imperiali, B. *Org. Biomol. Chem.* **2004**, *2*, 1965.

15. Kamal, J. K. A.; Zhao, L.; Zewail, A. H. *Proc. Natl. Acad. Sci. U. S. A.* **2004**, *101*, 13411.

16. Waka, Y.; Mataga, N.; Tanaka, F. *Photochem. Photobiol.* **1980**, *32*, 335.

17. Perochon, E.; Lopez, A.; Tocanne, J. F. *Biochemistry* **1992**, *31*, 7672.

18. Bernik, D. L.; Negri, R. M. *J. Colloid Interface Sci.* **1998**, *203*, 97.

19. Saxena, R.; Shrivastava, S.; Chattopadhyay, A. *J. Phys. Chem. B* **2008**, *112*, 12134.

20. Lippert, E. Z. *Elektrochem.* **1957**, *61*, 962.

21. Mataga, N.; Kaifu, Y.; Koizumi, M. *Bull. Chem. Soc. Jpn.* **1956**, *29*, 465.

22. Reichardt, C. *Chem. Rev.* **1994**, *94*, 2319.

23. Sinkeldam, R. W.; Tor, Y. *Org. Biomol. Chem.* **2007**, *5*, 2523.

24. Bucci, E.; Steiner, R. F. *Biophys. Chem.* **1988**, *30*, 199.

25. Yengo, C. M.; Berger, C. L. *Curr. Opin. Pharmacol.* **2010**, *10*, 731.

26. Yan, Y. L.; Marriott, G. *Curr. Opin. Chem. Biol.* **2003**, *7*, 635.

27. LiCata, V. J.; Wowor, A. J. In Biophysical Tools for Biologists – *In Vitro* Techniques: Applications of fluorescence Anisotropy to the study of Protein–DNA interactions; Academic Press: Amsterdam, **2008**; pp 243.

28. Anderson, B. J.; Larkin, C.; Guja, K.; Schildbach, J. F. *Methods Enzymol.* **2008**, *450*, 253.

29. Gilbert, S. D.; Batey, R. T., Riboswitches, Methods and Protocols – Monitoring RNA–Ligand Interactions Using Isothermal Titration Calorimetry. Humana Press: Totowa, NJ, **2009**.

30. Lentz, B. R. *Chem. Phys. Lipids* **1993**, *64*, 99.

31. Mykytczuk, N. C. S.; Trevors, J. T.; Leduc, L. G.; Ferroni, G. D. *Prog. Biophys. Mol. Biol.* **2007**, *95*, 60.

32. Davenport, L.; Targowski, P. *Biophys. J.* **1996**, *71*, 1837.

33. Santos, N. C.; Prieto, M.; Castanho, M. *Biochim. Biophys. Acta.* **2003**, *1612*, 123.

34. Jameson, D. M.; Ross, J. A. *Chem. Rev.* **2010**, *110*, 2685.

35. Gradinaru, C. C.; Marushchak, D. O.; Samim, M.; Krull, U. J. *Analyst* **2010**, *135*, 452.

36. Stern, O.; Volmer, M. *Phys. Z.* **1919**, *20*, 183.

37. Lehrer, S. S. *Biochemistry* **1971**, *10*, 3254.

38. Perrin, F. *C.R. Hebd. Seances Acad. Sci. 1924*, 178, **1978**.

39. Tinsley, R. A.; Walter, N. G. *RNA* **2006**, *12*, 522.

40. Silverman, A. P.; Kool, E. T. *Adv. Clin. Chem.* **2007**, *43*, 79.

41. Guo, J.; Ju, J.; Turro, N. *Anal. Bioanal. Chem.* **2012**, *402*, 3115.

42. Michalet, X.; Weiss, S.; Jager, M. *Chem. Rev.* **2006**, *106*, 1785.

43. Matyus, L.; Szollosi, J.; Jenei, A. *J. Photochem. Photobiol. B* **2006**, *83*, 223.

44. Rawat, S. S.; Kelkar, D. A.; Chattopadhyay, A. *Biophys. J.* **2004**, *87*, 831.

45. Munishkina, L. A.; Fink, A. L. *Biochim. Biophys. Acta.* **2007**, *1768*, 1862.

46. Heuck, A.; Johnson, A. *Cell Biochem. Biophys.* **2002**, *36*, 89.

47. Silvius, J. R.; Nabi, I. R. *Mol. Membr. Biol.* **2006**, *23*, 5.

48. Heberle, F. A.; Buboltz, J. T.; Stringer, D.; Feigenson, G. W. *Biochim. Biophys. Acta.* **2005**, *1746*, 186.

49. London, E.; Brown, D. A.; Xu, X. *Methods Enzymol.* **2000**, *312*, 272.

50. Almeida de, R. F. M.; Loura, L. M. S.; Prieto, M. *Chem. Phys. Lipids* **2009**, *157*, 61.

51. Dale, R. E.; Eisinger, J. *Biopolymers* **1974**, *13*, 1573.

52. Börjesson, K.; Preus, S.; El-Sagheer, A. H.; Brown, T.; Albinsson, B.; Wilhelmsson, L. M. *J. Am. Chem. Soc.* **2009**, *131*, 4288.

53. Förster, T. *Ann. Phys.* **1948**, *437*, 55.

54. Stryer, L. *Annu. Rev. Biochem.* **1978**, *47*, 819.

55. Merzlyakov, M.; Li, E.; Hristova, K. *Biointerphases* **2008**, *3*, FA80.

56. Li, P. T. X.; Vieregg, J.; Tinoco, I. *Annu. Rev. Biochem.* **2008**, *77*, 77.

57. Wilhelmsson, L. M. *Q. Rev. Biophys.* **2010**, *43*, 159.

58. Preus, S.; Wilhelmsson, L. M. *ChemBioChem* **2012**, *13*, 1990.

59. Kelbauskas, L.; Woodbury, N.; Lohr, D. *Biochem. Cell Biol.* **2009**, *87*, 323.

60. Schuler, B.; Eaton, W. A. *Curr. Opin. Struct. Biol.* **2008**, *18*, 16.

61. Ziv, G.; Haran, G. *J. Am. Chem. Soc.* **2009**, *131*, 2942.

62. Prinz, A.; Reither, G.; Diskar, M.; Schultz, C. *Proteomics* **2008**, *8*, 1179.

63. Molitoris, B. A.; Sandoval, R. M. *Adv. Drug Deliv. Rev.* **2006**, *58*, 809.

64. Benninger, R. K. P.; Hao, M.; Piston, D. W. *Rev. Physiol. Biochem. Pharmacol.* **2008**, *160*, 71.

65. Kapanidis, A. N.; Strick, T. *Trends Biochem. Sci.* **2009**, *34*, 234.

66. Bagatolli, L. A. *Biochim. Biophys. Acta.* **2006**, *1758*, 1541.

67. Diaspro, A.; Chirico, G.; Collini, M. *Q. Rev. Biophys.* **2005**, *38*, 97.

68. Ishikawa-Ankerhold, H. C.; Ankerhold, R.; Drummen, G. P. C. *Molecules* **2012**, *17*, 4047.

69. Levitt, J. A.; Matthews, D. R.; Ameer-Beg, S. M.; Suhling, K. *Curr. Opin. Biotechnol.* **2009**, *20*, 28.

70. Borst, J. W.; Visser, A. J. W. G. *Meas. Sci. Technol.* **2010**, *21*, 1.

71. Funatsu, T.; Harada, Y.; Tokunaga, M.; Saito, K.; Yanagida, T. *Nature* **1995**, *374*, 555.

72. Ha, T.; Enderle, T.; Ogletree, D. F.; Chemla, D. S.; Selvin, P. R.; Weiss, S. *Proc. Natl. Acad. Sci. U. S. A.* **1996**, *93*, 6264.

73. Funatsu, T.; Harada, Y.; Higuchi, H.; Tokunaga, M.; Saito, K.; Ishii, Y.; Vale, R. D.; Yanagida, T. *Biophys. Chem.* **1997**, *68*, 63.

74. Ashkin, A. *Proc. Natl. Acad. Sci. U. S. A.* **1997**, *94*, 4853.

75. Tinoco, I.; Gonzalez, R. L. *Genes Dev.* **2011**, *25*, 1205.

76. Mukhopadhyay, S.; Deniz, A. *J. Fluorescence* **2007**, *17*, 775.

77. Huang, B.; Bates, M.; Zhuang, X. *Annu. Rev. Biochem.* **2009**, *78*, 993.

78. Borgia, A.; Williams, P. M.; Clarke, J. *Annu. Rev. Biochem.* **2008**, *77*, 101.

79. Pljevaljcic, G.; Millar, D. P. *Methods Enzymol.* **2008**, *450*, 233.

80. Yang, H., *Curr. Opin. Chem. Biol.* **2010**, *14*, 3.

81. Selvin, P. R.; Ha, T. Single-Molecule Techniques, A Laboratory Manual. Cold Spring Harbor Laboratory Press: Cold Spring Harbor, NY, **2008**.

82. Hinterdorfer, P.; Oijen, A. Handbook of Single-Molecule Biophysics. Springer: New York, **2009**.

83. Bustamante, C. *Annu. Rev. Biochem.* **2008**, *77*, 45.

84. Krüger, A. C.; Hildebrandt, L. L.; Kragh, S. L.; Birkedal, V. *Methods Cell Biol.* **2013**, *113*, 1.

85. Zhao, R.; Rueda, D. *Methods* **2009**, *49*, 112.

86. Wilson, T. J.; Nahas, M.; Araki, L.; Harusawa, S.; Ha, T.; Lilley, D. M. J. *Blood Cells Mol. Dis.* **2007**, *38*, 8.

87. Klostermeier, D. *Biochem. Soc. Trans.* **2011**, *39*, 611.

88. Leuba, S.; Anand, S.; Harp, J.; Khan, S. *Chromosome Res.* **2008**, *16*, 451.

89. Lamichhane, R.; Solem, A.; Black, W.; Rueda, D. *Methods* **2010**, *52*, 192.

90. Deindl, S.; Zhuang, X. *Methods Enzymol.* **2012**, *513*, 59.

91. Buning, R.; van Noort, J. *Biochimie* **2010**, *92*, 1729.

92. Shimomura, O.; Johnson, F. H.; Saiga, Y. *J. Cell. Comp. Physiol.* **1962**, *59*, 223.

93. Heim, R.; Cubitt, A. B.; Tsien, R. Y. *Nature* **1995**, *373*, 663.

94. Tsien, R. Y. *Annu. Rev. Biochem.* **1998**, *67*, 509.

95. Ballou, B.; Ernst, L. A.; Waggoner, A. S. *Cur. Med. Chem.* **2005**, *12*, 795.

96. Almutairi, A.; Guillaudeu, S. J.; Berezin, M. Y.; Achilefu, S.; Frechet, J. M. J. *J. Am. Chem. Soc.* **2008**, *130*, 444.

97. Jobsis, F. F. *Science* **1977**, *198*, 1264.

98. Chance, B. *Ann. N. Y. Acad. Sci.* **1998**, *838*, 29.

99. Frangioni, J. V. *Curr. Opin. Chem. Biol.* **2003**, *7*, 626.

100. Berezin, M. Y.; Lee, H.; Akers, W.; Achilefu, S. *Biophys. J.* **2007**, *93*, 2892.

101. Sasaki, E.; Kojima, H.; Nishimatsu, H.; Urano, Y.; Kikuchi, K.; Hirata, Y.; Nagano, T. *J. Am. Chem. Soc.* **2005**, *127*, 3684.

102. Williams, D. C.; Soper, S. A. *Anal. Chem.* **1995**, *67*, 3427.

103. McWhorter, S.; Soper, S. A. *Electrophoresis* **2000**, *21*, 1267.

104. Kricka, L. J.; Fortina, P. *Clin. Chem.* **2009**, *55*, 670.

105. Kimura, R. H.; Cheng, Z.; Gambhir, S. S.; Cochran, J. R. *Cancer Res.* **2009**, *69*, 2435.

106. Kim, K.; Lee, M.; Park, H.; Kim, J. H.; Kim, S.; Chung, H.; Choi, K.; Kim, I. S.; Seong, B. L.; Kwon, I. C. *J. Am. Chem. Soc.* **2006**, *128*, 3490.

107. Michalet, X.; Pinaud, F. F.; Bentolila, L. A.; Tsay, J. M.; Doose, S.; Li, J. J.; Sundaresan, G.; Wu, A. M.; Gambhir, S. S.; Weiss, S. *Science* **2005**, *307*, 538.

108. Shu, X. K.; Royant, A.; Lin, M. Z.; Aguilera, T. A.; Lev-Ram, V.; Steinbach, P. A.; Tsien, R. Y. *Science* **2009**, *324*, 804.

109. Ntziachristos, V. *Annu. Rev. Biomed. Eng.* **2006**, *8*, 1.

110. Zipfel, W. R.; Williams, R. M.; Webb, W. W. *Nat. Biotechnol.* **2003**, *21*, 1369.

111. Helmchen, F.; Denk, W. *Nat. Methods* **2005**, *2*, 932.

112. Niesner, R. A.; Hauser, A. E. *Cytometry A* **2011**, *79A*, 789.

113. Yao, S.; Belfield, K. D. *Eur. J. Org. Chem.* **2012**, *2012*, 3199.

114. Akers, W.; Lesage, F.; Holten, D.; Achilefu, S. *Mol. Imaging* **2007**, *6*, 237.

115. Goiffon, R. J.; Akers, W. J.; Berezin, M. Y.; Lee, H.; Achilefu, S. *J. Biomed. Opt.* **2009**, *14*, 020501.

116. Rao, J. H.; Dragulescu-Andrasi, A.; Yao, H. Q. *Curr. Opin. Biotechnol.* **2007**, *18*, 17.

117. Amiot, C. L.; Xu, S. P.; Liang, S.; Pan, L. Y.; Zhao, J. X. J. *Sensors* **2008**, *8*, 3082.

118. Hilderbrand, S. A.; Weissleder, R. *Curr. Opin. Chem. Biol.* **2009**, *14*, 71.

119. Elson, E. L. In Fluorescence Fluctuation Spectroscopy: Brief Introduction to Fluorescence Correlation Spectroscopy; Tetin, S. Y., Ed.; Elsevier Academic Press Inc: San Diego, **2013**, pp 11–41 and all other chapters in this volume.

120. Bulseco, D. A.; Wolf, D. E. *Methods Cell Biol.* **2013**, *114*, pp 489.

121. Hensel, M.; Klingauf, J.; Piehler, J. *Biol. Chem.* **2013**, *394*, 1097.

122. Lin, Y. C.; Phua, S. C.; Lin, B.; Inoue, T. *Curr. Opin. Chem. Biol.* **2013**, *17*, 663.

2

NATURALLY OCCURRING AND SYNTHETIC FLUORESCENT BIOMOLECULAR BUILDING BLOCKS

RENATUS W. SINKELDAM AND YITZHAK TOR

Department of Chemistry and Biochemistry, University of California, San Diego, CA, USA

2.1 INTRODUCTION

Most common biomolecules and their building blocks lack appreciable emission. When emissive, as in the case of certain fluorescent amino acids (e.g., phenylalanine and tryptophan), their excitation and emission energies are relatively high and found in the UV range. As a result, their utility in biophysical studies, high-throughput assays and imaging applications can be rather limited. This has prompted the development of functional and emissive surrogates. This chapter, bridging our opening discussion and the more focused chapters to follow, concisely discusses the main contributions in this area.

Designer fluorescent probes should ideally resemble their natural counterparts in terms of their molecular size and shape, while retaining their inherent function. We refer to such probes as being isomorphic. This feature, of course, presents a fundamental predicament, as structural modifications aiming to alter the electronic features of a chromophore, inevitably also impact its basic physical properties as well as interactions with its environment and other biomolecules. Nevertheless, elegant advances have been made in this field. After we highlight the naturally occurring fluorescent

Fluorescent Analogs of Biomolecular Building Blocks: Design and Applications, First Edition.
Edited by Marcus Wilhelmsson and Yitzhak Tor.
© 2016 John Wiley & Sons, Inc. Published 2016 by John Wiley & Sons, Inc.

biomolecular building blocks, we concisely summarize the main derivatives developed as emissive surrogates of the major families of biomolecular building blocks.

2.2 NATURALLY OCCURRING EMISSIVE BIOMOLECULAR BUILDING BLOCKS

Whether viewed from a utility or design perspective, it is most inspiring to take notice of the inherently fluorescent building blocks selected by Nature. It is probably safe to state that it was not their fluorescence but rather their structural properties that made them pass Nature's selection criteria. Regardless, their emissive properties are a fortunate coincidence that provide scientists with molecular tools to study the biomolecules containing such building blocks. Interestingly, each family of biomolecules (i.e., proteins, nucleic acids, and lipids) has at least one known naturally occurring emissive building block (Fig. 2.1).

Best known and frequently utilized are the protein building blocks tryptophan (**1**) and tyrosine (**2**) (Fig. 2.1, Table 2.1). The rather unfavorable fluorescence properties of phenylalanine (**3**) limit its use. Tyrosine, on the other hand, enjoys reasonable fluorescence quantum yield with a pH-sensitive emission maximum that shifts from 310 to 340 nm upon deprotonation. Tyrosine lacks sensitivity toward polarity, a trait for which tryptophan, besides its robust fluorescence quantum yield, is well known for.[8] Due to its relative lipophilic character, tryptophan is often buried inside the hydrophobic protein interior. Protein unfolding exposes tryptophan to the polar aqueous environment, causing a shift in its emission maximum from 309 to 355 nm.[1] As discussed in numerous reviews, besides folding/unfolding experiments, applications include protein dynamics and ligand binding.[1, 9–11]

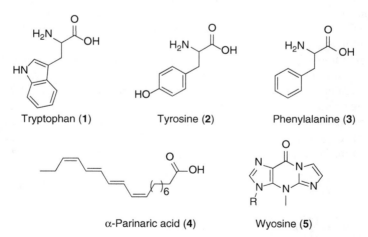

Figure 2.1 Naturally occurring fluorescent biomolecular building blocks.

TABLE 2.1 Selected Spectroscopic Properties of Naturally Occurring Fluorescent Biomolecular Building Blocks[a]

#	Name	Solvent	λ_{abs} (ε)	λ_{em}	Φ_{fl}	τ
1	Tryptophan	Buffer pH 7	279	355	0.01–0.4	
2	Tyrosine	Buffer pH 7	275	310[b]	0.14	3.3–3.8
3	Phenylalanine	Buffer pH 7	258	282	0.024	
4	α-Parinaric acid	Methanol	319, 304(79)	432	0.017	1.3
		Decane	321, 306(74)	432	0.054	5.2
5	Wyosine	Buffer pH 7	235(32), 295(7.4)[c]	450[d]	0.044[d]	

[a]Values for λ, and ε are given in nm and 10^3 M^{-1} cm^{-1}, respectively. Values for amino acids **1**, **2**, and **3** are obtained from different sources.[1-3]
[b]If deprotonated $\lambda_{em} = 340$ nm.[4]
[c]Spectral data, especially the long-wavelength absorption maximum, is pH sensitive.[5] Earlier findings for λ_{abs} in unbuffered water of 235 nm ($\varepsilon = 17.6$) and 294 ($\varepsilon = 4.2$) are seemingly contradicting with tabulated values.[6]
[d]Values for the nucleobase determined in aqueous 0.01 M Tris-HCl, containing 0.1 M NaCl, 10 mM Mg^{2+} at pH 7.5.[7]

A naturally occurring fluorescent membrane constituent is α-parinaric acid (**4**) (Fig. 2.2). This conjugated polyunsaturated fluorescent fatty acid was isolated for the first time from Parinari laurinum in 1933, and 20 years later identified as the (Z),(E),(E),(Z)-isomer.[12] Comparison of the spectral data in methanol and decane, two solvents of dramatically different polarity,[13] reveal almost identical absorption and emission maxima (Table 2.1). This can be attributed to the absence of a strong dipole moment in the ground and excited state as a result of the aliphatic hydrocarbons that cap the π-system. There are, however, significant differences in fluorescence quantum yield and fluorescence lifetime for the two solvents (Table 2.1).[12, 14] After treatment with iodine, the all trans-β-parinaric acid was obtained and spectroscopically characterized in the late 1970s,[15] followed by application as a fluorescent probe in the study of synthetic phospholipid membranes.[16]

Interestingly, adding limited amounts of polyunsaturated fatty acids can stabilize artificial phospholipid membranes, whereas larger amounts can destabilize them.[17] Other naturally occurring polyenes, although not native membrane constituents, that have been used in early membrane studies for their lipophilicity and emissive properties are retinol, retinal, and other cartenoids.[14, 18] For the same reasons, the macrolide antibiotics filipin and amphotericin also found use as fluorescent membrane probes.[19] The last two, however, are rather large, complex, and known to induce cell lyses.

The four canonical nucleobases that make up all nucleic acids virtually lack appreciable fluorescent properties with their high energy absorption, very low quantum yield, and short excited state lifetimes.[20-24] Despite numerous posttranscriptionally modified ribonucleosides,[25] the nucleic acids research field is perhaps the most deprived of naturally occurring fluorescent analogs of its building blocks. Yet, among them there is one, the fluorescent nucleoside wyosine (**5**) (Fig. 2.1) and its emissive derivatives wybutosine and wybutoxosine that share the same chromophore. Their biosynthetic pathway was recently deciphered,[26] and their natural occurrence has been established in baker's yeast tRNA^{Phe},[7, 27–29] Torula yeast,[30] rat and bovine liver,[31, 32] and plants.[33] Wyosine's glycosidic bond is exceptionally susceptible to hydrolysis, which proved to be a major hurdle in its isolation from natural sources.[6] Despite its potential, its chemical instability and lack of a Watson–Crick hydrogen bonding face likely explains why wyosine has not been recognized as a potential fluorescent nucleoside surrogate for the study of nucleic acids.

All naturally occurring fluorescent biomolecular building blocks share one critical structural commonality: an extended (aromatic) π-system (Fig. 2.1). They therefore offer inspiration and a blueprint for fluorescent probe designers. Introduction or extension of an existing π-system bathochromically shifts the absorption maximum allowing for a π–π* excitation, which is often followed by a radiative π*–π decay process. Hence, such a modification is an integral part of any design aimed to endow nonemissive natural biomolecular building blocks with appreciable fluorescence properties. Interestingly, the efficiency of the emissive process is strongly dependent on structural rigidity, nature of the substituents, and environmental factors (e.g., polarity, pH, viscosity). This provides probe designers with the opportunity to tailor the fluorescence response of the probe toward environmental characteristics.[34] For instance, sensitivity to pH can be controlled by inclusion or exclusion of basic or acidic sites.[35, 36] Introduction of a strong push–pull system by judicial placement of donor (electron releasing) and acceptor (electron withdrawing) moieties typically results in responsiveness to environmental polarity.[37–39] Sensitivity to viscosity, or molecular crowding, is virtually absent in the most rigid fluorophores that lack single-bond linkages between π-systems. Conversely, introduction of such a linkage constructs a "molecular rotor" and likely imparts a fluorescence probe with enhanced sensitivity to viscosity.[40] Besides the synthetic hurdles accompanied with such alterations, the real design challenge is to limit the structural modification to a minimum, often a prerequisite to ensure interchangeability of the natural building block with its fluorescent surrogate.

2.3   SYNTHETIC FLUORESCENT ANALOGS OF BIOMOLECULAR BUILDING BLOCKS

Due to the vast amount of scientific literature available, the following sections cannot and are not aimed to be comprehensive. Rather, they are intended to illustrate the various structural designs that have been explored to impart nonemissive natural biomolecular building blocks with desirable fluorescent properties. To this end, the scope, with a focus on isomorphicity, of fluorescent analogs of membrane constituents, amino acids, and nucleosides is given. Readers will, however, be directed to reviews that significantly elaborate on selected topics. The depicted structures herein, along with their tabulated basic spectral properties, allow for interesting comparison of their structure–photophysical properties relationship (SPPR).

2.3.1   Synthetic Emissive Analogs of Membranes Constituents

Biological membranes, crucial for sustaining cellular integrity and function, consist of a complex mixture of membrane constituents including proteins, phospholipids, and fatty acids. The amphiphilic nature of the last two constituents, a polar head group with a lipophilic tail, is the primary driving force that shapes the architecture of the biological membrane in aqueous environments.[41–43] Lipid bilayers are amenable to multiple kinds of fluorescent labeling as has been discussed in multiple reviews.[4, 44–52] Strategies can be categorized by polar head group labeling, apolar chain-end labeling, on- and in-chain labeling, and noncovalent labeling. The last approach typically exploits inherently lipophilic, and hence hydrophobic, dyes including 1-ethylpyrene,[53] diphenyl hexatriene (DPH) (**6**) (Fig. 2.2, Table 2.2),[60–62] methyl-9-anthroate,[63, 64] 4-(dicyanovinyl)julolidine,[65, 66] and steroidal skeletons resembling aminodesoxyequilenin.[67, 68] Exposed to an aqueous environment containing lipid bilayers, such dyes will either precipitate or dissolve in the apolar interior of the membrane according to a system-specific partition coefficient (K). Despite the presence of organized domains, for example, rafts and superlattices,[69, 70] the typical fluid nature of the membrane interior, however, provides ample mobility and thus limits knowledge and control over the dye's exact position. To improve its positioning, one of the benzene rings of diphenyl hexatriene (**6**) has been functionalized with the polar trimethylammonium to anchor it to the polar head group region.[71]

More common approaches exploit phospholipid head group, chain-end, or in- and on-chain functionalization, thereby effectively determining the probe's immediate environment and thus its applications. The lack of a suitable scaffold for synthetic extension or expansion to impart membrane constituents with fluorescence properties limits the strategy to the inclusion of known fluorophores. Located at the polar head group of a lipid, the fluorophore will be in immediate contact with the polar extracellular matrix allowing the probing of processes at the cell surface. A widely used example is the dansyl-labeled phosphatidyl ethanolamine (DPE) (**7**) (Fig. 2.2).[72] Its sensitivity to polarity was exploited for the study of protein–lipid interactions[73, 74] in addition to local polarity[72, 75] and fluidity[75] of biological membranes. Other probes in this category include phospholipid head groups labeled with coumarin,[76] nitrobenzoxadiazole,[77] or rhodamine B.[78]

To explore the interior of membranes, the chains of fatty acids and phospholipids have been functionalized with fluorophores. The on-chain approach is exemplified by 12-(9-anthroyloxy) stearic acid (12-AS) (**8**)[72] and its analog 12-(9-anthroyloxy) stearic acid anthraquinone.[63, 79] This design has the advantage that it allows for controlled positioning of the probe in proximity to or remote from the polar head group thereby facilitating the study of membrane polarity,[72] fluidity,[63, 79] and protein–lipid interactions.[80] Free rotation enabled by the linker, however, can complicate spectroscopic analysis, which likely explains the limited popularity of this approach. Conjugation of the fluorophore at the chain-end of a phospholipid or fatty acid positions it deep in the apolar interior of the membrane. An example of such a design

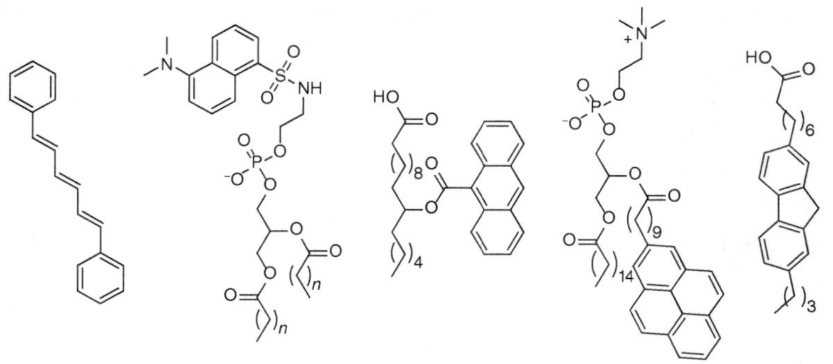

6 7 8 9 10

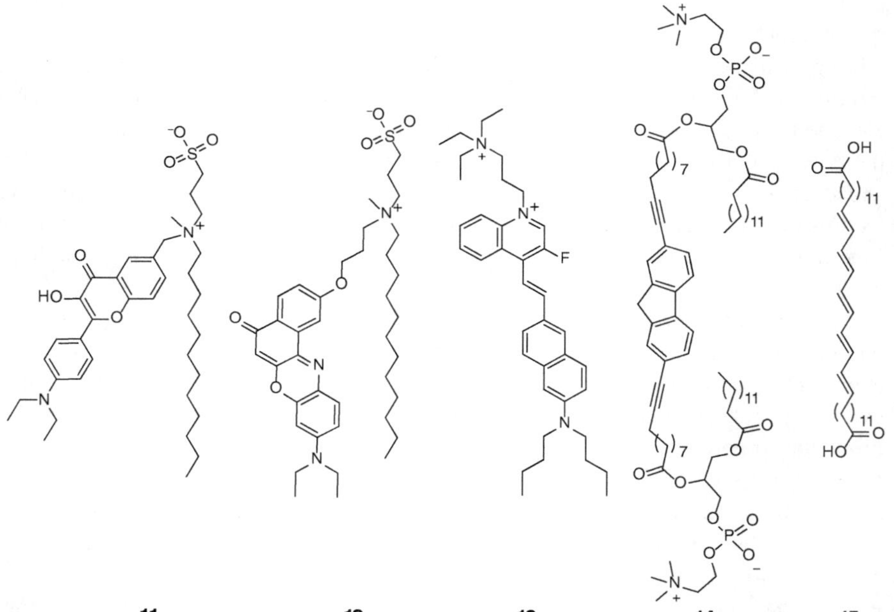

11 12 13 14 15

Figure 2.2 Examples of noncovalent, polar head group, on-chain, chain-end, and in-chain fluorescent membrane probes.

that enjoys very high fluorescence quantum yield is a pyrene chain-end labeled phosphatidylcholine (pyrene-PC) (**9**) (Fig. 2.2, Table 2.2).[53, 81, 82] This probe has been extensively used to study membrane fluidity,[81] the effect of cholesterol on membrane properties,[83] membrane microdomains,[84–86] protein–lipid interactions,[87] and membrane permeability.[88] This design strategy does, however, potentially suffers from "looping-back," where a probe aimed to be located at the inner membrane folds back closer to the polar head group due to the flexibility of the aliphatic chain.[89] This drawback can be overcome with an in-chain approach controlling positioning

TABLE 2.2 Spectroscopic Properties of Selected Synthetic Fluorescent Membrane Constituents[a]

#	Name	Solvent	λ_{abs} (ε)	λ_{em}	Φ_{fl}	τ
6	DPH[b]	Hexane	352, 370	430	0.64	15.7
7	DPE[c]	Methanol	346(3.6)	514		
8	12-AS	Methanol	362(7.8)	458	0.071	1.6
		Hexane		446		10.5
9	Pyrene-PC	Methanol	342(37)[d]	376[d]	0.65[e]	410[e]
10	C8A-FL-C4	Methanol	270(38), 297, 309	319	0.65	
11	3HF[f]	Ethanol	431	521/570		
		Hexane	396	423/554	0.14	
12	NR[g]	Buffer	521	657	0.002	
		Dioxane	526	592	0.74	
14	BAexFluorPC[h]	DMPC	308, 329	334		1.3
15	trans-PDA	Chloroform	353(92), 335(95), 320(60)	474	0.14[i]	

[a]Values for λ and ε are given in nm and 10^3 M^{-1} cm^{-1}, respectively.
[b]Data from Bachilo et al.[54] and Palmer et al.[55]
[c]Data from London et al.[92]
[d]Data from Sinkeldam et al.[50]
[e]Data in ethanol from Hermetter et al.[48]
[f]Data for 4-diethylamino-3-hydroxyflavone, λ_{em} data represent emission from normal and tautomeric state after ESIPT (N*/T*), respectively.[56, 57]
[g]Buffer is phosphate buffer of pH 7.4.
[h]Only a spectroscopic study in dimyristoylphosphatidylcholine (DMPC) vesicles is reported, λ_{abs} and λ_{em} are extracted from graphs, and only the most contributing τ is given.[58]
[i]Quantum yield in DMPC vesicles.[59]

of the fluorophore in the membrane by limiting its mobility. Hence, probes such as fluorene-labeled fatty acid (C8A-FL-C4) **10**[90, 91] (Fig. 2.2, Table 2.2) have been used in depth analysis of membranes with a focus on phospholipid topology[89, 92, 93] in addition to membrane localization and penetration of, for example, cholesterol[94–96] and membrane-bound proteins and peptides.[97]

Other examples of fluorescent aromatic hydrocarbons used include anthracene and ethynyl-extended anthracene,[98] and vinyl-extended dihydrophenanthrene.[99] Incorporation of 3-hydroxyflavone, 3HF, (**11**) and Nile Red, NR, (**12**)[100] illustrate the use of fluorescent aromatic heterocycles (Fig. 2.2, Table 2.2).[101, 102] The former, 3-hydroxyflavone (**11**), can undergo excited state intramolecular proton transfer (ESIPT)[56, 103] and is sensitive to surface charge and hydration.[101] Nile Red (**12**) is polarity sensitive as well but, in contrast to **11**, cannot undergo ESIPT due to the absence of acidic protons.[100] Both **11** and **12** bind specifically to the outer membrane leaflet, posses spectral properties sensitive to changes in lipid order, and can detect cellular apoptosis.[100–102] Like Nile Red (**12**), a recently reported family of environmentally sensitive quinolinium-based membrane probes show remarkable redshifted emission.[104] Among them, quinolinium **13** reveals a lipid order-dependent long-wavelength emission maximum ranging from ~660 to almost 700 nm with an emission profile tailing over 800 nm (Fig. 2.2).

Regardless of favorable spectroscopic properties, arguably the best position control of a membrane probe can be achieved using the symmetrical bolaamphiphile design where the in-chain fluorophore is anchored in the membrane by two polar head groups limiting both longitudinal and transverse maneuverability. This elaborated design is illustrated with the ethynyl-extended fluorene (**14**).[58] When it comes to isomorphic design principles, the most representative class of probes are the polyene membranes substituents. Their successful design is validated and strongly inspired by aforementioned naturally occurring α-parinaric acid (**4**) (Fig. 2.1). Superior to designs with bulky fluorophores, such polyene-containing phospholipids have been used as probes in live cells.[105] In all trans-pentanoic diacid (trans-PDA) (**15**), as in **14**, the bolaamphiphile design locks the polyene probe in place while its ultraslim fluorophore limits membrane perturbation to an absolute minimum (Fig. 2.2).[106] Its high molar absorptivity values, emission in the visible part of the electromagnetic spectrum, and sufficient quantum yield (Table 2.2) were used to study probe dynamics and fluidity of lipid bilayers.[59]

2.3.2 Synthetic Emissive Analogs of Amino Acids

Besides the emissive aromatic residues of tryptophan (**1**), tyrosine (**2**), and phenylalanine (**3**) (Fig. 2.1, Table 2.1), the naturally occurring amino acids have high-energy UV absorption maxima and lack fluorescence. Imparting the native amino acids with desirable fluorescence properties is a challenge since protein function relies heavily on correct folding thereby limiting the modifications that can be tolerated without detrimental structural or functional repercussions. Material-related applications have, however, exploited modified residues. Noncanonical amino acids[107] and their fluorescent amino acids substitutes have appeared in numerous overview articles.[50, 108–113]

Despite the already useful features of tryptophan (**1**) and tyrosine (**2**), a desirable enhancement of their spectral properties would include a redshifted absorption spectrum to allow for selective excitation, which is especially valuable in proteins that already contain one or more tryptophan residues. Synthetic derivatization of tryptophan's (**1**) aromatic core is a straightforward approach and has led to a variety of subtly modified tryptophan mimics, all sharing its shape and size but with altered spectroscopic properties.[50, 111, 114, 115] One notable modification is 7azaTrp (**16**) (Fig. 2.3, Table 2.3).[115, 121, 122] Replacing the benzene core in tryptophan for a pyridine core causes a notable redshift of the absorption and emission maximum to 291 and 391 nm, respectively. Despite the lower quantum yield compared to native tryptophan, 7azaTrp (**16**) has been incorporated in β-galactosidase,[123] membrane protein EIImtI,[115] and used for DNA–protein binding studies.[114] Tryptophan remains an inspiring starting point for new designs as is illustrated very recently with analogs comprised of an expanded ring system such as 1*H*-pyrrolo[3,2-*c*]isoquinoline. They possess a larger Stokes shifts and their absorption and emission maxima are distinct from tryptophan.[124]

Instead of exploiting tryptophan, a general strategy to impart amino acids with fluorescent properties is a straightforward attachment of known fluorophores. Side-chain modification gives access to virtually limitless number of fluorescent amino acids.[50]

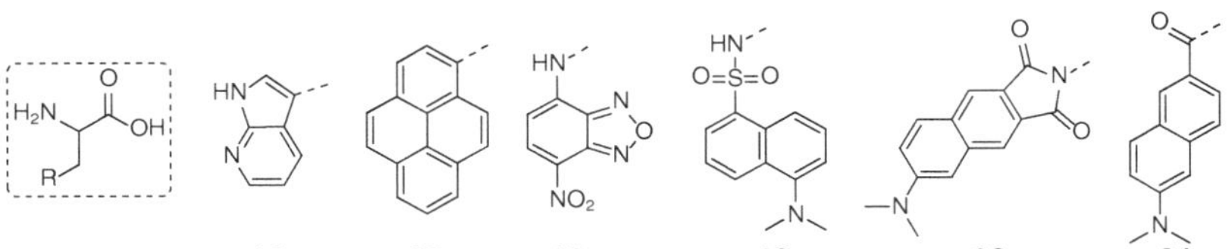

Figure 2.3 Examples of emissive amino acid analogs 7azaTrp (**16**), 1PyrAla (**17**), NBDAla (**18**), 51dansylAla (**19**), 6DMNA (**20**), and Aladan (**21**).

TABLE 2.3 Spectroscopic Properties of Selected Synthetic Fluorescent Amino Acids[a]

#	Name	Solvent	λ_{abs} (ε)	λ_{em}	Φ_{fl}	τ
16	7azaTrp	Water	291	391	0.01	1.24
		Methanol	297	366	0.01	
17	1PyrAla[b]	Ethanol	241(79.4), 272, 343	376	0.65	
18	NBDAla[c]	Ethanol	264, 330, 462(19.7)	532	0.38	
19	51DansylAla[d]	Methanol	335(4.0)	518	0.23	
		Dioxane	335(4.1)	479	0.54	
20	6DMNA[e]	Water	388	592	0.002	
		Dioxane	372	498	0.22	
21	Aladan[f]	Water	364(14.5)	531		
		Cyclohexane	342	401		

[a]Values for λ and ε are given in nm and 10^3 M^{-1} cm^{-1}.
[b]Values for pyrene.[48, 116]
[c]Data for 7-benzylamino-4-nitrobenz-2-oxa-1,3-diazole.[117]
[d]Values for 5-(dimethylamino)-*N*-methylnaphthalene-1-sulfonamide.[118]
[e]Values for model compound 6DMN-GlyOMe.[119]
[f]Values for Prodan.[120]

A plethora of derivatives have been reported, here exemplified by alanine mimic 1PyrAla (**17**)[125] (Fig. 2.3). Although analog **17** is endowed with the desirable high quantum yield of the hydrocarbon fluorophore pyrene, it lacks a significantly redshifted emission maximum (Table 2.3).

Shifting the emission to lower energies can frequently be introduced by judicial positioning of an electron-releasing moiety (donor or D) and an electron-withdrawing moiety (acceptor or A) yielding chromophores that are often referred to as "charge transfer," "push–pull," or D–A chromophores. Importantly, this electronic feature, augmenting the chromophore's polarization, is typically accompanied by an enhanced sensitivity to environmental polarity, a general concept that has been widely applied in fluorescent probe design including fluorescent amino acids.[50, 113]

A classical example of such a chromophore is 4-amino-7-nitro-2,1,3-benzoxadiazole (NDB). Interestingly, compounds containing a nitro group are often assumed to be nonemissive, making the first report in the late 1960s of the fluorescence properties of NBD containing glycine, with NBD attached to the N-terminus, a rare exception.[126] Almost a decade later, NBD alanine, also functionalized at the N-terminus, was subjected to rigorous investigation of its sensitivity to solvent polarity establishing that the quantum yield decreases, molar absorptivity increases, and both absorption and emission maximum undergo a redshift with increasing solvent polarity.[127] NBD's spectral responsiveness has led to applications in protein and membrane studies.[50] Much later, NBDAla (18) was developed, having alanine's methyl group replaced by NBD enabling substitution of alanine by its emissive surrogate 18, enjoying a high fluorescence quantum yield and redshifted emission maximum in ethanol of 0.38 and 532 nm, respectively (Fig. 2.3, Table 2.3).[128] Similar to 18, 51dansylAla (19), containing the well-known push–pull chromophore dansyl, also displays a significant polarity-dependent emission maximum while maintaining a robust quantum yield under common conditions (Fig. 2.3, Table 2.3).[123, 129] It has been used in folding/unfolding,[129] polarity,[130] and binding[131, 132] studies. Analogously, dansyl has also been exploited to impart lysine with emissive properties.[123]

Other related push–pull designs include 6DMNA (20)[119] and the Prodan-modified alanine Aladan (21)[133] (Fig. 2.3, Table 2.3). The former, 6DMNA (20), possessing a polarity-sensitive emission maximum ranging from 498 nm in dioxane to 592 nm in water, has been studied after incorporation in a central position in a hexapeptide[130] and in peptide–protein binding studies.[119] Prodan-containing Aladan (21), featuring a 130-nm difference in emission maximum going from apolar cyclohexane (401 nm) to polar water (531 nm), has been used to estimate local polarity in proteins[133] despite its alleged destabilizing effect.[134] It must be noted that the large polarity-dependent shift in emission maxima for Aladan (21) as well as 6DMNA (20) is also accompanied by significant shifts in absorption maximum thereby limiting the potential polarity sensitivity if expressed as a function of Stokes shift ($\nu_{abs} - \nu_{em}$). In contrast, polarity hardly has an effect on the absorption maximum of the dansyl chromophore (19). The polarity sensitivity of its emission maximum, however, is comparatively modest (Table 2.3).

It is important to note that even though the nomenclature of some of the aforementioned fluorescent amino acids implies that they are alanine mimics, their properties deviate significantly from the structural dimensions and polarity of the native amino acid. Hence, substitution of alanine for any of these emissive surrogates may adversely influence folding or stability upon incorporation into peptides or proteins, which in turn might affect the function. When isomorphicity is a prerequisite, the tryptophan mimics, here represented by 7azaTrp (16), are arguably among the most desirable fluorescent amino acid surrogates.

2.3.3 Synthetic Emissive Analogs of Nucleosides

Pioneering research in the 1960s and 1970s formed not only the foundation for modern fluorescent probe development for the study of membranes and proteins but also for the study of nucleosides. Merely 16 years after the unraveling of the double helix structure of DNA,[135] the inherent lack of fluorescent properties of the

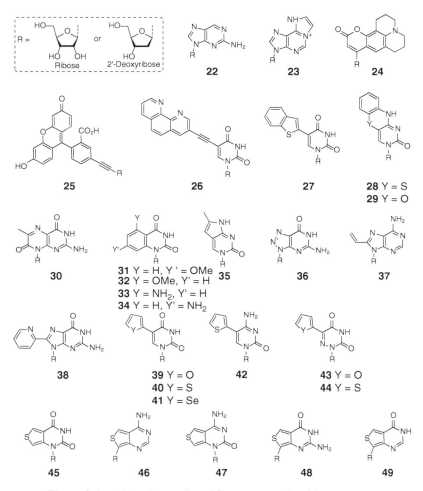

Figure 2.4 Selected examples of fluorescent nucleoside surrogates.

native nucleosides was overcome with the disclosure of 2-aminopurine, 2AP (**22**), an adenosine mimic (Fig. 2.4).[136] This report by Stryer *et al.* in 1969 was followed up by Leonard in 1972, who reported on another fluorescent adenosine surrogate: ethenoadenosine, ethenoA (**23**).[137] Although both adenosine analogs, they critically differ in design. Remarkable, and likely unpredictable, repositioning of the exocyclic amine from the 6 in A to 2 position in 2AP dramatically enhances the heterocycle's emission. Its spectral properties include a redshifted absorption maximum at 303 nm allowing for selective excitation, and an emission profile centering around 370 nm with a most desirable fluorescent quantum yield of 68% in water (Table 2.4). The emissive properties of ethenoA (**23**), on the other hand, stem from the expanded aromatic ring system providing a slightly less redshifted absorption maximum at

TABLE 2.4 Spectroscopic Properties of Selected Synthetic Fluorescent Nucleosides[a]

#	Name	Solvent/Condition	λ_{abs} (ε)	λ_{em}	Φ_{fl}	τ
22	2AP	Water	303 (6.8)	370	0.68	7.0
23	ethenoA	Buffer pH 7	275 (6.0), 294 (3.1)[b]	415	0.6	20
24	Coumarin	Buffer pH 7.2	400	515		7.4
25	exFLdU	Buffer pH 7.2	320 (23.9), 492 (57.0)	520	0.53	
26	exPhenU	Water	330 (20.0), 345 (18.0)[b]	408	0.16	
27	BTU	Water	318	458	0.035	1.04
		Dioxane	322	435	0.060	0.45
28	tC	Buffer 7.5	375 (4.0)	500	0.17	3.7
29	tC^O	Buffer 7.5	360 (9.0)	465	0.30	3.4
30	6-MI	Buffer pH 7.5	340	431	0.70	6.4
31	3-MeOQ	Water	320	395	0.16	
		Dioxane	314	362		
32	5-MeOQ	Water	305	357	0.08	
33	5-AQ	Water	349	445	0.42	
		Dioxane	350	407		
34	7-AQ	Water	316	361	0.04	
		Dioxane	316	336		
35	Pyrrolo-dC	Buffer pH 7.0	350 (5.9)	460	0.2	
36	8-AzaG	Neutral, pK_a 8.05	256 (12.9)	347	<0.01	
		Monoanion	278 (11.7)	362	0.55	
37	8-vinyldA	Buffer pH 7.5	290 (12.6)	382	0.66	4.7
38	8-PyrdG	Water	300 (20)	415	0.02	
39	5-FurdU	Water	316 (11.0)	431	0.03	1.0
40	5-ThiophdU	Water	332 (1.1)	455	0.20	4.9
		Dioxane	335 (1.3)	415	0.80	5.4
41	5-SelU	Water	325	454	0.014	0.4
		Dioxane	331	437	0.025	0.27
42	5-ThiopheneC	Water	310	443	0.020	
		Dioxane	309	421	0.006	
43	5-Fur6AzaU	Water	320 (1.0)	443	0.05	3.1
		Dioxane	327 (1.2)	414	0.60	5.6
44	5-Thioph6AzaU	Water	332 (1.1)	455	0.20	4.9
		Dioxane	335(1.3)	415	0.80	5.4
45	thU	Water	304(3.16)	409	0.41	11.5
		Dioxane	304 (3.50)	378	0.04	1.0
46	thA	Water	341 (7.44)	420	0.21	3.9
		Dioxane	345 (7.83)	411	0.14	3.2
47	thC	Water	320 (4.53)	429	0.41	15.2
		Dioxane	326 (4.21)	422	0.01	5.0
48	thG	Water	321 (4.15)	453	0.46	14.8
		Dioxane	333 (4.53)	424	0.50	13.0
49	thI	Buffer pH 7.4	315 (4.8)	391		

[a] Values for λ, ε, and τ are given in nm, 10^3 M^{-1} cm^{-1}, and ns, respectively.
[b] Only the two longest wavelength absorption maxima are given.

294 nm, but a significantly redshifted emission that peaks in the visible at 415 nm with an equally impressive quantum yield of 60%.

These early emissive nucleosides unintentionally illustrate that useful fluorescent properties can be reconciled with a minimally modified structure, a design principle that would three decades later resurface to be termed isomorphic.[138] Prior to the deliberate exploration of isomorphic designs, other approaches were, and still are, investigated to obtain nucleosides with emissive properties. Many have been reviewed,[50, 139–148] but only a few will be highlighted here to demonstrate their structural diversity. Parallel to the generation of fluorescent mimics of membrane constituent and amino acids, known fluorophores have been used as attachments to or substitutions of the native nucleobases. The latter approach does immediately benefit from the fluorophore's favorable spectroscopic properties but at the expense of inadequate structural resemblance to the native nucleobases, most notably the lack of a WC hydrogen bonding face. Many polycyclic aromatic hydrocarbons have been used including pyrene, perylene, anthracene, stilbene, terthiophene. This category is exemplified here by coumarin (**24**) (Fig. 2.4),[149] which was incorporated across an abasic site in duplex DNA[150] and used in several studies on DNA solvation and dynamics.[151–154] Other such nucleoside mimics have been used to study enzyme–substrate recognition,[155] in direct monitoring of damage and repair in duplex DNA,[156] and detection of nucleases.[157] Incorporated in oligomers comprised entirely of nonnatural emissive nucleoside mimics, they have been used for the real-time tracking of biological systems,[158] and monitoring of DNA base excision repair.[158] Outside of typical nucleic acid research, the sequence-dependent spectroscopic properties of water-soluble oligomers comprising these emissive nucleoside mimics[159] were used for the sensing of food spoilage[160] and are expected to contribute to the areas of solar energy conversion and biomedical detection.[159, 161]

Instead of replacing the nucleobase by a fluorophore, a number of analogs have been reported, where known fluorophores, including pyrene, perylene, naphthalene, anthracene, and fluorene, were attached to natural nucleobases as discussed in several reviews.[50, 142, 146, 148, 162, 163] Advantages of this approach include an intact hydrogen bonding face and the opportunity to electronically conjugate the fluorophore to the π-system of the native nucleobase. The latter gives an extended π-system likely to induce a redshifted absorption maximum. More importantly, local environmental perturbations at the nucleobase location possibly resonate throughout the entire π-system to be reflected by spectral changes. Examples according to this design of modified adenosine (A), guanosine (G), cytidine (C), and thymidine (T)/uridine (U) are abundant and can be found in the aforementioned reviews.

The ethynyl moiety, exploiting the robust Pd-mediated Sonogashira coupling, is frequently used to link the native nucleobase with the fluorophore as illustrated by fluorescein-modified T-mimic exFLdU (**25**) (Fig. 2.4).[164] The extended π-system and attachment of fluorescein convert nonemissive T in a fluorescent T-mimic with redshifted absorption and emission maxima and a robust fluorescence quantum yield (Table 2.4). Interestingly, studies showed that only further extended versions could be enzymatically incorporated in oligonucleotides.[164] Similarly, phenanthroline was introduced to give exPhenU (**26**) (Fig. 2.4, Table 2.4).[165] Its polarity-sensitive spectral properties facilitated discrimination between perfect base-pairing and mismatches. The chelating ability of **26** enabled formation of Ru^{II} and Os^{II} metal complexes, and their significantly different spectroscopic properties[166] were used to study donor–acceptor interactions in oligonucleotides.[167] Very recently published work on extension of the uridine π-system through a ethynyl linker by attaching a polarity-sensitive naphthalimide unit,[168] recognized for its polarity sensitivity[169] and also used as a fluorescent amino acid analog,[119, 130] illustrates the continued popularity of this approach. Numerous examples, where the fluorophore is directly conjugated to the nucleobase, have been reported (see the aforementioned reviews). This approach is illustrated here by the more recently reported benzothiophene-conjugated uridine, BTU (**27**).[170] The polarity sensitivity and sufficient quantum yield made this uridine analog electable for the study of oligonucleotides dynamics in artificial cell models.[171] An analogous design featuring a benzofuran modification was used for the detection of abasic sites in RNA.[172]

Instead of extending the native π-system by conjugation to know fluorophores, the π-system can also be expanded with additional aromatic rings. Examples of this rigid planar design include the aforementioned ethenoA (**23**),[137] and tC (**28**) (Fig. 2.4). The latter was first reported in 1995 for antisense applications,[173] and later recognized for its nonperturbing and fluorescent properties[174] and almost context-independent emission maximum and fluorescent quantum yield.[175] Its close relative tC^O (**29**), with an oxygen instead of sulfur bridge, also enjoys favorable spectroscopic properties (Fig. 2.4, Table 2.4).[176] Comparison of tC with tC^O reveals how small structural modifications, in this case an S to O substitution, can substantially influence the spectroscopic properties (Table 2.4). Nucleoside mimics tC (**28**) and tC^O (**29**) have been used in enzymatic incorporations using the Klenow fragment,[177] while incorporation of tC has also been studied with *Escherichia coli* DNA polymerase.[178] Both C-surrogates have been used for structural measurements in DNA using Förster resonance energy transfer (FRET).[179–181] Other applications include the use of tC^O in fluorescence-dependent immunoassays,[182] and the use tC in photo-induced DNA duplex destabilization.[183] A derivative of tC, bearing an aminoethoxy unit, has been used for the fluorescence detection of 8-oxoG.[184–186]

A motif rich in heteroatoms is represented by pteridines, which contain structural elements akin to the expanded nucleobases but with dimensions resembling isomorphic nucleosides.[139, 145] A member of this group is 6-MI (**30**), a nonperturbing emissive guanosine analog with an emission maximum that peaks in the visible and a very impressive fluorescent quantum yield of 70% (Fig. 2.4, Table 2.4). Noteworthy is that the incorporation of **30** into oligonucleotides appears to be benign,[187] but other pteridines are reported to cause a sequence-dependent destabilizing effect in oligonucleotides.[188]

Analogous in size and structure to pteridines are quinazoline-derived nucleobases. Their design relies on the fusion of an electron-rich aromatic ring with the electron-withdrawing pyrimidine core to obtain uridine mimics with redshifted absorption and emission maxima.[189–192] Introduction of electron-releasing exocyclic substituents, aimed to enlarge the chromophore's dipole, enable tailoring of the fluorophore's fluorescent properties as is illustrated by 3-MeOQ (**31**),[189] 5-MeOQ

(32),[191] 5-AQ (33),[190] 7-AQ (34)[192] (Fig. 2.4). They possess sufficient-to-good fluorescence quantum yields and are polarity sensitive (Table 2.4), and their native hydrogen bonding face and ribose positioning make them suitable uridine surrogates for perfect pairing with adenine.

The fluorescent uridine analog **31** was incorporated in a truncated version of the prokaryotic ribosomal decoding site, also called the A-site. As a fluorescence donor in a FRET assay, 3-MeOQ (**31**) reported real time on the binding of designer antibiotics to the A-site enabling determination of their binding affinity.[189] To assess the difference in binding affinity of designer antibiotics for prokaryotic and eukaryotic decoding sites, 5-MeOQ (**32**) was incorporated in the prokaryotic A-site as a fluorescence donor in a complex double FRET competition assay with the eukaryotic A-site.[191] Tryptophan's emission overlap with the absorption of 5-AQ (**33**) and the latter's emission in the visible was exploited in a FRET assay to monitor a model system of the binding of HIV-1 to the Rev Response Element.[190] Incorporated into oligomeric dsDNA, fluorescent uridine analog 7-AQ (**34**) displayed sensitivity to mismatched pairing and G-specific fluorescence enhancement.[192]

Although pyrrolo-dC (**35**) can be viewed as an expanded nucleoside, its minimal modification and structural likeness to cytidine make it a good example of an isomorphic fluorescent nucleoside mimic with favorable properties (Fig. 2.4, Table 2.4).[193, 194] It has been used in structure monitoring of RNA,[195] the study of T7 RNA polymerase,[193] as an aptamer sensor for 8-oxoG,[196] among other applications.[50] The rather remarkable fluorescent quantum yield for its size prompted the development of various derivatives, each with unique spectroscopic qualities.[197–201]

To minimize potential perturbation upon incorporation into oligonucleotides, it is obvious, yet essential, to limit the structural modification as is exemplified by 8-Aza-guanosine, 8-AzaG (**36**).[202] This minimal modification of guanosine, a C8 to N substitution, results in a pH-sensitive emission,[203, 204] enabling the study of RNA's catalytic function after enzymatic incorporation into oligomeric RNA.[205, 206] An example of an adenosine mimic with minimal π-system extension to impart it with fluorescent properties is represented by 8-vinyl-6-aminopurine, 8-vinyldA, (**37**) (Fig. 2.4).[207, 208] Aside from its redshifted absorption and emission maxima, this A-mimic features a most impressive quantum yield of 66% in buffer at pH 7.5 (Table 2.4). Like 2AP, however, 8-vinyldA's fluorescence quantum yield drops upon incorporation into oligonucleotides where it may, depending on the sequence, cause slight duplex destabilization.[207] Its favorable spectroscopic properties were exploited in applications including single nucleotide polymorphism.[209] Interestingly, applying the same 8-vinyl modification to guanosine furnishes 8-vinyldG[210] with equally desirable fluorescent properties, which were exploited in the study of G-quadruplexes[210] and peptide nucleic acids.[211] The research on G-quadruplexes also draws from the development of fluorescent nucleoside analogs,[212] with 8-pyridine functionalized G, 8-PydrG (**38**) as another example (Fig. 2.4, Table 2.4).[213–215] Its selective chelation of various metals as a bidentate ligand was studied in DNA as well.[216]

The attachment of furan in the five position of 2'-deoxyuridine gives 5-FurandU (**39**) imparting its practically nonemissive parent nucleoside with useful spectroscopic properties that were employed to detect abasic sites after incorporation into duplex DNA (Fig. 2.4, Table 2.4).[138] More than three decades after the discovery of 2AP (**22**) and ethenoA (**23**), 5-FurdU (**39**) illustrated once again that benign structural modifications can lead to favorable fluorescence properties. For its polarity-sensitive emission maximum and sufficient fluorescence quantum yield, 5-FurdU (**39**) was used for the detection of abasic sites[138] and estimation of major groove polarity in double-stranded DNA.[37] The ribose triphosphate derivative, possessing practically the same fluorescence properties, was recognized by T7 RNA polymerase and successfully incorporated into RNA constructs.[217, 218] The unpredictable relationship between molecular structure and photophysical properties is illustrated by 5-ThiophdU (**40**), a thiophene analog of **39** with a significantly depressed fluorescent quantum yield (Fig. 2.4, Table 2.4). Examples **39** and **40** led to the synthesis and spectroscopic analysis of other fluorescent purine and pyrimidine mimics bearing 5-membered aromatic heterocycles.[219, 220] Noteworthy pyrimidine examples include the recently reported selenophene-modified uridine 5-SelU (**41**) used for RNA-ligand binding studies[221] and thiophene-modified cytidine (**42**), which can discriminate between an opposite positioned G, 8-oxoG, or T in duplex DNA using fluorescence spectroscopy.[222]

In addition to conjugation of 5-membered heterocycles to uridine, a slight modification of its core has been explored as well. Uridine mimics 5-Fur6AzaU (**43**) and 5-Thioph6AzaU (**44**) comprise the previously used 5-membered heterocycles in addition to a C6-to-N substitution in the uracil core (Fig. 2.4).[38] The enhanced push–pull effect renders this 6-aza motif polarity sensitive and, compared to **39** and **40**, the C6-to-N modification results in a significant increase in the fluorescent quantum yield (Table 2.4). In addition, the 6-aza modification alters the pK_a of $N - 3$ making **43** and **44**, or their 2'deoxy ribose analogs, suggesting potential utility as hybridization probes after incorporation into RNA or DNA, respectively. Hence, as also seen for tC (**28**) and tC^O (**29**), the alteration of a single atom in the molecular structure can have a notable impact on the spectral properties. The 6-azaU motif has recently been further extended to provide visibly emitting and tunable nucleosides.[39]

Nucleosides **27**, **38–44**, and similar derivatives all have in common a single-bond linkage between the nucleobase and the conjugated 5-membered heterocycle. This so-called rotor element is known to impart fluorophores with sensitivity to viscosity or molecular crowding.[40, 66, 223, 224] Although rarely explored in the context of fluorescent nucleosides, this design indeed shows sensitivity toward the viscosity or crowdedness of its environment.[38, 40, 221] Recently, the fluorescence properties of selected examples of this design have been studied under 2-photon excitation conditions.[225]

The first report on isomorphic fluorescent nucleosides in 2005 by Tor *et al.* presented a design, illustrated by **39** and **40**, that reconciled isomorphic design criteria with useful fluorescent properties.[138] This work was followed up with thieno uridine,

[th]U (**45**), a new design based on the fusion of thiophene with the pyrimidine nucleobase (Fig. 2.4). This planar and rigid nucleobase boasts most favorable fluorescence properties including a polarity-sensitive emission maximum and an excellent fluorescence quantum yield in apolar as well as polar media (Table 2.4). Studies on the [th]U-triphosphate with T7 RNA polymerase using various templates showed that [th]U was amenable to enzymatic incorporation further establishing the potential of this design.[226, 227] The value of [th]U as a fluorescent probe was demonstrated as a sensor for mismatched pairing,[227] for the signaling of toxic ribosome-inactivating proteins activity.[228] The confirmation of [th]U's versatility spurred the development of a fluorescent nucleoside alphabet, also consisting of [th]A (**46**), [th]C (**47**), and [th]G (**48**), all derived from the thieno[3,4-*d*]-pyrimidine heterocycle (Fig. 2.4).[229] As was determined for [th]U, the photophysical analysis of the other alphabet members showed absorption maxima over 300 nm and emission in the visible with modest to good quantum yields for all (Table 2.4). Utilization of these exciting fluorescent nucleosides is in full swing. Recently, the translation of fluorescent [th]G-modified RNA could be monitored by changes in fluorescence intensity stemming from an mRNA codon–anticodon interaction at the ribosomal decoding center.[230] It was also found that T7 RNA polymerase can initiate and maintain transcription with [th]GTP replacing GTP, yielding fully modified and highly emissive transcripts.[231]

This emissive alphabet received attention from theoreticians as well who calculated the acid dissociation constant of [th]G by using density functional theory,[232] and the excited state properties of all alphabet members.[233] In addition to the canonical nucleoside analogs, an emissive analog of the naturally occurring and nonemissive inosine, [th]I (**49**) materialized, also with robust spectral properties (Table 2.4). Inosine is the product of the catabolic deamination of adenosine by adenosine deaminase, a chief enzyme of purine metabolism.[234–239] The distinct spectral properties of [th]I and [th]A enabled, for the first time, monitoring of the enzymatic deamination in real time using absorption and fluorescence spectroscopy and development of a high-throughput assay for the discovery and evaluation of ADA inhibitors.[240]

2.4   SUMMARY AND PERSPECTIVE

The selected examples of designer fluorescent biomolecular building blocks shown here, whether or not inspired by naturally occurring counterparts, demonstrate that (1) insights into the SPPR enable new and fluorescent probe designs with tunable properties, (2) minimal (isomorphic) structural modifications can be reconciled with favorable spectral properties, and (3) these probes are invaluable for the ongoing unraveling of the structure and function of biomolecules. The development and application of fluorescent biomolecular building blocks is ongoing, vibrant, and creative. Without a doubt, the future looks bright!

REFERENCES

1. Eftink, M.; Brand, L.; Johnson, M. L. *Methods Enzymol.* **1997**, *278*, 221.
2. Gokel, G. W. Dean's Handbook of Organic Chemistry. 2nd ed.; McGraw-Hill: New York, **2004**.
3. Ross, J. A.; Jameson, D. M. *Photochem. Photobiol. Sci.* **2008**, *7*, 1301.
4. Munishkina, L. A.; Fink, A. L. *Biochim. Biophys. Acta.* **2007**, *1768*, 1862.
5. Bazin, H.; Zhou, X. X.; Glemarec, C.; Chattopadhyaya, J. *Tetrahedron Lett.* **1987**, *28*, 3275.
6. Nakatsuka, S.-i.; Ohgi, T.; Goto, T., *Tetrahedron Lett.* **1978**, *19*, 2579.
7. Maelicke, A.; Von Der Haar, F.; Sprinzl, M.; Cramer, F. *Biopolymers* **1975**, *14*, 155.
8. Pierce, D. W.; Boxer, S. G. *Biophys. J.* **1995**, *68*, 1583.
9. Eftink, M.; Shastry, M. C. R.; Brand, L.; Johnson, M. L. *Methods Enzymol.* **1997**, *278*, 258.
10. Royer, C. A. *Chem. Rev.* **2006**, *106*, 1769.
11. Engelborghs, Y. *J. Fluorescence* **2003**, *13*, 9.
12. Eckey, E. W.; Miller, L. P. Vegetable Fats and Oils. Reinhold: New York, **1954**.
13. Reichardt, C. *Chem. Rev.* **1994**, *94*, 2319.
14. Radda, G. K.; Smith, D. S. *FEBS Lett.* **1970**, *9*, 287.
15. Sklar, L. A.; Hudson, B. S.; Petersen, M.; Diamond, J. *Biochemistry* **1977**, *16*, 813.
16. Sklar, L. A.; Hudson, B. S.; Simoni, R. D. *Biochemistry* **1977**, *16*, 819.
17. Hac-Wydro, K.; Wydro, P. *Chem. Phys. Lipids* **2007**, *150*, 66.
18. Chance, B. *Biomembranes* **1975**, *7*, 33.
19. Bittman, R.; Chen, W. C.; Anderson, O. R. *Biochemistry* **1974**, *13*, 1364.
20. Sprecher, C. A.; Johnson, W. C. *Biopolymers* **1977**, *16*, 2243.
21. Callis, P. R. *Annu. Rev. Phys. Chem.* **1983**, *34*, 329.
22. Peon, J.; Zewail, A. H. *Chem. Phys. Lett.* **2001**, *348*, 255.
23. Onidas, D.; Markovitsi, D.; Marguet, S.; Sharonov, A.; Gustavsson, T. *J. Phys. Chem. B* **2002**, *106*, 11367.
24. Cohen, B.; Crespo-Hernandez, C. E.; Kohler, B. *Faraday Discuss.* **2004**, *127*, 137.
25. Limbach, P. A.; Crain, P. F.; McCloskey, J. A. *Nucleic Acids Res.* **1994**, *22*, 2183.
26. de Crécy-Lagard, V.; Brochier-Armanet, C.; Urbonavičius, J.; Fernandez, B.; Phillips, G.; Lyons, B.; Noma, A.; Alvarez, S.; Droogmans, L.; Armengaud, J.; Grosjean, H. *Mol. Biol. Evol.* **2010**, *27*, 2062.
27. RajBhandary, U. L.; Chang, S. H.; Stuart, A.; Faulkner, R. D.; Hoskinson, R. M.; Khorana, H. G. *Proc. Natl. Acad. Sci. U. S. A.* **1967**, *57*, 751.
28. Thiebe, R.; Zachau, H. G. *Eur. J. Biochem.* **1968**, *5*, 546.
29. Nakanishi, K.; Furutachi, N.; Funamizu, M.; Grunberger, D.; Weinstein, I. B. *J. Am. Chem. Soc.* **1970**, *92*, 7617.
30. Itaya, T.; Kanai, T.; Sawada, T. *Chem. Pharm. Bull.* **2002**, *50*, 547.
31. Nakanishi, K.; Blobstein, S.; Funamizu, M.; Furutachi, N.; Van Lear, G.; Grunberger, D.; Lanks, K. W.; Weinstein, I. B. *Nat. New Biol.* **1971**, *234*, 107.
32. Blobstein, S. H.; Grunberger, D.; Weinstein, I. B.; Nakanishi, K. *Biochemistry* **1973**, *12*, 188.

33. Feinberg, A. M.; Nakanishi, K.; Barciszewski, J.; Rafalski, A. J.; Augustyniak, H.; Wiewiorowski, M. *J. Am. Chem. Soc.* **1974**, *96*, 7797.

34. See one of the following chapters by Sinkeldam, R. W. and Tor, Y.

35. Sinkeldam, R. W.; Marcus, P.; Uchenik, D.; Tor, Y. *ChemPhysChem* **2011**, *12*, 2260.

36. Moody, E. M.; Brown, T. S.; Bevilacqua, P. C. *J. Am. Chem. Soc.* **2004**, *126*, 10200.

37. Sinkeldam, R. W.; Greco, N. J.; Tor, Y. *ChemBioChem* **2008**, *9*, 706.

38. Sinkeldam, R. W.; Hopkins, P. A.; Tor, Y. *ChemPhysChem* **2012**, *13*, 3350.

39. Hopkins, P. A.; Sinkeldam, R. W.; Tor, Y. *Org. Lett.* **2014**, *16*, 5290.

40. Sinkeldam, R. W.; Wheat, A. J.; Boyaci, H.; Tor, Y. *ChemPhysChem* **2011**, *12*, 567.

41. Gorter, E.; Grendel, F. *J. Exp. Med.* **1925**, *41*, 439.

42. Singer, S. J.; Nicolson, G. L. *Science* **1972**, *175*, 720.

43. Alberts, B.; Johnson, A.; Lewis, J.; Raff, M.; Roberts, K.; Walter, P. Molecular Biology of the Cell. 4th ed.; Garland Science: New York, **2002**.

44. Radda, G. K.; Vanderkooi, J. *Biochim. Biophys. Acta* **1972**, *265*, 509.

45. Brand, L.; Gohlke, J. R. *Annu. Rev. Biochem.* **1972**, *41*, 843.

46. Azzi, A. *Q. Rev. Biophys.* **1975**, *8*, 237.

47. Maier, O.; Oberle, V.; Hoekstra, D. *Chem. Phys. Lipids* **2002**, *116*, 3.

48. Rasmussen, J. A. M.; Hermetter, A. *Prog. Lipid Res.* **2008**, *47*, 436.

49. Cairo, C. W.; Key, J. A.; Sadek, C. M. *Curr. Opin. Chem. Biol.* **2010**, *14*, 57.

50. Sinkeldam, R. W.; Greco, N. J.; Tor, Y. *Chem. Rev.* **2010**, *110*, 2579.

51. Loura, L. M. S.; Ramalho, J. P. P. *Molecules* **2011**, *16*, 5437.

52. Klymchenko, A. S.; Kreder, R. *Chem. Biol.* **2014**, *21*, 97.

53. Waka, Y.; Mataga, N.; Tanaka, F. *Photochem. Photobiol.* **1980**, *32*, 335.

54. Bondarev, S. L.; Bachilo, S. M. *J. Photochem. Photobiol., A* **1991**, *59*, 273.

55. Cehelnik, E. D.; Cundall, R. B.; Lockwood, J. R.; Palmer, T. F. *J. Phys. Chem.* **1975**, *79*, 1369.

56. Klymchenko, A. S.; Demchenko, A. P. *Phys. Chem. Chem. Phys.* **2003**, *5*, 461.

57. Klymchenko, A. S.; Ozturk, T.; Demchenko, A. P. *Tetrahedron Lett.* **2002**, *43*, 7079.

58. Starck, J. P.; Nakatani, Y.; Ourisson, G. *Tetrahedron* **1995**, *51*, 2629.

59. Acuna, A. U.; Amat-Guerri, F.; Quesada, E.; Velez, M. *Biophys. Chem.* **2006**, *122*, 27.

60. Lentz, B. R.; Barenholz, Y.; Thompson, T. E. *Biochemistry* **1976**, *15*, 4521.

61. Shinitzk, M.; Inbar, M. *J. Mol. Biol.* **1974**, *85*, 603.

62. Andrich, M. P.; Vanderkooi, J. M. *Biochemistry* **1976**, *15*, 1257.

63. Thulborn, K. R.; Sawyer, W. H. *Biochim. Biophys. Acta* **1978**, *511*, 125.

64. Thulborn, K. R.; Treloar, F. E.; Sawyer, W. H. *Biochem. Biophys. Res. Commun.* **1978**, *81*, 42.

65. Haidekker, M. A.; Brady, T. P.; Lichlyter, D.; Theodorakis, E. A. *Bioorg. Chem.* **2005**, *33*, 415.

66. Haidekker, M. A.; Theodorakis, E. A. *Org. Biomol. Chem.* **2007**, *5*, 1669.

67. Badley, R. A.; Schneider, H.; Martin, W. G. *Biochemistry* **1973**, *12*, 268.

68. Kellner, B. M. J.; Cadenhead, D. A. *Biochim. Biophys. Acta* **1978**, *513*, 301.

69. Simons, K.; Ikonen, E. *Nature* **1997**, *387*, 569.

70. Somerharju, P.; Virtanen, J. A.; Cheng, K. H. *Biochim. Biophys. Acta.* **1999**, *1440*, 32.

71. Prendergast, F. G.; Haugland, R. P.; Callahan, P. J. *Biochemistry* **1981**, *20*, 7333.

72. Waggoner, A. S.; Stryer, L. *Proc. Natl. Acad. Sci. U. S. A.* **1970**, *67*, 579.

73. Liu, J.; Blumenthal, K. M. *J. Biol. Chem.* **1988**, *263*, 6619.

74. Liu, J. W.; Blumenthal, K. M. *Biochim. Biophys. Acta* **1988**, *937*, 153.

75. Bernik, D. L.; Negri, R. M. *J. Colloid Interface Sci.* **1998**, *203*, 97.

76. Soh, N.; Makihara, K.; Ariyoshi, T.; Seto, D.; Maki, T.; Nakajima, H.; Nakano, K.; Imato, T. *Anal. Sci.* **2008**, *24*, 293.

77. Chattopadhyay, A. *Chem. Phys. Lipids* **1990**, *53*, 1.

78. Lapinski, M. M.; Blanchard, G. J. *Chem. Phys. Lipids* **2007**, *150*, 12.

79. Blatt, E.; Sawyer, W. H. *Biochim. Biophys. Acta* **1985**, *822*, 43.

80. Lenard, J.; Wong, C. Y.; Compans, R. W. *Biochim. Biophys. Acta* **1974**, *332*, 341.

81. Jones, M. E.; Lentz, B. R. *Biochemistry* **1986**, *25*, 567.

82. Encinas, M. V.; Lissi, E. A.; Alvarez, J. *Photochem. Photobiol.* **1994**, *59*, 30.

83. Bondar, O. P.; Rowe, E. S. *Biochim. Biophys. Acta.* **1998**, *1370*, 207.

84. Somerharju, P. J.; Virtanen, J. A.; Eklund, K. K.; Vainio, P.; Kinnunen, P. K. J. *Biochemistry* **1985**, *24*, 2773.

85. Tang, D.; Chong, P. L. G. *Biophys. J.* **1992**, *63*, 903.

86. Chong, P. L. G.; Tang, D.; Sugar, I. P. *Biophys. J.* **1994**, *66*, 2029.

87. Nicolay, K.; Hovius, R.; Bron, R.; Wirtz, K.; Dekruijff, B. *Biochim. Biophys. Acta* **1990**, *1025*, 49.

88. Langner, M.; Hui, S. W. *Biochim. Biophys. Acta.* **2000**, *1463*, 439.

89. Lala, A. K. *Chem. Phys. Lipids* **2002**, *116*, 177.

90. Lala, A. K.; Koppaka, V. *Biochemistry* **1992**, *31*, 5586.

91. Lala, A. K.; Dixit, R. R.; Koppaka, V.; Patel, S. *Biochemistry* **1988**, *27*, 8981.

92. Asuncion-Punzalan, E.; Kachel, K.; London, E. *Biochemistry* **1998**, *37*, 4603.

93. Kaiser, R. D.; London, E. *Biochemistry* **1998**, *37*, 8180.

94. Kutchai, H.; Chandler, L. H.; Zavoico, G. B. *Biochim. Biophys. Acta* **1983**, *736*, 137.

95. Blatt, E.; Sawyer, W. H.; Ghiggino, K. P. *Aus. J. Chem.* **1983**, *36*, 1079.

96. Vincent, M.; Deforesta, B.; Gallay, J.; Alfsen, A. *Biochem. Biophys. Res. Commun.* **1982**, *107*, 914.

97. Uemura, A.; Kimura, S.; Imanishi, Y. *Biochim. Biophys. Acta* **1983**, *729*, 28.

98. Quesada, E.; Ardhammar, M.; Norden, B.; Miesch, M.; Duportail, G.; Bonzi-Coulibaly, Y.; Nakatani, Y.; Ourisson, G. *Helv. Chim. Acta* **2000**, *83*, 2464.

99. Ventelon, L.; Charier, S.; Moreaux, L.; Mertz, J.; Blanchard-Desce, M. *Angew. Chem. Int. Ed.* **2001**, *40*, 2098.

100. Kucherak, O. A.; Oncul, S.; Darwich, Z.; Yushchenko, D. A.; Arntz, Y.; Didier, P.; Mély, Y.; Klymchenko, A. S. *J. Am. Chem. Soc.* **2010**, *132*, 4907.

101. Darwich, Z.; Klymchenko, A. S.; Kucherak, O. A.; Richert, L.; Mély, Y. *Biochim. Biophys. Acta.* **2012**, *1818*, 3048.

102. Shynkar, V. V.; Klymchenko, A. S.; Kunzelmann, C.; Duportail, G.; Muller, C. D.; Demchenko, A. P.; Freyssinet, J.-M.; Mely, Y. *J. Am. Chem. Soc.* **2007**, *129*, 2187.

103. Chou, P. T.; Martinez, M. L.; Clements, J. H. *J. Phys. Chem.* **1993**, *97*, 2618.

104. Kwiatek, J. M.; Owen, D. M.; Abu-Siniyeh, A.; Yan, P.; Loew, L. M.; Gaus, K. *Plos One* **2013**, *8*, e52960.

105. Kuerschner, L.; Ejsing, C. S.; Ekroos, K.; Shevchenko, A.; Anderson, K. I.; Thiele, C. *Nat. Methods* **2005**, *2*, 39.

106. Quesada, E.; Acuna, A. U.; Amat-Guerri, F. *Angew. Chem. Int. Ed.* **2001**, *40*, 2095.

107. Connor, R. E.; Tirrell, D. A. *J. Macromol. Sci., Polym. Rev.* **2007**, *47*, 9.

108. Sisido, M. *Prog. Polym. Sci.* **1992**, *17*, 699.

109. Hohsaka, T.; Sisido, M. *Curr. Opin. Chem. Biol.* **2002**, *6*, 809.

110. Twine, S. M.; Szabo, A. G. *Biophotonics, Pt A* **2003**, *360*, 104.

111. Twine, S. M.; Szabo, A. G. *Methods Enzymol.* **2003**, *360*, 104.

112. Katritzky, A. R.; Narindoshvili, T. *Org. Biomol. Chem.* **2009**, *7*, 627.

113. Krueger, A. T.; Imperiali, B. *ChemBioChem* **2013**, *14*, 788.

114. Ross, J. B. A.; Szabo, A. G.; Hogue, C. W. V. *Methods Enzymol.* **1997**, *278*, 151.

115. Broos, J.; Gabellieri, E.; Biemans-Oldehinkel, E.; Strambini, G. B. *Protein Sci.* **2003**, *12*, 1991.

116. Hirayama, K. Handbook of Ultraviolet and Visible Absorption Spectra of Organic Compounds. Springer-Verlag: New York, **1967**.

117. Kenner, R. A.; Aboderin, A. A. *Biochemistry* **1971**, *10*, 4433.

118. Sinkeldam, R. W.; Tor, Y. *Org. Biomol. Chem.* **2007**, *5*, 2523.

119. Vazquez, M. E.; Blanco, J. B.; Imperiali, B. *J. Am. Chem. Soc.* **2005**, *127*, 1300.

120. Weber, G.; Farris, F. J. *Biochemistry* **1979**, *18*, 3075.

121. De Filippis, V.; De Boni, S.; De Dea, E.; Dalzoppo, D.; Grandi, C.; Fontana, A. *Protein Sci.* **2004**, *13*, 1489.

122. Lotte, K.; Plessow, R.; Brockhinke, A. *Photochem. Photobiol. Sci.* **2004**, *3*, 348.

123. Steward, L. E.; Collins, C. S.; Gilmore, M. A.; Carlson, J. E.; Ross, J. B. A.; Chamberlin, A. R. *J. Am. Chem. Soc.* **1997**, *119*, 6.

124. Talukder, P.; Chen, S.; Arce, P. M.; Hecht, S. M. *Org. Lett.* **2014**, *16*, 556.

125. Egusa, S.; Sisido, M.; Imanishi, Y. *Chem. Lett.* **1983**, 1307.

126. Ghosh, P. B.; Whitehouse, M. W. *Biochem. J.* **1968**, *108*, 155.

127. Lancet, D.; Pecht, I. *Biochemistry* **1977**, *16*, 5150.

128. Turcatti, G.; Nemeth, K.; Edgerton, M. D.; Meseth, U.; Talabot, F.; Peitsch, M.; Knowles, J.; Vogel, H.; Chollet, A. *J. Biol. Chem.* **1996**, *271*, 19991.

129. Summerer, D.; Chen, S.; Wu, N.; Deiters, A.; Chin, J. W.; Schultz, P. G. *Proc. Natl. Acad. Sci. U. S. A.* **2006**, *103*, 9785.

130. Loving, G.; Imperiali, B. *J. Am. Chem. Soc.* **2008**, *130*, 13630.

131. Torok, K.; Cowley, D. J.; Brandmeier, B. D.; Howell, S.; Aitken, A.; Trentham, D. R. *Biochemistry* **1998**, *37*, 6188.

132. Zuhlke, R. D.; Pitt, G. S.; Deisseroth, K.; Tsien, R. W.; Reuter, H. *Nature* **1999**, *399*, 159.

133. Cohen, B. E.; McAnaney, T. B.; Park, E. S.; Jan, Y. N.; Boxer, S. G.; Jan, L. Y. *Science* **2002**, *296*, 1700.

134. Sundd, M.; Robertson, A. D. *Nat. Struct. Biol.* **2002**, *9*, 500.

135. Watson, J. D.; Crick, F. H. *Nature* **1953**, *171*, 737.

136. Ward, D. C.; Reich, E.; Stryer, L. *J. Biol. Chem.* **1969**, *244*, 1228.

137. Secrist, J. A.; Barrio, J. R.; Leonard, N. J. *Science* **1972**, *175*, 646.

138. Greco, N. J.; Tor, Y. *J. Am. Chem. Soc.* **2005**, *127*, 10784.

139. Hawkins, M. E. *Cell Biochem. Biophys.* **2001**, *34*, 257.

140. Rist, M. J.; Marino, J. P. *Curr. Org. Chem.* **2002**, *6*, 775.

141. Ranasinghe, R. T.; Brown, T. *Chem. Commun.* **2005**, 5487.

142. Wilson, J. N.; Kool, E. T. *Org. Biomol. Chem.* **2006**, *4*, 4265.

143. Asseline, U. *Curr. Org. Chem.* **2006**, *10*, 491.

144. Wagenknecht, H. A. *Ann. N.Y. Acad. Sci.* **2008**, *1130*, 122.

145. Hawkins, M. E.; Brand, L.; Johnson, M. L. *Methods Enzymol.* **2008**, *450*, 201.

146. Dodd, D. W.; Hudson, R. H. E. *Mini Rev. Org. Chem.* **2009**, *6*, 378.

147. Wilhelmsson, L. M. *Q. Rev. Biophys.* **2010**, *43*, 159.

148. Tanpure, A. A.; Pawar, M. G.; Srivatsan, S. G. *Isr. J. Chem.* **2013**, *53*, 366.

149. Coleman, R. S.; Madaras, M. L. *J. Org. Chem.* **1998**, *63*, 5700.

150. Brauns, E. B.; Madaras, M. L.; Coleman, R. S.; Murphy, C. J.; Berg, M. A. *J. Am. Chem. Soc.* **1999**, *121*, 11644.

151. Brauns, E. B.; Madaras, M. L.; Coleman, R. S.; Murphy, C. J.; Berg, M. A. *Phys. Rev. Lett.* **2002**, *88*, 158101.

152. Somoza, M. M.; Andreatta, D.; Murphy, C. J.; Coleman, R. S.; Berg, M. A. *Nucleic Acids Res.* **2004**, *32*, 2494.

153. Andreatta, D.; Perez Lustres, J. L.; Kovalenko, S. A.; Ernsting, N. P.; Murphy, C. J.; Coleman, R. S.; Berg, M. A. *J. Am. Chem. Soc.* **2005**, *127*, 7270.

154. Berg, M. A.; Coleman, R. S.; Murphy, C. J. *Phys. Chem. Chem. Phys.* **2008**, *10*, 1229.

155. Matray, T. J.; Kool, E. T. *Nature* **1999**, *399*, 704.

156. Lee, S. H.; Wang, S.; Kool, E. T. *Chem. Commun.* **2012**, *48*, 8069.

157. Jung, J.-W.; Edwards, S. K.; Kool, E. T. *ChemBioChem* **2013**, *14*, 440.

158. Wang, S.; Guo, J.; Ono, T.; Kool, E. T. *Angew. Chem. Int. Ed.* **2012**, *51*, 7176.

159. Teo, Y. N.; Wilson, J. N.; Kool, E. T. *J. Am. Chem. Soc.* **2009**, *131*, 3923.

160. Kwon, H.; Samain, F.; Kool, E. T. *Chem. Sci.* **2012**, *3*, 2542.

161. Teo, Y. N.; Kool, E. T. *Chem. Rev.* **2012**, *112*, 4221.

162. Okamoto, A.; Saito, Y.; Saito, I. *J. Photochem. Photobiol., C* **2005**, *6*, 108.

163. Schmucker, W.; Wagenknecht, H.-A. *Synlett* **2012**, *23*, 2435.

164. Thoresen, L. H.; Jiao, G. S.; Haaland, W. C.; Metzker, M. L.; Burgess, K. *Chem. Eur. J.* **2003**, *9*, 4603.

165. Hurley, D. J.; Seaman, S. E.; Mazura, J. C.; Tor, Y. *Org. Lett.* **2002**, *4*, 2305.

166. Hurley, D. J.; Tor, Y. *J. Am. Chem. Soc.* **2002**, *124*, 3749.

167. Hurley, D. J.; Tor, Y. *J. Am. Chem. Soc.* **2002**, *124*, 13231.

168. Tanpure, A. A.; Srivatsan, S. G. *ChemBioChem* **2014**, 1309.

169. Saha, S.; Samanta, A. *J. Phys. Chem. A* **2002**, *106*, 4763.

170. Pawar, M. G.; Srivatsan, S. G. *Org. Lett.* **2011**, *13*, 1114.

171. Pawar, M. G.; Srivatsan, S. G. *J. Phys. Chem. B* **2013**, *117*, 14273.

172. Tanpure, A. A.; Srivatsan, S. G. *ChemBioChem* **2012**, *13*, 2392.

173. Lin, K. Y.; Jones, R. J.; Matteucci, M. *J. Am. Chem. Soc.* **1995**, *117*, 3873.

174. Engman, K. C.; Sandin, P.; Osborne, S.; Brown, T.; Billeter, M.; Lincoln, P.; Nordén, B.; Albinsson, B.; Wilhelmsson, L. M. *Nucleic Acids Res.* **2004**, *32*, 5087.

175. Sandin, P.; Wilhelmsson, L. M.; Lincoln, P.; Powers, V. E.; Brown, T.; Albinsson, B. *Nucleic Acids Res.* **2005**, *33*, 5019.

176. Sandin, P.; Börjesson, K.; Li, H.; Mårtensson, J.; Brown, T.; Wilhelmsson, L. M.; Albinsson, B. *Nucleic Acids Res.* **2008**, *36*, 157.

177. Sandin, P.; Stengel, G.; Ljungdahl, T.; Borjesson, K.; Macao, B.; Wilhelmsson, L. M. *Nucleic Acids Res.* **2009**, *37*, 3924.

178. Walsh, J. M.; Bouamaied, I.; Brown, T.; Wilhelmsson, L. M.; Beuning, P. J. *J. Mol. Biol.* **2011**, *409*, 89.

179. Börjesson, K.; Preus, S.; El-Sagheer, A. H.; Brown, T.; Albinsson, B.; Wilhelmsson, L. M. *J. Am. Chem. Soc.* **2009**, *131*, 4288.

180. Preus, S.; Wilhelmsson, L. M. *ChemBioChem* **2012**, *13*, 1990.

181. Preus, S.; Kilså, K.; Miannay, F.-A.; Albinsson, B.; Wilhelmsson, L. M. *Nucleic Acids Res.* **2013**, *41*, e18.

182. Sellrie, F.; Lenz, C.; Andersson, A.; Wilhelmsson, L. M.; Schenk, J. A. *Talanta* **2014**, *124*, 67.

183. Preus, S.; Jonck, S.; Pittelkow, M.; Dierckx, A.; Karpkird, T.; Albinsson, B.; Wilhelmsson, L. M. *Photochem. Photobiol. Sci.* **2013**, *12*, 1416.

184. Nasr, T.; Li, Z.; Nakagawa, O.; Taniguchi, Y.; Ono, S.; Sasaki, S. *Bioorg. Med. Chem. Lett.* **2009**, *19*, 727.

185. Nakagawa, O.; Ono, S.; Li, Z.; Tsujimoto, A.; Sasaki, S. *Angew. Chem. Int. Ed.* **2007**, *46*, 4500.

186. Nakagawa, O.; Ono, S.; Tsujimoto, A.; Li, Z.; Sasaki, S. *Nucleos. Nucleot. Nucl.* **2007**, *26*, 645.

187. Driscoll, S. L.; Hawkins, M. E.; Balis, F. M.; Pfleiderer, W.; Laws, W. R. *Biophys. J.* **1997**, *73*, 3277.

188. Hawkins, M. E.; Pfleiderer, W.; Balis, F. M.; Porter, D.; Knutson, J. R. *Anal. Biochem.* **1997**, *244*, 86.

189. Xie, Y.; Dix, A. V.; Tor, Y. *J. Am. Chem. Soc.* **2009**, *131*, 17605.

190. Xie, Y.; Maxson, T.; Tor, Y. *J. Am. Chem. Soc.* **2010**, *132*, 11896.

191. Xie, Y.; Dix, A. V.; Tor, Y. *Chem. Commun.* **2010**, *46*, 5542.

192. Xie, Y.; Maxson, T.; Tor, Y. *Org. Biomol. Chem.* **2010**, *8*, 5053.

193. Liu, C. H.; Martin, C. T. *J. Mol. Biol.* **2001**, *308*, 465.

194. Berry, D. A.; Jung, K. Y.; Wise, D. S.; Sercel, A. D.; Pearson, W. H.; Mackie, H.; Randolph, J. B.; Somers, R. L. *Tetrahedron Lett.* **2004**, *45*, 2457.

195. Tinsley, R. A.; Walter, N. G. *RNA* **2006**, *12*, 522–529.

196. Roy, J.; Chirania, P.; Ganguly, S.; Huang, H. *Bioorg. Med. Chem. Lett.* **2012**, *22*, 863.

197. Hudson, R. H. E.; Ghorbani-Choghamarani, A. *Org. Biomol. Chem.* **2007**, *5*, 1845.

198. Hudson, R. H. E.; Choghamarani, A. G. *Nucleos. Nucleot. Nucl.* **2007**, *26*, 533.

199. Hudson, R. H. E.; Ghorbani-Choghamarani, A. *Synlett* **2007**, 870.

200. Gerrard, S. R.; Edrees, M. M.; Bouamaied, I.; Fox, K. R.; Brown, T. *Org. Biomol. Chem.* **2010**, *8*, 5087.

201. Noé, M. S.; Ríos, A. C.; Tor, Y. *Org. Lett.* **2012**, *14*, 3150.

202. Roblin, R. O.; Lampen, J. O.; English, J. P.; Cole, Q. P.; Vaughan, J. R. *J. Am. Chem. Soc.* **1945**, *67*, 290.

203. Wierzchowski, J.; Wielgus-Kutrowska, B.; Shugar, D. *Biochim. Biophys. Acta.* **1996**, *1290*, 9.

204. Wierzchowski, J.; Ogiela, M.; Iwańska, B.; Shugar, D. *Anal. Chim. Acta* **2002**, *472*, 63.

205. Da Costa, C. P.; Fedor, M. J.; Scott, L. G. *J. Am. Chem. Soc.* **2007**, *129*, 3426.

206. Liu, L.; Cottrell, J. W.; Scott, L. G.; Fedor, M. J. *Nat. Chem. Biol.* **2009**, *5*, 351.

207. Gaied, N. B.; Glasser, N.; Ramalanjaona, N.; Beltz, H.; Wolff, P.; Marquet, R.; Burger, A.; Mely, Y. *Nucleic Acids Res.* **2005**, *33*, 1031.

208. Kenfack, C. A.; Burger, A.; Mely, Y. *J. Phys. Chem. B* **2006**, *110*, 26327.

209. Kenfack, C. A.; Piemont, E.; Ben Gaied, N.; Burger, A.; Mely, Y. *J. Phys. Chem. B* **2008**, *112*, 9736.

210. Nadler, A.; Strohmeier, J.; Diederichsen, U. *Angew. Chem. Int. Ed.* **2011**, *50*, 5392.

211. Müllar, S.; Strohmeier, J.; Diederichsen, U. *Org. Lett.* **2012**, *14*, 1382.

212. Vummidi, B. R.; Alzeer, J.; Luedtke, N. W. *ChemBioChem* **2013**, *14*, 540.

213. Dumas, A.; Luedtke, N. W. *ChemBioChem* **2011**, *12*, 2044.

214. Dumas, A.; Luedtke, N. W. *Nucleic Acids Res.* **2011**, *39*, 6825.

215. Dumas, A.; Luedtke, N. W. *J. Am. Chem. Soc.* **2010**, *132*, 18004.

216. Dumas, A.; Luedtke, N. W. *Chemistry* **2012**, *18*, 245.

217. Srivatsan, S. G.; Tor, Y. *J. Am. Chem. Soc.* **2007**, *129*, 2044.

218. Srivatsan, S. G.; Tor, Y. *Tetrahedron* **2007**, *63*, 3601.

219. Greco, N. J.; Tor, Y. *Tetrahedron* **2007**, *63*, 3515.

220. Pesnot, T.; Tedaldi, L. M.; Jambrina, P. G.; Rosta, E.; Wagner, G. K. *Org. Biomol. Chem.* **2013**, *11*, 6357.

221. Pawar, M. G.; Nuthanakanti, A.; Srivatsan, S. G. *Bioconjug. Chem.* **2013**, *24*, 1367.

222. Greco, N. J.; Sinkeldam, R. W.; Tor, Y. *Org. Lett.* **2009**, *11*, 1115.

223. Förster, T.; Hoffmann, G. *Z. Phys. Chem.* **1971**, *75*, 63.

224. Law, K. Y. *Chem. Phys. Lett.* **1980**, *75*, 545.

225. Lane, R. S. K.; Jones, R.; Sinkeldam, R. W.; Tor, Y.; Magennis, S. W. *ChemPhysChem* **2014**, *15*, 867.

226. Srivatsan, S. G.; Tor, Y. *Chem. Asian J.* **2009**, *4*, 419.

227. Srivatsan, S. G.; Weizman, H.; Tor, Y. *Org. Biomol. Chem.* **2008**, *6*, 1334.

228. Srivatsan, S. G.; Greco, N. J.; Tor, Y. *Angew. Chem. Int. Ed.* **2008**, *47*, 6661.

229. Shin, D.; Sinkeldam, R. W.; Tor, Y. *J. Am. Chem. Soc.* **2011**, *133*, 14912.

230. Liu, W.; Shin, D.; Tor, Y.; Cooperman, B. S. *ACS Chem. Biol.* **2013**, *8*, 2017.

231. McCoy, L. S.; Shin, D.; Tor, Y. *J. Am. Chem. Soc.* **2014**, *136*, 15176.

232. Lee, Y.-J.; Jang, Y. H.; Kim, Y.; Hwang, S. *Bull. Korean Chem. Soc.* **2012**, *33*, 4255.

233. Gedik, M.; Brown, A. *J. Photochem. Photobiol., A* **2013**, *259*, 25.

234. Wilson, D. K.; Rudolph, F. B.; Quiocho, F. A. *Science* **1991**, *252*, 1278.

235. Kinoshita, T.; Nakanishi, I.; Terasaka, T.; Kuno, M.; Seki, N.; Warizaya, M.; Matsumura, H.; Inoue, T.; Takano, K.; Adachi, H.; Mori, Y.; Fujii, T. *Biochemistry* **2005**, *44*, 10562.

236. Kinoshita, T.; Nishio, N.; Nakanishi, I.; Sato, A.; Fujii, T. *Acta Crystallogr. D. Biol. Crystallogr.* **2003**, *59*, 299.

237. Cristalli, G.; Costanzi, S.; Lambertucci, C.; Lupidi, G.; Vittori, S.; Volpini, R.; Camaioni, E. *Med. Res. Rev.* **2001**, *21*, 105.

238. Glazer, R. I. *Cancer Chemother. Pharmacol.* **1980**, *4*, 227.

239. Odwyer, P. J.; Wagner, B.; Leylandjones, B.; Wittes, R. E.; Cheson, B. D.; Hoth, D. F. *Ann. Intern. Med.* **1988**, *108*, 733.

240. Sinkeldam, R. W.; McCoy, L. S.; Shin, D.; Tor, Y. *Angew. Chem.* **2013**, *125*, 14276.

3

POLARIZED SPECTROSCOPY WITH FLUORESCENT BIOMOLECULAR BUILDING BLOCKS

BO ALBINSSON AND BENGT NORDÉN

Chemistry and Chemical Engineering/Chemistry and Biochemistry, Chalmers University of Technology, Gothenburg, Sweden

3.1 TRANSITION MOMENTS

A fundamental property for the understanding and quantitative interpretation of any absorption and fluorescence measurement is the quantum mechanical concept known as the electric dipole transition moment of the transitions studied:

$$\boldsymbol{\mu}_{0j} = \langle j|\boldsymbol{\mu}|0\rangle \tag{3.1}$$

where $|0\rangle$ and $|j\rangle$ refer to ground and excited states, respectively, and $\boldsymbol{\mu}$ is the electric dipole operator $\sum q_i \mathbf{r}_i$, with q_i the charge at position $\mathbf{r}_i$, summation over all electrons and atomic nuclei of the molecule. $\boldsymbol{\mu}_{0j}$ has vector properties whose direction in the molecular frame is determined by the properties of the respective molecular wave functions of the $|0\rangle$ and $|j\rangle$ states, that is, for the electronic states the symmetry properties of the respective molecular orbitals involved in the transition. The transition moment may be regarded as a molecular antenna by which light can be absorbed through electric field interaction. It determines the fast (sub fs) response to photon–matter interaction whether it is annihilation (absorption) or creation

Fluorescent Analogs of Biomolecular Building Blocks: Design and Applications, First Edition.
Edited by Marcus Wilhelmsson and Yitzhak Tor.
© 2016 John Wiley & Sons, Inc. Published 2016 by John Wiley & Sons, Inc.

(luminescence) of a photon. Both absorption and fluorescence can be related to transition probability, and intensity of absorption/emission is proportional to

$$|\langle j|\mathbf{\mu}|0\rangle \cdot \mathbf{E}|^2 \propto \cos^2\phi|\mathbf{\mu}_{0j}|^2 \tag{3.2}$$

with ϕ the angle between the photon electric field $\mathbf{E}$ and the transition moment. $|\mathbf{\mu}_{0j}|^2$ is the dipole strength, proportional to the area under the corresponding absorption (emission) band. In the following, we shall consider linear dichroism (LD), magnetic circular dichroism (MCD), and Förster resonance energy transfer (FRET) and how these phenomena are related to and may report on transition moment orientations and on molecular structure.

3.2   LINEAR DICHROISM

Linear dichroism (for reviews and textbooks on polarized light spectroscopy, see Refs. [1–4]) is the difference between absorption of light polarized with the electric field vector parallel to two orthogonal reference directions, the laboratory-fixed axes Z and Y:

$$\text{LD} = A_Z - A_Y \propto \left(\cos^2\phi_Z - \cos^2\phi_Y\right)|\mathbf{\mu}_{0j}|^2 \tag{3.3}$$

where ϕ_Z and ϕ_Y are the angles between $\mathbf{\mu}_{0j}$ and Z and Y, respectively. For a sample of macroscopically oriented molecules with uniaxial distribution relative to axis Z:

$$\left\langle \cos^2\phi_Y \right\rangle = \frac{1}{2}\left\langle \sin^2\phi_Z \right\rangle \tag{3.4}$$

so that

$$\text{LD} = A_Z - A_y \propto \frac{1}{2}\left(3\left\langle \cos^2\phi_Z \right\rangle - 1\right)|\mathbf{\mu}_{0j}|^2 \tag{3.5}$$

The absorbance of the corresponding isotropic sample is

$$A_{\text{iso}} = \frac{1}{3}\left(A_Z + 2A_y\right) \propto \frac{1}{3}|\mathbf{\mu}_{0j}|^2 \tag{3.6}$$

and the dimensionless quantity, "reduced" linear dichroism, LD_r:

$$\text{LD}_r = \text{LD}/A_{\text{iso}} = \frac{3}{2}\left(3\left\langle \cos^2\phi_Z \right\rangle - 1\right) \tag{3.7a}$$

Obviously $-1.5 \leq \text{LD}_r \leq +3$ corresponding, respectively, to perpendicular and parallel orientations of the transition moment with respect to the unique axis. At the "magic" angle 54.7° LD vanishes, as well as for perfectly isotropic distributions for which $\left\langle \cos^2\phi_Z \right\rangle = \frac{1}{3}$. For DNA and other "universal-joint" flexible macromolecules orienting like a rod, Equation 3.7a can be complemented with a single orientation

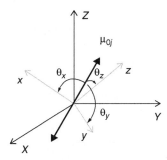

Figure 3.1 Schematic showing the molecular fixed (lower case x, y, z) and laboratory coordinate systems (upper case X, Y, Z). The projection of the transition moment, μ_{0j}, on the molecule axis is shown by the angles θ_x, θ_y, and θ_z.

factor, S, describing the degree of polymer alignment (i.e., orientation of the molecular fixed unique long axis).

$$\mathrm{LD}_r = \frac{3}{2} S \left(3 \left\langle \cos^2\theta_z \right\rangle - 1 \right) \quad 0 \le S \le 1 \tag{3.7b}$$

where the angle θ_z is the projection of the transition moment on the molecular fixed long axis (z, see Fig. 3.1).

For a single molecule, one has from Equation 3.3 the maximum and minimum LD proportional to $+$ and $-|\mu_{0j}|^2$, and the reduced LD (since for an isotropic distribution $\left\langle \cos^2\phi_Z \right\rangle = \left\langle \cos^2\phi_Y \right\rangle = 1/3$) is then

$$\mathrm{LD}_r = 3 \left(\cos^2\phi_Z - \cos^2\phi_Y \right) \tag{3.8}$$

Here, obviously $-3 \le \mathrm{LD}_r \le +3$, corresponding to the alignment of transition moment parallel to the Y or Z axis, respectively.

For the uniaxial ensemble of molecules, an expression useful for many applications is[1,3]

$$\begin{aligned}\mathrm{LD}_r &= 3 \left(S_{zz}\varepsilon_z + S_{yy}\varepsilon_y + S_{xx}\varepsilon_x \right) / \left(\varepsilon_z + \varepsilon_y + \varepsilon_x \right) \\ &= 3 \left(S_{zz}\cos^2\theta_z + S_{yy}\cos^2\theta_y + S_{xx}\cos^2\theta_x \right)\end{aligned} \tag{3.9}$$

where the Saupe order parameters are defined with respect to molecule-fixed axes x, y, z and angles θ_u ($u = x, y,$ or z) are the projections of the transition moment onto these molecular axes so that (Fig. 3.1)

$$S_{xx} \le S_{yy} \le S_{xx} \quad S_{uu} = \frac{1}{2} \left(3 \left\langle \cos^2\Theta_{Zu} \right\rangle - 1 \right) \quad u = x, y, \text{ or } z \tag{3.10}$$

Here Θ_{Zu} is the angle that molecular axis u makes with the laboratory axis Z around which the distribution is uniaxial. ε_z, ε_y, and ε_x are extinction coefficients for light polarized parallel to the respective molecule axes $z, y,$ and x. For orientations in polymer or liquid crystalline hosts, the longest dimension of a molecule generally takes the highest value of S_{uu}, which defines the z-axis. Due to symmetry, for a planar molecule, the normal to the molecular plane defines one of the three axes, usually the x-axis. The choice of the axes y and z in the molecular plane is determined by at what angle $\langle \cos^2 \Theta_{Zu} \rangle$ has its maximum (this is the z-axis) and its minimum (the y-axis).

From the definition of S_{uu} in Equation 3.10, it follows that the three order parameters are not independent because

$$S_{xx} + S_{yy} + S_{xx} = 0 \qquad (3.11)$$

Two simple examples of use of these relations are the applications to determine transition moment directions in planar molecules. Obviously, for an out-of-plane polarized transition: $\theta_x = 0$ and $\theta_z = \theta_y = \pi/2$, so only the third term in Equation 3.9 will be nonzero. Since S_{xx} must be negative, LD is also negative. The first application of this principle to planar molecules oriented in stretched poly(ethylene) matrix was for the detection of out-of-plane polarized $n \rightarrow \pi^*$ transitions in pyridine, pyrazine, and s-triazine.[5,6]

In case we search for the direction of an in-plane polarized transition, we have correspondingly that $\theta_x = \pi/2$, and the third term of Equation 3.9 will vanish. Here, obviously $\theta_z = \pi/2 - \theta_y$ and only a single angle needs to be determined. However, because of a sign ambiguity of the quadratic form of Equation 3.9, two solutions, $+$ and $-\theta_z$, are both possible. Various ways to solve the sign problem have been developed over the years. One is to systematically introduce and vary the position of electronically "inert" substituent groups (such as methyl substituents) that will affect and rotate the preferred in-plane axes z and y in a qualitatively predictable way, and thus the emerging solutions for θ_z. If one may assume that the substituent does not significantly perturb the electronic states and transition studied, the observed increase or decrease of the appearing θ_z value may often be used to exclude one of the two solutions to the quadratic equation. Another method is to exploit additional information available from fluorescence anisotropy (see Eqs 3.20 and 3.21) that relates the directions of exciting and emitting transition moments. Many examples of the use of both of these methods are available in the literature in connection with the determination of transition moment directions in natural as well as synthetic nucleobases.[7–12]

In order to determine the absolute transition moment directions in a low-symmetry (planar) molecule, the order parameters need to be known as well as the direction of the molecular orientation axis. This can be achieved by measuring the LD of vibrational transitions for which sometimes the direction of the corresponding transition moment is known. Alternatively, since there are many IR-allowed vibrational transitions in a medium-sized planar molecule of low symmetry, one can make fair estimates of the order parameters from the maximum and minimum LD_r-values of the in-plane polarized transitions. In addition, the out-of-plane polarized vibrational transitions all should have the same, negative LD_r-value, thus providing information for the order parameter describing the orientation of the out-of-plane axis. This procedure was successfully applied for the determination of the orientation parameters for indole derivatives[8,13], purines,[9,11] and nonnatural DNA bases.[10,12]

In one case, the combination of luminescence anisotropy and linear dichroism, together with information about the direction in which a ligand is facing a substrate, could solve the optical and structural sign ambiguities: first resolving the absorption tensor of a substitution-inert ruthenium-oligo-pyridyl complex, that is, determining the components of polarized absorption along symmetry axes as a function of wavelength.[14] Secondly, this information could be used to determine the binding geometry of the complex bound to DNA, to detect a small but significant clockwise rotation with respect to a given reference axis. It was found that two (oppositely twisted) enantiomeric forms of the complex were rotated in the same clockwise direction, an effect due to an inherent tilt of DNA bases: clockwise when viewed into the minor groove, an effect assigned to a local A-conformation of the double helix.[14,15]

LD when measured using polarization–modulation technique is a sensitive way to study molecular alignment in macroscopically oriented systems. A prestretched sheet of poly(ethylene) (PE) or poly(propylene) (PP) can thus produce highly resolved LD spectra from molecules – even for small molecules – that are allowed to diffuse into the anisotropic host.[16] The orientation distribution is to a good approximation uniaxial, so Equation 3.9 applies, or variants of the equation that include the alignment of the polymer chains as a separable parameter.[17] An early LD experiment on naphthalene – which is iso-π–electronic with indole (see Section 3.7) – showed quite different LD spectra in PE and PP. This is not an effect due to any significant perturbation of the chromophore but to different orientation distributions: the strongly overlapping vibronic absorption features of the first two transitions, corresponding to a long-axis polarized spectrum, $\varepsilon_z(\lambda)$, overlapping with a short-axis polarized spectrum, $\varepsilon_y(\lambda)$, give different $LD(\lambda) = S_{zz}\varepsilon_z(\lambda) + S_{yy}\varepsilon_y(\lambda)$ simply because of different sets of values of S_{zz} and S_{yy}, whereas the resolved $\varepsilon_z(\lambda)$ and $\varepsilon_y(\lambda)$ spectra are unchanged. The example is interesting since the less robust chromophore indole, by contrast, seems more prone to alter also its vibronic absorption envelopes and effective transition moment directions when subject to substituent perturbations, a fact important to consider in biophysical applications of tryptophan chromophore (*vide infra*).

The study of LD of ensembles of molecules requires that a macroscopic orientation can be somehow achieved. For example, DNA can be subject to hydrodynamic alignment in a Couette flow cell.[2] Also electrophoretic orientation can be studied with LD, with the alignment caused by steric interactions with the gel fibers during the migration of the macromolecule through the gel.[18] Lipid bilayer membranes and solutes therein may be studied using lamellar lyotropic lipid bilayer systems, which will spontaneously align between parallel silica windows.[19] Alternatively, lipid vesicles (liposomes), which when subject to shear forces become deformed and aligned, may be used to orient small solute molecules as well as proteins in their lipid bilayers.[20–23] A special case of lipid bilayer model membranes are "bicelles," which may be aligned in strong magnetic fields or by hydrodynamic flow.[24] In principle, also electric field orientation is an option, but with physiological electrolyte concentrations, such experiments are generally associated with disturbing heat dissipation. Photoselection, finally, by which the excitation pulse selects molecules with their transition moments preferentially parallel to the electric field of light, may be used for the study of fast (ps) conformational or solvent rearrangement processes by applying fast laser spectroscopy.[25]

3.3 MAGNETIC CIRCULAR DICHROISM

In addition to LD, another polarized absorption property that is often very useful for assignment applications of electronic transitions is MCD, which is defined as the differential absorption between opposite forms of circularly polarized light (i.e., circular dichroism) due to the presence of a static magnetic field directed parallel to the propagation direction of the light beam. For an electronically allowed, nondegenerate transition between (ground) state $|0\rangle$ and (excited) state $|f\rangle$, the magnitude of MCD is given within first-order perturbation theory by the so-called B-term as a sum of contributions from other (nondegenerate) excited states $|i\rangle$:

$$B\,(0 \to f) = \sum_{i \neq 0} (\langle i| - i\mathbf{m}\,|0\rangle \cdot \langle 0|\,\boldsymbol{\mu}\,|f\rangle \times \langle f|\,\boldsymbol{\mu}\,|i\rangle)\big/_{(E_i - E_0)}$$

$$+ \sum_{i \neq f} (\langle f| - i\mathbf{m}\,|j\rangle \cdot \langle 0|\,\boldsymbol{\mu}\,|\mathrm{f}\rangle \times \langle i|\,\boldsymbol{\mu}\,|0\rangle)\big/_{(E_i - E_f)} \qquad (3.12)$$

where

$$\mathbf{m}_{0j} = \langle j| - i\mathbf{m}|0\rangle \qquad (3.13)$$

is the magnetic dipole transition moment with the magnetic dipole operator $\mathbf{m} = \Sigma q_i \mathbf{r}_i \times \mathbf{p}_i$ with $\mathbf{r}_i$ and $\mathbf{p}_i$ position and momentum vectors, respectively, for all particles and q_i their charges.

MCD is generally (just like CD) a very weak effect for electronic transitions as a result of small values for magnetic dipole transition moments and also due to the large energy separation in the denominator of the first sum of terms, between ground and excited states. However, $(E_i - E_f)$ in the denominator of the second sum of terms may lead to stronger effects for the case of energetically close-lying excited states. This was for the first time demonstrated with thiophene (and homologous five-membered heterocycles), where the presence of two near-degenerate $\pi - \pi^*$ states was serendipitously discovered by their exceptionally strong bisignate MCD spectrum.[6] Thus, as a result of orthogonal (in-plane) polarizations of the two electric dipole transition moments $\langle 0|\boldsymbol{\mu}|f\rangle$ and $\langle 0|\boldsymbol{\mu}|i\rangle$ of the two first $\pi \to \pi *$ transitions in thiophene, the vector product $\langle 0|\boldsymbol{\mu}|f\rangle \times \langle i|\boldsymbol{\mu}|0\rangle$ becomes large. This vector product is directed orthogonal to the plane of the chromophore and, therefore, yields a large scalar product value with the magnetic dipole transition moment $\langle f| - i\mathbf{m}|j\rangle$; similarly, polarized perpendicular to the plane in case the hypothetical transition $j \to f$ is magnetic dipole allowed and corresponds to rotation of charge within the molecular plane.

MCD is an asset in combination with LD and fluorescence anisotropy for the deciphering of transition moment directions in biophysically important chromophores, and many examples of applications are found among nucleobases as well as protein aromatic chromophores, where in particular the presence of close-lying electric dipole allowed transitions with nonparallel polarizations can be detected.[7–9,11–13]

3.4   FÖRSTER RESONANCE ENERGY TRANSFER (FRET)

The phenomenon of FRET is due to through-space dipolar coupling between two chromophores, the electronically excited "donor" chromophore D and an "acceptor" chromophore with potential of becoming excited. It may be regarded as the electric correspondence to the magnetic Nuclear Overhauser Effect in NMR, the transfer of spin polarization between atomic nuclei. The donor D is a chromophore initially in the vibrational ground state 0 of its electronically excited state i, which we denote as state $|i, \, 0\rangle$. The deexcitation to its electronic ground state, and vibrational excited state u, $|0, \, u\rangle$, is represented by the electric dipole transition moment:

$$\mathbf{\mu}_{0i} = \langle 0, \, u|\mathbf{\mu}|i, \, 0 \rangle = \mathbf{\mu}_{D} \qquad (3.14)$$

Correspondingly, the acceptor chromophore, A, may be excited from its original electronic and vibrational ground state $|0, \, 0\rangle$ to an electronic and vibrational (vibronic) excited state $|j, \, v\rangle$, an excitation process represented by a transition moment

$$\mathbf{\mu}_{0j} = \langle 0, \, 0|\mathbf{\mu}|j, \, v \rangle = \mathbf{\mu}_{A} \qquad (3.15)$$

The transfer does not involve any photon fields but is due to direct Coulombic interactions between the D and A chromophores. The probability (rate) for the transfer is given by

$$\langle (j, \, v)(0, \, u)|V|(i, \, 0)(0, \, 0) \rangle^{2} \qquad (3.16)$$

with wavefunctions for the initial and final states as the products written to the right and to the left, respectively, and V the Coulomb interaction operator.

Neglecting orbital overlap and therefore electron exchange terms at sufficiently large donor–acceptor separations and assuming the dipole approximation, V is given by the relative orientations and the separation of the transition moments of donor and acceptor chromophore as

$$V \propto |\mathbf{\mu}_{D}||\mathbf{\mu}_{A}|(\cos\theta_{DA} - 3\cos\theta_{D}\cos\theta_{A})/4\pi\varepsilon_{0}R_{DA}^{3} \qquad (3.17)$$

with μ_{D} and μ_{A} being the transition moments as defined above, θ_{DA} the angle between the two transition moments and θ_{D} and θ_{A} the angles formed by the transition moments with respect to a line connecting their centers, and R_{DA} the distance separating the transition moments along this line.

From Equations 3.16 and 3.17, we have that the FRET rate is proportional to

$$(\cos\theta_{DA} - 3\cos\theta_{D}\cos\theta_{A})^{2}/R_{DA}^{6} \qquad (3.18)$$

which provides a basis for the quantitative interpretation of FRET in terms of distance separation and/or relative orientations of the donor and acceptor chromophores.

3.5 FLUORESCENCE ANISOTROPY

A property corresponding to the absorption anisotropy (linear dichroism) for emission is the fluorescence anisotropy:

$$r = \frac{I_{vv} - I_{vh}G}{I_{vv} + 2I_{vh}G} \qquad (3.19)$$

where the first index refers to the polarization setting of the polarizer of the excitation source and the second to the emission polarizer. The factor, $G = I_{hv}/I_{hh}$, corrects for any polarization bias inherent in the emission detection.

For a transition with pure (nonoverlapping) polarization in a molecule that is not rotating during the excited state lifetime, the fluorescence anisotropy is given by

$$r_{0i} = \frac{1}{5} \left(3 \left\langle \cos^2 \chi_i \right\rangle - 1 \right) \qquad (3.20)$$

where χ_i is the angle that the emitting transition moment makes to the absorbing transition moment for the ith electronic transition. For an absorption spectrum with overlapping bands, one has to take into account the degree by which different states become excited (i.e., overlapping transitions) at a given wavelength, so one has

$$r_0(\lambda) = \frac{1}{5} \frac{\sum_i A_i(\lambda) \left(3 \left\langle \cos^2 \chi_i \right\rangle - 1 \right)}{\sum_i A_i(\lambda)} \qquad (3.21)$$

where $A_i(\lambda)$ is the absorption of transition i. This is a useful relation by which the relative polarizations of different transitions can be combined with LD data for determining both the absolute directions of transition moments in the molecular frame and the angular orientations (including signs of angles) of the chromophores in various structures (see Section 3.2).

3.6 FLUORESCENT NUCLEOBASES

The normal abundant nucleobases, the four Watson–Crick bases adenine, guanine, cytosine, and thymine are all practically nonfluorescent at ambient temperature and will therefore not be discussed at any length in this chapter. The lack of fluorescence is an effect of efficient nonradiative deexcitation processes, which make the excited state lifetimes very short (ps). Indeed, one could consider the efficient deactivation property a result of Darwinian evolution that has selected chromophores which when exposed to strong solar UV radiation exhibit minimal photochemical degradation or modification. Thus, it may be characteristic that one of the very few naturally fluorescent nucleobases, the "Wye base" does never occur in DNA context but only in the naturally short-lived RNA, such as in the anticodon loop of tRNA. The UV absorption spectrum of the Wye base is due to four transitions whose transition moments are spread over all directions of the compass and therefore useful for various polarized light applications.[7]

A more artificial fluorescent DNA base is $1,N^6$-ethenoadenine, "$\varepsilon - A$," a three-ring purine base that can be produced in situ by the reaction of chloro- or bromoacetaldehyde with DNA whereby adenines become modified to varying yields. The fluorescence properties are brilliant, but the occurrence of wavelength-dependent excitation and several lifetimes suggest the presence of a tautomeric equilibrium that can obscure probe applications.[10] Ethenoadenine has been used in DNA complexes with recombinase RecA in order to discriminate nucleobase absorption ($\varepsilon - A$ having an absorption band above 300 nm) from normal DNA and from tryptophan and tyrosine absorption of the protein; also fluorescence-detected LD has been applied to probe the base orientation.[26,27]

A frequently used artificial fluorescent DNA base analog that appears to faithfully incorporate itself into DNA stacks, though probably with less clean base pairing, is the 2-aminopurine chromophore. Its near-UV spectrum can be resolved into five transitions with polarizations spread in various directions of the molecular plane, and also an indication of an out-of-plane $n \rightarrow \pi^*$ transition.[12]

The tricyclic cytosine analog, "tC," originally designed for Janus–basepair applications in PNA contexts,[28] but serendipitously found to be brightly fluorescent by Wilhelmsson, will be dealt with in detail in chapter 10 of this book.[29–31]

3.7   FLUORESCENT PEPTIDE CHROMOPHORES

The amino acid tryptophan, with indole as its photoactive aromatic chromophore (Fig. 3.2), is the most frequently used natural fluorescent probe in biophysical contexts, due to its relatively high fluorescence quantum yield and well-resolved absorption profile in the near-UV. Indole has two strongly overlapping electronic transitions in the lowest near-UV absorption band. Since indole is an N-heterocycle with 10 π-electrons distributed on 9 second-row atoms, the nomenclature for naming the electronic states was derived from the corresponding pure hydrocarbons, namely, naphthalene. The two lowest singlet excited electronic states are thus given the Platt symbols L_a and L_b and the two next B_a and B_b. Transitions to the two L-states are weakly allowed, whereas transitions to the B states are much stronger, giving an absorption spectrum with a weak band between 300 and 240 nm and a strong band

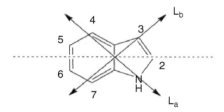

Figure 3.2   Transition moment direction in indole for the two lowest electronic transitions. The substitution positions are marked with numbers and the pseudo symmetry axis as a dashed line.

between 230 and 200 nm. In order to experimentally resolve the absorption spectrum of indole into its "pure" electronic transitions and at the same time determine the corresponding transition moment directions, LD in the UV and IR regions on stretched polyethylene samples combined with MCD and fluorescence anisotropy was measured.[8,13] Already from the quite strong wavelength dependence of the reduced linear dichroism and fluorescence anisotropy spectra over the lowest absorption band, it was clear that two electronic transitions heavily overlap in this region. From linear combination of the respective polarized absorption and excitation spectra components, the pure $A \rightarrow L_a$ and $A \rightarrow L_b$ absorption components were determined. In this procedure, information about the relative (from fluorescence anisotropy) and absolute (from LD) transition moment directions were obtained. In order to settle the absolute transition moment directions, estimates of the order parameters and molecular orientation axis were obtained from IR LD measurements. For indole, the N–H stretch is particularly helpful since it is a pure vibrational transition that does not mix with any other vibrations (i.e., it is a group vibration) and thus has a well-defined and predictable direction within the molecular frame. As mentioned above (Section 3.2), due to the quadratic dependence on the directional cosines, two solutions for the transition moment directions are always possible. This was resolved by comparing the results from indole derivatives substituted with fairly electronically inert methyl and methoxy substituents in the 3- and 5-positions. These substituents cause the molecule to orient slightly differently in the stretched polymer, making a choice of absolute (including sign) directions possible. These experimental results have later been supported by high-level electronic calculations,[32] and there is today a general agreement that in indole and tryptophan (which is related to 3-methylindole) the two lowest electronic transitions are oriented with a large relative angle (close to 90°) and with a projection close to 45° on the pseudosymmetry axis (Fig. 3.2). The $A \rightarrow L_b$ transition has vibrational structure, makes a positive angle with the pseudosymmetry axis, and it is the lowest electronic transition for 5-substituted indoles. The $A \rightarrow L_a$ transition is unstructured, Franck–Condon forbidden, makes a negative angle with the pseudosymmetry axis, and is the lowest electronic transition in 3-methylindole and tryptophan. For indole, the order of the two lowest electronic transitions is uncertain, but at least in a polar environment the $A \rightarrow L_a$ transition is lowest.

When the indole chromophore is substituted in other positions than 5 (which is para to the nitrogen) on the six-membered ring, the transition moment direction of the $A \rightarrow L_b$ transition is affected.[8] It appears as if the $A \rightarrow L_b$ transition is localized on the six-membered ring and resembles the corresponding transition in a nitrogen-substituted benzene. Para-substitution (i.e., in the 5-position) does not change the direction, whereas ortho- (7-position) or meta- (4- or 6-position) substitution turns the transition moment in the expected direction in perfect analogy with substituted benzenes. It is interesting to note that the $A \rightarrow L_a$ transition is unaffected by substitution in the six-membered ring. This is consistent with the partial charge transfer character of this transition (involving the double bond of the five-membered ring). Upon substitution in the 2- or 3-position, the $A \rightarrow L_a$ transition redshifts, which is the reason for the L_a state to be lower in energy for these derivatives.

Next after tryptophan, due to its abundance and use as a fluorescent aromatic amino acid probe, comes probably tyrosine. Its spectrum is characterized by the first two transitions in near-UV, at 230 and 280 nm, polarized parallel and perpendicular to the 1,4 axis of the parent chromophore phenol. As shown from studies of 1,4-methylphenol (para-cresol) in stretched poly(vinyl alcohol) and poly(ethylene) matrix and in alcohol and cyclohexane solvents, the energies (and vibrational fine-structure) of the spectral components are quite sensitive to polar environment, probably mainly due to hydrogen bonding involving the OH group. By contrast, the transition moment directions appear robust so that the absorption envelopes have essentially pure polarizations under the respective bands independent of environment (L. Fornander *et al.*, unpublished result).

3.8   SITE-SPECIFIC LINEAR DICHROISM (SSLD)

A relatively new methodology for studying protein structure in solution or other noncrystalline environment is through Site-Specific Linear Dichroism by Molecular Replacement (SSLD-MR). This application of polarized light spectroscopy was first demonstrated for the fibrous complex between DNA and recombination enzyme RecA.[33] The principle may be described as follows. The LD of the wild-type complex is compared with that of a mutant in which one aromatic amino acid (e.g., a tryptophan or a tyrosine) has been replaced by a spectroscopically "silent" chromophore, that is, one with insignificant absorption in the wavelength region of interest (e.g., threonine or phenylalanine). If the relative degrees of orientation (parameter S in Eq. 3.7b) can be calibrated somehow with enough precision, the differential spectrum LD(wild)/S(wild) – LD(mutant)/S(mutant) will exactly correspond to the LD of the replaced residue and will thus contain information about its orientation, generally in terms of two in-plane transition moments of the aromatic chromophore. By studying a battery of mutants, and for each of them having verified their remaining biological activity or other measurables in support for an unperturbed structure, a series of angular coordinates can be deduced. Such a set of data may then be used in a molecular modeling simulation to build up a three-dimensional structure of the protein complex.[2,27,34]

A special case of SSLD is when fluorescent molecules are exploited, either as naturally occurring building blocks in a biomolecular structure or when implanted as replacement for nonfluorescent residues. An advantage is then that the SSLD spectrum of the selected chromophore may be obtained as a fluorescence-detected LD (FLD) signal without need for subtraction of wild-type LD spectrum and extra calibration of S.[3]

3.9   SINGLE-MOLECULE FLUORESCENCE RESONANCE ENERGY TRANSFER (smFRET)

Fluorescence studies using site-specific incorporation of the artificial fluores-cent nucleobase "tC" have been used to study conformational kinetics of DNA polymerase.[35] At single-molecule level, using a fluorophore label with high brightness, the assembly pathway and DNA unwinding activity have been monitored for the bacteriophage T4 helicase – primase (primosome) complex.[36] The helicase substrates have been surface-immobilized model DNA replication forks "internally" labeled in the duplex region with opposed donor/acceptor (iCy3/iCy5) chromophore pairs in the lagging and leading strands. The time dependence of the smFRET signals can be monitored during the unwinding process, and the helicase rates and processivities measured as a function of GTP concentration. Such single-molecule experiments can provide a detailed real-time visualization of the assembly pathway and duplex DNA unwinding activity and the results can be correlated with more indirect equilibrium and steady-state results obtained in bulk solution studies.

3.10   SINGLE-MOLECULE FLUORESCENCE-DETECTED LINEAR DICHROISM (smFLD)

Polarization-resolved single-molecule experiments have until recently only been applied to study reorientation dynamics in the time domain 10 ms or longer. However, using MHz polarization modulation technique, together with a postacquisition data analysis technique, time resolution in the submillisecond regime is possible, able to use for observing DNA breathing and its role in the binding and assembly mechanism of a replication helicase.[36]

As shown in Figure 3.3, a DNA replication fork construct can be "internally" labeled with the FRET donor–acceptor chromophores iCy3 and iCy5.[37] These cyanine dyes are rigidly incorporated into the sugar phosphate backbone with each fluorophore replacing an opposed DNA base at each label position. With the polarized excitation setup outlined in Figure 3.3b, the smFRET of the iCy3 and iCy5 system and the smFLD of the iCy3 chromophore may be simultaneously monitored.

The principle of these measurements is based on the idea that a single-oriented DNA replication fork construct will absorb light with different probabilities depending on the polarization of the laser field. By rapidly modulating the laser polarization, and recording the phase of the modulation for every detected signal photon, it is possible to monitor the time-dependent projection of the absorbing electric transition dipole moment onto arbitrarily defined laboratory axes. While the smFRET signal is sensitive to the relative separation and relative orientation of the iCy3 and iCy5 chromophores, the smFLD signal is directly probing the iCy3 orientation in the laboratory frame. As shown by Phelps *et al.*, both signals are affected by DNA breathing.[36]

The smFLD method is a useful complement to smFRET since it can help to avoid misinterpretation of false smFRET signals. For example, if the acceptor fluorophore of a coupled donor–acceptor FRET pair was to undergo a photo-bleach event (permanent or temporary), there would be a drop in acceptor intensity coincident with a rise in donor intensity. An acceptor photo-bleach event can thus be misinterpreted as a sudden separation between the donor–acceptor pair. This ambiguity can be addressed by simultaneously monitoring the smFLD signal, because a FRET conversion event

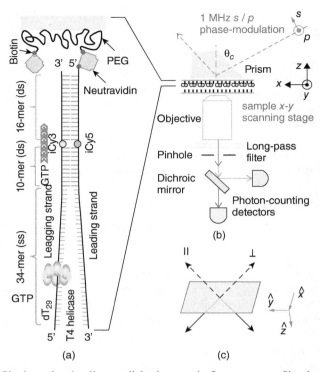

Figure 3.3 Single-molecule linear dichroism and fluorescence. Simultaneous single molecule sub-millisecond measurement of Fluorescence-detected Linear Dichroism (smFLD) and Förster Resonance Energy Transfer (smFRET). (a) Position of FRET pair to detect opening of the replication fork construct as helicase binds to the d(T)29 loading sequence of the lagging strand. (b) Total Internal Reflection Fluorescence (TIRF) microscope setup excitation scheme and detection method. The polarization of the exciting beam is modulated between orthogonal linear polarizations at 1 MHz. The p polarization component of the exciting electric field points in the direction of the y axis, and the s polarization component is lying within the x-z-plane. From Ref. [36]. Reproduced with permission from PNAS, USA.

due to donor–acceptor pair separation will be accompanied by a significant change in the smFLD signal, whereas an acceptor photo-bleach event will not.

Superimposed on the transiently fluctuating LD – which when averaged over time is zero for a system without macroscopic orientation – there may also be an LD due to some geometric confinement (in the example in Figure 3.3 due to covalent attachment to the TIRF surface). One may thus envisage several interesting applications of the smFLD technique by which the effect of confinement can be dynamically resolved to provide information, for example, about internal flexibility of a biomacromolecule as well as about its ambient interactions. Other interesting potential future applications along these lines may include sensors for screening genetic errors or gene sequencing devices based on PNA base pairing.[38,39]

The recently discovered 50% elongation as a stable conformation of overstretched DNA is another interesting topic worthy of addressing at single-molecule level using FLD and FRET. In particular, a question to resolve is whether three base pairs (or bases) spontaneously form a stack, followed by a bigger gap, as proposed from energetic arguments by Bosaeus *et al.*, or whether the stretch is somehow homogeneous.[40]

REFERENCES

1. Norden, B. *Appl. Spect. Rev.* **1978**, *14*, 157.
2. Norden, B.; Kubista, M.; Kurucsev, T. *Q. Rev. Biophys.* **1992**, *25*, 51.
3. Nordén, B.; Rodger, A.; Dafforn, T. Linear Dichroism and Circular Dichroism. A Textbook on Polarized-Light Spectroscopy. Royal Society of Chemistry London, **2010**.
4. Michl, J.; Thulstrup, E. W. Spectroscopy with Polarized Light. VCH: New York, **1986**.
5. Norden, B. *Chem. Phys. Lett.* **1973**, *23*, 200.
6. Norden, B.; Hakansson, R.; Pedersen, P. B.; Thulstrup, E. W. *Chem. Phys.* **1978**, *33*, 355.
7. Albinsson, B.; Kubista, M.; Sandros, K.; Norden, B. *J. Phys. Chem.* **1990**, *94*, 4006.
8. Albinsson, B.; Norden, B. *J. Phys. Chem.* **1992**, *96*, 6204.
9. Albinsson, B.; Norden, B. *J. Am. Chem. Soc.* **1993**, *115*, 223.
10. Holmen, A.; Albinsson, B.; Norden, B. *J. Phys. Chem.* **1994**, *98*, 13460.
11. Holmen, A.; Broo, A.; Albinsson, B.; Norden, B. *J. Am. Chem. Soc.* **1997**, *119*, 12240.
12. Holmen, A.; Norden, B.; Albinsson, B. *J. Am. Chem. Soc.* **1997**, *119*, 3114.
13. Albinsson, B.; Kubista, M.; Norden, B.; Thulstrup, E. W. *J. Phys. Chem.* **1989**, *93*, 6646.
14. Lincoln, P.; Broo, A.; Norden, B. *J. Am. Chem. Soc.* **1996**, *118*, 2644.
15. Lincoln, P.; Norden, B. *J. Phys. Chem. B* **1998**, *102*, 9583.
16. Norden, B. *Chem. Scripta* **1975**, *7*, 167.
17. Norden, B. *J. Chem. Phys.* **1980**, *72*, 5032.
18. Norden, B.; Elvingson, C.; Jonsson, M.; Akerman, B. *Q. Rev. Biophys.* **1991**, *24*, 103.
19. Norden, B.; Lindblom, G.; Jonas, I. *J. Phys. Chem.* **1977**, *81*, 2086.
20. Ardhammar, M.; Lincoln, P.; Norden, B. *Proc. Natl. Acad. Sci. U. S. A.* **2002**, *99*, 15313.
21. Ardhammar, M.; Mikati, N.; Norden, B. *J. Am. Chem. Soc.* **1998**, *120*, 9957.
22. Caesar, C. E. B.; Esbjorner, E. K.; Lincoln, P.; Norden, B. *Biochemistry* **2006**, *45*, 7682.
23. Quesada, E.; Ardhammar, M.; Norden, B.; Miesch, M.; Duportail, G.; Bonzi-Coulibaly, Y.; Nakatani, Y.; Ourisson, G. *Helv. Chim. Acta* **2000**, *83*, 2464.
24. Kogan, M.; Beke-Somfai, T.; Norden, B. *Chem. Commun.* **2011**, *47*, 7356.
25. Onfelt, B.; Lincoln, P.; Norden, B.; Baskin, J. S.; Zewail, A. H. *Proc. Natl. Acad. Sci. U. S. A.* **2000**, *97*, 5708.
26. Takahashi, M.; Kubista, M.; Norden, B. *J. Mol. Biol.* **1989**, *205*, 137.
27. Morimatsu, K.; Takahashi, M.; Norden, B. *Proc. Natl. Acad. Sci. U. S. A.* **2002**, *99*, 11688.
28. Ray, A.; Norden, B. *FASEB J.* **2000**, *14*, 1041.
29. Wilhelmsson, L. M.; Sandin, P.; Holmen, A.; Albinsson, B.; Lincoln, P.; Norden, B. *J. Phys. Chem. B* **2003**, *107*, 9094.
30. Engman, K. C.; Sandin, P.; Osborne, S.; Brown, T.; Billeter, M.; Lincoln, P.; Norden, B.; Albinsson, B.; Wilhelmsson, L. M. *Nucleic Acids Res.* **2004**, *32*, 5087.

31. Sandin, P.; Wilhelmsson, L. M.; Lincoln, P.; Powers, V. E. C.; Brown, T.; Albinsson, B. *Nucleic Acids Res.* **2005**, *33*, 5019.

32. SerranoAndres, L.; Roos, B. O. *J. Am. Chem. Soc.* **1996**, *118*, 185.

33. Hagmar, P.; Norden, B.; Baty, D.; Chartier, M.; Takahashi, M. *J. Mol. Biol.* **1992**, *226*, 1193.

34. Reymer, A.; Frykholm, K.; Morimatsu, K.; Takahashi, M.; Norden, B. *Proc. Natl. Acad. Sci. U. S. A.* **2009**, *106*, 13248.

35. Stengel, G.; Gill, J. P.; Sandin, P.; Wilhelmsson, L. M.; Albinsson, B.; Norden, B.; Millar, D. *Biochemistry* **2007**, *46*, 12289.

36. Phelps, C.; Lee, W.; Jose, D.; von Hippel, P. H.; Marcus, A. H. *Proc. Natl. Acad. Sci. U. S. A.* **2013**, *110*, 17320.

37. Lee, W.; Jose, D.; Phelps, C.; Marcus, A. H.; von Hippel, P. H. *Biochemistry* **2013**, *52*, 3157.

38. Carlsson, C.; Jonsson, M.; Norden, B.; Dulay, M. T.; Zare, R. N.; Noolandi, J.; Nielsen, P. E.; Tsui, L. C.; Zielenski, J. *Nature* **1996**, *380*, 207.

39. Wittung-Stafshede, P.; Rodahl, M.; Kasemo, B.; Nielsen, P.; Norden, B. *Colloids Surf. A.* **2000**, *174*, 269.

40. Bosaeus, N.; El-Sagheer, A. H.; Brown, T.; Smith, S. B.; Akerman, B.; Bustamante, C.; Norden, B. *Proc. Natl. Acad. Sci. U. S. A.* **2012**, *109*, 15179.

4

FLUORESCENT PROTEINS: THE SHOW MUST GO ON!

GREGOR JUNG

Biophysical Chemistry, Saarland University, Saarbruecken, Germany

4.1 INTRODUCTION

The usage of Autofluorescent Proteins revolutionized all life sciences, and its development was consequently awarded by the Nobel Prize for Chemistry in 2008.[1–3] Many reviews and even books were written about this topic.[4–8] The intention of the chapter is to give an overview of the overwhelmingly rich photochemical and spectroscopic features of the proteins and to briefly discuss their usage with regard to bioanalysis. Again, the Nobel Prize for Chemistry in 2014 might confirm the ongoing meaning of Fluorescent Protein technology: The role of Fluorescent Proteins for super-resolution microscopy is highlighted in a recent compilation.[9]

4.2 HISTORICAL SURVEY

At the beginning of this successful scientific story, there was the curiosity of a researcher. Being interested in the origin of the bioluminescence of the Pacific jellyfish *Aequorea victoria*, Shimomura isolated the protein Aequorin and, later on, the so-called Green Fluorescent Protein (*av*GFP) from this animal.[1, 10] It turned out that Aequorin is excited in a bioluminescent reaction.[11] It is likely not wrong to say that most of the interest in the following almost 30 years was spent on Aequorin as its blue luminescence is dependent on calcium ions.[12] The usage of the protein and

Fluorescent Analogs of Biomolecular Building Blocks: Design and Applications, First Edition.
Edited by Marcus Wilhelmsson and Yitzhak Tor.

its cofactors in cell cultures allowed for following changes of Ca^{2+} concentration. The molecular function of GFP appeared just to act as an energy transfer acceptor that converts the blue chemiluminescence with a low quantum yield into green fluorescence with high quantum yield.[13] Only some chemical characterization of this "boring" GFP was made over three decades.[14] Despite the knowledge that its chromophore structure is different from that of other fluorescent proteins, that is, it consists of an oligopeptide, only cloning the encoding DNA sequence could prepare the ground for the forthcoming revolution.[15] The first seminal transformation of bacteria and the transfection of the nematode *Caenorhabditis elegans* by Chalfie *et al.* proved that the DNA sequence is sufficient to produce specifically labeled nerve cells in living organisms.[16] The total synthesis of the peptide sequence of 238 amino acids by purely chemical means finally could show that, indeed, only the peptide sequence is sufficient to generate the green luminescence.[17] No further enzymatic catalysis or the action of chaperons is required for chromophore formation. Shortly afterward, the three-dimensional structure was resolved by X-ray crystallography and showed the so-called β-barrel as tertiary structure with some tendency for dimerization (Fig. 4.1a).[18, 19] The chromophoric tripeptide is located in the center as part of a central α-helix representing the axis of the can-like structure. Eleven β-strands surround the axis and thus shield the chromophoric moiety from the outer surrounding, that is, the solvent. The bimodal absorption of wild-type (wt) GFP could be explained by the existence of two basic chromophore forms, the neutral (RH) form and the anionic chromophore (R⁻) form (Fig. 4.1b).[20, 21] It was also clear that strong interactions of the chromophore with its vicinity must exist as its synthetic counterpart does not fluoresce at room temperature.[22] The features of Fluorescent Proteins therefore can only be understood by the interplay between the intrinsic chromophore properties and the surrounding supermolecular architecture, as will be discussed in the following chapters. They are also the key to modify the emission color by mutagenesis. This approach vastly was exploited by Tsien and coworkers for creating further Fluorescent Proteins with blue, blue-green, and green-yellowish emission, that is, Blue (BFPs), Cyan (CFPs), and Yellow Fluorescent Proteins (YFPs), respectively.[3] Color tuning could be further expanded not until chromoproteins and further Fluorescent Proteins were detected in corals and other relatives to the jellyfishes.[23–25] Especially the tetrameric protein *ds*Red, which was derived from the coral genus *Discosoma sp.*, turned out to be another milestone in the development of Fluorescent Protein technology[26, 27]; it was monomerized by multiple amino acid exchanges[28] and, afterward, successfully converted to a family of Yellow to Red Fluorescent Proteins (RFPs), the so-called mFruit proteins, again by Tsien and coworkers.[29] Researchers in the life sciences have nowadays a huge palette of genetically encodable fluorescence colors for multicolor labeling at their disposal.[30] Fluorescent Proteins proved to be useful workhorses for other biosciences including the study of protein dynamics by kinetic, denaturation, and other biophysical methods.[31–33] Expanding the genetic code, biochemists were able to incorporate unnatural amino acids into the scaffold to obtain a Gold Fluorescent Protein.[34, 35] Finally, Fluorescent Proteins are often used for educational purposes as well.[36–38]

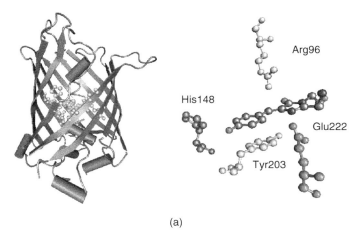

(a)

(b)

Figure 4.1 Structural and spectroscopic fundamentals of *av*FP Tyr66. (a) Left: tertiary structure of eYFP with the chromophore (green) in the center of the β-barrel. The interaction with its nearby amino acids (right: chromophore atoms color coded as shown in Figure 4.5; surrounding amino acid in orange) determines its spectroscopic properties. Note the π–π stacking of the phenolic moieties of Tyr203 with the chromophore. (b) From a chemical point of view, the absorption properties of *av*FP Tyr66 are determined by the equilibrium between the neutral chromophore (RH) and the anionic chromophore (R⁻). This equilibrium is influenced by the surrounding (see also Fig. 4.3) within the protein barrel. (*See color plate section for the color representation of this figure.*)

4.3 PHOTOPHYSICAL PROPERTIES

4.3.1 Absorption Properties and Color Hue Modification (Fig. 4.2)

From a chemical point of view, the anionic chromophoric unit in GFP merely resembles an oxonol- or merocyanine-type dye, with a grain of salt.[39] A rather basic

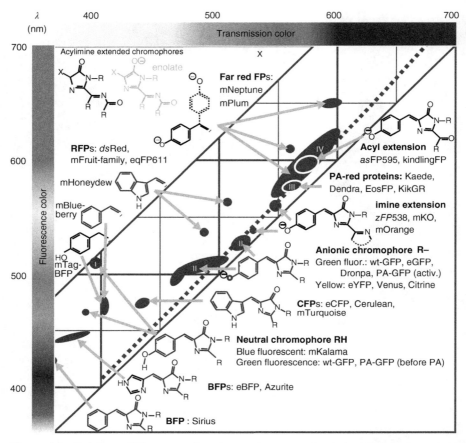

Figure 4.2 Rainbow of Fluorescent Proteins. Typical absorption and emission colors of the underlying chromophore structures are depicted (upper-left corner: acylimine-derived chromophores; lower-right triangle: other chromophores). Color tuning within one chromophore class, that is, the shape and extension of the ellipses, is achieved by interactions with the chromophore surrounding. The dashed line corresponds to a constant Stokes shift of $500 \, cm^{-1}$; see Section 4.3.3 for further information. Details for some selected areas can be found in other figures: I → Figure 4.9a; II → Figures 4.3 and 4.6; III → Figure 4.9d; IV → Figure 4.4. The ESPT-exhibiting protein mKeima is not depicted due to its large spectral changes ($\lambda_{exc} = 440 \, nm$, $\lambda_{em} = 620 \, nm$; see Section 4.4.1). (*See color plate section for the color representation of this figure.*) Data taken from Refs [53, 54, 211] and references therein.

description of their photophysical properties uses two mesomeric structures, which is also valid for the well-known indicator dye phenolphthalein ($\lambda_{max,abs} = 555 \, nm$) or benzaurin ($\lambda_{max,abs} = 567 \, nm$) (Fig. 4.3a).[40] Their absorption maxima are faithfully reproduced by quantum-chemical computations.[39] Due to the slightly smaller imidazolinone moiety compared to the phenyl rings in the organic dyes, the absorption maximum of the anionic chromophore is thus expected to be blueshifted by about 60 nm, that is, to $\lambda_{max} = 500–510$ nm. Experimentally, YFPs exhibit the strongest redshift of proteins with R^- chromophores with an absorption maximum at $\lambda_{max} = 524$ nm (0–0 transition at cryogenic temperatures).[41] Only about 10 nm of the redshift compared to the absorbance in wt-GFP can be assigned to $\pi–\pi$ interactions of Tyr203 and the chromophore (Fig. 4.1a).[42] Any distortion toward a more asymmetric charge distribution must lead to a blueshift. This can be achieved by neutralizing R^- to RH, but also the short-wavelength absorption of BFPs and CFPs, raised by amino acid substitutions at position 66, can be interpreted in the same way. The same tendency is also experienced by more subtle changes of the environment: removal of the hydrogen bonds from quinone-stabilizing Arg96 by mutation induces a distinct hypsochromic shift,[43, 44] stabilization of the negative charge at the phenolic moiety favors one of the mesomeric structures and, thus, shortens the absorption wavelength as well (Fig. 4.3b). These considerations explain the difference between the absorption maxima of the A, B, and I-state of wt-GFP.[45] Hence, the I-state possesses spectroscopic features that are close to the above-mentioned naïve estimation derived from benzaurin. Another bathochromic shift of 6–10 nm is achieved by the mutation Ser205Val, which further destabilizes the benzenoid structure.[46] The finally obtained PhiYFP, a naturally occurring YFP derivative, therefore exhibits an absorption maximum close to the low-temperature limit.[46, 47]

In other Fluorescent Proteins, however, the conjugation length is increased along amino acid 65. There is a remarkable redshift in absorption of roughly 65 nm going from λ_{max} of the I-state in GFP to λ_{max} of *as*FP just by addition of one carbonyl function to the chromophoric unit. In fact, the obtained chromophore has a conjugation length such as benzaurin, and the absorption maxima also almost coincide. The strong electron-accepting capability of the carbonyl function withdraws electrons from the phenolic moiety thus enhancing the symmetry of the electron distribution. Consequently, carbonyl groups in conjugation with the extended π-system (acyl-extended chromophore) lead to a stronger redshift than imino-group expansions as in *zFP*538. Similar considerations also hold for the acylimine-extended chromophores (upper-left triangle in Fig. 4.2). An explanation, similar to the above treatment of the B- and I-state, was recently given to explain the color tuning in RFPs.[48] Remarkably, mTag-BFP exhibits similar blue fluorescence such as mBlueberry despite a distinctly shorter conjugation length.[49] Here, the imidazolinone moiety undergoes keto–enol tautomerism, and a putative enolate structure generates a symmetric chromophore (see gray structure in Fig. 4.2).[50]

Although qualitatively useful in understanding the absorbance of the Far-RFPs, the presented application of the polymethine dye concept should be regarded with suspicion as it vastly ignores the local environment. *ds*Red and its monomer mRFP, for example, absorb at a shorter wavelength than *as*FP despite a more extended chromophore. Moreover, mutagenesis can assist in redshifting the absorption spectrum by about 30 nm as in mCherry.[29] It was shown for the mFruit family that already the internal electric field, established by the local protein environment around the chromophore and therefore varying from mutant to mutant, can tune the absorbance by 50 nm (Fig. 4.4).[51] A twofold effect on the absorption

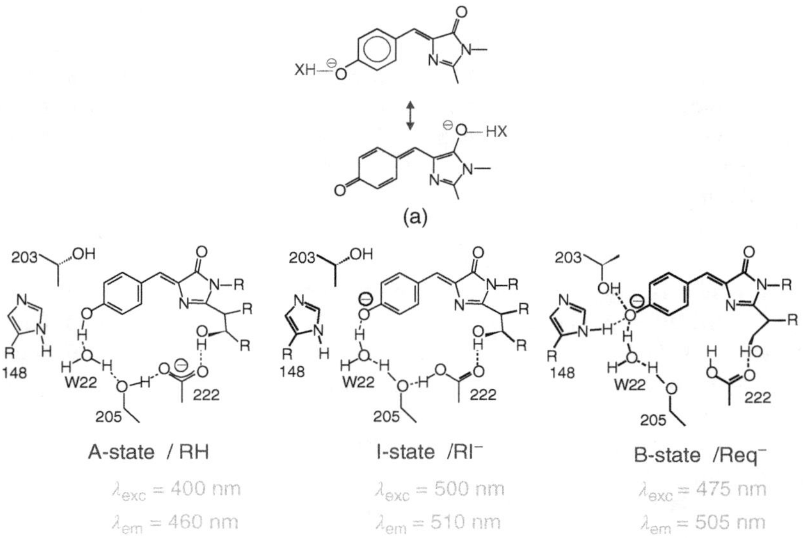

(a)

A-state / RH

λ_{exc} = 400 nm

λ_{em} = 460 nm

I-state /RI⁻

λ_{exc} = 500 nm

λ_{em} = 510 nm

B-state /Req⁻

λ_{exc} = 475 nm

λ_{em} = 505 nm

(b)

Figure 4.3 Influence of hydrogen bonding on spectroscopic properties. (a) Description of the electronic states by the mesomeric effect between the benzenoid (upper structure) and quinonoid (lower structure) formulas. Their respective contributions to the resonance structure are influenced by hydrogen bonding. From Ref. [40]. (b): In wt-GFP, at least three different states can be distinguished by spectroscopic means, that is, the A-state, the photocycle intermediate I (see Fig. 4.9a) and the predominant B-state.[122, 123, 212] Only the first state unambiguously coincides with the chemically defined neutral chromophore RH state, whereas the last two states have the anionic chromophore in common.[20, 21] The absorption redshift from the B-state to the I-state is mostly an effect of the destabilization of the negative charge at the phenolic moiety.[45] Another bathochromic shift of 6–10 nm is observed when stabilizing Ser205 is replaced by Val.[46] The hydrogen-bonding network makes the difference.

wavelength is noticed: on the one hand, increasing the electric field within the protein barrel shifts the absorption wavelength to the red due to the interaction with the permanent dipole moments of the ground (S_0) and excited state (S_1), that is, lowering of the energy gap; on the other hand, it diminishes the effective change of the permanent dipole moments, $\Delta\mu_0$, due to polarizability, which is reduced in the excited state. Consequently, a quadratic dependence of the transition energy from the internal electric field is derived. Quantum-chemical computations identified the relevant surrounding amino acids Lys and Arg.[48] From the combination of symmetrizing hydrogen bonding and the presumable help of π–π interactions and a final shift to λ_{max} = 608 nm in mGrape3 is established.[52] It should be noted that the presented overall classification cannot take all peculiarities into account due to the manifold of discovered chemical structures, cf. Refs [53, 54] for more chromophoric structures.

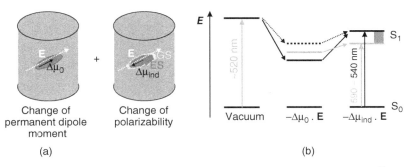

Figure 4.4 Spectral changes in the mFruit-family due to quadratic Stark-effect.[74] (a) The internal electric field **E**, which is influenced by the mutation pattern, is interacting with the change of permanent dipole moment upon excitation, $\Delta\mu_0$, that is, the difference between the static dipoles of the ground (S_0) and the excited state (S_1). The net change of permanent dipole moment, however, is modified by the dipole moment that is induced by the electric field, $\Delta\mu_{ind}$. $\Delta\mu_{ind}$ is opposed to $\Delta\mu_0$ because of the reduced polarizability in the excited state, as indicated by the clouds on the right. (b) The hypothetic vacuum transition wavelength[213] is bathochromically shifted by the linear Stark effect, that is, the larger **E**, the larger is the energetic stabilization (solid black line vs dashed black line). Due to the converse direction of $\Delta\mu_{ind}$, which itself linearly depends on **E**, the energy gap between S_0 and S_1 is raised again, that is, altogether a quadratic dependence. The strongest redshift is therefore observed for mutants with moderate internal electric fields (gray lines). However, the linear Stark effect is relevant for two-photon excitation cross sections.[57]

Besides the absorption maximum, also the molar extinction coefficient is of great importance. Under the assumption of constant oscillator strength for identical chromophores in various protein environments, the extinction coefficient depends on the width of the absorption spectrum, that is, mainly how strongly the geometry of the molecule is changing upon excitation, and, secondly, of the percentage of molecules which absorb. Slight changes due to the protein surrounding are discussed.[48] The author of this chapter keeps on arguing that the molar extinction coefficient, as it is often listed, is a useless quantity in a photophysical sense:[55–57] it does not allow for calculating molecular properties such as the Einstein coefficient for absorption, the magnitude of the transition dipole moment, or even Förster radii for energy transfer. The reason is that it is strongly related to the chromophore formation (or maturation) efficiency, which will be explained in the following section. A convenient way to correct for this, if enough material is present, is denaturation and subsequent absorption spectrometry.[58]

4.3.2 Chromophore Formation

Autocatalytic chromophore formation is best understood for avGFP.[59] The chromophore consists of three amino acids, of which Gly67 is conserved and Tyr66 can be replaced by other aromatic amino acids yielding BFPs or CFPs, and Ser65. The latter position offers the largest possibility for variations. The triad itself does not spontaneously transform into the chromophore but requires the specific protein environment. At first, the correct architecture must be fixed bringing the nitrogen atom of Gly67 close to the carbonyl carbon atom of amino acid 65 for heterocycle formation.[60] The following steps are catalyzed by Arg96 and Glu222.[61] Their role is emphasized as they are highly conserved in naturally occurring Fluorescent Proteins, and only few mutagenic substitutions at that position are reported.[44, 62, 63] Among them, Glu222His appears promising as the photolabile carboxylate (see Section 4.4.3) is replaced by the stable histidine while the high expression yield and, thus, the catalysis is maintained.[64]

Oxidation of the side chain of amino acid 66 is required for forming a colored protein. While it was accepted for a long time that chromophore formation follows the *cyclization–dehydration–oxidation* mechanism (Fig. 4.5a),[65] kinetic experiments, also with deuterated amino acids, support the reversed succession of the last two processes (*cyclization–oxidation–dehydration*, Fig. 4.5b).[66, 67] Also the maturation of RFPs, where an additional oxidation step has to proceed, is still under debate.[54] While the prediction, in which the red chromophore is generated from a green one (Fig. 4.5a), leads to the development of a fluorescent timer for monitoring the onset of expression,[68] the development of blue-to-red timer is only compatible with a blue, mTagBFP-like intermediate.[50] Recent kinetic experiments postulated a branching in *ds*Red toward green and red chromophore (Fig. 4.5b and c).[69] Despite this ambiguity, glutamic acid and arginine residues are ubiquitously placed close enough to the chromophore heterocycle to fulfill their catalytic action in the maturation (although the primary sequence of the abundant number of proteins and, thus, the amino acid numbering is different). In a protein ensemble exhibiting conformational heterogeneity and with the reversibility of some reaction steps in mind, one could easily imagine that individual proteins rest in one or the other state thus reducing the amount of completely formed chromophore.

There are experimental observations that chromophore formation may be incomplete in many Fluorescent Proteins. In the first crystallographic analysis of wt-GFP, an electron density corresponding to a fraction of about 30% was observed, which was interpreted as a still hydrated chromophore.[18] Only slightly less amount of immature proteins (~20%) was found by counting single molecules.[70] Substitution of catalyzing amino acids also distinctly reduces the fraction of absorbing species.[43] Experimental hints to incomplete maturation also exist for RFPs. The published optical spectra of *ds*Red still contain residual absorption of the green absorbing "precursor."[26] Single-molecule experiments unraveled that in *ds*Red, the green-to-red chromophore ratio is in the range of 1:1.2–1:1.5, which corresponds to only 55–60% completed red chromophore and which reproduced previous mass spectrometric findings.[27, 71] This value might be even further reduced due to the later-found alternative pathways in the chromophore formation in *ds*Red (Fig. 4.5b and c). Recently, fluorescence cross-correlation spectroscopy was undertaken with necessary care and proved that eGFP in a tandem construct with mCherry fully matures, in agreement with an earlier study,[43] whereas the red absorbing counterpart is far from being completed.[72] Hence, the determined extinction coefficients might be inadequate for usage in photophysics.[57, 73] In biological applications, the situation

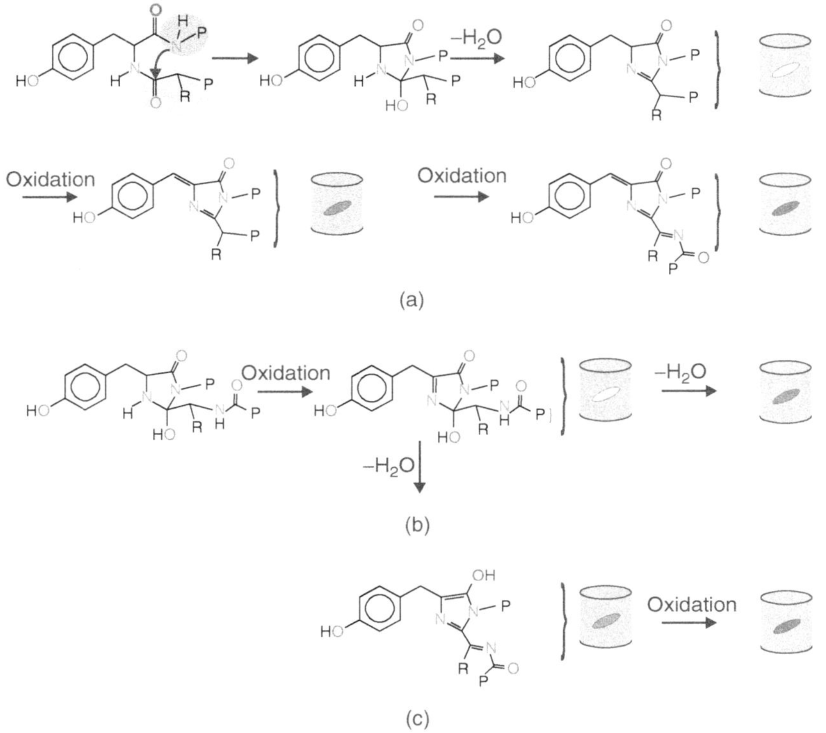

(a)

(b)

(c)

Figure 4.5 Simplified models of chromophore formation in Green and Red Fluorescent Proteins under discussion. The spectroscopically observable, fluorescent states are symbolized by barrels with correspondingly colored ellipses. Various protonation and deprotonation steps are omitted for the sake of simplicity. (a) In the original model, the cyclization of the tripeptide (green: Gly67; yellow: Ser65 in *av*FP) is followed by dehydration, thus forming the imidazolinone moiety. Subsequent oxidation yields the green fluorescent state; further oxidation can turn the proteins into Red Fluorescent Proteins. (b) The model in (a) is challenged by more recent kinetic experiments, which reverses the latter chromophore formation steps.[66, 67] (c) Especially, the current investigation of the chromophore formation in Red Fluorescent Proteins shows a blue fluorescent intermediate that likely exhibits the chromophore structure of mTagBFP. Further oxidation completes the red chromophore. In the branching model (b/c), the green and the red fluorescent states are dead ends in the chromophore genesis. It should be noted that the originally proposed branching model by Strack *et al.*[69] postulates two oxidation steps before dehydration, which leads to an oxidized mTagBFP-chromophore as hypothetically colored intermediate. Further ring-structure formation or acylimine hydrolysis causes the large variety of chromophore structures.[49, 53, 171] (*See color plate section for the color representation of this figure.*)

might be even worse because of continuous protein synthesis and degradation. Extensive studies, however, are lacking as those proteins that are stuck at an earlier stage of the chromophore formation are spectroscopically silent. There is a great demand for the development of tools with which suchlike phenomena can be quantified especially in microscopy.

4.3.3 Fluorescence Color and Dynamics

For most Fluorescent Proteins possessing one of the anionic chromophores, the emission hue is close to the color of the excitation light (Fig. 4.2). Two-photon excitation is possible as well and is strongly connected to the factors that determine the absorption color.[57, 74] A typical Stokes shift, that is, the energy difference between absorption and emission maxima, lies there in the range of roughly 400–600 cm^{-1} (see the dashed line in Fig. 4.2). Astonishingly in the actual case, the Fluorescent Proteins are known to exhibit distinct changes of the permanent dipole moment $\Delta\mu_0$ in the range of 5–7 D.[75–78] Dyes with comparable changes of the electron distribution are used as solvatochromic probes.[79] One can conclude that the observed, small Stokes shifts reflect the protein stiffness that is required to maintain fluorescence (see below and Fig. 4.7b). The measured energy differences are therefore close to minimum Stokes shifts observed for dyes in apolar liquids, where only intramolecular geometry changes and unspecific solvent responses contribute.[79]

Smaller deviations from the sketched empirical relation are observed when comparing the Stokes shifts of the B-state dominated proteins with those exhibiting the I-state (Fig. 4.6). The mirror symmetry rule between absorption and emission of the anionic species is still fulfilled (Fig. 4.6a). The vibrational shoulder of the spectra at ambient conditions is shifted by about 1350–1450 cm^{-1}, as can be read from low-temperature experiments.[41, 80] The underlying internal coordinate q likely belongs to the bicyclic chromophore framework,[40, 81, 82] but a clear assignment of vibrational modes below 1500 cm^{-1} is difficult.[83] The stronger the vibrational shoulder is, that is, the broader the absorption or emission spectra are, the larger is the Stokes shift (Fig. 4.6b). On the basis of the above-introduced model of polymethine dyes, one can argue that the charge distribution becomes more homogeneous upon excitation. The stronger the quinonoid mesomeric structure (Fig. 4.3a) contributes already in the ground state, the smaller are the required structural changes along q during relaxation in the excited state. The latter add to the relaxation energy λ, that is, the energetic equivalent of half of the experimental Stokes shift $\Delta\lambda$ (Fig. 4.6c).

Although the given explanation of relaxation along only one coordinate q might explain the differences in the Stokes shift between the B- and I-state, additional modes with strongly differing potentials between the ground and excited states must be considered for assessing the complete relaxation.[81, 84, 85] This is especially true for the large Stokes shifts of Blue Fluorescent Proteins, associated with low fluorescence quantum yields. Exhaustive investigation including Stark-effect experiments, resonance Raman scattering, low-temperature studies in combination with theoretical support were not exhaustively attempted yet. Larger Stokes shifts are observed as well in the far-red-emitting Fluorescent Proteins.[77] Here, particular amino acids

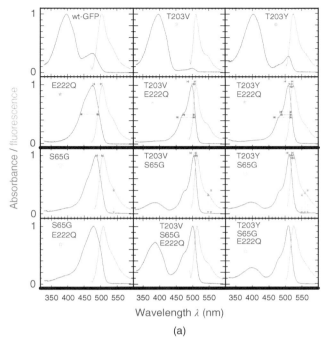

(a)

Figure 4.6 Spectroscopic variation in a set of 12 mutants derived from wt-GFP (*G. Jung, J. Wiehler; unpublished results*). (a) Normalized absorption and emission spectra of various GFP mutants (one-letter notation for the amino acids). The width of the absorption spectra (horizontal blue arrows), exemplified for the second row (Glu222Gln-mutants), is related to the Stokes shift (horizontal red arrows). The same trend is observed for the height of the shoulder in the emission spectrum (vertical green arrows) as exemplified in the third row (Ser65Gly-mutants). The arrows are each transferred from the left spectra for a better comparison. (b) The area of the absorption spectrum, normalized to the R^- absorption maximum, exhibits a linear correlation with the Stokes shift, that is, twice the reorganization energy λ. The area is a quantitative measure of the width in the absorption spectrum (see a, second row). The intrinsic Stokes shift bias of $400–500\,cm^{-1}$ is plotted in Figure 4.2. The symbols refer to the mutants in (a). (c) The strongest influence on the Stokes shift is observed for varying amino acid at position 203. The putative explanation is based on the oxonol-like properties of the anionic chromophore R^- (see Fig. 4.3a). Structural changes to which the normal coordinate q belongs reflect the bond alterations according to the mesomeric concept.[40] In the excited state, the negative charge is more homogeneously smeared over the chromophore than in the ground state. The stronger the charge stabilization at the phenolic oxygen atom is in the electronic ground state S_0, the stronger is the structural change upon excitation into S_1 along q, and thus, the Franck–Condon factors for absorption and emission (indicated by vertical blue and green arrows). Moreover, the larger is also the reorganization energy λ in the linear electron-vibration coupling regime. The charge stabilization is most distinct in the Thr203 mutants (cf. to the B-state in Fig. 4.3b) and less pronounced in the Thr203Val mutants (cf. to the I-state in Fig. 4.3b). The π–π stacking can explain the slightly larger Stokes shift in the YFPs (Thr203Tyr mutants), as it partially stabilizes the negative charge at the phenolic moiety. (*See color plate section for the color representation of this figure.*)

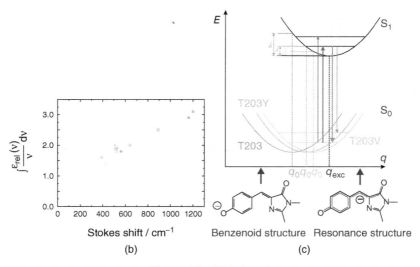

Figure 4.6   (*Continued*)

with hydrogen bonding to the chromophore react on large changes of the permanent dipole moment by reorientation during the excited-state lifetime.[86] Such geometrical relaxation, which was not known from other anionic chromophore species with lower Stokes shifts (Fig. 4.7), is supposed to stabilize the excited state with respect to the ground state. Generally, acylimine extended chromophores exhibit distinctly larger Stokes shifts (Fig. 4.2).

A major energetic difference between excitation and emission can also result from excited-state reactions (see Section 4.4.1). It is worth mentioning that this latter experimental observation exemplifies the ambiguity of the definition of *Stokes shift* as the energy difference between the absorption maximum and the emission maximum, since no clue to the causality is provided. Another example is given in CFPs, where a distinct absorption maximum at $\lambda_{abs} \sim$ 450–455 nm is only slightly less intense than the absorption maximum at $\lambda_{abs} \sim$ 435 nm thus exaggerating the estimated relaxation energy based on the spectroscopic maxima.

The capability of proteins to fluoresce is the molecular property that makes them so valuable in the life sciences. The chromophore itself, being isolated from proteins or synthesized by chemical means, is fluorescent only at cryogenic temperatures.[87, 88] The fluorescence intensity depends on biochemical parameters such as the maturation efficiency (see Section 4.3.2) but also on the photophysics of the chromophore. The radiative lifetime, which is the upper limit of the fluorescence lifetime, can be quite accurately calculated once the *real* molecular extinction coefficient of the absorbing species is known.[73] Excitation and emission must involve the same electronic states for the validity of the underlying Strickler–Berg relation.[89] Again, *av*GFP and its derivatives are most intensely investigated.[55] The values for τ_{rad} of *av*GFP-derived proteins are in the range of 4.5 ns.[90, 91] Three mechanisms are deemed responsible for the fast radiationless decay from the excited state, that is, two single-bond rotations

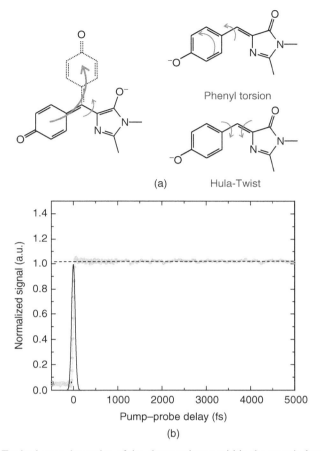

(a)

Phenyl torsion

Hula-Twist

(b)

Figure 4.7 Excited-state dynamics of the chromophores within the protein barrel. (a) Three different movements discussed in internal conversion. The one-bond flip on the left requires large free space and is therefore suppressed within the protein. The one-bond twist around the phenyl single bond (phenyl torsional deformation) is presumably the main pathway for internal conversion.[81, 84] It strongly depends on the double-bond character of the single bond and is therefore influenced by the stabilization of the negative charge at the phenolic moiety (see also Fig. 4.3).[55] The influence of the concerted two-bond flip, the so-called Hula-Twist movement, can be studied in variants, where the competing one-bond rotation is widely disabled as in Thr203Val variants. Deviations from chromophore planarity accelerate internal conversion thus reducing fluorescence quantum yields.[98] (b) Pump-probe experiment ($\lambda_{pump} = 525$ nm, $\lambda_{probe} = 560$ nm) on eYFP (*G. Jung, G. Fleming; unpublished results*). The (stimulated emission) signal rises within the instrumental response function (black curve) and decays slowly with the fluorescence lifetime. No additional contribution from the protein environment, reacting on the change of permanent dipole moment $\Delta\mu_0$,[75] is detected. The missing solvation dynamics is assigned to the protein stiffness around the chromophore.[214] Additional dynamics due to conformational changes of nearby, hydrogen-bonded amino acids can be detected in proteins with a large Stokes shift such as mPlum.[77] Other ultrafast changes of the transient absorption or transient grating signal may result from vibrational cooling after ESPT[215] or from coherent excitation of low-frequency vibrational modes,[216] which might also indicate phenyl torsional movements.[81]

and one concerted two-bond flip, the so-called Hula-Twist (Fig. 4.7a).[92] The space requirements of the rotation around the double bond close to the heterocycle make this process unlikely to occur in GFP. The other one-bond rotation, the phenyl torsional movement was made responsible for the rapid internal conversion in BFPs.[93] The extent to which the latter movement contributes to the radiationless decay of the excited state in GFP and YFP[81, 84] depends on the mesomeric structures introduced in Fig. 4.3a[55]: The stronger electrons are pulled to the phenolic moiety, the stronger is the single-bond character and, hence, the faster is the decay. This qualitative explanation is exemplified in comparison of the fluorescence lifetimes of the B- and I-state of *av*FP[94, 95] but also in comparison of fluorinated chromophores embedded in eGFP and eYFP compared to their parent counterparts.[73] The Hula-Twist movement, which preferentially occurs in restricted geometries, is thought to be responsible for the low fluorescence quantum yields in Ser65Gly mutants and some photoconverted proteins (see Section 4.4.3).[55] Molecular dynamics simulations point to the contribution of the Hula-Twist even in the bright Fluorescent Proteins.[96, 97] It also gained interest as mechanism for isomerization reactions (see Section 4.4.2).[98] It seems, however, that the protein surrounding of the chromophore is stiff enough to effectively suppress these large amplitude motions when the fluorescence quantum yield is high.[81, 84, 85]

Concerning the extended chromophores of RFPs, one could imagine that more decay pathways are operative. In addition, the energy gap law predicts faster internal conversion in RFPs. This combination of two effects might explain why the maximum brightness for the variety of Fluorescent Proteins is found in the yellow-green region,[53, 99] where the most symmetric charge distributions are conceivable. To the best of my knowledge, systematic investigations are still lacking.

4.3.4   Directional Properties along with Optical Transitions

Besides the extent by which the permanent dipole moment changes upon excitation, the transition dipole moment is important for the description of the optical properties of the Fluorescent Proteins. While the former influences the Stokes shift as well as the transition energy, the magnitude of the latter is a measure of the extinction coefficient and, via the Strickler–Berg relation, connected to the radiative lifetime.[73] The angle between both vectors can be determined by Stark-effect experiments; it amounts to ~20° in the anionic species in wt-GFP[75] and to 13° in *ds*Red.[76]

Reorientation of the transition dipole moment between absorption and emission leads to depolarization, that is, loss of the fluorescence anisotropy. However, the observation of strongly polarized emission of GFP crystals, excited with polarized light, indicated conservation of the transition direction.[100] Other experiments have proved that the transition dipole moments for absorption and emission are collinear in many Fluorescent Proteins.[101–103] The fluorescence anisotropy of monomeric Fluorescent Proteins decays with a time constant of approximately 16 ns, which corresponds to the rotational correlation time of the protein barrel.[101, 102, 104] Time-resolved anisotropy measurements with Fluorescent Proteins are therefore valuable tools in bioanalysis (see Section 4.3.5). Due to the excellent polarization-maintaining properties, fluorescence anisotropy measurements are worthwhile for studying oligomerization.[105, 106]

The absolute direction of the transition dipole moment with respect to the molecular structure can be determined by recording absorption spectra of oriented samples[107] or by using UV–Vis/IR pump-probe experiments. Here, the transition dipole moment is measured with respect to the orientation of the heterocycle's carbonyl bond as internal reference.[108] Furthermore, this procedure seems to be more reliable than the analysis of single-crystal dichroism. As the transition dipole moments of most Fluorescent Proteins are not yet experimentally verified, scientists have to rely on quantum-chemical computations (Fig. 4.8a).[109]

4.3.5   Energy Transfer and Energy Migration

So far, the description of the photophysical properties was focused on monomeric proteins. However, some naturally occurring proteins such as *ds*Red are obligatory oligomers.[110, 111] Due to the compact size of the barrel with a diameter of 2.4 nm and a height of 4.2 nm, the distances therein fall in the range of efficient energy transfer (FRET) (Fig. 4.8b).[112, 113] Ultrafast FRET was observed in *ds*Red, where the residual absorbing green fraction served as the energy donor and the red absorbing population acted as energy acceptor.[114] Due to the various distances and existing angles between the transition dipole moments, the disappearance of the green luminescence and the appearance of the red luminescence in such systems obey a multiexponential decay and rise behavior, respectively, the shortest time constant even being too fast to be detected by time-correlated single-photon counting methods.[104] An easier interpretation is achieved when tandem-constructs or dimers are investigated (Fig. 4.8c).[72, 105, 115] The latter were found in some of the crystallographic structures of *av*FPs. The dimerization constant on the order of 10–20 mM^{-1} at physiological pH[116] was exploited for studying the energy migration, that is, FRET between identical chromophores.[103] Here, the decay of the fluorescence anisotropy from 0.4 to 0.28 occurs with a time constant of 2.2 ps pointing to transition dipole moments tilted by less than 30°. Lower values of κ^2 that describes the mutual orientation of similar chromophores were determined in an independent study.[108] Both experiments show that the commonly accepted value for $\kappa^2 = 2/3$ is misleading for tightly bound FRET pairs. Moreover, it was pointed out recently that each FRET pair might exhibit different orientation factors κ^2.[109] Also calculations for pairs in tetrameric *ds*Red show deviations from $\kappa^2 = 2/3$ despite unidentified orientation of the transition dipole moments.[113] The availability of appropriate experimental setups for the determination of geometric factors, however, might be limited due to the expected time constants.

The preferred approach for quantifying analytes is spectral two-channel detection of a FRET pair with a sensor moiety in between.[117] On the basis of published Förster-radii, several donor–acceptor pairs appear useful, and countless examples for bioanalytical FRET applications are proposed (Fig. 4.8b).[53, 118] The dynamic range of such assays, however, often appears limited: there is some FRET without any analyte and less than 100% FRET efficiency with the analyte in saturation.[119] The response

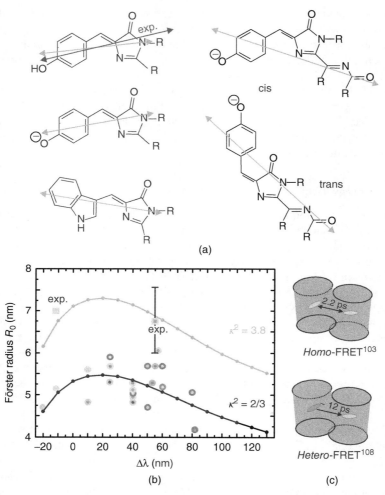

(a)

(b)   (c)

Figure 4.8 Transition dipole moments and their coupling in Förster resonance energy transfer (FRET). (a) Orientation of the transition dipole moments of various chromophores, based on recently published computations (gray arrows; see Ref. [109]). Trustworthy experimental data are sparse and only available for the neutral GFP chromophore (blue arrow).[108] An angle deviation of ~7° with respect to the reference carbonyl bond is found between calculation and experiment. (b) Calculated Förster radii (round symbols, taken from Ref. [53, 118]) and experimental values (square symbols, see (c)) for various pairs are plotted against the difference of the wavelength maxima between the donor emission and the acceptor absorbance. *Homo*-FRET, also called energy migration, is found at negative $\Delta\lambda$ values. The colors of the dots (interior: donor; contour line: acceptor) refer to the fluorescence colors in Figure 4.2. The trend of FRET values can be calculated by manually redshifting the absorption spectrum of GFP-mutant Ser65Thr against its emission spectrum. This is done for two different orientational factors κ^2 (black and gray line). (*See color plate section for the color representation of this figure.*) (c) Only few experiments were performed to experimentally determine the relative orientation of two chromophores in sticking FRET pairs. From the FRET time constant and the change of the fluorescence anisotropy, κ^2 and the Förster-radius can be calculated. κ^2 is considerably larger than expected for an averaged orientation, that is, $\kappa^2 \sim 3.8$[103] and $\kappa^2 \sim 2.6$.[108, 109] R_0 recalculation for Ref. [108] was done by the author.

to physiological changes is therefore challenging in such a situation. One reason certainly is that the chromophore formation efficiency of the acceptor, such as in the pair eGFP and mCherry,[72] can be far below 100% under physiological conditions: pairs with a silent acceptor cannot experience any change of the FRET efficiency upon conformational changes. But also the Förster-radius R_0 might be misjudged if the extinction coefficient of the acceptor is underestimated. Estimates on the basis of Fig. 4.8b imply minimum chromophore distances of 6 nm or more, that is, >1 nm above R_0, for completely disabled FRET in any pair. Lower values of R_0, which could be beneficial for shorter maximum distances, are achieved by reducing the fluorescence yield of the donor alone, by separating the emission of the donor and absorbance of the acceptor further, or by reducing the absorbance of the acceptor. It is therefore not surprising that the development of physiologically useful Ca^{2+}-indicators, which was the first established sensor, took more than 10 years of improvements.[120]

4.4 PHOTOCHEMICAL REACTIONS

4.4.1 Excited-state Proton Transfer (ESPT)

The interplay of the embedded chromophore with its surrounding not only influences the fluorescence properties but also leads to excited-state reactions, which are unknown for the chromophore itself.[121] The first photochemical reaction in GFP that attracted the interest of physicochemists was the excited-state proton transfer (ESPT) (Fig. 4.9a).[122, 123] Early on, there were doubts about the origin of the absorption maximum at around 400 nm leading to green fluorescence; note, for comparison, the rather similar experimental data, redshifted by 50–80 nm, for FRET in dsRed.[114] A distinct isotope effect in the reaction kinetics, an explicit temperature dependence of the process as well as the missing near-UV absorption in some mutants, excluded the role of higher excited states thus favoring a proton transfer reaction. To the same degree, to which the reaction is decelerated by external factors, blue emission at around 460 nm occurs. But also RFPs can exhibit ESPT with these characteristic features.[124, 125] The reaction in mKeima ($\lambda_{exc} = 440$ nm, $\lambda_{em} = 620$ nm), for example, is approximately twice as fast as in wt-GFP.[124]

It is known for several aromatic alcohols that their acidity is increased upon excitation.[126] The pK_a value of the GFP chromophore is lowered from 8.2 to roughly 0 - 2 in the excited state and allows for proton transfer to nearby bases.[127, 128] The main pathway for proton transfer in GFP connects the phenolic moiety with Glu222 via a bound water molecule and Ser205,[19–21, 129] but some more proton wires were deciphered.[130, 131] The previously introduced I-state (see Section 4.3.1, Fig. 4.3) is left after the proton transfer, hence causing the eponymous green fluorescence. Its denomination is based on its *intermediate* character: back to the ground state, the proton is reshuffled from the glutamic acid to the chromophore establishing a Förster-photocycle. Only mutagenesis and low-temperature experiments could stabilize the I-state.[45, 80] The time constant for the reprotonation of the chromophore was determined by pump–dump–probe spectroscopy.[132] It should be noted that the

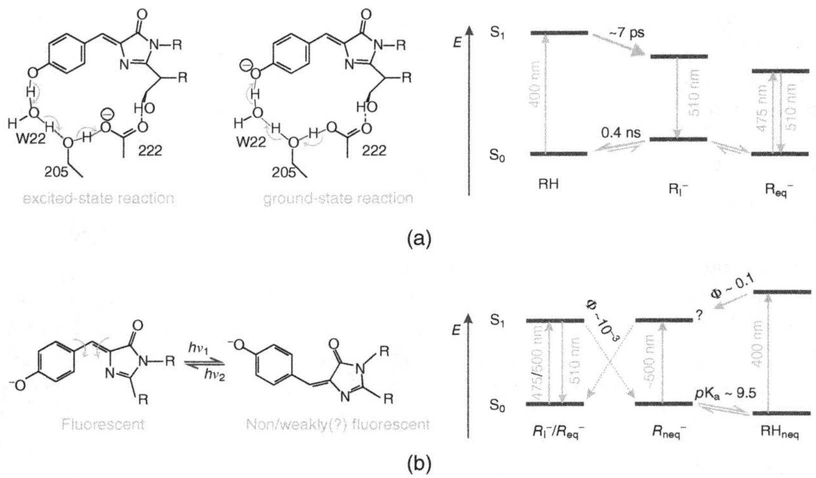

(a)

(b)

Figure 4.9 Simplified photochemical pathways in Fluorescent Proteins (left: chemical models; right: spectroscopic state models). (a) Excited-state proton transfer (ESPT) in wt-GFP moves a proton along the proton chain from the phenolic moiety of the chromophore toward the final acceptor Glu222. In the ground state, the anionic chromophore is reprotonated due to the higher acidity of the glutamic acid. Time constants are taken from Refs [122, 132]. Other examples of ESPT reactions are found, for example, in mKeima leading to a red fluorescent chromophore.[124, 125] The B-state in wt-GFP that is R_{eq}- is accessed on the ground-state energy surface. (b) Cis–trans isomerization to a nonequilibrated, mostly dark R_{neq}- form in Fluorescent Proteins presumably occurs via the Hula-Twist movement. As the isomeric form possesses similar electronic properties, back-transfer to the initial fluorescent form is achieved by the excitation wavelength as well. A constant population of the isomeric state under photostationary condition indicates this mechanism (see also Fig. 4.12).[135] However, the pK_a value of the isomeric state can be distinctly higher than the initial state, thus shifting the main absorbance of the dark state.[144] This reversible photoconversion or photoactivation allows double-resonance experiments ($\lambda_{exc} = 400$ and $475\,nm$) as used in super-resolution microscopy (see also Fig. 4.13). It is unlikely that reconversion with $\lambda_{exc} = 400\,nm$ proceeds via ESPT as no fluorescence is observed during the activation process.[217] The scheme also applies to Red Fluorescent Proteins, however, with bathochromic shifts $\Delta\lambda \sim 50\,nm$ due to their elongated chromophores.[145] Values approximated from Refs [134, 144]. (c) The irreversible photoconversion, or photoactivation, leads to decarboxylation of Glu222, thus removing its acidity. Hence, reprotonation of the anionic chromophore is prevented, and the absorbance is shifted toward the anionic chromophore state. Excitation into higher excited states is even more efficient (S_n). Direct photoactivation of the anionic chromophore R$^-$ states was proved by several authors,[161, 169] but photoactivation light microscopy (PALM) with addressing the neutral chromophore RH is especially useful.[156] Based on quantum yields measurements, there is some evidence for a direct photoconversion from the excited RH state thus bypassing ESPT.[168] Values taken from Refs [155, 161, 166, 168]. (d) Another pathway of irreversible photoconversion, which shifts the fluorescence color from green to red, is found in Fluorescent Proteins such as mKaede and Dendra with histidine next to the green chromophore. Photochemical cleavage of the peptide backbone is accompanied by extension of the conjugation. It can be concluded from the wavelength dependence of photoactivation that the initial state is a weakly populated neutral chromophore state. No evidence is found yet that the reaction proceeds via ESPT. Photoactivation via the "green" anionic chromophore state was detected by single-molecule methods.[198, 219] Electron-transfer reactions likely are involved.[220]

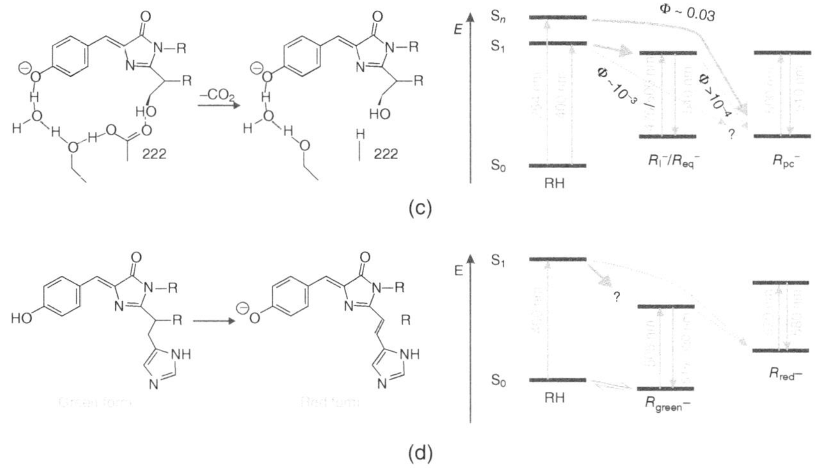

(c)

(d)

Figure 4.9   (*Continued*)

knowledge of the underlying reaction has some practical relevance: the interruption of the hydrogen-bonding network by replacing the final acceptor by Glu222Gln or inhibiting the proton transfer leads to Blue Fluorescent Proteins stemming from the emission of the neutral chromophore.[55, 133]

4.4.2   Isomerization Reactions: Reversible Photoswitching

The most basic reaction that occurs in Fluorescent Proteins is the light-driven cis–trans isomerization (Fig. 4.9b).[134] It was detected by fluorescence correlation spectroscopy where a constant fraction of a dark state was populated upon irradiation.[135] A dark population, where only the population kinetics is intensity dependent, reflects photostationary conditions where both on- and off-transitions are light-induced (see also Section 4.6.2). The observed behavior resembles the isomerization dynamics in cyanine dyes, which exhibit floppy double bonds.[136] From the decay time constant of the autocorrelation function, the quantum yield for the isomerization can be calculated, which is on the order of 10^{-3}–10^{-4}.[137] However, there is no direct proof that indeed a whole isomerization takes place in these experiments. In addition to vibrational spectroscopic investigations,[83] only X-ray crystallography, which was intended to explain the photoswitching behavior, provided evidence that indeed isomerization around the exocyclic methylene bridge can occur.[138–140] Cis–trans isomeric forms of Fluorescent Proteins' chromophores can be both fluorescent with distinctly different optical properties, even in the same protonation state.[141, 142] Observation of fluorescence is more a matter of how accurately the chromophore can be pressed into planarity, that is, how efficiently the surrounding can adapt to the isomeric state.[98, 143]

Most interestingly, the pK_a value of the isomeric state might be shifted even above the value of the free chromophore.[144, 145] In other words, the fluorescent anionic chromophore is in a photostationary equilibrium with the isomeric neutral state over a pH range from 7 to 9. In consequence, the effective absorption spectrum of the dark state is blueshifted under physiological conditions,[146] and fluorescence excited with blue-green light ($\lambda_{exc} \sim 475\,nm$) can be restored with low levels of light, which addresses the neutral chromophore state ($\lambda_{exc} \sim 400\,nm$).[147–150] The switching process itself appears to be more efficient by addressing the neutral chromophore than the anionic state,[137, 144] which might reflect the higher tendency of the former to undergo isomerization reactions (see Section 4.3.3). Once the susceptibility to undergo isomerization is further lowered, for example, by introducing the bulky amino acid Tyr at position 203 or by replacing Glu222 with Gln, the double-resonance excitation scheme can be exploited for photoswitching (see also Section 4.6.3).[144, 150, 151] Although being best understood for GFP-like chromophoric systems, similar observations and explanations also hold for RFPs such as *as*FP (Kindling FP) and IrisFP with the respectively redshifted wavelengths.[140, 145, 152]

4.4.3 Photoconversion: Irreversible Bond Rupture

Besides the presented reversible reactions, irreversible reactions can take place.[153] Already in the first experiments on GFP with intense near-UV irradiation, it was noticed that a complete reversal of photoswitching was not achieved.[122, 154] Mass spectrometry and X-ray crystallography unraveled that a light-driven decarboxylation of Glu222 takes place with a low quantum yield (Fig. 4.9c).[155] The meaning in the context of the reversible ESPT-photocycle is that the acidic compound for the chromophore reprotonation is removed, thus disturbing the hydrogen-bonding network. This was exploited in a seminal application, the so-called photoactivation: here, a GFP mutant bearing Thr203His, which is largely shifted toward the neutral chromophore state, is photochemically ($\lambda_{exc} \sim 400\,nm$) converted into an anionic form.[156] The change of the absorbance is read out by the wavelength with which green fluorescence can be excited, that is, by fluorescence excitation spectroscopy: At the beginning, only near-UV generates green fluorescence via ESPT, but after photoactivation, green fluorescence can also be excited with blue-green light ($\lambda_{exc} = 475–500\,nm$). It turned out that the underlying chemical process, that is, the decarboxylation of a nearby glutamic acid, is observed in other Fluorescent Proteins as well.[157–159]

Moreover, the light-driven decarboxylation can also show up as a reduction of the fluorescence lifetime.[55, 160, 161] The light-induced reduction of the fluorescence lifetime was also detected for other Fluorescent Proteins, but it is not a characteristic signature.[162–164] Furthermore, it is also not solved to date whether all observed changes of the fluorescence lifetime upon extended illumination result from irreversible or reversible photoreactions. Time-resolved experiments confirmed previous experiments indicating that higher excited states exhibit higher quantum yields of photoconversion.[165] Also, the excited states of the anionic chromophore can be the starting point for this photochemical process.[166, 167] The latter observation is understandable considering that the residence time in the excited anionic state is longer by two orders of magnitude than the dwell time in the excited neutral state with a few picoseconds before it undergoes ESPT. It is therefore not clear yet whether excitation of RH induces the photoactivation via ESPT or in a photochemical side reaction although quantum yield measurements suggest the latter.[168] Photoconversion experiments with deuterated GFP samples, where decarboxylation should be favored due to the decelerated ESPT reaction,[122, 123] could enlighten the exact pathway.

Not only nearby amino acids can be the target of photochemistry but also the conjugation of the chromophore could be extended (Fig. 4.9d). This photochemically driven photoactivation turns green-emitting fluorophores into red-emitting fluorophores by a subsequent oxidation reaction. This conversion is recognized for several Fluorescent Proteins, where the first amino acid in the chromophore triade is a histidine as in Kaede and IrisFP,[153] but suchlike spectral shifts were also detected in single-molecule experiments on *av*FP-derived proteins.[169] It is interesting to note that the mentioned green-to-red conversion for GFP was reported even before red chromophores were found in nature.[170] The mechanism of this photoactivation is still not understood in detail. Application of the reversible and the irreversible photoactivation is found in super-resolution microscopy (see Section 4.6.3).[30, 152] A unique benefit of the irreversible photoconversion is the ability to follow proteins, which can be time- and spatially tagged.[53, 171]

4.4.4   Other Photochemical Reactions

Besides the above-described processes, which found most attention in Fluorescent Proteins, other reactions might occur. Oxygen sensitization indicates that intersystem crossing in Fluorescent Proteins occurs.[172, 173] The generation of further reactive oxygen species was described, which might turn the so-called KillerRed into a pivotal phototoxic agent in the life sciences.[174, 175] Last but not least, the reasons for ubiquitous photobleaching still need to be elucidated. Quantum yields for the irreversible photodestruction of GFP-R$^-$ species were consistently found $>10^{-5}$ with slightly higher values for the GFP than for YFPs.[144, 151, 176, 177] Some attempts to stabilize GFP in sugar crystals were undertaken.[178]

4.5   ION SENSITIVITY

4.5.1   Ground-State Equilibria of Protonation States

Already the archetypal wt-GFP, isolated from the jellyfish, exhibits two absorptions of the neutral form and the anionic chromophore form (see Fig. 4.1). Most astonishingly, the population ratio between these two forms is unaffected by the external pH value over a large range despite a pK_a value >8 of the isolated chromophore.[127] The pH dependence is only introduced when the hydrogen-bonding network around the chromophore is altered, for example, by mutations or by photochemical reactions (see also Section 4.4.3). pK_a values then span the range between 6 and 8.[179, 180] A closer inspection revealed that not only ionization of the chromophore but also the acid–base equilibrium of a nearby protonation site, presumably Glu222, is required to modulate the changes of the optical spectra.[181] Both moieties are connected via the hydrogen-bonding chain, which also operates in ESPT (Fig. 4.9a). This finding explains why the ionization of the amino acid counteracts the buildup of the negatively charged state of the chromophore. It also provides rationales for the experimental observations that binding of anions or disturbance of the hydrogen-bonding network affects the equilibrium between the two protonation states of the chromophore.[182–184]

As the chromophore is shielded from the outer solvent by the surrounding amino acids, the transfer of the protons from the exterior to neutralize, for example, by buffer molecules, cannot be realized directly. A *gatekeeper*, that is, an amphoteric moiety passing the proton from outside onto the chromophore, was postulated from fluctuation experiments that were performed in dependence of the buffer concentration.[185] Later on, further kinetic experiments starting with nonequilibrium conditions supported the identification of His148 as the primary proton acceptor.[186] It should be noted here that pH changes might also affect the chromophore conformation.[141, 187]

4.5.2 Quenching by Small Ions

Besides absorbance alterations, other changes of spectroscopic features might be useful for analytical purposes. As the fluorescence lifetime amounts to a few nanoseconds, only static quenching, that is, complex formation in the electronic ground state, can be exploited for quantification.[79] Some small anions were found to bind to the interior of the protein barrel.[182] Their binding is weak ($K_D \gg$ mM) and, therefore, can only be analyzed at the upper edge of the physiological window, especially at high pH values.[183] A change of the fluorescence lifetime was not detected. The practical value is certainly limited due to the experimental observation that there is not a unique binding site for anions but at least two of them in the chromophore pocket.[188]

Much stronger affinities are required for measuring the concentrations of multivalent metal ions. Most effort has been made to develop Ca^{2+}-sensitive constructs primarily based on FRET.[120, 189] Heavy metal binding sites were introduced in Fluorescent Proteins as well or were found in naturally occurring chromoproteins.[190, 191] The attachment of suchlike metal ions results in a reduction of the fluorescence intensity, or, rarely, a fluorescence enhancement is achieved by removing some of the internal conversion mechanisms.[192] Recently, the His6-tag, which is introduced for purification by affinity chromatography, was exploited for monitoring the uptake of Cu^{2+} by plant root cells (Fig. 4.10).[193] The specificity was obtained due to its strong blue color, a unique feature among all other competitive ions. Moreover, the ambiguous fluorescence intensity diminishment could be circumvented since the fluorescence lifetime was used as readout.[194] Other self-calibrating approaches are based on FRET constructs.[195]

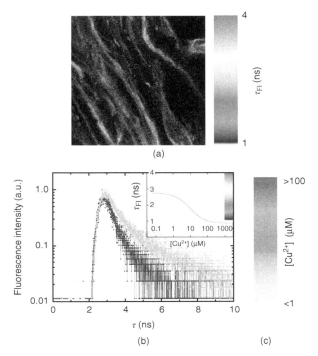

Figure 4.10 Visualization of Cu^{2+}-uptake into plant root cells by means of a sensor GFP and fluorescence lifetime imaging microscopy (FLIM).[193] (a) *Arabidopsis thaliana* root cells expressing GFP-variant Ser65Thr were incubated with 100 μM Cu^{2+} and FLIM was performed. Quenching of the GFP fluorescence due to FRET with Cu^{2+} as (dark) acceptor lead to a reduction of the fluorescence lifetime τ_{Fl}.[194] (b) The green areas in (a) display an averaged $\langle \tau_{Fl} \rangle \sim$ 1.8 ns, which is reduced to $\langle \tau_{Fl} \rangle \sim 1.2$ ns in the blue areas. From the calibration titration (inset), a rough estimate of free $[Cu^{2+}]$ over two orders of magnitude can be made (c). For an exact quantification, however, effects of the surrounding have to be considered.[218] (*See color plate section for the color representation of this figure.*)

4.6 RELATION MICROSCOPY–SPECTROSCOPY FOR FLUORESCENT PROTEINS

4.6.1 Brightness Alteration from Cuvette to Microscopic Experiments

The brightness of fluorescently tagged samples depends on, apart from considering instrumental settings and the local concentration, molecular, physicochemical parameters as introduced in Section 4.3. The proportionality according to these relationships might be suspended due to photobleaching or due to an incomplete chromophore formation. All mentioned parameters can also be deciphered from ensemble cuvette experiments. Further deviation from the irradiation intensity-signal proportionality might stem from saturation of the electronic transition. However, the latter effect is diminished if the use of strongly focusing optics or intense light sources is avoided.

By using lasers for excitation and strongly collimating microscope objectives, even a small light power can result in enormous intensities. Intensities on the order of $100\,kW/cm^2$ can be achieved within the diffraction-limited focus of a conventional laser pointer operating at a power of 1 mW. Calculations on the basis of typical absorption cross sections of 2–3 $Å^2$ at λ_{max} yield approximately 5×10^7 excitations/s.[137] Without any doubt, from this rough estimate, it is obvious that rare events with probabilities of $>10^{-4}$ can be detrimental to the brightness, for example, in confocal microscopy.[56] Here, typical pixel dwell times of 2 µs, up to 100 pixels per diffraction limited spot and averaging over several frames result in doses that impact the behavior of the Fluorescent Proteins.[161] Figure 4.11 exemplifies the impact of several imperfections on the expected brightness of dsRed in microscopy. One can conclude that the described phenomena complicate quantitative analyses.[163, 164] Therefore, methods that do not rely on fluorescence intensities are preferential, but also alternatives such as lifetime-based methods require careful interpretation.

4.6.2   Lessons from Microspectrometry

Some of the aforementioned light-driven reactions (see Section 4.4), such as irreversible photoconversion,[154, 160, 169, 170] isomerization reactions,[135, 196] or photoactivation,[147–150] were postulated or observed by microscopic experiments before structural investigations have proved the underlying mechanisms. The rate constants of such rare events can be obtained by fluorescence correlation spectroscopy.[137] The most widespread application of FCS is the analysis of the diffusional properties, and photobleaching can be analyzed from reduction of the apparent diffusional time.[177] Chemical reasons for fluctuations are light-driven transient population of isomeric dark states[90, 135] and reversible reactions such as protonation.[185, 197] As the residence time of the molecules within a confocal volume is limited to ~1 ms (see however Ref. [162]), only light-driven processes with quantum yields $\gg 10^{-6}$ can be analyzed in FCS. For slower reactions on a timescale beyond 1 ms or consecutive reactions, single-molecule methods are preferential. Another advantage of the latter methods is that the chronology of rare events is easily detectable,[198] whereas FCS averages fluctuations over a large number of molecules.

The nature of dark or metastable states usually remains hidden or speculative in microscopic experiments as those experiments mostly rely on fluorescence spectroscopy. Despite sometimes small fluorescence quantum yields of those states, fluorescence emission spectra reveal spectral shifts, which are attributed to isomerization reactions[158, 160, 196] or photoconversion.[169] Lifetime measurements of single molecules allow for sorting different populations.[199] Finally, changes of the excitation spectra can be concluded by varying the excitation wavelength in FCS (see Fig. 4.12). Double-resonance experiments allowed for detecting the neutral chromophore state in the photophysics of some anionic chromophore species.[146, 147, 150] A unifying potential scheme in its revisited form is shown in Figure 4.9b.[144] It replaces a previous

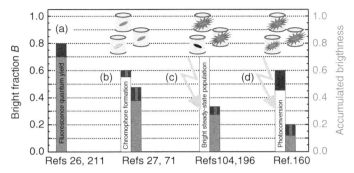

Figure 4.11 Relative brightness in cuvette experiments compared with the brightness in microscopic experiments, exemplified for *ds*Red. (a) The conventionally measured brightness is composed of the fluorescence quantum yield and the extinction coefficient, which reflects also the chromophore formation (see Fig. 4.5). (b) In *ds*Red, almost half of the chromophores are green.[27, 71] In these reports, however, trapping in other spectroscopically silent state as, for example, the blue fluorescent state might be overseen. Accordingly, even lower values of chromophore formation were detected in mCherry and mRFP.[72] (c) The constant dark-state population under photostationary conditions (see Fig. 4.9b) further reduces the actual output of fluorescence.[104] Even lower fractions of molecules in the bright state were detected.[196] (d) Irreversible light-induced decarboxylation of Glu215 (similar to Glu222 in *av*FP[158]) is associated with a reduction in the fluorescence lifetime from 3.6 to 1.5 ns,[160] and therefore, a proportional reduction in the fluorescence quantum yield. Concomitantly, a change of the emission wavelength is observed, which is attributed to a trans-isomeric state.[158, 196] The last two light-induced processes (c and d) are only observed in microscopy due to the distinctly higher excitation intensities. Considering all experimental observations, the initial molecular brightness as determined by quantum yield measurements in cuvettes is reduced by at least a factor of 4–6 in microscopy. Quantitative values for other proteins are vastly missing (see Ref.[56]).

version and is valid to describe double-resonance experiments on some GFP variants bearing the Glu222Gln mutation (Fig. 4.13).[137, 147]

4.6.3   Tools for Advanced Microscopic Techniques

There is no doubt that the discovery and the development of the Autofluorescent Proteins has had a great impact on all areas of life sciences. Fluorescently labeled organisms enabled experiments that could not be foreseen 20 years ago.[2] Countless articles, book chapters, and so on describe the wealth of applications of Fluorescent Proteins.[4–9] Also genetically encoded fluorescence evolved to replace classical fluorescence dyes for measuring ion-concentration or pH values *in vivo*.[119, 193, 200, 201] Two-photon excitation of Fluorescent Proteins[57, 202] second-harmonic generation[203] as well as the development of far-red-emitting fluorophores paves the way for imaging thicker specimen.[52] Another advantage is the labeling stoichiometry of gene products, which is not as easily achieved by classical fluorescence labeling.[204] Two-color experiments are exploited to enhance the sensitivity of imaging experiments.[205, 206]

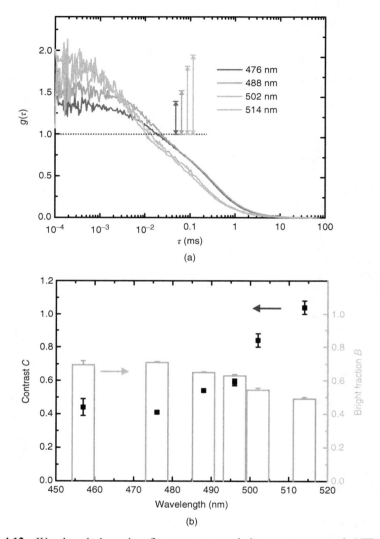

Figure 4.12 Wavelength-dependent fluorescence correlation spectroscopy of eYFP (λ_{abs} = 514 nm, λ_{em} = 527 nm) at similar excitation intensities (*S. Veettil, G. Jung; unpublished results*). (a) Autocorrelation functions $g(\tau)$ of YFP, normalized to one molecule, show an intensity-independent decay on the microsecond timescale due to cis–trans isomerization.[135] With blueshifting the excitation wavelength, the amplitude of this fast decay, that is, contrast C, decreases, as indicated by the arrows. The reduction of the apparent diffusion time on the sub-ms timescale going from λ_{exc} = 514 nm to λ_{exc} = 476 nm is interpreted as photobleaching due to the higher number of excitation cycles the closer to the excitation maximum.[177] (b) As both the forward transition into and backward transition from the dark state are light-driven (see Fig. 4.9b), the contrast C scales roughly with the relative population of molecules in the dark state.[136] The bright fraction is smaller close to the excitation maximum in this particular case (see also Fig. 4.11). Therefore, a blueshifted absorption spectrum of the dark state can be concluded from the photostationary conditions. (*See color plate section for the color representation of this figure.*)

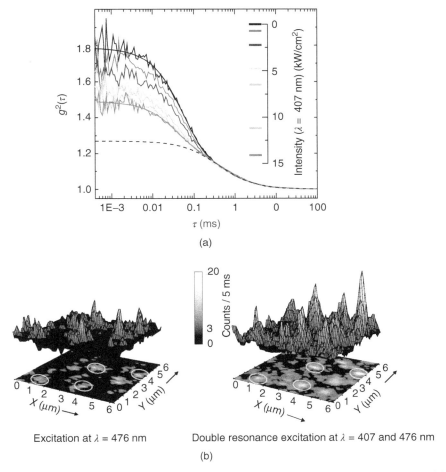

Figure 4.13 Double resonance experiments with $\lambda_{exc,1} = 476$ nm and $\lambda_{exc,2} = 407$ nm with *av*FP bearing the mutation Glu222Gln. (a) In FCS-experiments of the anionic chromophore R^-, the light-driven population of the dark state, that is, the amplitude of the colored curve with respect to the diffusional part (dashed line), is reduced by depopulation of the neutral chromophore absorbing $\lambda_{exc} = 400$ nm (cf. Fig. 4.9b). (b) The brightness of single molecules can be enhanced (left → right) by double-resonance excitation as the dark fluorescence trap is depleted by $\lambda_{exc,2} = 407$ nm. However, the benefit of double-resonance excitation for super-resolution microscopy[149] was overseen at that time. Reproduced from Ref. [147] with permission. (*See color plate section for the color representation of this figure.*)

However, subcellular localization of fluorescence was restricted due to the optical resolution of roughly $\lambda/2 > 200$ nm in the visible range. Only during the past decade, several methods were developed to break the diffraction limit in optical resolution. While predominantly physical methods such as stimulated emission depletion microscopy (STED-microscopy) can be executed with Autofluorescent Proteins as well,[207] other super-resolution techniques on the basis of single-molecule methods exploit the specific photochemical reactions discussed in Section 4.4.[134, 171] Photoactivated localization microscopy (PALM) exploits the irreversible change of the excitation properties by photoconversion or changes of the covalent chromophore structure.[208] Reversible switching between fluorescent on–off states as in the isomerization reaction is exploited in reversible switchable optical fluorescence transitions (RESOLFTs).[209] Quite recently, decoupling of switching and fluorescence readout was achieved.[210] These are beautiful examples in which basic experimental findings are turned into exploitable tools by biotechnological engineering, which were, at least partially, awarded by the Nobel Prize for Chemistry in 2014.

4.7 PROSPECTS AND OUTLOOK

Life sciences are revolutionized by the discovery and development of Autofluorescent Proteins, but their rich chemistry inspired biophysicists to investigate them by various kinds of spectroscopic techniques. Specifically tailored to scientific problems, a plethora of proteins allows for addressing the basic analytical question *when, where, what*: Using timer proteins or photoactivation, researchers can mark the temporal zero-point in dynamics and kinetics; microscopy with photoswitchable proteins maps their location down to an accuracy of nanometers; tagging of gene sequences selects the gene product. With the whole rainbow of fluorescence colors at our disposal, multicolor experiments are feasible. Interactions can be studied by FRET, and transportation is visualized by time-lapse microscopy. There are, however, still open questions such as "how much of a certain compound is existing (in time and space)." Here, the (photo)-stability and the brightness of proteins, especially of those in the red spectral region, are regarded as limitations. Further developments of Fluorescent Protein technology are required despite yet particular successful approaches of quantifying second messengers. I anticipate that for at least some of these applications the ubiquitous photochemical reactions, which guided the invention of new microscopic tools, have to be overcome. It took more than 30 years from the very early experiments to the *Green Revolution*, and the development of Fluorescent Proteins still has to go on.

ACKNOWLEDGMENTS

The author thanks C. Spies for help in the preparation of some figures. Generous financial support is provided by the German Science Foundation (DFG; grant JU650/2-2 and 5-1).

REFERENCES

1. Shimomura, O. *Angew. Chem. Int. Ed.* **2009**, *48*, 5590.
2. Chalfie, M. *Angew. Chem. Int. Ed.* **2009**, *48*, 5603.

3. Tsien, R. *Angew. Chem. Int. Ed.* **2009**, *48*, 5612.

4. Chalfie, M.; Kain, S., Eds. Green Fluorescent Protein – Properties, Applications and Protocols; 2nd ed.; Wiley-Interscience: NJ, **2006**.

5. Zimmer, M. Glowing Genes: A Revolution in Biotechnology. Prometheus: Amherst, NY, **2005**.

6. Pieribone, V.; Gruber, D. A Glow in the Dark: The Revolutionary Science of Biofluorescence. Belknap Press: Cambridge, MA, **2005**.

7. Jung, G., Ed. Fluorescent Proteins I – From Understanding to Design: Springer Series on Fluorescence 11. Springer: Heidelberg, Germany, **2012**.

8. Jung, G., Ed. Fluorescent Proteins II – Application of Fluorescent Protein Technology: Springer Series on Fluorescence 12. Springer: Heidelberg, Germany, **2012**.

9. Tinnefeld, P.; Eggeling, C.; Hell, S., Eds. Far-Field Optical Nanoscopy: Springer Series on Fluorescence 14. Springer: Heidelberg, Germany, **2015**.

10. Shimomura, O.; Johnson, F.; Saiga, Y. *J. Cell. Comp. Physiol.* **1962**, *62*, 223.

11. Head, J.; Inouye, S.; Teranishi, K.; Shimomura, O. *Nature* **2000**, *405*, 372.

12. Shimomura, O.; Johnson, F. *Proc. Natl. Acad. Sci. U. S. A.* **1978**, *75*, 2611.

13. Morise, H.; Shimomura, O.; Johnson, F.; Winant, J. *Biochemistry* **1974**, *13*, 2656.

14. Cody, C.; Prasher, D.; Westler, W.; Prendergast, F.; Ward, W. *Biochemistry* **1993**, *32*, 1212.

15. Prasher, D.; Eckenrode, V.; Ward, W.; Prendergast, F.; Cormier, M. *Gene* **1992**, *111*, 229.

16. Chalfie, M.; Tu, Y.; Euskirchen, G.; Ward, W.; Prasher, D. *Science* **1994**, *263*, 802.

17. Nishiuchi, Y.; Inui, T.; Nishio, H.; Bodi, J.; Kimura, T.; Tsuji, F.; Sakakibara, S. *Proc. Natl. Acad. Sci. U. S. A.* **1998**, *95*, 13549.

18. Ormö, M.; Cubitt, A.; Kallio, K.; Gross, L.; Tsien, R.; Remington, J. *Science* **1996**, *273*, 1392.

19. Yang, F.; Moss, L.; Phillips, G. *Nat. Biotechnol.* **1996**, *14*, 1246.

20. Palm, G.; Zdanov, A.; Gaitanaris, G.; Stauber, R.; Pavlakis, G.; Wlodawer, A. *Nat. Struct. Biol.* **1997**, *4*, 361.

21. Brejc, K.; Sixma, T.; Kitts, P.; Kain, S.; Tsien, R.; Ormö, M.; Remington, J. *Proc. Natl. Acad. Sci. U. S. A.* **1997**, *94*, 2306.

22. Kojima, S.; Ohkawa, H.; Hirano, T.; Maki, S.; Niwa, H.; Ohashi, M.; Inouye, S.; Tsuji, F. *Tetrahedron Lett.* **1998**, *39*, 5239.

23. Matz, M.; Fradkov, A.; Labas, Y.; Savitsky, A.; Zaraisky, A.; Markelov, M.; Lukyanov, S. *Nat. Biotechnol.* **1999**, *17*, 969.

24. Wiedenmann, J.; D'Angelo, C.; Nienhaus, U. in Ref. 8; pp. 3–33.

25. Labas, Y.; Gurskaya, N.; Yanushevich, Y.; Fradkov, A.; Lukyanov, S.; Matz, M. *Proc. Natl. Acad. Sci. U. S. A.* **2002**, *99*, 4256.

26. Baird, G.; Zacharias, D.; Tsien, R. *Proc. Natl. Acad. Sci. U. S. A.* **2000**, *97*, 11984.

27. Gross, L.; Baird, G.; Hoffman, R.; Baldridge, K.; Tsien, R. *Proc. Natl. Acad. Sci. U. S. A.* **2000**, *97*, 11990.

28. Campbell, R.; Tour, O.; Palmer, A.; Steinbach, P.; Baird, G.; Zacharias, D.; Tsien, R. *Proc. Natl. Acad. Sci. U. S. A.* **2002**, *99*, 7877.

29. Shaner, N.; Campbell, R.; Steinbach, P.; Giepmans, B.; Palmer, A.; Tsien, R. *Nat. Biotechnol.* **2004**, *22*, 1567.

30. Day, R.; Davidson, M. *Chem. Soc. Rev.* **2009**, *38*, 2887.

31. Mickler, M.; Dima, R.; Dietz, H.; Hyeon, C.; Thirumalai, D.; Rief, M. *Proc. Natl. Acad. Sci. U. S. A.* **2007**, *104*, 20268.

32. Hsu, S.; Blaser, G.; Jackson, S. *Chem. Soc. Rev.* **2009**, *38*, 2951.

33. Seifert, M.; Ksiazek, D.; Azim, K.; Smialowski, P.; Budisa, N.; Holak, T. *J. Am. Chem. Soc.* **2002**, *124*, 7932.

34. Bae, J.; Rubini, M.; Jung, G.; Wiegand, G.; Seifert, M.; Azim, K.; Kim, J.; Zumbusch, A.; Holak, T.; Moroder, L.; Huber, R.; Budisa, N. *J. Mol. Biol.* **2003**, *328*, 1071.

35. Hoesl, M.; Merkel, L.; Budisa, N. in Ref. 7; pp. 99–130.

36. Fisher, H.; Mintz, C. *J. Ind. Microbiol. Biotechnol.* **2000**, *24*, 323.

37. Ruller, R.; Silva-Rocha, R.; Silva, A.; Cruz Schneider, M.; Ward, R. *Biochem. Mol. Biol. Educ.* **2011**, *39*, 21.

38. Giron, M.; Salto, R. *Biochem. Mol. Biol. Educ.* **2011**, *39*, 309.

39. Olson, S.; McKenzie, R. *Chem. Phys. Lett.* **2010**, *492*, 150.

40. Laino, T.; Nifosi, R.; Tozzini, V. *Chem. Phys.* **2004**, *298*, 17.

41. Creemers, T.; Lock, A.; Subramaniam, V.; Jovin, T.; Völker, S. *Chem. Phys.* **2002**, *275*, 109.

42. Kummer, A.; Wiehler, J.; Rehaber, H.; Kompa, C.; Steipe, B.; Michel-Beyerle, M. *J. Phys. Chem. B* **2000**, *104*, 4791.

43. Sniegowski, J.; Phail, R.; Wachter, R. *Biochem. Biophys. Res. Commun.* **2005**, *332*, 657.

44. Wood, T.; Barondeau, D.; Hitomi, C.; Kassmann, C.; Tainer, J.; Getzoff, E. *Biochemistry* **2005**, *44*, 16211.

45. Wiehler, J.; Jung, G.; Seebacher, C.; Zumbusch, A.; Steipe, B. *ChemBioChem* **2003**, *4*, 1164.

46. Pakhomov, A.; Martynov, V. *Biochem. Biophys. Res. Commun.* **2011**, *407*, 230.

47. Shagin, D.; Barsova, E.; Yanushevich, Y.; Fradkov, A.; Lukyanov, K.; Labas, Y.; Semenova, T.; Ugalde, J.; Meyers, A.; Nunez, J.; Widder, E.; Lukyanov, S.; Matz, M. *Mol. Biol. Evol.* **2004**, *21*, 841.

48. List, N.; Olsen, J.; Jensen, H.; Steindal, A.; Kongsted, J. *J. Phys. Chem. Lett.* **2012**, *3*, 3513.

49. Subach, O.; Malashkevich, V.; Zencheck, W.; Morozova, K.; Piatkevich, K.; Almo, S.; Verkhusha, V. *Chem. Biol.* **2010**, *17*, 333.

50. Pletnev, S.; Subach, F.; Dauter, Z.; Wlodawer, A.; Verkhusha, V. *J. Am. Chem. Soc.* **2010**, *132*, 2243.

51. Drobizhev, M.; Tillo, S.; Makarov, N.; Hughes, T.; Rebane, A. *J. Phys. Chem. B* **2009**, *113*, 12860.

52. Lin, M.; McKeown, M.; Ng, H.; Aguilera, T.; Shaner, N.; Campbell, R.; Adams, S.; Gross, L.; Ma, W.; Alber, T.; Tsien, R. *Chem. Biol.* **2009**, *16*, 1169.

53. Chudakov, D.; Matz, M.; Lukyanov, S.; Lukyanov, K. *Physiol. Rev.* **2010**, *90*, 1103.

54. Subach, F.; Verkhusha, V. *Chem. Rev.* **2012**, *112*, 4308.

55. Jung, G.; Wiehler, J.; Zumbusch, A. *Biophys. J.* **2005**, *88*, 1932.

56. Jung, G.; Zumbusch, A. *Microsc. Res. Tech.* **2006**, *69*, 175.

57. Drobizhev, M.; Makarov, N.; Tillo, S.; Hughes, T.; Rebane, A. *Nat. Methods* **2011**, *8*, 393.

58. Ward, W. in Ref. 4; pp. 39–65.

59. Craggs, T. *Chem. Soc. Rev.* **2009**, *38*, 2865.

60. Branchini, B.; Nemser, A.; Zimmer, M. *J. Am. Chem. Soc.* **1998**, *120*, 1.

61. Barondeau, D.; Putnam, C.; Kassmann, C.; Tainer, J.; Getzoff, E. *Proc. Natl. Acad. Sci. U. S. A.* **2003**, *100*, 12111.

62. Sniegowski, J.; Lappe, J.; Patel, H.; Huffman, H.; Wachter, R. *J. Biol. Chem.* **2005**, *280*, 26248.

63. Ehrig, T.; O'Kane, D.; Prendergast, F. *FEBS Lett.* **1995**, *367*, 163.

64. Auerbach, D.; Klein, M.; Franz, S.; Carius, Y.; Lancaster, R. *ChemBioChem* **2014**, *15*, 1404.

65. Barondeau, D.; Tainer, J.; Getzoff, E. *J. Am. Chem. Soc.* **2006**, *128*, 3166.

66. Pouwels, L.; Zhang, L.; Chan, N.; Dorrestein, P.; Wachter, R. *Biochemistry* **2008**, *47*, 10111.

67. Zhang, L.; Patel, H.; Lappe, J.; Wachter, R. *J. Am. Chem. Soc.* **2006**, *128*, 4766.

68. Terskikh, A.; Fradkov, A.; Ermakova, G.; Zaraisky, A.; Tan, P.; Kajava, A.; Zhao, X.; Lukyanov, S.; Matz, M.; Kim, S.; Weissman, I.; Siebert, P. *Science* **2000**, *290*, 1585.

69. Strack, R.; Strongin, D.; Mets, L.; Glick, B.; Keenan, R. *J. Am. Chem. Soc.* **2010**, *132*, 8496.

70. Ulbrich, M.; Isacoff, E. *Nat. Methods* **2007**, *4*, 319.

71. Garcia-Parajo, M.; Koopman, M.; van Dijk, E.; Subramaniam, V.; van Hulst, N. *Proc. Natl. Acad. Sci. U. S. A.* **2001**, *98*, 14392.

72. Foo, Y.; Naredi-Rainer, N.; Lamb, D.; Ahmed, S.; Wohland, T. *Biophys. J.* **2012**, *102*, 1174.

73. Jung, G.; Brockhinke, A.; Gensch, T.; Hötzer, B.; Schwedler, S.; Veettil, S. in Ref. 7; pp. 69–97.

74. Drobizhev, M.; Tillo, S.; Makarov, N.; Hughes, T.; Rebane, A. *J. Phys. Chem. B* **2009**, *113*, 855.

75. Bublitz, G.; King, B.; Boxer, S. *J. Am. Chem. Soc.* **1998**, *120*, 9370.

76. Lounis, B.; Deich, J.; Rosell, F.; Boxer, S.; Moerner, W. *J. Phys. Chem. B* **2001**, *105*, 5048.

77. Abbyad, P.; Childs, W.; Shi, X.; Boxer, S. *Proc. Natl. Acad. Sci. U. S. A.* **2007**, *104*, 20189.

78. Drobhizev, M.; Makarov, N.; Hughes, T.; Rebane, A. *J. Phys. Chem. B* **2007**, *111*, 14051.

79. Lakowicz, J.; Principles of Fluorescence Spectroscopy, 3rd ed.; Springer: NY, **2006**.

80. Seebacher, C.; Deeg, F.; Bräuchle, C.; Wiehler, J.; Steipe, B. *J. Phys. Chem. B* **1999**, *103*, 7728.

81. Martin, M.; Negri, F.; Olivucci, M. *J. Am. Chem. Soc.* **2004**, *126*, 5462.

82. Fang, C.; Frontiera, R.; Tran, R.; Mathies, R. *Nature* **2009**, *462*, 200.

83. Luin, S.; Tozzini, V. in Ref. 7; pp. 133–169.

84. Stavrov, S.; Solntsev, K.; Tolbert, L.; Huppert, D. *J. Am. Chem. Soc.* **2006**, *128*, 1540.

85. Litvinenko, K.; Meech, S. *Phys. Chem. Chem. Phys.* **2004**, *6*, 2012.

86. Shu, X.; Wang, L.; Colip, L.; Kallio, K.; Remington, J. *Protein Sci.* **2009**, *18*, 460.

87. Kummer, A.; Kompa, C.; Niwa, H.; Hirano, T.; Kojima, S.; Michel-Beyerle, M. *J. Phys. Chem. B* **2002**, *106*, 7554.

88. Litvinenko, K.; Webber, N.; Meech, S. *J. Phys. Chem. A* **2003**, *107*, 2616.

89. Strickler, S.; Berg, R. *J. Chem. Phys.* **1962**, *37*, 814.

90. Heikal, A.; Hess, S.; Webb, W. *Chem. Phys.* **2001**, *274*, 37.

91. Goedhart, J.; van Weeren, L.; Hink, M.; Vischer, N.; Jalink, K.; Gadella, T. *Nat. Methods* **2010**, *7*, 137.

92. Weber, W.; Helms, V.; McCammon, A.; Langhoff, P. *Proc. Natl. Acad. Sci. U. S. A.* **1999**, *96*, 6177.

93. Kummer, A.; Wiehler, J.; Schüttrigkeit, T.; Berger, B.; Steipe, B.; Michel-Beyerle, M. *ChemBioChem* **2002**, *3*, 659.

94. Striker, G.; Subramaniam, V.; Seidel, C.; Volkmer, A. *J. Phys. Chem. B* **1999**, *103*, 8612.

95. Cotlet, M.; Hofkens, J.; Maus, M.; Gensch, T.; Van der Auweraer, M.; Michiels, J.; Dirix, G.; Van Guyse, M.; Vanderleyden, J.; Visser, A.; DeSchryver, F. *J. Phys. Chem. B* **2001**, *105*, 4999.

96. Megley, C.; Dickson, L.; Maddalo, S.; Chandler, G.; Zimmer, M. *J. Phys. Chem. B* **2009**, *113*, 302.

97. Lelimousin, M.; Noirclerc-Savoye, M.; Lazareno-Saez, C.; Paetzold, B.; Le Vot, S.; Chazal, R.; Macheboeuf, P.; Field, M.; Bourgeois, D.; Royant, A. *Biochemistry* **2009**, *48*, 10038.

98. Maddalo, S.; Zimmer, M.; *Photochem. Photobiol.* **2006**, *82*, 367.

99. Dedecker, P.; De Schryver, F.; Hofkens, J. *J. Am. Chem. Soc.* **2013**, *135*, 2387.

100. Inouye, S.; Shimomura, O.; Goda, M.; Shribak, M.; Tran, P. *Proc. Natl. Acad. Sci. U. S. A.* **2002**, *99*, 4272.

101. Volkmer, A.; Subramaniam, V.; Birch, D.; Jovin, T. *Biophys. J.* **2000**, *78*, 1589.

102. Borst, J.; Hink, M.; van Hoek, A.; Visser, A. *J. Fluoresc.* **2005**, *15*, 153.

103. Jung, G.; Ma, Y.; Prall, B.; Fleming, G. *ChemPhysChem* **2005**, *6*, 1628.

104. Heikal, A.; Hess, S.; Baird, G.; Tsien, R.; Webb, W. *Proc. Natl. Acad. Sci. U. S. A.* **2000**, *97*, 11996.

105. Gautier, I.; Tramier, M.; Durieux, C.; Coppey, J.; Pansu, R.; Nicolas, J.; Kemnitz, K.; Coppey-Moisan, M. *Biophys. J.* **2001**, *80*, 3000.

106. Bader, A.; Hoetzl, S.; Hofman, E.; Voortman, J.; van Bergen en Henegouwen, P.; van Meer, G.; Gerritsen, H. *ChemPhysChem* **2011**, *12*, 475.

107. Rosell, F.; Boxer, S. *Biochemistry* **2003**, *42*, 177.

108. Shi, X.; Basran, J.; Seward, H.; Childs, W.; Bagshaw, C.; Boxer, S. *Biochemistry* **2007**, *46*, 14403.

109. Ansbacher, T.; Srivastava, H.; Stein, T.; Baer, R.; Merkx, M.; Shurki, A. *Phys. Chem. Chem. Phys.* **2012**, *14*, 4109.

110. Wall, M.; Socolich, M.; Ranganathan, R. *Nat. Struct. Biol.* **2000**, *7*, 1133.

111. Yarbrough, D.; Wachter, R.; Kallio, K.; Matz, M.; Remington, M. *Proc. Natl. Acad. Sci. U. S. A.* **2001**, *98*, 462.

112. Sanchez-Mosteiro, G.; Koopman, M.; van Dijk, E.; Hernando, J.; van Hulst, N.; Garcia-Parajo, M. *ChemPhysChem* **2004**, *5*, 1782.

113. Schleifenbaum, F.; Blum, C.; Elgass, K.; Subramaniam, V.; Meixner, A. *J. Phys. Chem. B* **2008**, *112*, 7669.

114. Schüttrigkeit, T.; Zachariae, U.; von Feilitzsch, T.; Wiehler, J.; von Hummel, J.; Steipe, B.; Michel-Beyerle, M. *ChemPhysChem* **2001**, *2*, 325.

115. Koushik, S.; Chen, H.; Thaler, C.; Puhl, H.; Vogel, S. *Biophys. J.* **2006**, *91*, L99.

116. Zacharias, D.; Violin, J.; Newton, A.; Tsien, R. *Science* **2002**, *296*, 913.

117. Zeug, A.; Woehler, A.; Neher, E.; Ponimaskin, E. *Biophys. J.* **2012**, *103*, 1821.

118. Patterson, G.; Piston, D.; Barisas, G. *Anal. Biochem.* **2000**, *284*, 438.

119. Miyawaki, A.; Lopis, J.; Heim, R.; McCaffery, M.; Adams, J.; Ikura, M.; Tsien, R. *Nature* **1997**, *388*, 882.

120. Gensch, T.; Kaschuba, D. in Ref. 8; pp. 125–161.

121. Meech, S. in Ref. 7; pp. 41–68.

122. Chattoraj, M.; King, B.; Bublitz, G.; Boxer, S. *Proc. Natl. Acad. Sci. U. S. A.* **1996**, *93*, 8362.

123. Lossau, H.; Kummer, A.; Heinecke, R.; Pöllinger-Dammer, F.; Kompa, C.; Bieser, G.; Jonsson, T.; Silva, C.; Yang, M.; Youvan, D.; Michel-Beyerle, M. *Chem. Phys.* **1996**, *213*, 1.

124. Henderson, N.; Osborn, M.; Koon, N.; Gepshtein, R.; Huppert, D.; Remington, J. *J. Am. Chem. Soc.* **2009**, *131*, 13212.

125. Piatchevich, K.; Malashkevich, V.; Almo, S.; Verkhushka, V. *J. Am. Chem. Soc.* **2010**, *132*, 10762.

126. Agmon, N. *J. Phys. Chem. A* **2005**, *109*, 13.

127. Scharnagl, C.; Raupp-Kossmann, R. *J. Phys. Chem. B* **2004**, *108*, 477.

128. Baranov, M.; Lukyanov, K.; Borissova, A.; Shamir, J.; Kosenkov, D.; Slipchenko, L.; Tolbert, L.; Yampolsky, I.; Solntsev, K. *J. Am. Chem. Soc.* **2012**, *134*, 6025.

129. Lill, M.; Helms, V. *Proc. Natl. Acad. Sci. U. S. A.* **2002**, *99*, 2778.

130. Stoner-Ma, D.; Jaye, A.; Ronayne, K.; Nappa, J.; Meech, S.; Tonge, P. *J. Am. Chem. Soc.* **2008**, *130*, 1227.

131. Leiderman, P.; Huppert, D.; Agmon, N. *Biophys. J.* **2006**, *90*, 1009.

132. Kennis, J.; Larsen, D.; van Stokkum, I.; Vengris, M.; van Thor, J.; van Grondelle, R. *Proc. Natl. Acad. Sci. U. S. A.* **2004**, *101*, 17988.

133. Ai, H.; Shaner, N.; Cheng, Z.; Tsien, R.; Campbell, R. *Biochemistry* **2007**, *46*, 5904.

134. Bourgeois, D.; Adam, V. *IUBMB Life* **2012**, *64*, 482.

135. Schwille, P.; Kummer, S.; Heikal, A.; Moerner, W.; Webb, W. *Proc. Natl. Acad. Sci. U. S. A.* **2000**, *97*, 151.

136. Widengren, J.; Schwille, P. *J. Phys. Chem. A* **2000**, *104*, 6416.

137. Jung, G.; Bräuchle, C.; Zumbusch, A. *J. Chem. Phys.* **2001**, *114*, 3149.

138. Andresen, M.; Wahl, M.; Stiel, A.; Gräter, F.; Schäfer, L.; Trowitzsch, S.; Weber, G.; Eggeling, C.; Grubmüller, H.; Hell, S.; Jakobs, S. *Proc. Natl. Acad. Sci. U. S. A.* **2005**, *102*, 13070.

139. Henderson, N.; Ai, H.; Campbell, R.; Remington, J. *Proc. Natl. Acad. Sci. U. S. A.* **2007**, *104*, 6672.

140. Andresen, M.; Stiel, A.; Trowitzsch, S.; Weber, G.; Eggeling, C.; Wahl, M.; Hell, S.; Jakobs, S. *Proc. Natl. Acad. Sci. U. S. A.* **2007**, *104*, 13005.

141. Violot, S.; Carpentier, P.; Blanchoin, L.; Bourgeois, D. *J. Am. Chem. Soc.* **2009**, *131*, 10356.

142. Pletneva, N.; Pletnev, V.; Shemiakina, I.; Chudakov, D.; Artemyev, I.; Wlodawer, A.; Dauter, Z.; Pletnev, S. *Protein Sci.* **2011**, *20*, 1265.

143. Brakemann, T.; Weber, G.; Andresn, M.; Groenhof, G.; Stiel, A.; Trowitzsch, S.; Eggeling, C.; Grubmüller, H.; Hell, S.; Wahl, M.; Jakobs, S. *J. Biol. Chem.* **2010**, *285*, 14603.

144. Bizzarri, R.; Serresi, M.; Cardarelli, F.; Abbruzzetti, S.; Campanini, B.; Viappiani, C.; Beltram, F. *J. Am. Chem. Soc.* **2010**, *132*, 85.

145. Gayda, S.; Nienhaus, K.; Nienhaus, U. *Biophys. J.* **2012**, *103*, 2521.

146. Habuchi, S.; Dedecker, P.; Hotta, J.; Flors, C.; Ando, R.; Mizuno, H.; Miyawaki, A.; Hofkens, J. *Photochem. Photobiol. Sci.* **2006**, *5*, 567.

147. Jung, G.; Wiehler, J.; Steipe, B.; Bräuchle, C.; Zumbusch, A. *ChemPhysChem* **2001**, *2*, 392.

148. Jung, G.; Mais, S.; Zumbusch, A.; Bräuchle, C. *J. Phys. Chem. A* **2000**, *104*, 873.

149. Habuchi, S.; Ando, R.; Dedecker, P.; Verheijen, W.; Mizuno, H.; Miyawaki, A.; Hofkens, J. *Proc. Natl. Acad. Sci. U. S. A.* **2005**, *102*, 9511.

150. Dickson, R.; Cubitt, A.; Tsien, R.; Moerner, W. *Science* **1997**, *388*, 355.

151. Nifosi, R.; Ferrari, A.; Arcangeli, C.; Tozzini, V.; Pellegrini, V.; Beltram, F. *J. Phys. Chem. B* **2003**, *107*, 1679.

152. Nienhaus, U.; Nienhaus, K.; Wiedenmann, J. in Ref. 7; pp. 241–263.

153. Van Thor, J. in Ref. 7; pp. 183–216.

154. Yokoe, H.; Meyer, T. *Nat. Biotechnol.* **1996**, *14*, 1252.

155. van Thor, J.; Gensch, T.; Hellingwerf, K.; Johnson, L. *Nat. Struct. Biol.* **2002**, *9*, 37.

156. Patterson, G.; Lippincott-Schwartz, J. *Science* **2002**, *297*, 1873.

157. McAnaney, T.; Zeng, W.; Doe, C.; Bhanji, N.; Wakelin, S.; Pearson, D.; Abbyad, P.; Shi, X.; Boxer, S.; Bagshaw, C. *Biochemistry* **2005**, *44*, 5510.

158. Habuchi, S.; Cotlet, M.; Gensch, T.; Bednarz, T.; Haber-Pohlmeier, S.; Rozenski, J.; Dirix, G.; Michiels, J.; Vanderleyden, J.; Heberle, J.; DeSchryver, F.; Hofkens, J. *J. Am. Chem. Soc.* **2005**, *127*, 8977.

159. Royant, A.; Noirclerc-Savoye, M.; *J. Struct. Biol.* **2011**, *174*, 385.

160. Bowen, B.; Woodbury, N. *Photochem. Photobiol.* **2003**, *77*, 362.

161. Jung, G.; Werner, M.; Schneider, M. *ChemPhysChem* **2008**, *9*, 1867.

162. Düser, M.; Zarrabi, Y. Zimmermann, B.; Dunn, S.; Börsch, M. *Proc. SPIE* **2006**, *6092*, 60920.

163. Tramier, M.; Zahid, M.; Mevel, J.; Masse, M.; Coppey-Moisan, M. *Microsc. Res. Tech.* **2006**, *69*, 933.

164. Hoffmann, B.; Zimmer, T.; Klöcker, N.; Kelbauskas, L.; König, K.; Benndorf, K.; Biskup, C. *J. Biomed. Opt.* **2008**, *13*, 031205.

165. Langhojer, F.; Dimler, F.; Jung, G.; Brixner, T. *Biophys. J.* **2009**, *96*, 2763.

166. Bell, A.; Stoner-Ma, D.; Wachter, R.; Tonge, P. *J. Am. Chem. Soc.* **2003**, *125*, 6919.

167. Testa, I.; Mazza, D.; Barozzi, S.; Faretta, M.; Diaspro, A. *Appl. Phys. Lett.* **2007**, *91*, 133902.

168. van Thor, J.; Sage, T. *Photochem. Photobiol. Sci.* **2006**, *5*, 597.

169. Blum, C.; Meixner, A.; Subramaniam, V. *Biophys. J.* **2004**, *87*, 4172.

170. Elowitz, M.; Surette, M.; Wolf, P.; Stock, J.; Leibler, S. *Curr. Biol.* **1997**, *7*, 809.

171. Shcherbakova, D.; Subach, O.; Verkhusha, V. *Angew. Chem. Int. Ed.* **2012**, *51*, 10724.

172. Jimenez-Banzo, A.; Nonell, S.; Hofkens, J.; Flors, C. *Biophys. J.* **2008**, *94*, 168.

173. Ragas, X.; Cooper, L.; White, J.; Nonell, S.; Flors, C. *ChemPhysChem* **2011**, *12*, 161.

174. Carpentier, P.; Violot, S.; Blanchoin, L.; Bourgeois, D. *FEBS Lett.* **2009**, *583*, 2839.

175. Vegh, R.; Solntsev, K.; Kuimova, M.; Cho, S.; Liang, Y.; Loo, B.; Tolbert, L.; Bommarius, A. *Chem. Commun.* **2011**, *47*, 4887.

176. Harms, G.; Cognet, L.; Lommerse, P.; Blab, G.; Schmidt, T. *Biophys. J.* **2001**, *80*, 2396.

177. Veettil, S.; Budisa, N.; Jung, G. *Biophys. Chem.* **2008**, *136*, 38.

178. Kurimoto, M.; Subramony,P.; Gurney, R.; Lovell, S.; Chmielewski, J.; Kahr, B. *J. Am. Chem. Soc.* **1999**, *121*, 6952.

179. Wachter, R.; Elsliger, M.; Kallio, K.; Hanson, G.; Remington, J. *Structure* **1998**, *6*, 1267.

180. Elsliger, M.; Wachter, R.; Hanson, G.; Kallio, K.; Remington, J. *Biochemistry* **1999**, *38*, 5296.

181. Bizzarri, R.; Nifosi, R.; Abbruzetti, S.; Rocchia, W.; Guidi, S.; Arosio, D.; Garau, G.; Campanini, B.; Grandi, E.; Ricci, F.; Viappiani, C.; Beltram, F. *Biochemistry* **2007**, *46*, 5494.

182. Jayaraman, S.; Haggie, P.; Wachter, R.; Remington, J.; Verkman, A. *J. Biol. Chem.* **2000**, *275*, 6047.

183. Arosio, D.; Garau, G.; Ricci, F.; Marchetti, L.; Bizzarri, R.; Nifosi, R.; Beltram, F. *Biophys. J.* **2007**, *93*, 232.

184. Bizzarri, R. in Ref. 8; pp. 59–97.

185. Widengren, J.; Terry, B.; Rigler, R. *Chem. Phys.* **1999**, *249*, 259.

186. Abbruzzetti, S.; Grandi, E.; Viappiani, C.; Bologna, S.; Campanini, B.; Raboni, S.; Bettati, S.; Mozzarelli, A. *J. Am. Chem. Soc.* **2005**, *127*, 626.

187. Malo, G.; Pouwels, L.; Wang, M.; Weichsel, A.; Montfort, W.; Rizzo, M.; Piston, D.; Wachter, R. *Biochemistry* **2007**, *46*, 9865.

188. Bregestovski, P.; Arosio, D. in Ref. 8; pp. 99–124.

189. Baubet, V.; Le Mouellic, H.; Campbell, A.; Lucas-Meunier, E.; Fossier, P.; Brulet, P. *Proc. Natl. Acad. Sci. U. S. A.* **2000**, *97*, 7260.

190. Richmond, T.; Takahashi, T.; Shimikhada, R.; Bernsdorf, J. *Biochem. Biophys. Res. Commun.* **2000**, *268*, 462.

191. Balint, E.; Petres, J.; Szabo, M.; Orban, C.; Szilagyi, L.; Abraham, B. *J. Fluoresc.* **2013**, *23*, 273.

192. Barondeau, D.; Kassmann, C.; Tainer, J.; Getzoff, E. *J. Am. Chem. Soc.* **2002**, *124*, 3522.

193. Hötzer, B.; Ivanov, R.; Brumbarova, T.; Bauer, P.; Jung, G. *FEBS J.* **2012**, *279*, 410.

194. Hötzer, B.; Ivanov, R.; Altmeier, S.; Kappl, R.; Jung, G. *J. Fluoresc.* **2011**, *21*, 2143.

195. Vinkenborg, J.; Koay, M.; Merkx, M. *Curr. Opin. Chem. Biol.* **2010**, *14*, 231.

196. Malvezzi-Campeggi, F.; Jahnz, M.; Heinze, K.; Dittrich, P.; Schwille, P. *Biophys. J.* **2001**, *81*, 1776.

197. Haupts, U.; Maiti, S.; Schwille, P.; Webb, W. *Proc. Natl. Acad. Sci. U. S. A.* **1998**, *95*, 13573.

198. Blum, C.; Subramaniam, V. in Ref. 7; pp. 217–240.

199. Cotlet, M.; Goodwin, P.; Waldo, G.; Werner, J. *ChemPhysChem* **2006**, *7*, 250.

200. Serresi, M.; Bizzarri, R.; Cardarelli, F.; Beltram, F. *Anal. Bioanal. Chem.* **2009**, *393*, 1123.

201. Arosio, D.; Ricci, F.; Marchetti, L.; Gualdani, R.; Albertazzi, L.; Beltram, F. *Nat. Methods* **2010**, *7*, 516.

202. Blab, G.; Lommerse, P.; Cognet, L.; Harms, G.; Schmidt, T. *Chem. Phys. Lett.* **2001**, *350*, 71.

203. De Meulenaere, E.; Asselberghs, I.; de Wergifosse, M.; Botek, E.; Spaepen, S.; Champagne, B.; Vanderleyden, J.; Clays, K. *J. Mater. Chem.* **2009**, *19*, 7514.

204. Foo, Y.; Korzh, V.; Wohland, T. in Ref. 8; pp. 213–248.

205. Marriott, G.; Mao, S.; Sakata, T.; Ran, J.; Jackson, D.; Petchprayoon, C.; Gomez, T.; Warp, E.; Tulyathan, O.; Aaron, H.; Isacoff, E.; Yan, Y. *Proc. Natl. Acad. Sci. U. S. A.* **2008**, *105*, 17789.

206. Chirico, G.; Collini, M.; D'Alfonso, L.; Caccia, M.; Daglio, S.; Campanini, B. in Ref. 8; pp. 35–55.

207. Willig, K.; Kellner, R.; Medda, R.; Hein, B.; Jakobs, S.; Hell, S. *Nat. Methods* **2006**, *3*, 721.

208. Betzig, E.; Patterson, G.; Sougrat, R.; Lindwasser, W.; Olenych, S.; Bonifacino, J.; Davidson, M.; Lippincott-Schwartz, J.; Hess, H. *Science* **2006**, *313*, 1642.

209. Hofmann, M.; Eggeling, C.; Jakobs, S.; Hell, S. *Proc. Natl. Acad. Sci. U. S. A.* **2005**, *102*, 17565.

210. Brakemann, T.; Stiel, A.; Weber, G.; Andresen, M.; Testa, I.; Grotjohann, T.; Leutenegger, M.; Plessmann, U.; Urlaub, H.; Eggeling, C.; Wahl, M.; Hell, S.; Jakobs, S. *Nat. Biotechnol.* **2011**, *29*, 942.

211. Nifosi, R.; Tozzini, V. in Ref. 7; pp. 3–40.

212. Creemers, T.; Lock, A.; Subramaniam, V.; Jovin, T.; Völker, S. *Nat. Struct. Biol.* **1999**, *6*, 557.

213. Nifosi, R.; Luo, Y. *J. Phys. Chem. B* **2007**, *111*, 14043.

214. Mandal, D.; Tahara, T.; Meech, S. *J. Phys. Chem. B* **2004**, *108*, 1102.

215. Kondo, M.; Heisler, I.; Stoner-Ma, D.; Tonge, P.; Meech, S. *J. Am. Chem. Soc.* **2010**, *132*, 1452.

216. Cinelli, R.; Tozzini, V.; Pellegrini, V.; Beltram, F.; Cerullo, G.; Zavelani-Rossi, M.; De Sylvestri, S.; Tyagi, M.; Giacca, M. *Phys. Rev. Lett.* **2001**, *86*, 3439.

217. Ando, R.; Flors, C.; Mizuno, H.; Hofkens, J.; Miyawaki, A. *Biophys. J.* **2007**, *87*, L97.

218. Suhling, K.; Siegel, J.; Phillips, D.; French, P.; Lévêque-Fort, S.; Webb, S.; Davis, D. *Biophys. J.* **2002**, *83*, 3589.

219. Durisic, N.; Laparra-Cuervo, L.; Sandoval-Alvarez, A.; Borbely, J.; Lakadamyali, M. *Nat. Methods* **2014**, *11*, 156.

220. Bogdanov, A.; Mishin, A.; Yampolsky, I.; Belousov, V.; Chudakov, D.; Subach, F.; Verkhusha, V.; Lukyanov, S.; Lukyanov, K. *Nat. Chem. Biol.* **2009**, *5*, 459.

5

DESIGN AND APPLICATION OF AUTOFLUORESCENT PROTEINS BY BIOLOGICAL INCORPORATION OF INTRINSICALLY FLUORESCENT NONCANONICAL AMINO ACIDS

PATRICK M. DURKIN AND NEDILJKO BUDISA

Department of Chemistry, Berlin Institute of Technology/TU Berlin, Biocatalysis Group, Berlin, Germany

5.1 INTRODUCTION

Eukaryotic and prokaryotic cells contain a variety of compounds that fluoresce upon excitation with UV light. Proteins and peptides, with aromatic amino acids, are intrinsically fluorescent when excited with UV light. Some enzymatic cofactors (e.g., FMN, NAD, porphyrins) are also intrinsically fluorescent, adding to the protein fluorescence. These structures all possess aromatic ring structures that absorb UV light – a key feature required for fluorescence. These come in the form of tryptophan, tyrosine, and phenylalanine, each of which has characteristic absorption and emission peaks shown as shown in Table 5.1.

Tryptophan displays much stronger fluorescence than both tyrosine and phenylalanine. However, the fluorescent properties of tryptophan are solvent dependent; as solvent polarity decreases, the spectrum shifts to shorter wavelengths and increases in intensity. For this reason, tryptophan residues in hydrophobic domains of folded proteins exhibit a spectral shift of 10–20 nm, and this phenomenon has been exploited to

Fluorescent Analogs of Biomolecular Building Blocks: Design and Applications, First Edition.
Edited by Marcus Wilhelmsson and Yitzhak Tor.
© 2016 John Wiley & Sons, Inc. Published 2016 by John Wiley & Sons, Inc.

TABLE 5.1   Fluorescence Characteristics of Free Amino Acids: Tryptophan, Tyrosine, and Phenylalanine

Amino Acid	Lifetime (ns)	Absorption		Fluorescence	
		Wavelength (nm)	Molar Absorptivity $(M^{-1} cm^{-1})$	Wavelength (nm)	Quantum Yield
Tryptophan	2.6	280	5600	348	0.20
Tyrosine	3.6	274	1400	303	0.14
Phenylalanine	6.4	257	200	282	0.04

investigate protein denaturation.[1] The natural abundance of tryptophan is ~1.4% of all amino acid residues in human proteins (significantly less than phenylalanine and tyrosine, 3.9% and 3.2%, respectively), yet it has the strongest fluorescence of the three, making this natural optical probe an important residue for fluorescent proteins. Tyrosine has a similar excitation wavelength to that of tryptophan, but a distinctly different emission wavelength. Quenching of tyrosine fluorescence via resonance energy transfer has been observed when a tryptophan residue is proximal to the tyrosine residue. Fluorescence can also be quenched in this case by ionization of the aromatic hydroxyl group. Phenylalanine is very weakly fluorescent and can only be observed in the absence of both tryptophan and tyrosine.

Despite the data above, it is often difficult to predict the fluorescence characteristics of an aromatic residue, as this is dependent on its structural context in a particular protein molecule. Comparing the folded and unfolded states, the quantum yield may be increased or decreased by folding; thus, fluorescence can be increased or decreased by folding. This change in fluorescence intensity can be used as a probe for perturbations of both secondary and tertiary structures of proteins and peptides. However, the wavelength range of emitted light (i.e., emission profile) is a better indication of the environment of the fluorophore, for example, tryptophan residues that are exposed to water have maximal fluorescence (λ_{max}) at a wavelength of ~340–350 nm, whereas totally buried residues fluoresce have λ_{max} ~330 nm.[2]

Green Fluorescent Protein (GFP) is a naturally occurring protein, which exhibits a bright green fluorescence upon exposure to light of the blue-UV range. Its main excitation absorption is found at 395 nm, with a minor absorption at 475 nm. Its emission peak is at 509 nm and it has a quantum yield of 0.79.[3] Its structure is comprised of 238 amino acid residues (~27 kDa), which form a β-barrel type formation with a hydrophobic core (Fig. 5.1).

This hydrophobic core contains a covalently bound fluorophore, 4-(p-hydroxybenzylidene)imidazolidin-5-one (p-HBI), which arises from a posttranslational condensation and oxidation reaction of three residues: Ser65–Tyr66–Gly67.[5-7] The biosynthetic pathway for the formation of the p-HBI moiety via condensation of these residues and subsequent oxidation was proposed by Heim et al.[5] and is based upon the observations that expression of recombinant GFP in *Escherichia coli*

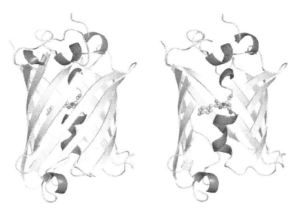

Figure 5.1 Structure of GFP (PDB entry 1gfl) displaying β-barrel structure, which buries the fluorophore within the hydrophobic cavity leading to fluorescence. From Ref. [4].

is possible without the use of any specific enzymes from the natural source *Aequorea victoria*. Also imidazolin-5-ones undergo autoxidative formation of double bonds at the 4-position completing the fluorophore,[5] and when GFP is expressed in *E. coli* under anaerobic conditions, it is nonfluorescent and indistinguishable from native GFP by on a denaturing SDS-gel.

The fluorescence of this protein is highly dependent on the chemical environment and the protonation state of the fluorophore, and the folding state of the protein, in particular, the β-barrel structure plays a key role.[8] The β-barrel provides a hydrophobic core, excluding water from the fluorophore and allowing several proximal residues (Glu222, Thr203, His148, Gln94, and Arg96) to form hydrogen bonds with the fluorophore (Fig. 5.2). If the *p*-HBI moiety is exposed to water, for example, when the protein is denatured by treatment with guanidine, a complete quenching of fluorescence is observed.[11] The tyrosine moiety is also deprotonated in the active form, and when this is protonated fluorescence is quenched.[12]

The key point here is that the *p*-HBI moiety is required for fluorescence; however, the environment in which this is situated dictates the observed fluorescence wavelength and intensity. With this in mind, researchers developed a wide array of variants of GFP through single point mutations; one of the first examples was the single mutation (S65T) that resulted in increased photostability and fluorescence of GFP, and a shift of the major excitation peak to 488 nm while keeping the peak emission at 509 nm making this much more practical as an optical marker as these optical characteristics matched well with commonly available FITC filter sets.[13] This was followed by the point mutation of a phenylalanine residue to leucine (F67L) leads to enhanced GFP (EGFP) which exhibits a 37 °C folding efficiency allowing use of this protein in mammalian cells.

The color of emission can also be modified; blue fluorescent proteins (EBFP, EBFP2, and Azurite) and yellow fluorescent proteins (YFP, Venus, and Citrine) are all mutations of GFP (Fig. 5.3). These changes in color can be brought about

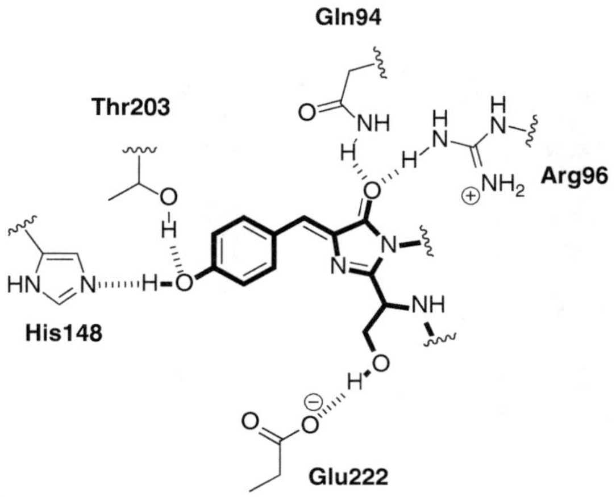

Figure 5.2   Structure of the chromophore of GFP and the proximal residues.[9,10] The hydrogen bonds between these residues and the *p*-HBI chromophore are pivotal in the fluorescent properties of the protein. When the chromophore moiety is exposed to water, fluorescence is quenched. From Ref. [11].

by direct modification of the fluorophore, for example, a (Y66H) mutation leads to blue fluorescence, or by a change in the environment the fluorophore is placed, for example, a (T203Y) mutation leads π-stacking of the tyrosine residue and the *p*-HBI giving rise to a redshift.[14] Incorporation of noncanonical α-amino acids (those which are not naturally proteinogenic) into this chromophore can also lead to interesting fluorescence properties. A good example of this is gold fluorescent protein (GdFP), which arises from the incorporation of 4-amino-tryptophan residue, displays a large redshift in its emission maximum that is found at 572 nm. Thus, it has a Stokes shift of ~100 nm, much larger than any mutant containing canonical amino acids (ca. 10-40 nm).[15]

GFP has found widespread application in the field of cellular biology as a fluorescent marker when incorporated into other proteins, in the form of a fusion protein. This is achieved by incorporating the gene encoding for GFP into the gene of the protein to be studied. This provides a biocompatible autofluorescent marker into a desired protein that would normally be expressed in the cell and can be detected without the need of extraneous cofactor to allow fluorescence to occur. This has been used in imaging studies in microscopy, Förster resonance energy transfer (FRET) studies for conformational analysis, and protein–protein interaction,[20] as well as an efficient reporter for expression of proteins. For the most part, this methodology is intracellularly compatible; that is, the incorporation of the GFP gene into *E. coli* to form a fusion protein does not necessarily have a detrimental effect on the function of the protein compared to the wild-type protein.

Though it should be noted that due to the size of GFP (27 kDa), its application can be somewhat limited when studying smaller proteins and peptides. With these types of systems in mind, the design of small fluorescent building blocks that can

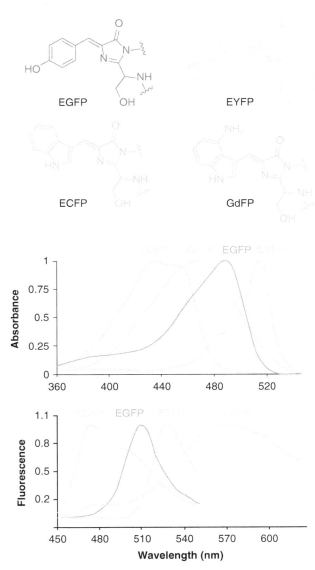

Figure 5.3 Chromophores of several GFP mutants and their associated absorbance and fluorescence profiles. Each of the fluorophores possesses the *p*-HBI functionality but has distinctly different fluorescence profiles due to the interaction of the tyrosine and tyrosine analogs and the surrounding protein matrix.[14–17] See Table 5.2 for data regarding fluorescence properties. Reprinted from Journal of Molecular Biology, 2003, 328 (5), 1071–1081, with permission from Elsevier.

TABLE 5.2 Fluorescence Characteristics of GFP and Various Analogs[14,18,19]

Mutation	Common Name	λ_{ex} (ε)	λ_{em} (QY)	Relative Fluorescence Intensity @37°C
Class 1, wild type				
None or Q80R	Wild Type	395-397 (25-30) 470-475 (9.5-14)	504 (0.79)	6
F99S, M153T, V163	Cycle 3	397 (30) 475 (6.5-8.5)	506 (0.79)	100
Class 2, phenolate anion				
S65T	EGFP	489 (52-58)	509-511 (0.64)	12
F64L, S65T		488 (55–57)	507-509 (0.60)	20
F64L, S65T, V163A		488 (42)	511 (0.58)	54
S65T, S72A, N149K, M153T, I167T	Emerald	487 (57.5)	509 (0.68)	100
Class 3, neutral phenol				
S202F, T203I	H9	399 (20)	511 (0.60)	13
T203I, S72A, Y145F	H9-40	399 (29)	511 (0.64)	100
Class 4, phenolate anion stacked with π-electron system (yellow fluorescent proteins)				
S65G, S72A, T203F		512 (65.5)	522 (0.70)	6
S65G, S72A, T203H		508 (48.5)	518 (0.78)	12
S65G, V68L, Q69K, S72A, T203Y	10C Q69K	516 (62)	529 (0.71)	50
S65G, V68L, S72A, T203Y	10C	514 (83.4)	527 (0.61)	58
S65G, S72A, K79R, T203Y	Topaz	514 (94.5)	527 (0.60)	100
Class 5, indole in chromophore (cyan fluorescent proteins)				
Y66W		436	485	–
Y66W, N146I, M153T, V163A	W7	434 (23.9)	476 (0.42)	61
F64L, S65T, Y66W	W1B or	434 (32.5)	505	80
N146I, M153T, V163A	ECFP	452	476 (0.40)	–
S65A, Y66W, S72A, N146I, M153T, V163A	W1C	435 (21.2)	495 (0.39)	100

be introduced into proteins and peptides is highly desirable – particularly when considering fluorescent amino acids that can be easily incorporated within the backbone of a peptide.

5.2   DESIGN AND SYNTHESIS OF FLUORESCENT BUILDING BLOCKS IN PROTEINS

The desired property of fluorescence can be introduced into proteins by either attaching an extrinsic fluorescent label or by the use of aromatic amino acids within the protein sequence.

The advantage of using an extrinsic fluorescent label is that one does not necessarily need to alter the sequence itself in order to achieve a fluorescent protein. Merely by "tagging" a protein with a label, it is possible to afford a wide variety of absorption or emission maxima for almost any sequence with little or no impact on the protein activity. The difficulty arises when designing the fluorescent label to be specific for a particular site of a protein, which usually must be on the outer surface of the protein.

The use of the fluorescent proteinogenic (canonical) amino acids can often be limited; these residues will already be present within other proteins in an *in vivo* setting; thus, specific excitation of the fluorescent residue may not be possible. So, novel noncanonical fluorescent amino acids are highly desirable. The fluorescent amino acid format has several advantages over the labeling solution. Firstly, a wide range of synthetic amino acids with fluorescent properties can be synthesized with relative ease, and these amino acids can be used in manual peptide synthesis (either solid or solution phase), and it is also possible to achieve ribosomal synthesis by careful reprogramming of biological systems (see Section 5.2.4). Secondly, the fluorescence characteristics of the amino acid can be finely tuned by "atomic mutation" without changing the gross structure of the protein in question; for example, substituting nitrogen atoms in place of a CH rather than changing the entire structure of a side-chain residue, it can be possible to bring about large changes in fluorescence.

Regardless of the method of introduction, there are many factors that must be considered when designing a fluorescent reporter. The fluorescent moiety should ideally have a high fluorescence quantum yield, sufficiently large Stokes shift and sensitivity to the environment in which it is located, that is, solvent polarity or hydrophobic/hydrophilic environments within a protein matrix. The last point is more important for the amino acid–based approach, whereas extrinsic labels are generally more sensitive to solvent conditions.

5.2.1   Extrinsic Fluorescent Labels

Fluorescent labels covalently attached to proteins or peptides via side-chain residues can be used to bring about fluorescence in an otherwise poorly or nonfluorescent protein. The challenge that arises from this method is that the fluorophore must be designed such that it can be chemically ligated to a protein/peptide sequence specifically at a residue or region, while only slightly perturbing the structure of the protein to be investigated.

Specific incorporation of fluorophores can be addressed by functionalizing the amino acid residue prior incorporation; this is generally trivial for manually synthesized peptides, but more challenging for ribosomally synthesized proteins, as these effectively become noncanonical α-amino acids (ncAAs). This means that without developing a specific tRNA synthetase to charge the tRNA with the amino acid bearing a fluorescent label (see Section 5.2.4); thus, extrinsic fluorescent labeling can often only be achieved after the protein has been synthesized.

This means the label must be specific for a single residue or a group of residues present in the protein. This is not always possible, for example, the side-chain functionalities present in proteins are somewhat limited, that is, many residues are not amenable to ligation, for example, glycine, leucine, and isoleucine. Those that are amenable are too similar in reactivity to easily distinguish (e.g., serine and threonine, or aspartic and glutamic acid). The number of residues in a protein can be large, and the folding of the protein may also bury the residue to be functionalized. The potential for labeling to occur at the wrong position would mean it would not be possible to be applied to an *in vivo* system. To overcome this, one must design a system whereby the fluorescent label "recognizes" the correct position of binding within the protein (i.e., via multivalent interaction of label and protein), for example, a particular sequence of amino acids that can be incorporated in the protein that the label binds to. A good example of how this can be achieved is work by Griffin *et al.*[21]

The tag/binding domain had to be designed such that (1) the ligand (tag) should be of low molecular weight and membrane permeable (ideally <700 Da), (2) the binding domain should be a short sequence of naturally occurring amino acids, and (3) the ligand should bind tightly and specifically to the receptor.

The FlAsH-EDT$_2$ system was developed on this basis and is comprised of a ligand–receptor system reliant upon a sulfur–arsenic interaction. This was synthesized efficiently from fluorescein mercuric acetate, arsenic chloride, and ethane dithiol (EDT) (Scheme 5.1). The binding domain for this ligand was based upon a α-helical domain with four cysteine residues placed at the i, $i+1$, $i+4$, and $i+5$ positions of the α-helix, which can be attached into a protein at any desired point. The four thiol groups of the cysteine residues then displace the EDT of the dithioarsolane to form a fluorescent complex specific for proteins bearing the tetracysteine binding domain. Reportedly, the ligand has relatively few binding sites in nontransfected mammalian cells but binds to the designed peptide domain with high selectivity (nanomolar or lower dissociation constant). An added bonus of this ligand is that it remains nonfluorescent until it binds its target, whereupon it becomes strongly fluorescent.

5.2.2 Intrinsic Fluorescent Labels

The canonical amino acid Trp is one of the most suitable targets for the engineering and design of protein fluorescence owing to its low abundance in natural sequences and high relevance for protein stability, function, and spectroscopic features. In addition, it has a large fluorescence quantum yield and a wide scope for chemical

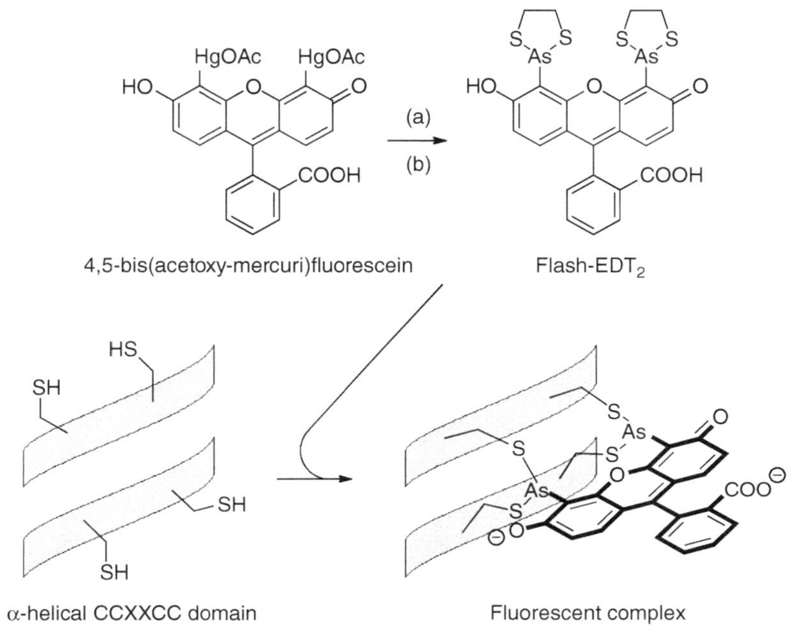

Scheme 5.1 Synthesis and incorporation of the FlAsH-EDT$_2$ system into a tetracysteine-containing peptide or protein is shown. The binding of the fluorescent ligand is dependent on the four cysteine residues binding of the peptide to the bis-arsenic functionality of the ligand. Reagents and conditions: (a) AsCl$_3$, Pd(OAc)$_2$, DIPEA, NMP; (b) EDT, aqueous acetone.[21]

modification. Altering the structure of tryptophan can be achieved in a facile manner, either by total synthesis or by chemically modifying the amino acid itself.

Analogs of tryptophan (and the other fluorescent amino acids) were originally used as probes for the determination of metabolic pathways and mechanisms by which proteins are synthesized.[22] However, due to limitations in protein expression technology at the time, incorporation of the analogs was often capricious and so interest in this field waned. With the development of expression systems, with tightly regulated protein expression, came an increase in the reliability of incorporation and renewed interest in this area (see Section 5.2.4).

The investigation of the photophysics of a variety of functionalized tryptophans and their related indoles revealed that characteristic excitation and emission wavelengths can be obtained from these analogs. In particular, aza- and hydroxytryptophans have drawn significant interest. Work by Szabo showed that 5-hydroxytryptophan (5HW) could be used as a fluorescent marker in protein–protein and protein–nucleic acid interactions.[23] This analog showed a redshift allowing a selective excitation of the 5HW chromophore in the presence of other native proteins and nucleic acids. Azatryptophans and their related indoles have been studied by Petrich and coworkers,[24] and showed a redshift compared to tryptophan. In Table 5.3,

TABLE 5.3 Fluorescence Characteristics of Tryptophan and Synthetic Analogs

Amino Acid	Lifetime (ns)	Absorption		Fluorescence	
		Wavelength (nm)	Molar Absorptivity $(M^{-1} cm^{-1})$	Wavelength (nm)	Quantum Yield
Tryptophan	2.6	280	5600	348	0.20
7-Aza-tryptophan	0.75	290	6000	403	0.016
5-Fluoro-tryptophan	2.7	285	5400	360	0.14
5-Hydroxy-tryptophan	3.6	277	4800	339	0.256

Measurements recorded at 20 °C in 100 mM Tris HCl at pH 7.5.

several common probes used in fluorescence studies and their fluorescence properties are listed.

5.2.2.1 Synthesis of Tryptophans: As a result of these investigations, general methods for the synthesis of tryptophans have become a hot topic within the synthetic chemistry community; a comprehensive overview of this area would be beyond the scope of this chapter; however, a flavor of the methods available for the synthesis of tryptophans is provided here.

Some of the simplest methods for the synthesis of tryptophans are the direct electrophilic substitutions of the indole moiety for example, nitration, amination, fluorination, and hydroxylation. Exploiting the nucleophilic character of the 3-poisition of indole, it is possible to synthesize tryptophans expediently by substitution of the parent indoles at this position. This can be achieved either by "classical" synthesis, enzymatically or by cross-coupling methodologies.

The classical route employs Mannich conditions followed by substitution with an amido malonate residue and then hydrolysis (Scheme 5.2). This has the drawback that enantioselectivity is not present in this route, and a late-stage enzymatic resolution (by *N*-acylation and then treatment with an acylase) must be employed to achieve this.

Enzymatic synthesis can be employed using tryptophan synthase and L-serine, as shown in Scheme 5.3. The benefit of this method is that it is stereochemically simpler to undertake as the stereochemistry is set in the serine (which can be isolated from natural sources in high enanantiopurity). The synthase, usually present as a dimer, comprises two subunits: the α-subunit catalyzes the formation of indole and glyceraldehyde-3-phosphate from a retro-aldol cleavage of indole-3-glycerol phosphate; the β-subunit is responsible for the condensation of L-serine and indole to form tryptophan and pyridoxal phosphate. This synthase turns out to be fairly general, allowing a variety of indoles to be converted into substituted tryptophans. This is attributed to the large binding pocket in the active site of the β-subunit allowing many variants to be synthesized in this manner.[25]

The benefit of these two approaches is their generality; a wide array of indoles can be synthesized simply, on large scale, for example, Fischer indole synthesis, and then directly functionalized to the desired tryptophan. However, for those indoles not

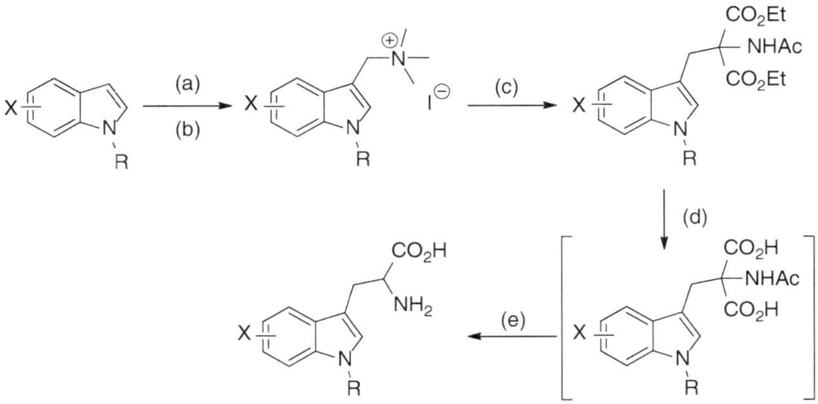

Scheme 5.2 Reagents and conditions: $CH_2(NMe_2)_2$, p-formaldehyde, AcOH; (b) MeI, EtOH; (c) $AcNH(CO_2Et)_2$, NaOEt, EtOH; (d) 10% KOH, EtOH; (e) HCl, THF.

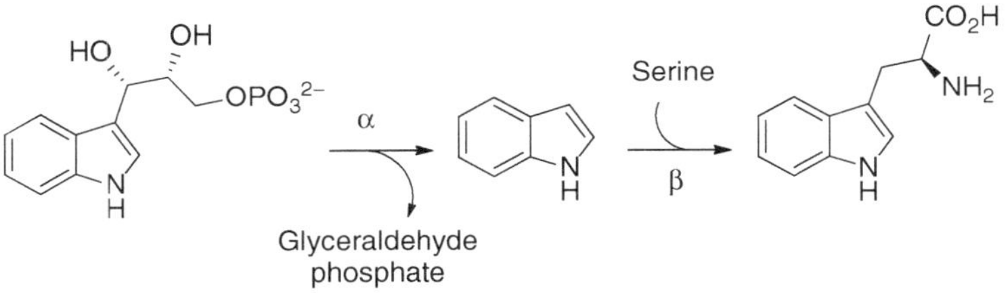

Scheme 5.3 The biosynthetic pathway for tryptophan using tryptophan synthase. The α-subunit is responsible for the formation of the indole and glyceraldehyde phosphate, and the β-subunit condenses indole with serine to form tryptophan. The second step can be exploited to synthesize a wide array of analogous tryptophans by substituting indole for indole analogs.[25]

amenable to enzymatic synthesis or conditions employed in the classical route, it is possible to use cross coupling to synthesize tryptophans. A C—H activation methodology coupling an indole and dehydroalanine derivative by means of a palladium-(II) catalyst is a method reported by Murakami and coworkers[26] Alternatively, one could employ a Suzuki coupling with bromodehydroalanine derivatives and the boronic acid/ester of the desired indole (Scheme 5.4).

After coupling, the resultant compounds can then be reduced by catalytic hydrogenation – employing a chiral rhodium catalyst affords the desired enantiomer of the tryptophan in good *e.e.* and yields.[27]

It is also possible to synthesize tryptophans from materials other than the parent indole; indeed, a wide variety of synthetic tryptophans can be synthesized employing the Schöllkopf chiral auxiliary.[28]

The chiral auxiliary for the synthesis of L-tryptophans is synthesized expediently by the condensation of D-valine with glycine ethyl ester and subsequent treatment with Meerwein's reagent (Scheme 5.5). This chiral auxiliary is attached to a

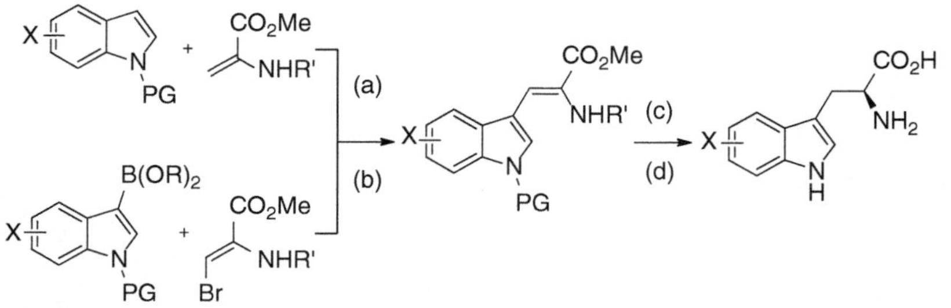

Scheme 5.4 Reagents and conditions: (a) PdCl$_2$, NaOAc, AcOH; (b) Pd(PPh$_3$)$_4$, NaOEt, benzene; (c) H$_2$, [(COD)Rh(R,R)-Et-DuPHOS]$^+$TfO$^-$, MeOH-CH$_2$Cl$_2$; (d) Deprotection via appropriate method, for example, hydrolysis.[27]

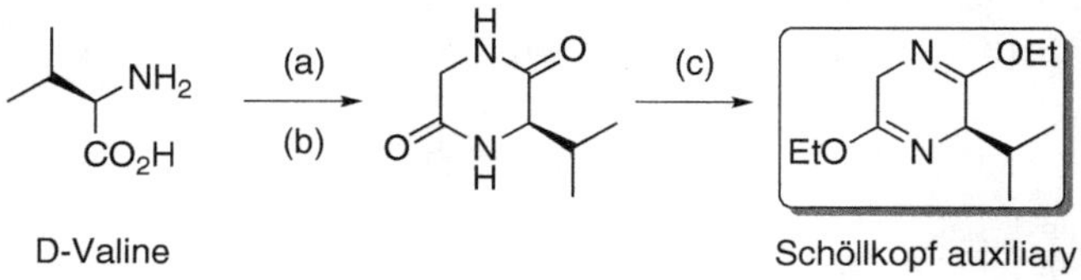

Scheme 5.5 Reagents and conditions: (a) triphosgene, THF; (b) glycine ethyl ester, NEt$_3$, THF; (c) Et$_3$O$^+$BF$_4$$^-$, CH$_2Cl_2$.

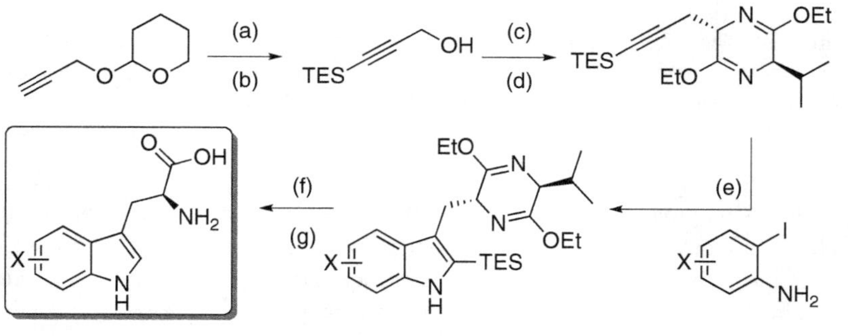

Scheme 5.6 Reagents and conditions: (a) n-BuLi, TESCl, THF; (b) p-TSA, MeOH, reflux; (c) (PhO)$_2$POCl, KOH, Et$_2$O; (d) nBuLi, Schöllkopf auxiliary; (e) Pd(OAc)$_2$, Na$_2$CO$_3$, LiCl, DMF, iodoaniline; (f) TBAF, THF; (g) aq. HCl then NaOH, EtOH. From Ref. [28].

functionalized propargyl alcohol, which is then coupled with an iodoaniline species via a tandem cross-coupling amination reaction – a variant of the Larock indole synthesis.[29] The regiochemistry and diastereoselectivity of the reaction is controlled by the size of the silyl-moiety and the chiral auxiliary (Scheme 5.6). After removal of the silyl group and chiral auxiliary by hydrolytic methods, the product tryptophan is isolated in good yields and high e.e.[28]

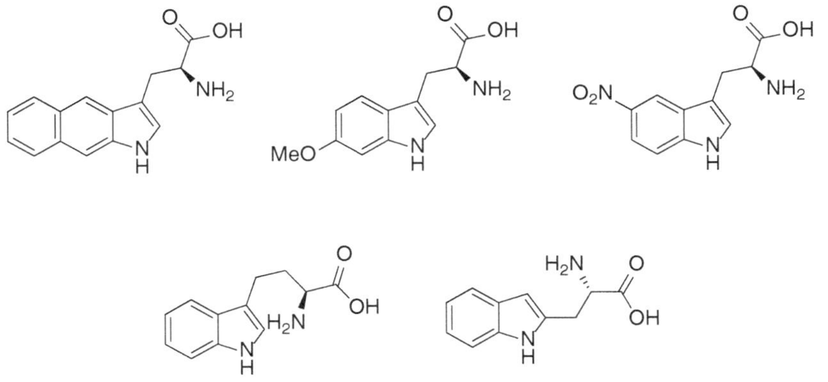

Figure 5.4 Examples of tryptophan analogs synthesized using the Schöllkopf chiral auxiliary.[28] (From Ref. [23].

Either enantiomer of the tryptophan can be synthesized using this method, by using the appropriate enantiomer of valine for the auxiliary. The drawback here is that for L-tryptophan the expensive D-valine is required; however, the chiral auxiliary is recovered after the hydrolytic step and can be recycled. A wide variety of iodoanilines are amenable to this methodology – from electron-rich or electron-poor compounds to annulated, homologated, and regioisomeric products all being synthesized in this manner (Fig. 5.4).

5.2.2.2 Endocyclic modifications of Tryptophan: Of greater interest in this field is the ability to maintain structural similarity and steric size of the fluorescent building block, such that perturbations to the overall structure of the protein to be studied are minimized. Incorporation of a heteroatom into the ring structure of tryptophan is a key example of how one can achieve this. Modification of the aromatic ring system, by replacing a carbon atom by nitrogen affords a range of different tryptophans (and indoles) which, as described in the previous section, would give rise to desirable fluorescence properties.

Indeed, azaindoles have been extensively probed from a photochemical stance, with a variety of studies investigating the role of hydrogen bonding between the azaindoles and bulk water, and the photochemically excited states of these hydrates.[23,30] The behavior of theses substrates in proton transfer mechanisms can lead to complicated optical properties and lower quantum yields of fluorescence. The measurement of the fluorescent properties of the azaindoles and related azatryptophans though has shown that 4-azaindole and 7-azaindole are, in particular, interesting substrates and the related tryptophans could be used as fluorescent probes for protein studies (Table 5.4).

5.2.2.3 Azulenylalanine: The bicyclic system azulene has found use as a spectroscopic marker in a variety of systems owing to its intense blue color and distinct

TABLE 5.4   **Fluorescence Characteristics of Tryptophan and Aza Analogs**

Substance	$\lambda_{ab,max}$ (nm)	$\lambda_{em,max}$ (nm)	Quantum Yield Φ_F
Tryptophan	280	348	0.20
7-Azaindole	288	386	0.03
	279	402	–
7-Azatryptophan	290	397	–
4-Azaindole	288	418	0.04
4-Azatryptophan	289	425	–
Indole	269	348	0.14

The various optical characteristics of indoles and their related tryptophans are shown in the table. The key features that are desirable are either a distinct absorption or emission along with a good quantum yield – that is, a strong fluorescence that can be excited or detected specifically.[1,2,31–33]

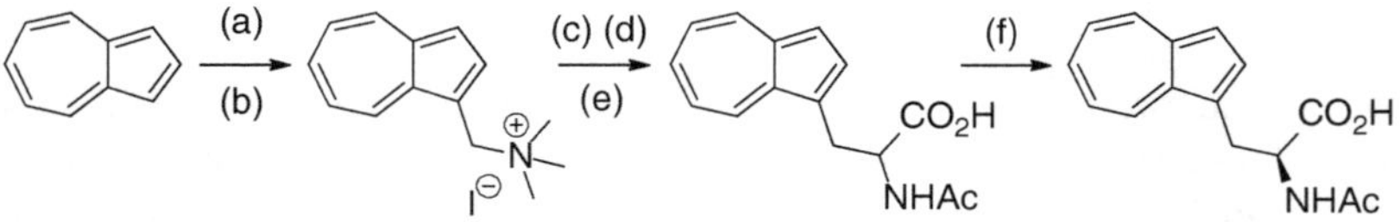

Scheme 5.7   Reagents and conditions: (a) $CH_2(NMe_2)_2$, formaldehyde, AcOH, CH_2Cl_2; (b) MeI, EtOH; (c) $AcNH(CO_2Et)_2$, NaOEt, EtOH, reflux; (d) 10% KOH, EtOH, reflux; (e) 0.2 M HCl, THF, reflux; (f) acylase I from *Aspergillus melleus*, 0.1 M pH 7.5 phosphate buffer, 37 °C.

fluorescence properties. This analog is approximately isosteric with Trp and has been used as a fluorescent marker within a peptide setting,[34] and also in conjunction with a stable radical *N*-oxide bearing amino acid (structurally similar to TEMPO, which acts as a fluorescence quencher) as structural probes for hexapeptides.[35] An effective synthesis of an amino acid bearing the azulenyl side-chain residue, β-(1-azulenyl)-L-alanine (Aal) was reported by Moroder and coworkers,[34] synthesized by direct functionalization of azulene (Scheme 5.7). The elegant synthesis of the enantiomerically pure amino acid was achieved in six steps, starting from azulene itself. The final step of the synthesis is an enzymatic resolution employing acylase I from *Aspergillus melleus*. The free amino acid can then be protected appropriately for use in polymer-supported or solution-based peptide synthesis.

    The fluorescence spectrum of Aal excitation shows two major absorption maxima at 276 and 339 nm and shows emission at 381 nm (Fig. 5.5). This presents the possibility of using this ncAA as a fluorescent probe in the presence of tryptophan due to its distinct excitation wavelength. However, the relative fluorescence intensity of Aal, when excited at 339 nm, is lower than that of tryptophan (by a factor of about 3).[34]

5.2.2.4 *Coumarin-Based Systems:* Coumarins (benzopyrones) represent a large section of commercially available dyes for fluorescence purposes, most of which

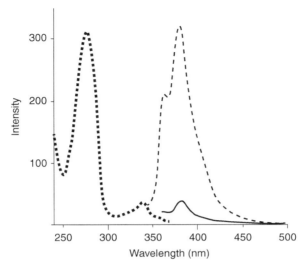

Figure 5.5   Fluorescence spectra of β-(1-azulenyl)-L-alanine; c = 10^{-6} M in 0.1 M K_3PO_4 buffer, pH 7.5; dotted line: excitation (λ_{max} = 276 nm and 339 nm); dashed line: emission (λ_{ex} = 276 nm, λ_{max} = 381 nm); solid line: emission (λ_{ex} = 339 nm, λ_{max} = 381 nm).   From Ref. [34]. Reprinted with permission from John Wiley and Sons.

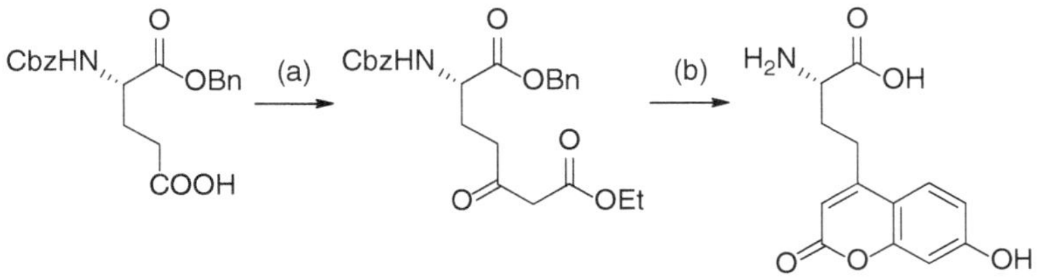

Scheme 5.8   Reagents and conditions: (a) N,N'-Carbonyldiimidazole, rt, 2 h, then ethyl magnesium malonate, rt, 40%; (b) resorcinol, $MeSO_3H$, rt, 50%.

display high emission quantum yield, photostability, and good solubility in many solvents.

In particular, 7-hydroxycoumarin has been used as this displays a high quantum yield, large Stokes shift and sensitivity to pH and solvent polarity – all of which are important for a good reporter moiety. So, an amino acid based upon the 7-hydroxycoumarin structure was synthesized by Garbay and coworkers[36] and later Schultz and coworkers,[37] by converting the side-chain acid of N-α-Cbz-L-glutamic acid α-benzyl ester into the β-keto ester, which was then condensed with resorcinol in methanesulfonic acid (Pechmann condensation)[36] (Scheme 5.8).

As mentioned previously, the fluorescence profile (Fig. 5.6) of this ncAA shows a distinct dependence upon pH, due to the acidic phenolic proton of this moiety. At

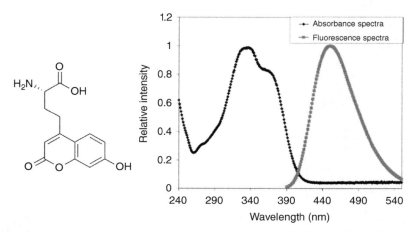

Figure 5.6 Fluorescence profile of coumarin-based amino acid developed by Schultz. The phenolate form displays an extinction coefficient of 17000 at 360 nm (right shoulder, black curve) and a quantum yield of 0.63. The emission maximum (line with filled square) is found at ~450 nm, in the gray region. From Ref. [37]. Reprinted with permission from Journal of the American Chemical Society.

pH 7.4, 100 mM sodium phosphate buffer, the coumarin is present in both phenol and phenolate forms (in approximately a 2:1 ratio), which makes this structure interesting not only from the point of fluorescence, but also for investigating pH of intracellular processes and proton transfer mechanisms.

5.2.2.5 DAN-Based System: Chromophores based on naphthalenes have been studied extensively, in particular *N*-dimethylaminonaphthalenes. The development of these types of chromophores arose from a study by Cohen *et al.*[38] The aim of this work was to develop an amino acid that can be used to probe the electrostatic interactions in proteins, in particular, the stabilizing effects that arise from these interactions. AlaDAN, a modified alanine appended with a 6-dimethylamino-2-acyl-naphthalene (DAN) moiety, was synthesized as shown in Scheme 5.9.

This synthesis is achieved effectively, using a chiral cinchodinium catalyst to control the stereochemistry of the glycine moiety (glycine-*t*Bu ester condensed with acetophenone). This intermediate amino acid can then be protected for solid-phase synthesis, or alternatively aminoacylated to the appropriate tRNA (in 5 steps) for use in nonsense suppression.

The DAN moiety undergoes large charge redistribution upon excitation, making this perfect for probing electrostatic or solvent interactions. Indeed, studies into *N*-acylated AlaDAN fluorescence show a distinct dependency upon solvent, indicating that in bulk solvent aprotic solvents lead to a strong blue fluorescence, whereas protic ones leads to red fluorescence.

Another use of DAN is shown in the case of the DANsyl group – a sulfonic acid–bearing DAN moiety. This is a more general fluorescent label than AlaDAN, in

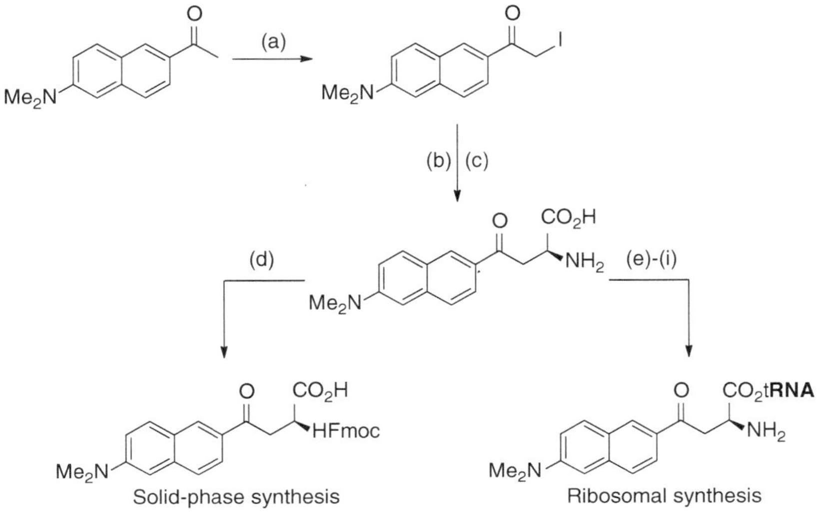

Scheme 5.9 Reagents and conditions: (a) LiHMDS, THF then I_2; (b) $Ph_2CNCH_2CO_2{}^tBu$, cinchonidinium catalyst, $CsOH.H_2O$, CH_2Cl_2; (c) TFA, ethane-1,2-dithiol; (d) Fmoc-OSuc, DMF; (e) allyl chloroformate, $NaHCO_3$, THF/H_2O; (f) CH_2ClCN, Et_3N, DMF; (g) pdCpA, DMF; (h) AcOH, N-methylmorpholine, DMF then $Pd(PPh_3)_4$; (i) *Tetrahymena thermophila* suppressor tRNA.[38]

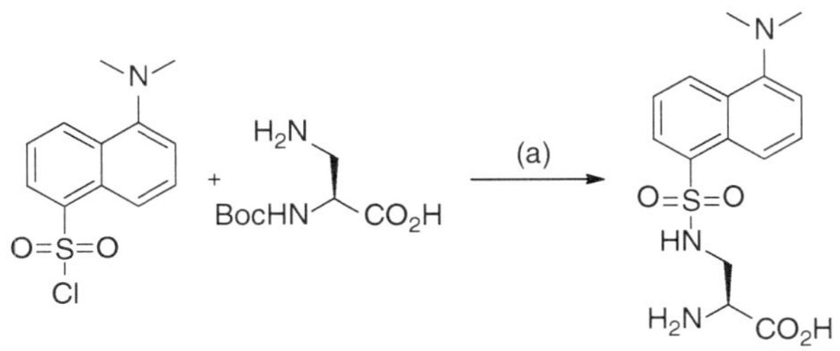

Scheme 5.10 DANsyl group can be used as an extrinsic fluorophore for amino acids bearing amine side chains. Reagents and conditions: (a) NEt_3, CH_2Cl_2 then TFA, H_2O. From Ref. [39].

that it can be attached to a variety of amine-side-chain-bearing amino acids, though the efficiency of incorporation of the amino acid is then dependent on the synthetic approach used.[39]

For example, DANsylated diaminopropionic acid (Scheme 5.10) can be genetically encoded efficiently in *Saccharomyces cerevisiae* by the use of an amber nonsense codon and an orthogonal tRNA/aminoacyl-tRNA synthetase (aaRS) pair (see Section 5.2.4.3).[39]

5.2.3 Extrinsic Labels Chemically Ligated using Cycloaddition Chemistry

The Huisgen cycloaddition reaction of an alkyne and an azide to form a 1,2,3-triazole has found application in a variety of settings. This is primarily due to the robust nature of this reaction and the ability to incorporate a range of "tags" onto virtually any system. The incorporation of fluorescent tags into proteins has indeed been achieved in this manner, employing alkyne-bearing proteins/peptides coupled with fluorescent moieties possessing azide functionalities.

Schultz and coworkers[40] reported the incorporation of both *O*-propargyltyrosine and *p*-azidophenylalanine into human superoxide dismutase-1 (SOD) using a modified *E. coli* tyrosyl-tRNA synthetase (see Section 5.2.4.3). These residues were coupled *in vitro* with either a DANsyl or fluorescein moiety under copper catalyzed conditions (Scheme 5.11).

This methodology exploits the use of both ncAAs and an extrinsic fluorescent label, with a high degree of flexibility with the azide donor of the "click reaction" coming from either the protein moiety (i.e., *p*-azidophenylalanine) or from the dye,

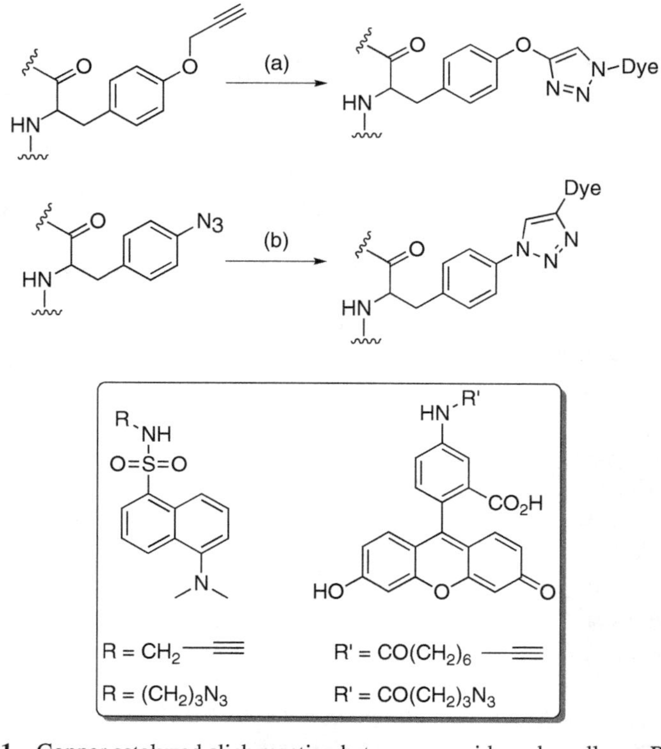

Scheme 5.11 Copper catalyzed click reaction between an azide and an alkyne. Reagents and conditions: (a) 0.01 mM protein in phosphate buffer, pH 8, 2 mM dye, 1 mM $CuSO_4$(aq.), 1 mg Cu wire.

meaning that one can tailor the dye system to needs of the protein. The main drawback of this procedure is the requirement of a copper catalyst, which often precludes this from use *in vivo*, though this can be overcome by using strained cyclooctynes rather than terminal alkynes[41–43] or by careful choice of chelating ligands for the copper species.[44,45] This process also requires large excess of copper salts and dye (100 and 200 equivalents respectively), which could limit its use on a larger scale (see Section 5.3.3).

5.2.4   Modification of the Genetic Code to Incorporate ncAAs

Generation of sequence diversity by classical protein engineering in the frame of 20 canonical amino acids has received a great deal of attention in past few decades. However, it is obvious that this set of building blocks prescribed by the universal genetic code does not span all dimensions of chemical variability that could be potentially advantageous to diversify the catalytic performance of enzymes. For this reason, synthetic noncanonical amino acids represent an ideal tool to supply proteins with novel and unusual functions for specific applications. Thus, the incorporation of ncAAs, from outside of the proteinogenic amino acids, allows novel functionalities and physical properties to be introduced into a peptide/protein structure. There are several methods by which ncAAs can be incorporated, which are outlined in the following sections.

5.2.4.1   *Auxotrophic Host Approach:*   The primary method of achieving protein expression *in vivo* is by using auxotrophic hosts. An auxotrophic host is engineered such that it is unable to synthesize a particular compound, for example, amino acid, meaning it is unable to grow without this compound (i.e., essential amino acids). By supplementing the growth medium with an analog, it is then possible to substitute this into proteins – for example, cells auxotrophic for phenylalanine can incorporate *p*-azido-phenylalanine.

   E. coli is normally employed for protein expression in this manner and a number of *E. coli* tryptophan auxotrophs are available for use – ideal for fluorescent analogs described in the previous sections. However, there are many factors that affect the level of analog incorporation, which are highly dependent on the expression system and the protein to be expressed, meaning systems are often designed for specific incorporations. The use of a stringent promoter is a must for these systems as a leaky promoter leads to expression of unmodified protein. However, with this approach, one must consider that all proteins in this system synthesized after induction will then incorporate the analog, which can have an adverse effect on biosynthetic pathways leading to a decrease in overall protein expression.

   There are two variants for auxotrophic expression: a two-step method reliant upon a rich medium to encourage rapid growth ($OD_{600} > 1.0$) followed by cell harvesting and changing to a minimal medium (with a short growth period to deplete residual tryptophan), then adding the analog and the promoter; *or* a one-step method with cells containing the expression plasmids are grown in minimal medium from the outset, with small quantities of tryptophan (~2% w/v), until a constant OD_{600} is achieved, and then just prior to induction the analog added directly to the medium. This second option has some advantages, such as avoiding the cell-harvesting step, which can lead to cell damage and lower yield in protein. Detailed descriptions, as well as protocols, of incorporation of noncanonical amino acids are available in the literature.[22,31,46,47]

In brief, routine bioexpression protocols based on an auxotrophic method enable an almost quantitative incorporation of amino acid analogs in proteins. Variants of parent proteins (congeners) are generated by supplementing amino acid auxotrophic *E. coli* host strains with the analogs during heterologous expression. This method exploits the natural substrate tolerance of the components of the translation apparatus toward isosteric ncAA analog. Experimentally, cell mass is accumulated first under noninducing conditions in the presence of limiting amounts of the canonical amino acid. As soon as the external supply of the canonical amino acid is consumed, the cells are supplemented with the analog and target protein expression is turned on. Proteins produced in this way can be termed as congeneric or congenic since they are originated from the identical gene, but they contain only a small fraction of amino acids exchanged with analogs in a residue-specific manner.

5.2.4.2 Amber Stop Codon Suppression – "Expanded Genetic Code": Site-directed incorporations of noncanonical amino acids into single recombinant proteins are highly relevant to address some specific biological problems (e.g., spectroscopic tag attachment). "Expanded genetic code" methodologies consider some termination or nontriplet coding units as "blanks" for novel addition in the amino acid repertoire. In the simplest scenario, one or two stop codons can be considered as "blank" and used uniquely to encode specific noncanonical amino acid. These experiments require the evolution of novel aaRSs capable to specifically charge cognate tRNA intracellularly. Such aaRS:tRNA pair should be orthogonal, that is, there should be no cross-reactivity between heterologous aaRSs and tRNAs with the natural host endogenous synthetases, amino acids and tRNAs. Schultz and coworkers generated *Methanocaldococcus jannaschii* TyrRS:tRNA$_{CUA}$^{Tyr} orthogonal pair in *E. coli* for position-specific incorporation.[48] This system proved to be especially useful for the incorporation of a wide array of useful noncanonical tyrosine and phenylalanine analogs or extended aromatic systems. Another system relies on the use of PylRS:tRNA$_{CUA}$^{Pyl} and a recent example of the evolution of multiple orthogonal prolyl-tRNA synthetase/tRNA pairs has been developed by Shultz and coworkers.[49]

An improved and more versatile orthogonal pair is based upon the naturally occurring pyrrolysyl-tRNA synthetase (PylRS), which incorporates pyrrolysine (Pyl) the "22nd amino acid" into proteins in some methanogenic bacteria.[50,51] The Pyl structure allows for the chemical manipulation of its aliphatic side chain giving rise to numerous analogs with interesting functionalities for incorporation into proteins. Not surprisingly, engineered PylRS:tRNAPyl pairs were made for incorporation of versatile lysine analogs into proteins at single and multiple positions of recombinant target proteins.[52] A thorough overview of this area is given in a review from our group.[53]

There are a few restrictions associated with this methodology; for successful translation of the ncAA, it must be associated with a codon that is not currently in use by the system (e.g., nonsense codon). Together, the tRNA, aminoacyl tRNA synthetase, and codon are called an orthogonal set. Orthogonality requires that the set does not display crosstalk with endogenous tRNA and synthetase sets, while still being functionally compatible with the ribosome and other associated translational apparatus, and the synthetase must also only recognize the orthogonal tRNA and the ncAA, that is, mutually orthogonal.

5.2.4.3 *Directed Evolution of Orthogonal Aminoacyl tRNA-Synthetases:* With an orthologous set in hand, it is possible to mutate the plasmid through error-prone PCR or through degenerate primers for the synthetase's active site to allow the tRNA to be charged with an ncAA. Selection for the desired mutation is achieved in a two-step process, positive and negative selection conditions, where cells are removed by toxic conditions brought about by expression of a gene or lack of expression.

A good example of how this is achieved is with positive selection made by expression of chloramphenicol acetyl transferase with a premature amber codon. In this step, the cells are exposed to the ncAA and a toxin, in this case an antibiotic chloramphenicol. Those cells that use aminoacylated orthogonal tRNA with the ncAA or cAA will override the amber codon, and will display chloramphenicol resistance. This step selects for the orthogonal tRNA. The second step is to remove the cells that used the cAA in the first step, this is achieved by inserting a plasmid containing a barnase gene (a toxic protein) with a premature amber codon into cells without the ncAA, removing all cells that did not specifically recognize the ncAA in the first step. Since the initial reports of this directed evolution, many further examples can be seen in the literature, for the incorporation of a wide variety of ncAAs.[54]

5.3 APPLICATION OF FLUORESCENT BUILDING BLOCKS IN PROTEINS

In this section, a few key examples of the methods described previously will be discussed, with a particular emphasis on their applications and limitations.

5.3.1 Azatryptophans

One of the earliest examples of a purified analog-containing protein was alkaline phosphatase.[31] This enzyme was expressed with the successful incorporation of either 7-azatryptophan or 2-azatryptophan. No change in analogous enzyme's activity was reported, but large changes in the absorbance and fluorescence spectra of the protein were observed. However, due to technological limitations in the expression of this type of system, the yields of modified protein were not particularly large.

A more recent example though is the use of azatryptophans in the Human Annexin A5 protein (anxA5).[32] The role of this protein is currently not fully understood, but it is believed to play a role in blood clotting processes. The incorporation of 4- and 5-azatryptophan in anxA5 can be achieved by feeding tryptophan-auxtotrophic *E. coli* with the appropriate azaindole (1.3 mM) and indole (1–5 μM), which allows the

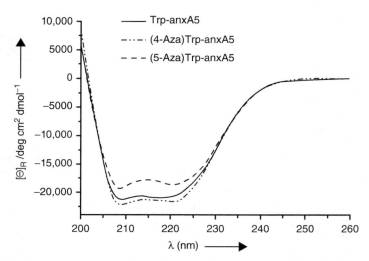

Figure 5.7 Secondary structure of (aza)Trp-anxA5 and the parent Trp-protein. Far-UV CD spectra from 200 to 260 nm were recorded at 4 °C. From Ref. [32]. Reprinted with permission from Proceedings of the National Academy of Sciences of the United States of America.

cells to grow and express the anxA5 analogs. It appeared that the indole was required in early stages of growth and without this the cells would not reach stationary phase ($OD_{600} \sim 2.0$–2.5). The replacement of a tryptophan residue, Trp187, for an (aza)Trp residue does not appear to change the secondary structure of the protein considerably (Fig. 5.7), while providing a characteristic absorption maximum at $\sim$325 nm and a large Stokes shift of $\sim$100 nm (Fig. 5.8).

Interestingly, this places the fluorescence emission in the visible range, and when comparing SDS-gel of the wild-type protein containing tryptophan and the analog, one can see upon exposure to a hand-held UV-light clear fluorescence in the case of the azatryptophan.

Table 5.5 shows the comparison of the fluorescence properties of anxA5 and the two ncAAs. Though the ncAAs both show decreased relative fluorescence to the wild-type protein, they both have a distinct redshift in emission maximum, making them fluoresce in the visible region. This property provides a useful spectroscopic tool for quantitative and qualitative analyses of these proteins.

5.3.2 FlAsH-EDT$_2$ Extrinsic Labeling System

The FlAsH-EDT$_2$ system was synthesized as previously described, and the tetra-cysteine binding domain (CCXXCC) was incorporated into the *Xenopus* calmodulin gene (a calcium-modulated messenger protein that is normally expressed in all eukaryotic cells), by mutation of the gene, replacing four residues of the N-terminal α-helix (Glu^6, Glu^7, Ala^{10}, and Glu^{11}) with cysteines.[21] This modified gene was then transiently transfected in HeLa cells, and after 36 h the cells were then exposed to

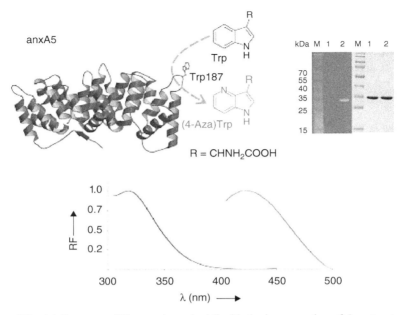

Figure 5.8 (a) Structure of Human Annexin A5 with the incorporation of 4-azatryptophan. (b) SDS-PAGE gel showing the comparison of the wild-type protein containing Trp187 residue (lane 1) and the modified protein containing 4-azatryptophan (lane 2) under UV conditions and Coomassie staining. (c) The absorption and fluorescence spectra of the 4-azatryptophan, $\lambda_{ex} = 280$ nm. From Ref. [32]. Reprinted with permission from Proceedings of the National Academy of Sciences of the United States of America. (*See color plate section for the color representation of this figure.*)

TABLE 5.5   Fluorescence Characteristics of anxA5 and Aza analogs

Substance	$\lambda_{ab,max}$ (nm)	$\varepsilon_{M,max}$ (M^{-1} cm^{-1})	$\lambda_{em,max}$ (nm)	Relative Fluorescence	Quantum Yield Φ_F
Indole	269	5264	348	1.00	0.220
4-azaindole	288	7262	418	0.30	0.077
5-azaindole	266	2663	402	0.16	0.099
Trp-anxA5	278	21105	318	1.00	–
(4-aza)Trp-anxA5	278	21020	423	0.46	–
(4-aza)Trp-anxA5	278	19730	414	0.24	–

Indole-analog fluorescent measurements were normalized to that of the indole, and accordingly for the natural anxA5 with its analogs. Measurements recorded at 20 °C in 100 mM Tris.HCl at pH 7.5.[32]

1 μM FlAsH-EDT$_2$ ligand and 10 μM EDT in Hank's balanced salt solution for 1 h at 25 °C. The results show that, after washing the cells to remove any free dye, the transfected HeLa cells showed a high degree of fluorescence after exposure to the ligand, whereas those without the binding domain calmodulin gene transfection showed greatly reduced fluorescence. The residual fluorescence in the nontransfected cells was attributed primarily to the mitochondrial fluorescence. The authors did note that the CCXXCC system can arise in naturally occurring proteins but made the argument that the sequence must be placed in a superficial position such that the FlAsH ligand can bind effectively; hence, their choice of an N-terminus of the α-helix of a relatively simple model protein. However, the relative ease of incorporation of this binding domain and the fact that the FlAsH-EDT$_2$ ligand is commercially available have made this a popular method of labeling various proteins *in vivo*, especially allowing nonfluorescent proteins to be used where fluorescence microscopy is employed as a visualization technique.

However, as already alluded to previously, an issue that can arise from the use of FlAsH-EDT$_2$ ligand is that the specificity of the binding of the arsenic ligand is not perfect and binding can be irreversible. This means that if the binding is not exclusively at the synthetic CCXXCC binding domain, toxic effects can arise in cellular processes, for example, when binding occurs in an important enzyme. Indeed, work by Beam and coworkers[55] showed that the FlAsH-EDT$_2$ ligand can bind at other sites in proteins that display CXXC motifs (albeit with lower affinity than the CCXXCC), meaning that labeling of the desired protein may not be entirely selective when endogenous proteins have similar domains to that of the synthetic binding domain. Despite this limitation, this system is still finding application in current research.[56]

5.3.3   Huisgen Dipolar Cycloaddition System

The azide–alkyne "click reaction" has found application in a wide variety of different systems and is a highly versatile and robust reaction.[57–59] As shown previously, this can be applied in a peptide/protein setting, by incorporating ncAAs that bear either an azide or an alkyne, which can then be coupled with the appropriate fluorescent partner.

Schultz used this system to incorporate DANsyl and fluorescein-based fluorescent moieties into human SOD-1.[40] This system is based upon the incorporation of *p*-azidophenylalanine or *O*-propargyltyrosine, which then can be coupled to a fluorescent moiety bearing an alkyne- or azide-functionality, providing a flexible approach for incorporating fluorescence labels and providing a choice of ncAA to be incorporated (Fig. 5.9).

The tRNA synthetases required to incorporate the ncAAs into the protein were generated from *E. coli* tyrosyl-tRNA synthetase by randomization of the codons for Tyr[37], Asn[126], Asp[182], Phe[183], and Leu[186]; these residues being picked on the basis of the crystal structure of the homologous synthetase from *Bacillus stearothermophilus*.[40]

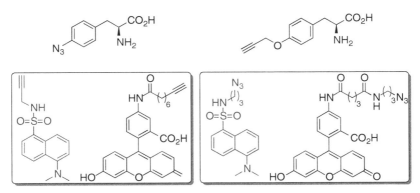

Figure 5.9 The two ncAAs, *p*-azidophenylalanine and *O*-propargyltyrosine, and their corresponding coupling partners based upon DAN or coumarin chromophores bearing an alkyne/azide handle.

The selection system for the desired synthetase used a transcriptional activator GAL4,[60,61] in which Thr[44] and Arg[110] codons are converted into amber nonsense codons (TAG).[62] Positive selection is then achieved by suppression of these codons in the MaV203:pGADGAL4-(2TAG) yeast strain, leading to full-length GAL4. This leads to the subsequent expression of *HIS3* and *URA3* reporter genes. These gene products are being particularly important in histidine and uracil auxotrophy, allowing the clones with the desired synthetase mutants in the presence of the ncAA to be selected. Negative selection to remove synthetases that charge endogenous amino acids is achieved by using growth medium containing 5-fluoroorotic acid and lacking the ncAA. The 5-fluoroorotic acid is converted into a toxic product by *URA3*, thereby removing those synthetases. After several rounds of selection, the synthetases were then identified and isolated, each showing a high degree of homology around the residues identified above to each other.[40] Several of the alkyne-based synthetases evolved showed the ability to charge both azide- and alkyne-ncAAs, whereas the azide-based synthetases were generally more selective, with only one example aminoacylating propargyltyrosine to its corresponding tRNA.

As an example application in fluorescent protein production, the codon for the permissive residue Trp[33] of SOD fused to a C-terminal 6xHis tag was mutated to an amber stop codon (TAG). With the 6xHis tag, it was then possible to purify the expressed protein, containing either *p*-azidophenylalanine or propargyltyrosine, by Ni-affinity chromatography. With the proteins in hand, the coupling of the fluorescent moiety (fluorescein or DANsyl structures) to the ncAA residue was then achieved by the [3+2] cycloaddition in the presence of copper (Scheme 5.12).

This reaction is, however, not without its limitations; the use of copper in this reaction means that it is not (generally) amenable to *in vivo* use. In this example, 100-fold excess of copper(I) sulfate and 200-fold excess of fluorescent coupling partner compared to protein are required, which is clearly far from optimal. However, this does represent an important tool for the labeling of proteins with a label, fluorescent or otherwise, by means of a [3+2] cycloaddition reaction.

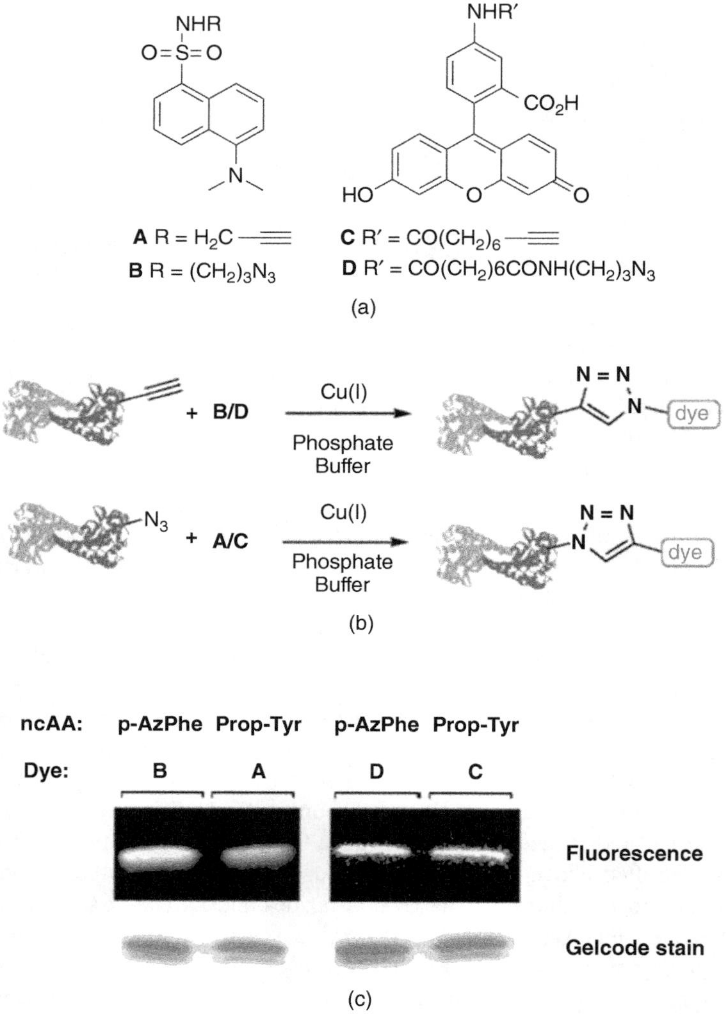

A R = H₂C———≡   **C** R′ = CO(CH₂)₆———≡
B R = (CH₂)₃N₃   **D** R′ = CO(CH₂)6CONH(CH₂)₃N₃

(a)

(b)

(c)

Scheme 5.12 Protein labeling by [3+2] cycloaddition: (a) synthesized dye labels A–D; (b) reaction between SOD and dye; (c) in-gel fluorescence scanning and Gelcode staining. From Ref. [40]. Reprinted with permission of American Chemical Society.

Much work in the area is currently being undertaken in an effort to generate conditions that are amenable to *in vivo* use (copper-free, biological pH, etc.). The use of strained cyclooctyne systems in the [3+2] cycloaddition reaction is becoming popular in recent publications.[41–43] Cyclooctynes have been chosen as they can be synthesized as amino acids in a facile manner that can be easily incorporated, and the inherent strain in the alkyne makes this more reactive toward azides, meaning the [3+2] cycloaddition reaction can occur without the use of copper (Fig. 5.10).

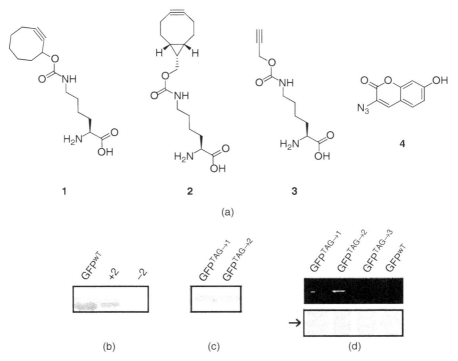

Figure 5.10 Example incorporation of alkyne-bearing amino acids into GFP for use in copper-free click reaction with an azide. (a) Structures of ncAAs 1, 2, 3 and fluorogenic azido coumarin 4. (b) Coomassie-stained SDS-PAGE gel of GFP$^{Y39TAG \rightarrow ncAA}$ in the presence (+) and absence (−) of 2 and comparison to wild-type GFP (GFPWT) and (c) GFP$^{Y39TAG \rightarrow 1}$ and GFP$^{Y39TAG \rightarrow 2}$ after purification with Ni-NTA chromatography. (d) Fluorescence image and corresponding Coomassie stain of labeling reaction with **4** inside *E. coli* cells expressing GFP$^{Y39TAG \rightarrow 1}$, GFP$^{Y39TAG \rightarrow 2}$, GFP$^{Y39TAG \rightarrow 3}$, and GFPWT (lanes 1–4, respectively; arrow indicates height of GFP band). From Ref. [43]. Reprinted with permission of WILEY-VCH Verlag GmbH & Co. KGaA, Weinheim.

As an alternative to the strained cyclooctynes, it is also possible to use copper chelating ligands limiting the cytotoxic effects associated with copper catalyzed reactions. A good example of this is the work by Ting[45] where a copper chelating ligand bearing based upon picolyl azide is coupled with 7-ethynyl coumarin in live cells in the presence of bis[(tertbutyltriazoyl)methyl]-[(2-carboxymethyltriazoyl)methyl]-amine (BTTAA)[63] or tris(3-hydroxypropyltriazolylmethyl)amine (THPTA),[44] which are biocompatible copper chelating agents.

5.4 CONCLUSIONS

As can be seen from the previous sections, a wide array of fluorescent labels is available for incorporation into proteins. The more generally applicable of these labels are based upon biomimetic ncAAs (aza-tryptophans, *p*-azidophenylalanine, etc.), which can be incorporated into proteins either by auxotrophic strain systems or by synthetic biological methods (orthogonal pair system). In general, GFP and its analogs have been used as good markers to monitor and study protein turnover, transport, and molecular interactions by using techniques such as FRET, fluorescence lifetime imaging, bimolecular fluorescence complementation, fluorescence recovery after photobleaching, or photoactivation. However, recently these have begun to fall from favor principally due to the limitation that the fusion proteins that are generated with target proteins might generate experimental artifacts. For example, fusion constructs might alter the location within the cell or introduce oligomer formation of the native protein. Similarly, the incorporation of the FlAsH-EDT$_2$ is also relatively limited for reasons previously described.

The choice of method of incorporating fluorescence into a protein is dependent on several factors: (1) nature of the fluorescent label, (2) nature of the protein into which the label is incorporated, and (3) the fluorescent properties desired for the protein. A comparison table of a selection of properties of the various labels described previously is summarized in Table 5.6.

The future of this field is in the incorporation of novel noncanonical amino acids into protein sequences that is, the expansion of the genetic code. Trp analogs (particularly the aza-analogs of Trp) have led to significant advancements in this area, showing it is possible to reprogram the biological systems involved in translation to generate novel fluorescent proteins, bringing about large changes in optical properties (e.g., comparing the Stokes shift of (4-aza)Trp to Trp) for a relatively small change in structure of the chromophore.[32] This point is crucial in investigating proteins and is clearly more favorable than using relatively large fusion proteins (such as GFP) or having to include large tags or binding domains (such as FlAsH).

5.5   PROSPECTS AND OUTLOOK

Synthetic biology that combines organic chemistry, physics, engineering, biochemistry, and biology has encouraged increased interest toward biocompatible

TABLE 5.6   Comparison of Methods for Incorporation of Fluorescence by Label

Substance	Mass (g/mol)	$\lambda_{ab,max}$ (nm)	$\lambda_{em,max}$ (nm)	Stokes shift (nm)
EGFP[a16]	~27,000	489	509	20
FlAsH-EDT$_2$[b21]	664	508	528	20
Coumarin[a64]	~250	395	470	75
(4-Aza)-Trp[32]	205	289	425	136
Trp	204	280	350	70
Tyr	181	274	304	30

[a]Relative masses of these substances are given for rough indication.
[b]Mass excludes the necessary incorporation of the synthetic binding site.

compounds as tools central to the study vital processes in the living cells as well as their rational design. Many useful tools capable of acting as biosensors or reporters in living cells are well known and established: classical fluorescent labels, isotopic markers, radioactive tracers, colorimetric biosensors, photo-switchable biomaterials, photochromic compounds, and electrochemical sensors. With the exception of radioactive labeling, the majority of these probes, to a certain extent, disturb the biological system in which they are used. In this context, the long-term goal of this field is the direct *in vivo* cotranslational generation of fluorescent proteins in living cells without the need for further posttranslational modification or chemical functionalization by externally added reagents. There are many potential developments to be undertaken and a brief outlook is given as follows.

5.5.1 Heteroatom-Containing Trp Analogs

DNA bases exhibit a combination of both exocyclic and endocyclic nitrogen atoms, charge transfer, and basicity. In particular, their exocyclic amino groups are key recognition sites in intra- and intermolecular interactions. Intriguingly, nitrogen-containing Trp analogs have many of the structural features of the purine nucleobases of DNA (adenine and guanine), and their synthetic analog 6-aminopurine, and share the capacity of the bases in pH-sensitive intramolecular charge transfer. Considering the investigation of novel heterocyclic building blocks, similar to azatryptophans, one can imagine that the use of purines and pyrimidines as the key component of the fluorophore. Endowing proteins with these "purine-Trp-analogs" will establish a new class of DNA/RNA-protein complexes that are able to mimic DNA or RNA, respectively ("DNA mimicry") and enable complex new interactions between proteins and proteins and between proteins and DNA/RNA. Furthermore, these substances may be targets for genetic code engineering by incorporating nonnatural base pairs into DNA or RNA.

Methylation of heteroatoms present within heteroaromatic amino acids based upon tryptophan (e.g., aza-, amino- and hydroxytryptophan) can be employed to disrupt hydrogen-bonding systems in proteins, while still offering potentially novel fluorescent markers within a protein setting. From preliminary studies, *N*-methyl-4-azaindole-based substituents offer good candidates for this.[65] It should be kept in mind that mono-methylated azaindole derivatives are generally unstable under the physiological conditions as it was described for 7-methyl-7-azaindole.[66,67] On the other hand, it is well known that even less stable amino acids such as telluromethionine were successfully incorporated into recombinant proteins.[68] In addition the 1,7-dimethyl-7-azaindole-based systems proved to be useful for these purposes although its low quantum yield could be problematic in attempt to design intrinsically fluorescent proteins.

5.5.2 Expanded Genetic Code – Orthogonal Pairs

The future development of fluorescent proteins lies primarily in the development of further novel amino acids with fluorescence properties suited to a wide variety of applications (*in vivo* labeling, chemical sensing, structural probes, etc.). With these novel amino acids, the development of multiple novel orthogonal pairs will be required for efficient in-cell expression of these in synthetic proteins. The development of novel aaRS:tRNA pairs is already receiving great interest from the synthetic biology community, and this input will doubtlessly allow for the incorporation of a wide range of noncanonical amino acids into proteins in the near future. While most of the development in this field has been the incorporation of fluorescence into a protein, the underlying science will have much wider implications. The expansion of the genetic code will not only allow for a multitude of fluorescent amino acids to be incorporated into almost any protein, but also allow one to design from the ground up and synthesize proteins not currently accessible to synthetic biologists. This will have a large impact on the future of drug design, catalysis, and the development of biocompatible materials suitable for medical applications. In summary, this exciting field will lead to the development of a wide variety of biologically relevant amino acids, and related proteins, and methodologies that will enable one to synthesize bespoke proteins with specific characteristics at will.

ACKNOWLEDGMENTS

The authors wish to thank the DFG and UniCAT cluster, TU Berlin, for financial support.

REFERENCES

1. Alston, R. W.; Urbanikova, L.; Sevcik, J.; Lasagna, M.; Reinhart, G. D.; Scholtz, J. M.; Pace, C. N. *Biophys. J.* **2004**, *87* (6), 4036–4047.

2. Wong, C.-Y.; Eftink, M. R. *Biochemistry* **1998**, *37* (25), 8938–8946.

3. Chalfie, M. *Photochem. Photobiol.* **1995**, *62* (4), 651–656.

4. Keller, R. GFP chromophore illustration. http://en.wikipedia.org/wiki/File:Gfp_and_fluorophore.png (accessed 2012 Sep 01).

5. Heim, R.; Prasher, D. C.; Tsien, R. Y. *Proc. Natl. Acad. Sci. U. S. A.* **1994**, *91* (26), 12501–12504.

6. Barondeau, D. P.; Putnam, C. D.; Kassmann, C. J.; Tainer, J. A.; Getzoff, E. D. *Proc. Natl. Acad. Sci. U. S. A.* **2003**, *100* (21), 12111–12116.

7. Cody, C. W.; Prasher, D. C.; Westler, W. M.; Prendergast, F. G.; Ward, W. W. *Biochemistry* **1993**, *32* (5), 1212–1218.

8. Nishiuchi, Y.; Inui, T.; Nishio, H.; Bódi, J.; Kimura, T.; Tsuji, F. I.; Sakakibara, S. *Proc. Natl. Acad. Sci. U. S. A.* **1998**, *95* (23), 13549–13554.

9. Niwa, H.; Inouye, S.; Hirano, T.; Matsuno, T.; Kojima, S.; Kubota, M.; Ohashi, M.; Tsuji, F. I. *Proc. Natl. Acad. Sci. U. S. A.* **1996**, *93* (24), 13617–13622.

10. Yang, F.; Moss, L. G.; Phillips, G. N. *Nat. Biotechnol.* **1996**, *14* (10), 1246–1251.

11. Jung, K.; Park, J.; Maeng, P.-J.; Kim, H. *Bull. Korean Chem. Soc.* **2005**, *26* (3), 413–417.

12. Ormö, M.; Cubitt, A. B.; Kallio, K.; Gross, L. A.; Tsien, R. Y.; Remington, S. J. *Science* **1996**, *273* (5280), 1392–1395.

13. Heim, R.; Cubitt, A. B.; Tsien, R. Y. *Nature* **1995**, *373* (6516), 663–664.

14. Tsien, R. Y. *Annu. Rev. Biochem.* **1998**, *67* (1), 509–544.

15. Hyun Bae, J.; Rubini, M.; Jung, G.; Wiegand, G.; Seifert, M. H. J.; Azim, M. K.; Kim, J.-S.; Zumbusch, A.; Holak, T. A.; Moroder, L.; Huber, R.; Budisa, N. *J. Mol. Biol.* **2003**, *328* (5), 1071–1081.

16. Palm, G. J.; Wlodawer, A. Spectral variants of green fluorescent protein. *Methods Enzymol.* **1999**; *302*, 378–394.

17. Budisa, N.; Pal, P.; Alefelder, S.; Birle, P.; Krywcun, T.; Rubini, M.; Wenger, W.; Bae, J.; Steiner, T., *Biol. Chem.* **2005**, *385* (2), 103–202.

18. Patterson, G. H.; Knobel, S. M.; Sharif, W. D.; Kain, S. R.; Piston, D. W., *Biophys. J.* **1997**, *73* (5), 2782–2790.

19. Cubitt, A. B.; Woollenweber, L. A.; Heim, R. *Methods Cell Biol.* **1998** *58*, 19–30.

20. Truong, K.; Ikura, M. *Curr. Opin. Struct. Biol.* **2001**, *11* (5), 573–578.

21. Griffin, B. A.; Adams, S. R.; Tsien, R. Y. *Science* **1998**, *281* (5374), 269–272.

22. Richmond, M. H. *Bacteriol. Rev.* **1962** *26* (4), 398–420.

23. Ross, J. B. A.; Szabo, A. G.; Hogue, C. W. V. *Methods Enzymol.* **1997** *278*, 151–190.

24. Chen, Y.; Gai, F.; Petrich, J. W. *J. Phys. Chem.* **1994**, *98* (8), 2203–2209.

25. Fukuda, D. S.; Mabe, J. A.; Brannon, D. R. *Appl. Environ. Microbiol.* **1971**, *21* (5), 841–843.

26. Yokoyama, Y.; Takahashi, M.; Takashima, M.; Kohno, Y.; Kobayashi, H.; Kataoka, K.; Shidori, K.; Murakami, Y. *Chem. Pharm. Bull.* **1994**, *42* (4), 832–838.

27. Prieto, M. N.; Mayor, S.; Lloyd-Williams, P.; Giralt, E. *J. Org. Chem..* **2009**, *74* (23), 9202–9205.

28. Ma, C.; Liu, X.; Li, X.; Flippen-Anderson, J.; Yu, S.; Cook, J. M. *J. Org. Chem.* **2001**, *66* (13), 4525–4542.

29. Larock, R. C.; Yum, E. K.; Refvik, M. D. *J. Org. Chem.* **1998**, *63* (22), 7652–7662.

30. Twine, S. M.; Szabo, A. G. **2003** *360*, 104–127.

31. Schlesinger, S., *J. Biol. Chem.* **1968**, *243* (14), 3877–3883.

32. Lepthien, S.; Hoesl, M. G.; Merkel, L.; Budisa, N. *Proc. Natl. Acad. Sci. U. S. A.* **2008**, *105* (42), 16095–16100.

33. Lepthien, S.; Wiltschi, B.; Bolic, B.; Budisa, N. *Appl. Microbiol. Biotechnol.* **2006**, *73* (4), 740–754.

34. Loidl, G.; Musiol, H.-J.; Budisa, N.; Huber, R.; Poirot, S.; Fourmy, D.; Moroder, L. *J. Pept. Sci.* **2000**, *6* (3), 139–144.

35. Venanzi, M.; Valeri, A.; Palleschi, A.; Stella, L.; Moroder, L.; Formaggio, F.; Toniolo, C.; Pispisa, B. *Biopolymers* **2004**, *75* (2), 128–139.

36. Brun, M.-P.; Bischoff, L.; Garbay, C. *Angew. Chem., Int. Ed. Engl.* **2004**, *43* (26), 3432–3436.

37. Wang, J.; Xie, J.; Schultz, P. G. *J. Am. Chem. Soc.* **2006**, *128* (27), 8738–8739.

38. Cohen, B. E.; McAnaney, T. B.; Park, E. S.; Jan, Y. N.; Boxer, S. G.; Jan, L. Y. *Science* **2002**, *296* (5573), 1700–1703.

39. Summerer, D.; Chen, S.; Wu, N.; Deiters, A.; Chin, J. W.; Schultz, P. G. *Proc. Natl. Acad. Sci. U. S. A.* **2006**, *103* (26), 9785–9789.

40. Deiters, A.; Cropp, T. A.; Mukherji, M.; Chin, J. W.; Anderson, J. C.; Schultz, P. G. *J. Am. Chem. Soc.* **2003**, *125* (39), 11782–11783.

41. Baskin, J. M.; Prescher, J. A.; Laughlin, S. T.; Agard, N. J.; Chang, P. V.; Miller, I. A.; Lo, A.; Codelli, J. A.; Bertozzi, C. R. *Proc. Natl. Acad. Sci. U. S. A.* **2007**, *104* (43), 16793–16797.

42. Plass, T.; Milles, S.; Koehler, C.; Schultz, C.; Lemke, E. A. *Angew. Chem., Int. Ed. Engl.* **2011**, *50* (17), 3878–3881.

43. Borrmann, A.; Milles, S.; Plass, T.; Dommerholt, J.; Verkade, J. M. M.; Wießler, M.; Schultz, C.; van Hest, J. C. M.; van Delft, F. L.; Lemke, E. A. *ChemBioChem* **2012**, *13* (14), 2094–2099.

44. Soriano del Amo, D.; Wang, W.; Jiang, H.; Besanceney, C.; Yan, A. C.; Levy, M.; Liu, Y.; Marlow, F. L.; Wu, P. *J. Am. Chem. Soc.* **2010**, *132* (47), 16893–16899.

45. Uttamapinant, C.; Tangpeerachaikul, A.; Grecian, S.; Clarke, S.; Singh, U.; Slade, P.; Gee, K. R.; Ting, A. Y. *Angew. Chem., Int. Ed. Engl.* **2012**, *51* (24), 5852–5856.

46. Sambrook, J.; Fritsch, E. F.; Maniatis, T. Molecular Cloning : A Laboratory Manual. Cold Spring Harbor Laboratory: Cold Spring Harbor, N.Y., **1989**.

47. Levine, M.; Tarver, H. *J. Biol. Chem.* **1951**, *192* (2), 835–850.

48. Wang, L.; Brock, A.; Herberich, B.; Schultz, P. G. *Science* **2001**, *292* (5516), 498–500.

49. Chatterjee, A.; Xiao, H.; Schultz, P. G., *Proc. Natl. Acad. Sci. U. S. A.* **2012**, *109* (37), 14841–14846.

50. Hao, B.; Zhao, G.; Kang, P. T.; Soares, J. A.; Ferguson, T. K.; Gallucci, J.; Krzycki, J. A.; Chan, M. K. *Chem. Biol.* **2004**, *11* (9), 1317–1324.

51. Polycarpo, C.; Ambrogelly, A.; Bérubé, A.; Winbush, S. M.; McCloskey, J. A.; Crain, P. F.; Wood, J. L.; Söll, D. *Proc. Natl. Acad. Sci. U. S. A.* **2004**, *101* (34), 12450–12454.

52. Li, W.-T.; Mahapatra, A.; Longstaff, D. G.; Bechtel, J.; Zhao, G.; Kang, P. T.; Chan, M. K.; Krzycki, J. A. *J. Mol. Biol.* **2009**, *385* (4), 1156–1164.

53. Hoesl, M. G.; Budisa, N. *Angew. Chem. Int. Ed. Engl..* **2011**, *50* (13), 2896–2902.

54. Liu, C. C.; Schultz, P. G. *Annu. Rev. Biochem.* **2010**, *79* (1), 413–444.

55. Stroffekova, K.; Proenza, C.; Beam, K. *Pflueg. Arch. Eur. J. Physiol.* **2001**, *442* (6), 859–866.

56. Rutkowska, A.; Haering, C. H.; Schultz, C. *Angew. Chem. Int. Ed.* **2011**, *50* (52), 12655–12658.

57. Huisgen, R.; Seidel, M.; Sauer, J.; McFarland, J.; Wallbillich, G. *J. Org. Chem.* **1959**, *24* (6), 892–893.

58. Kolb, H. C.; Finn, M. G.; Sharpless, K. B. *Angew. Chem., Int. Ed. Engl.* **2001**, *40* (11), 2004–2021.

59. Meldal, M.; Tornøe, C. W. *Chem. Rev.* **2008**, *108* (8), 2952–3015.

60. Keegan, L.; Gill, G.; Ptashne, M. *Science* **1986**, *231* (4739), 699–704.

61. Ptashne, M. *Nature* **1988**, *335* (6192), 683–689.

62. Chin, J. W.; Cropp, T. A.; Chu, S.; Meggers, E.; Schultz, P. G. *Chem. Biol.* **2003**, *10* (6), 511–519.

63. Besanceney-Webler, C.; Jiang, H.; Zheng, T.; Feng, L.; Soriano del Amo, D.; Wang, W.; Klivansky, L. M.; Marlow, F. L.; Liu, Y.; Wu, P. *Angew. Chem.* **2011**, *123* (35), 8201–8206.

64. Beatty, K. E.; Xie, F.; Wang, Q.; Tirrell, D. A. *J. Am. Chem. Soc.* **2005**, *127* (41), 14150–14151.

65. Merkel, L.; Hoesl, M. G.; Albrecht, M.; Schmidt, A.; Budisa, N. *ChemBioChem* **2010**, *11* (3), 305–314.

66. Waluk, J.; Pakuła, B.; Komorowski, S. J. *J. Photochem. Photobiol. A* **1987**, *39* (1), 49–58.

67. Chapman, C. F.; Maroncelli, M. *J. Phys. Chem.* **1992**, *96* (21), 8430–8441.

68. Budisa, N.; Karnbrock, W.; Steinbacher, S.; Humm, A.; Prade, L.; Neuefeind, T.; Moroder, L.; Huber, R. *J. Mol. Biol.* **1997**, *270* (4), 616–623.

6

FLUOROMODULES: FLUORESCENT DYE–PROTEIN COMPLEXES FOR GENETICALLY ENCODABLE LABELS

BRUCE A. ARMITAGE

Department of Chemistry and Molecular Biosensor and Imaging Center, Carnegie Mellon University, Pittsburgh, PA, USA

6.1 INTRODUCTION

The ability to study the inner workings of the cell has made tremendous advances since the introduction of green fluorescent protein (GFP) as a genetically encodable element that can be fused to a protein of interest, providing a covalently attached fluorescent tag for tracking protein localization, dynamics, and binding interactions.[1,2] GFP has also been used to indirectly label RNA, through fusion to an RNA-binding protein that recognizes a specific sequence tag.[3] Thus, GFP is an example of a *fluoromodule*, that is, a fluorescent component that can be used in a modular sense, as in fusion constructs. The palette of fluorescent proteins is diverse, ranging from the blue/cyan to red and providing an extremely useful catalog of fluoromodules for modern cell biology.[4]

GFP and other proteins used as genetically encoded tags can be classified as *inherently fluorescent proteins*, meaning that no exogenous agents are needed to produce fluorescence (Fig. 6.1a). In these proteins, the fluorescent component is formed via a series of chemical reactions that occur posttranslationally and involve solely the amino acid side chains of specific residues. Having all of the necessary ingredients genetically encoded certainly simplifies the use of inherently fluorescent proteins,

Fluorescent Analogs of Biomolecular Building Blocks: Design and Applications, First Edition.
Edited by Marcus Wilhelmsson and Yitzhak Tor.
© 2016 John Wiley & Sons, Inc. Published 2016 by John Wiley & Sons, Inc.

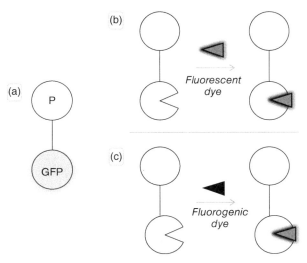

Figure 6.1   Schematic examples of genetically encoded tags for proteins. (a) Labeling with a constitutively fluorescent protein (e.g., GFP); (b) labeling with a nonfluorescent protein that subsequently binds to a constitutively fluorescent exogenous dye; (c) labeling with a nonfluorescent protein that subsequently binds to a fluorogenic exogenous dye.

but it also minimizes the amount of control over experimental conditions available to the user. For example, changing the color of a tagged protein usually requires a new fusion construct to be prepared in which the original fluoromodule is replaced by one of a different color. The user also has limited control over timing: transcription and translation are normally strongly coupled, making it difficult to actively dictate the time and/or spatial location at which the fluorescence signal should appear. Furthermore, there is a time lag between protein translation and maturation of the fluorophore in inherently fluorescent proteins. During this window, the tagged protein might be undergoing important biological processes, but these remain unobservable until the fluorophore maturation is complete.

The desire for greater control over fluorescence-based imaging experiments has led to the development of numerous fluoromodule technologies that combine genetically encoded *apomodules* with exogenously supplied fluorescent dyes.[5,6] In some cases, the dye is linked to a reactive group that results in formation of a covalent bond to the apomodule, whereas in other cases binding is noncovalent. Another distinction arises from the dye, which can be either constitutively fluorescent (Fig. 6.1b) or *fluorogenic*, meaning the fluorescence only appears after the dye binds to the apomodule (Fig. 6.1c). The latter case is often more appealing because it gives rise to low background fluorescence and unbound dye need not be washed out of a sample prior to imaging.

Several classes of covalent fluoromodules have been reported in recent years. These include the FlAsH/ReAsH,[7,8] SNAP[9] and Halo[10] tags, and biotin ligase[11] systems. FlAsH/ReAsH benefits from fluorogenic dyes and a relatively small reactive apomodule that minimizes potential artifacts that can arise from having large protein domains genetically fused to the protein of interest. However, a disadvantage of this system is the toxicity of the biarsenical dyes, necessitating administration of an "antidote" immediately after the labeling reaction has been performed. The SNAP and Halo tags do not have this feature, but their apomodules are considerably larger than the tetracysteine motif recognized by the biarsenical dyes. In addition, SNAP, Halo, and biotin ligase labeling typically rely on constitutively fluorescent dyes, meaning unbound dye needs to be washed out prior to imaging. Nevertheless, the value of covalent labeling is that, provided the reaction is specific and the kinetics are reasonable, low expression levels do not necessarily preclude efficient labeling, in contrast to a noncovalent labeling system, where imaging of low-abundance proteins demands a dye-apomodule with a very low K_D value.

The focus of this chapter is on noncovalent fluoromodules based on fluorogenic dyes and genetically encodable fluorogen-activating proteins (FAPs) based on single-chain antibody fragments (scFvs). Combining rational design and standard organic synthesis methods to obtain the dye components, a diverse scFv library and a powerful high-throughput screening method to rapidly isolate functional FAPs from the library has led to a veritable rainbow of fluoromodules that are finding increasing applications in the biological sciences and biotechnology.

6.2   FLUOROMODULE DEVELOPMENT AND CHARACTERIZATION

There have been several examples of noncovalent fluoromodules based on dye–protein complexes. Early examples of fluoromodules consisted of antibodies isolated from rabbits or mice after immunization with fluorescein[12] or julolidene[13] haptens. In later work, Wittrup and coworkers selected scFvs that bind to fluorescein from a yeast surface-display library.[14] Although this demonstrated the concept of using *in vitro* selection to identify protein partners for organic dyes, fluorescein typically exhibits bright fluorescence even in the unbound state and, in fact, can show quenching upon binding to certain proteins, undesirable properties for imaging applications.

In a series of papers, Lerner, Schultz, and coworkers reported the selection of monoclonal antibodies and scFvs that bind to stilbene-based dyes.[15] In fluid solution, *trans*-stilbene undergoes photoisomerization to the cis isomer, resulting in low fluorescence quantum yields.[16] However, rigid environments such as viscous solvents or frozen glasses are known to significantly enhance fluorescence by preventing the conformational motion needed for isomerization.[17] A similar mechanism is in effect for the protein hosts for *trans*-stilbene ligands: conformational restriction leads to enhanced fluorescence. By varying the substituents on the stilbene dye and screening several different selected proteins, a limited range of colors were obtained.[18] In one intriguing case, the stilbene was bound adjacent to a tryptophan residue deep within the protein, leading to formation of an exciplex with distinctive blue emission.[19]

Over the past several years, Cornish and coworkers have developed a robust, noncovalent fluoromodule system based on the high-affinity binding between

Escherichia coli dihydrofolate reductase and the folate analog trimethoprim (TMP).[6,20,21] Covalent linkage of various fluorescent dyes to the TMP ligand allows selective labeling of eDHFR fusion proteins. The value of this tagging technology has been validated in numerous contexts, ranging from intracellular super-resolution protein imaging[22] to single-molecule analysis of spliceosome assembly.[23]

Our approach to creating new protein-based fluoromodules relies on rational design of the fluorogenic dye and efficient functional selection of protein binding partners for the dyes. We have focused on two classes of dyes: triarylmethanes, such as malachite green, and unsymmetrical cyanines, such as thiazole orange. Upon excitation, these dyes typically undergo facilitated bond rotations that lead to nonradiative deactivation, resulting in very low fluorescence quantum yields in fluid solution.[24–26] However, conformational constraints imposed by binding to biopolymer hosts such as double-stranded DNA[27] or RNA aptamers[28,29] can lead to fluorescence enhancements of 2–3 orders of magnitude. The ease with which these dyes can be synthesized and spectrally tuned made them attractive candidates for fluoromodule construction.

The FAP component of the fluoromodule is impossible to design based on our current limited knowledge of protein folding. Therefore, we rely on a combinatorial library of scFv proteins first reported by Wittrup and coworkers.[14] The proteins are displayed on the surface of yeast, with each yeast displaying thousands of copies of an individual protein. (The total diversity of the library is approximately 10^9.) The great value of the yeast surface display format arises from the powerful selection method that it enables: mixing of a fluorogenic dye with the yeast library leads to fluorescent labeling of the surface of those yeast that express proteins capable of binding and activating the dye (Fig. 6.2). These yeast are readily detected and isolated by fluorescence-activated cell sorting (FACS). We routinely screen approximately 2.5×10^6 yeast in a single FACS experiment. In our initial reports, we first

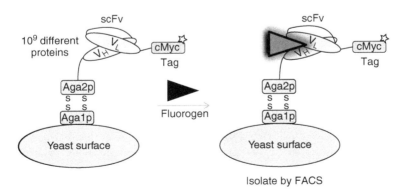

Figure 6.2 Design of FAP library and selection strategy. scFv proteins are expressed as fusions with a yeast cell surface protein. Addition of a fluorogenic dye results in fluorescent staining of yeast bearing complementary FAPs, which are easily sorted from the library using FACS.

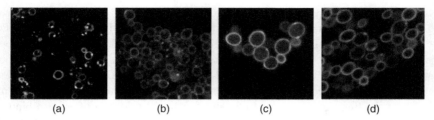

(a)   (b)   (c)   (d)

Figure 6.3 Yeast expressing cell surface fluorogen-activating scFv proteins after staining with OTB, TO, αCN-DIR or DIR (a–d). (*See color plate section for the color representation of this figure.*)

used two magnetic-bead-based affinity enrichments of the library before proceeding to the FACS screening.[30,31] However, we recently demonstrated that the naïve library can be taken directly to the FACS with a new fluorogen and reasonably high-affinity and high-quantum-yield FAPs can be obtained.[32] This simplification of the procedure not only eliminates the two affinity enrichment steps but also simplifies fluorogen synthesis as there is no longer a need to prepare a biotinylated version of the dye for immobilization on the magnetic bead. With this approach, promising FAPs can be obtained within 3–4 rounds of FACS screening, which requires only 2–3 weeks. Figure 6.3 shows imaging of several yeast-displayed FAPs after staining with the appropriate cognate dye.

6.2.1 Fluorogenic Dyes

There are many classes of fluorogenic dyes that can be activated by various changes in environmental conditions. Our efforts have focused on torsionally deactivated dyes that exhibit large increases in fluorescence quantum yield due to conformational restrictions. Several dyes from the triarylmethane and unsymmetrical cyanine classes of fluorogens have been used to select FAPs.

6.2.1.1 Malachite Green: In the first publication from our Center, a derivative of malachite green (MG, see Chart 6.1 for all fluorogen structures) was used as the fluorogen for selecting FAPs.[30] Although the triarylmethane dyes are not as versatile as the cyanines in terms of spectral tuning, MG exhibits an exceptionally low quantum yield in fluid solution, creating a potentially dark background against which fluoromodule fluorescence can be imaged. FAPs isolated for activating MG were found to have submicromolar K_D values and ϕ_f values of up to 0.25, corresponding to 10^4-fold enhancement above unbound fluorogen. As with RNA fluoromodules based on triarylmethane dyes,[28] the MG-FAPs exhibited bright red fluorescence.

6.2.1.2 Thiazole Orange: In the same report where MG-FAPs were described, the classic DNA/RNA stain thiazole orange[27] was also used to select scFv proteins.[30] A negatively charged alkylsulfonate group was appended to one heterocycle to reduce nonspecific nucleic acid binding, whereas an oligo PEG linker was attached to the

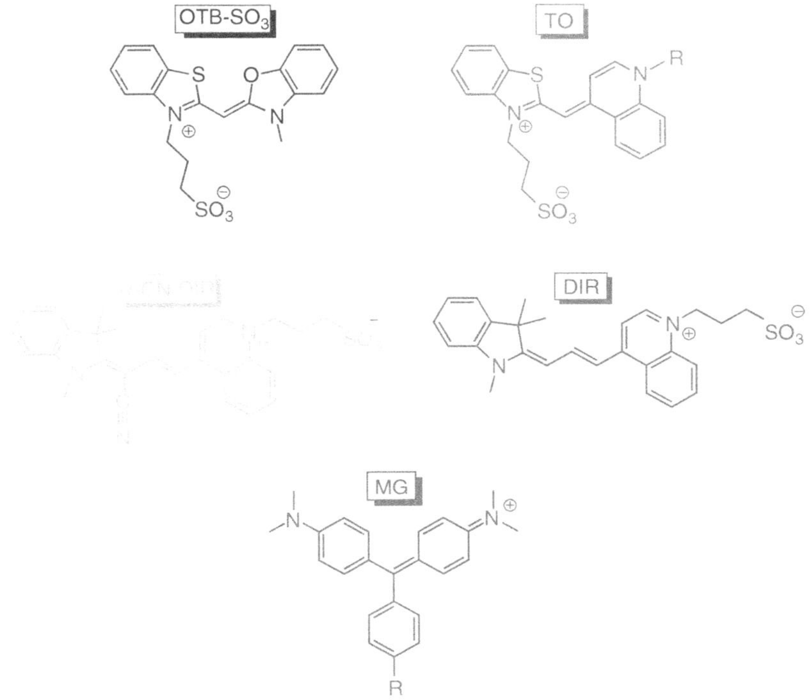

Chart 6.1 Fluorogenic dyes used in protein–dye fluoromodules. (R in TO and MG structures corresponds to a long, flexible substituent but is unnecessary for protein binding.)

other heterocycle for display of biotin used for initial magnetic bead enrichment steps. The K_D obtained for one of the FAPs is ca. 100-fold lower than for typical intercalation of TO into DNA while the ϕ_f is comparable. TO-FAPs offer green fluorescence.

6.2.1.3 Dimethylindole Red: An important step toward the development of our technology came from screening the scFv library for binding and activation of the new fluorogen *dimethylindole red* (DIR).[31] (Note that we named this dye based not on the color of the solid, as was done classically for dyes such as MG and TO, but rather on the color of its emission.) DIR was designed to have minimal nucleic acid binding due to the steric effect of the dimethylindole group, which hinders the pi-stacking interactions important to intercalative binding, as well as to electrostatic repulsions due to the propylsulfonate group. DIR and a biotinylated analog were originally synthesized for selection of RNA aptamers that can activate fluorescence from the dye.[29] These were readily used in the scFv screening method, leading to selection of several new FAPs, the most interesting of which, called "K7," binds to DIR with $K_D = 13$ nM to give a moderately efficient fluoromodule ($\phi_f = 0.3$). The same protein also bound and activated several other unsymmetrical cyanines. This includes dyes with variable heterocycles and conjugation lengths, meaning a single protein gives rise to a family of fluoromodules with spectra ranging from the blue to the near-infrared.

6.2.1.4 *Oxazole Thiazole Blue:* A blue fluoromodule could be created from the promiscuous FAP K7 and the monomethine dye PO-PRO-1, but the K_D and ϕ_f values were unimpressive. Therefore, we designed a new fluorogen with appropriate spectral properties. The dye, named oxazole thiazole blue (OTB-SO$_3$), was inspired by the previously reported dye Cyan 47[33] and exhibits reasonably high molar absorptivity in the violet region of the spectrum ($\varepsilon_{399} = 44{,}000\ M^{-1}\ cm^{-1}$), low fluorescence in fluid solution, low activation by double-stranded DNA, and good fluorescence enhancement in glycerol. The dye was used to select FAPs from the yeast-displayed scFv library, resulting in several promising proteins, three of which exhibited $K_D <$ 100 nM and $\phi_f > 80\%$.[32] This work also led to an important technological advance: OTB-SO$_3$ was added directly to the naïve library, which was then screened by FACS. Even though only ca. 2.5×10^6 cells were sorted, bright, high-affinity proteins were obtained after 4–6 rounds of selection. This indicates that the library has sufficient diversity to yield many independent solutions to the problem of fluorogen recognition. Furthermore, as stated above, the ability to bypass the magnetic bead–based sorting steps means that we no longer must synthesize biotinylated versions of the fluorogens before screening can commence.

6.2.1.5 *α-Cyano-DIR:* Although they offer the ability to fine-tune spectral properties over the entire visible/NIR range, unsymmetrical cyanine dyes can exhibit relatively low photostability. Following work by Toutchkine and Hahn,[34,35] we designed a substituted version of the red fluorogen DIR in which an electron-withdrawing cyano group was attached to the methine carbon adjacent (i.e., alpha) to the dimethylindole heterocycle.[36] Although we have not verified this for DIR, it is known that cyanine photobleaching can occur at least partly through oxidative attack by singlet oxygen (generated by the fraction of the dye excited state that intersystem crosses to the triplet state).[35,37] The new fluorogen, α-CN-DIR, bound to the promiscuous FAP K7 ($K_D = 60$ nM and $\phi_f = 0.16$), so new FAPs were not needed. As expected, the α-CN-DIR/K7 fluoromodule was approximately eightfold more photostable than the parent DIR/K7 fluoromodule. An added benefit was that α-CN-DIR/K7 exhibited a bright orange fluorescence due to the 60-nm blueshift of the cyano-substituted dye, providing an additional color to the fluoromodule palette.

6.2.1.6 *Bichromophoric Reagents for Improved Brightness:* In order to improve fluoromodule brightness, a bichromophoric reagent consisting of MG functionalized with up to four Cy3 dyes was synthesized.[38] These dendritic "dyedron" structures exhibit efficient Cy3 quenching by the appended MG, but no MG fluorescence until an MG-binding FAP is present. The high molar extinction coefficient arising from the four Cy3 dyes and efficient Cy3-MG energy transfer leads to nearly sevenfold greater brightness when Cy3 is excited than when MG is excited directly.

6.2.2 Fluorogen-Activating Protein (FAP) Optimization

Although we are unable to design *de novo* the FAP component of our fluoromodules, our FACS-based screening and enrichment method can yield proteins that give rise to high-affinity/high-quantum-yield complexes with their cognate dyes in a few weeks' time. Furthermore, once a promising FAP has been isolated, directed evolution and/or rational design can be used to improve the properties of the protein.

6.2.2.1 Directed Evolution: Affinity maturation of a FAP can be readily performed using mutagenic PCR to create a new library for selections, which are typically done at dye concentrations significantly below the K_D of the parent FAP. This method has been used to improve the affinity of MG-[30] and DIR-binding[39] FAPs more than 100-fold. Tighter binding can also lead to improved quantum yield, although this is not universally true.

6.2.2.2 Rational Design: The scFv proteins typically consist of heavy (V_H) and light (V_L) chains connected through a $(Gly_4Ser)_3$ linker. An interesting phenomenon that we have observed in several cases is the isolation of FAPs that utilize only one of the two domains (V_H or V_L) for dye binding.[30,39] The presence of single-domain proteins in the library is attributed to artifacts during library construction. However, we have also found cases where a two-domain FAP isolated from the library actually exhibits higher affinity after treatment with a protease enzyme that cleaves the polypeptide in the linker separating the two domains.[40] In these cases, the fluorogen most likely binds to an intermolecular dimer of the protein, relying on the individual active domains from each FAP. Once the active domain is identified, a synthetic dimer of that domain (e.g., $V_L - V_L$) can be constructed that exhibits further enhanced affinity.[39]

The question naturally arises of how a full-length $V_H - V_L$ scFv that does not strongly activate fluorescence manages to survive the selection process, which is strongly biased in favor of brightly activating FAPs. We hypothesize that relatively high surface expression levels of the scFvs permit binding of a single-dye molecule between two active domains of nearby scFvs. Thus, the fluorogen noncovalently cross-links two scFvs, leading to a functional, ternary complex. When the scFv is cloned and expressed as a soluble protein, the relatively low nano-/micromolar concentrations normally used for fluoromodule characterization disfavor ternary complex formation, leading to weak binding and fluorescence.

In addition to using experimental information such as weak solution fluorescence to assist in determining which FAPs should be deconstructed into single-domain proteins, a DIR-binding protein known as M8 provided an interesting alternative. When M8 was isolated from the library and sequenced, it was found to have an identical V_L domain as another protein isolated during the same selection, while the corresponding V_H domains exhibited only 48% homology, suggesting that dye-recognition activity resided in the V_L domain.[31] The full-length M8 protein exhibited virtually no fluorescence activation of DIR in solution in contrast to the yeast cell surface.[39] However, the isolated V_L domain successfully activated DIR fluorescence, demonstrating the ability to discard inactive partner domains for certain FAPs. The M8 V_L protein was then converted into a synthetic $V_L - V_L$ dimer with variable $(Gly_4Ser)_n$ ($n = 3–5$) linker lengths between the two V_L domains. In each case, the $K_D < 0.1\,nM$ while the quantum yields varied from 0.38 to 0.64. These results demonstrate the value of using rational engineering of proteins identified by the initial combinatorial library selection in order to optimize affinity and brightness.

6.2.3 Fluoromodule Recycling

Another interesting feature of the original fluoromodules was the finding that, under prolonged illumination, fluorescence intensity of a TO fluoromodule initially decayed but then stabilized at a level that depended on the TO concentration.[30] This indicates that bleached fluorogens can be replaced by pristine unbleached fluorogen from the surrounding medium in order to reconstitute fluorescence. This also illustrates another advantage of noncovalent fluoromodules: since the fluorogenic dyes are dark, excess dye can be present to allow *in situ* reconstitution of bleached fluoromodules, provided the excess dye does not bind nonspecifically to other cellular structures or induce its own photochemistry.

Recycling was also studied for the K7-DIR fluoromodule.[36] Starting with a 1:1 ratio of protein:dye (i.e., no excess dye), a solution was irradiated with visible light for 15 min, resulting in ca. 55% loss of fluorescence intensity. Addition of a second equivalent of dye resulted in recovery of fluorescence to 90% of the original value, but addition of more dye failed to increase the fluorescence beyond this value. This result indicated that the photobleaching pathway partitions between dye- and protein-localized damage. In the case of K7-DIR, approximately 80% of the damage was localized to the dye and could be recovered with fresh dye, whereas the remaining 20% was likely due to protein damage, which is not recoverable. (This ratio will likely vary with the specific protein and dye components of the fluoromodule.) The damage presumably involves reactive oxygen species such as singlet oxygen, although the mechanism has not been investigated in detail.

6.3 IMPLEMENTATION

6.3.1 Fusion Constructs for Protein Tagging

The original report on our fluoromodule technology demonstrated the ability to create fusion constructs of TO- or MG-binding FAPs with endogenous proteins, specifically the platelet-derived growth factor receptor (PDGFR), a transmembrane protein, such that the FAP would be displayed on the cell surface.[30] The fusion proteins were expressed in two mammalian cell lines and successfully stained with the green and red fluorogens. When cell-impermeant versions of the fluorogens were used, fluorescence was observed only at the cell periphery. However, cell-permeant fluorogens also stained elements of the secretory apparatus, indicating that newly synthesized protein in the process of being transported to the cell surface had successfully folded and captured fluorogenic dye. Thus, fusion constructs made from FAPs are analogous to GFP and other inherently fluorescent proteins.

Fusion of either TO- or MG-binding FAPs to the β_2-adrenergic receptor (β_2AR) allowed imaging and measurement of receptor internalization in mammalian cells in response to an agonist.[41,42] Constructs in which a FAP and GFP are expressed at the outer and inner surfaces, respectively, of the plasma membrane are also useful for normalizing imaging results to expression level, since the GFP fluorescence is not dependent on the presence of the fluorogenic dye. Fusion constructs involving FAPs and membrane proteins are showing considerable promise for use in screening assays for identifying drug candidates targeted to membrane receptors.

6.3.2 Protein Tagging and pH Sensing

β_2AR-FAP fusion constructs also were used to demonstrate the versatility of this technology as a biosensor in addition to a label: a tandem fluorogen consisting of TO-linked to a pH-sensitive Cy5 derivative was synthesized.[43] The tandem dye should bind to the cognate FAP under any conditions, but FRET to the Cy5 derivative should only occur at low pH, where the Cy5 component becomes protonated. Thus, internalized FAP-β_2AR exhibits red Cy5 emission due to FRET, while cell surface localized β_2AR remains green.

6.3.3 Super-Resolution Imaging

FAP-based fluoromodules have also been used for super-resolution nanoscopy, providing better images than is possible with standard confocal imaging.[44] A malachite green-binding FAP was genetically fused to actin for intracellular expression and exhibited approximately threefold better resolution in stimulated emission depletion (STED) nanoscopy compared with confocal imaging of the same samples.

In addition to demonstrating the use of scFv-based fluoromodules in the rapidly growing area of super-resolution imaging, this work demonstrated the ability to express functional FAPs in the cytoplasm of mammalian cells. The reducing environment of the cytoplasm is problematic for scFvs because of their reliance on disulfide bonds for structural stability. Therefore, cysteine residues in an MG-binding FAP were replaced by alanine using site-directed mutagenesis, and the resulting mutant was subjected to affinity maturation to obtain a functional, disulfide-free version that was expressed and folded stably as an actin fusion.

6.3.4 Protease Biosensors

The discovery that full-length $V_H - V_L$ FAPs can exhibit lower affinity and fluorescence than a corresponding isolated V_H or V_L domain suggested the creation of inactive or blocked FAPs that could serve as biosensors for protease activity.[40] The strategy involved fusing an active V_H or V_L domain to a counterpart domain that would block fluorogen binding. If the two domains are linked by a polypeptide sequence that can serve as a substrate for a protease enzyme, proteolytic activity can result in separation of the two domains, releasing the active domain to bind and activate fluorogen.

6.4 CONCLUSIONS

The fluoromodule technology unites two powerful capabilities: rational design and synthesis of fluorogenic dyes and rapid flow-based sorting of a yeast-displayed library to identify proteins that bind and activate fluorescence from the dyes. Multiple unique proteins that give mid-/low-nanomolar K_d and >10% quantum yields are usually obtained in less than a month for a variety of dyes. The proteins can also exhibit a wide range of promiscuity, offering considerable postselection flexibility in choosing the color of a particular fusion construct or biosensor.

6.5 PROSPECTS AND OUTLOOK

The biggest challenge facing further development of the fluoromodule technology is identifying a reliable path to intracellular expression of scFv-based FAPs, although preliminary efforts based on both protein and dye engineering are showing promise. Alternative scaffolds, (e.g., DARPin proteins[45]) that do not rely on disulfide bonds for stability and are known to be cytoplasmically functional are candidates for replacements for scFvs. Selection against cytoplasmic rather than surface-displayed libraries would also facilitate development of fluoromodules for intracellular applications. Nevertheless, the rapid progress in creating a catalog of genetically encodable FAPs and fluorogenic dye partners has already enabled numerous applications, where cell-surface or *in vitro* fluoromodules are desired.

ACKNOWLEDGMENTS

This work is the result of the contributions of many talented scientists, listed as coauthors on the publications cited below. Special acknowledgment is reserved for Alan S. Waggoner, Director of the Molecular Biosensor and Imaging Center, for conceiving and directing this project, and Peter B. Berget, for leading my research group into the fascinating world of proteins. Financial support from the US National Institutes of Health (grant U54 RR022241) is greatly appreciated.

REFERENCES

1. Chalfie, M.; Tu, Y.; Euskirchen, G.; Ward, W. W.; Prasher, D. C. *Science* **1994**, *263*, 802.
2. Tsien, R. Y. *Annu. Rev. Biochem.* **1998**, *67*, 509.
3. Bertrand, E.; Chartrand, P.; Schaefer, M.; Shenoy, S. M.; Singer, R. H.; Long, R. M. *Mol. Cell* **1998**, *2*, 437.

4. Shaner, N. C.; Steinbach, P. A.; Tsien, R. Y. *Nat. Methods* **2005**, *2*, 905.

5. Newman, R. H.; Fosbrink, M. D.; Zhang, J. *Chem. Rev.* **2011**, *111*, 3614.

6. Jing, C.; Cornish, V. W. *Acc. Chem. Res.* **2011**, *44*, 784.

7. Griffin, B. A.; Adams, S. R.; Tsien, R. Y. *Science* **1998**, *281*, 269.

8. Adams, S. R.; Campbell, R. E.; Gross, L. A.; Martin, B. R.; Walkup, G. K.; Yao, Y.; Llopis, J.; Tsien, R. Y. *J. Am. Chem. Soc.* **2002**, *124*, 6063.

9. Keppler, A.; Gendreizig, S.; Gronemeyer, T.; Pick, H.; Vogel, H.; Johnsson, K. *Nat. Biotechnol.* **2003**, *21*, 86.

10. Los, G. V.; Encell, L. P.; McDougall, M. G.; Hartzell, D. D.; Karassina, N.; Zimprich, C.; Wood, M. G.; Learish, R.; Ohana, R. F.; Urh, M.; Simpson, D.; Mendez, J.; Zimmerman, K.; Otto, P.; Vidugiris, G.; Zhu, J.; Darzins, A.; Klaubert, D. H.; Bulleit, R. F.; Wood, K. V. *ACS Chem. Biol.* **2008**, *3*, 373.

11. Chen, I.; Howarth, M.; Lin, W.; Ting, A. Y. *Nat. Methods* **2005**, *2*, 99.

12. Voss, E. W.; Lopatin, D. E. *Biochemistry* **1971**, *10*, 208.

13. Iwaki, T.; Torigoe, C.; Noji, M.; Nakanishi, M. *Biochemistry* **1993**, *32*, 7589.

14. Feldhaus, M. J.; Siegel, R. W.; Opresko, L. K.; Coleman, J. R.; Feldhaus, J. M.; Yeung, Y. A.; Cochran, J. R.; Heinzelman, P.; Colby, D.; Swers, J.; Graff, C.; Wiley, H. S.; Wittrup, K. D. *Nat. Biotechnol.* **2003**, *21*, 163.

15. Simeonov, A.; Matsushita, M.; Juban, E. A.; Thompson, E. H. Z.; Hoffman, T. Z.; Beuscher IV, A. E.; Taylor, M. J.; Wirsching, P.; Rettig, W.; McCusker, J. K.; Stevens, R. C.; Millar, D. P.; Schultz, P. G.; Lerner, R. A.; Janda, K. D. *Science* **2000**, *290*, 307.

16. Saltiel, J. *J. Am. Chem. Soc.* **1967**, *89*, 1036.

17. Saltiel, J.; D'Agostino, J. T. *J. Am. Chem. Soc.* **1972**, *94*, 6445.

18. Tian, F.; Debler, E. W.; Millar, D. P.; Deniz, A. A.; Wilson, I. A.; Schultz, P. G. *Angew. Chem. Int. Ed.* **2006**, *45*, 7763.

19. Debler, E. W.; Kaufmann, G. F.; Meijler, M. M.; Heine, A.; Mee, J. M.; Pljevaljcic, G.; Di Bilio, A. J.; Schultz, P. G.; Millar, D. P.; Janda, K. D.; Wilson, I. A.; Gray, H. B.; Lerner, R. A. *Science* **2008**, *319*, 1232.

20. Miller, L. W.; Cai, Y.; Sheetz, M. P.; Cornish, V. W. *Nat. Methods* **2005**, *2*, 255.

21. Calloway, N. T.; Choob, M.; Sanz, A.; Sheetz, M. P.; Milller, L. W.; Cornish, V. W. *ChemBioChem* **2007**, *8*, 767.

22. Wombacher, R.; Heidbreder, M.; van de Linde, S.; Sheetz, M. P.; Heilemann, M.; Cornish, V. W.; Sauer, M. *Nat Methods* **2010**, *7*, 717.

23. Hoskins, A. A.; Friedman, L. J.; Gallagher, S. S.; Crawford, D. J.; Anderson, E. G.; Wombacher, R.; Ramirez, N.; Cornish, V. W.; Gelles, J.; Moore, M. *Science* **2011**, *331*, 1289.

24. Oster, G.; Nishijima, Y. *J. Am. Chem. Soc.* **1956**, *78*, 1581.

25. Silva, G. L.; Ediz, V.; Armitage, B. A.; Yaron, D. *J. Am. Chem. Soc.* **2007**, *129*, 5710.

26. Duxbury, D. F. *Chem. Rev.* **1993**, *93*, 381.

27. Lee, L. G.; Chen, C.; Liu, L. A. *Cytometry* **1986**, *7*, 508.

28. Babendure, J. R.; Adams, S. R.; Tsien, R. Y. *J. Am. Chem. Soc.* **2003**, *125*, 14716.

29. Constantin, T.; Silva, G. L.; Robertson, K. L.; Hamilton, T. P.; Fague, K. M.; Waggoner, A. S.; Armitage, B. A. *Org. Lett.* **2008**, *10*, 1561.

30. Szent-Gyorgyi, C.; Schmidt, B. F.; Creeger, Y.; Fisher, G. W.; Zakel, K. L.; Adler, S.; Fitzpatrick, J. A.; Woolford, C. A.; Yan, Q.; Vasilev, K. V.; Berget, P. B.; Bruchez, M. P.; Jarvik, J. W.; Waggoner, A. *Nat. Biotechnol.* **2007**, *26*, 235.

31. Özhalici-Ünal, H.; Lee Pow, C.; Marks, S. A.; Jesper, L. D.; Silva, G. L.; Shank, N. I.; Jones, E. W.; Burnette, J. M., III; Berget, P. B.; Armitage, B. A. *J. Am. Chem. Soc.* **2008**, *130*, 12620.

32. Zanotti, K. J.; Silva, G. L.; Creeger, Y.; Robertson, K. L.; Waggoner, A. S.; Berget, P. B.; Armitage, B. A. *Org. Biomol. Chem.* **2011**, *9*, 1012.

33. Yarmoluk, S. M.; Lukashov, S. S.; Ogul'chansky, T. Y.; Losytskyy, M. Y.; Kornyushyna, O. S. *Biopolymers* **2001**, *62*, 219.

34. Toutchkine, A.; Nguyen, D.-V.; Hahn, K. M. *Org. Lett.* **2007**, *9*, 2775.

35. Toutchkine, A.; Kraynov, V.; Hahn, K. *J. Am. Chem. Soc.* **2003**, *125*, 4132.

36. Shank, N. I.; Zanotti, K. J.; Lanni, F.; Berget, P. B.; Waggoner, A. S.; Armitage, B. A. *J. Am. Chem. Soc.* **2009**, *131*, 12960.

37. Horiuchi, H.; Ishibashi, S.; Tobita, S.; Uchida, M.; Sato, M.; Toriba, K. -i.; Otaguro, K.; Hiratsuka, H. *J. Phys. Chem. B* **2003**, *107*, 7739.

38. Szent-Gyorgyi, C.; Schmidt, B. F.; Fitzpatrick, J. A. J.; Bruchez, M. P. *J. Am. Chem. Soc.* **2010**, *132*, 11103.

39. Senutovitch, N.; Stanfield, R. L.; Bhattacharyya, S.; Rule, G. S.; Wilson, I. A.; Armitage, B. A.; Waggoner, A. S.; Berget, P. B. *Biochemistry* **2012**, *51*, 2471.

40. Falco, C. N.; Dykstra, K. M.; Yates, B. P.; Berget, P. B. *Biotechnol. J.* **2009**, *4*, 1328.

41. Fisher, G. W.; Adler, S. A.; Fuhrman, M. H.; Waggoner, A. S.; Bruchez, M. P.; Jarvik, J. W. *J. Biomol. Screen.* **2010**, *15*, 703.

42. Holleran, J. P.; Brown, D.; Fuhrman, M. H.; Adler, S. A.; Fisher, G. W.; Jarvik, J. W. *Cytometry A* **2010**, *77A*, 776.

43. Grover, A.; Schmidt, B. F.; Salter, R. D.; Watkins, S. C.; Waggoner, A. S.; Bruchez, M. P. *Angew. Chem. Int. Ed.* **2012**, *51*, 4838.

44. Fitzpatrick, J. A. J.; Yan, Q.; Sieber, J. J.; Dyba, M.; Schwarz, U.; Szent-Gyorgyi, C.; Woolford, C. A.; Berget, P. B.; Waggoner, A. S.; Bruchez, M. P. *Bioconjug. Chem.* **2009**, *20*, 1843.

45. Hanes, J.; Schaffitzel, C.; Knappik, A.; Plückthun, A. *Nat. Biotechnol.* **2000**, *18*, 1287.

7

DESIGN OF ENVIRONMENTALLY SENSITIVE FLUORESCENT NUCLEOSIDES AND THEIR APPLICATIONS

SUBHENDU SEKHAR BAG

Department of Chemistry, Indian Institute of Technology, Guwahati, India

ISAO SAITO

Department of Materials Chemistry and Engineering, School of Engineering, Nihon University, Koriyama, Japan

7.1 INTRODUCTION

Unraveling the structure, functions, dynamics, and intermolecular interactions of biological macromolecules such as nucleic acids and proteins in many cases rely on the development of new fluorescence-based technology.[1] Fluorescence-based detection technologies are now widely used in numerous applications such as DNA sequencing,[2] *in situ* genetic analysis, single nucleotide polymorphisms (SNPs) typing through techniques such as fluorescence in situ hybridization (FISH),[3] and high-throughput screening (HTS)[4] as well as for many types of bioimaging. However, many of the biomolecules are complex in nature and do not show any inherent emissive properties in their intricate interactions. Excluding a few amino acids, ordinary biomacromolecules lack significant fluorescence properties. For example, the intrinsic fluorescence of the naturally occurring nucleic acid bases in DNA and RNA is extremely weak. These bases exhibit very short fluorescent

Fluorescent Analogs of Biomolecular Building Blocks: Design and Applications, First Edition.
Edited by Marcus Wilhelmsson and Yitzhak Tor.
© 2016 John Wiley & Sons, Inc. Published 2016 by John Wiley & Sons, Inc.

decay times in the range of a few picoseconds and do not provide much structural information since signals are normally averaged over all bases in the oligonucleotide sequence.

Therefore, the development of ideal fluorescent probes replacing the biomolecular building blocks, especially fluorescent DNA nucleoside analogs in having solvatofluorochromic properties for monitoring microenvironmental changes around nucleic acids is very important for understanding biological events associated with biomolecular interactions.[5] In particular, monitoring the change of local microenvironments such as dielectric properties of DNA surface and the inside DNA is highly important for understanding structures, functions, and dynamics of DNA and biomolecular interactions. In such a probing scenario, the fluorophore's emission property may be modulated in a way that the fluorescence can be enhanced or quenched and the emission can be red- or blue shifted, thereby enabling a visual observation of the fluorescence for studying structures, dynamics, and functions of biomacromolecules.

An ideal probe for monitoring various structures and dynamics of such biomolecules and its surroundings should be sensitive to its local microenvironment, including pH, viscosity, biological analytes, and solvent polarity, and should interact strongly with the biomolecules via electrostatic interaction, H-bonding, and other noncovalent interactions. Also, the design of small organic fluorophores having strong absorption and long emission wavelength has attracted much current attention because they can extract inner biomolecular information when conjugated to or complexed with a biomolecule. In this respect, highly suitable fluorescent analogs are mostly solvatochromic fluorophores that are being widely used as reporter probes for investigating chemical and biochemical phenomena because of their ability in sensing a small variation in dielectric constants within a biomolecular microenvironment.[6] Also, long-wavelength emission, especially emission in the visible region, is highly desirable; otherwise, the autofluorescence from biological macromolecules would inhibit the detection sensitivity.

In this chapter, we describe the characteristics of environmentally sensitive fluorescent (ESF) nucleosides and fluorescently labeled oligonucleotide probes developed in our laboratory that are useful for structural studies of DNA and RNA.

7.1.1 Solvatochromic Fluorophores

The phenomenon of solvent polarity-sensitive emission by a fluorophore is known as solvatochromism and the fluorophore is said to be solvatochromic.[7] For such fluorophores, the dipole moment of the excited state is greater than that of the ground state. Therefore, rearrangement of solvent molecules can lower the energy of the excited state prior to emission, resulting in a redshift of the emission maximum. Solvatochromic fluorophores exhibit emission properties such as emission wavelengths, quantum yields, and fluorescence lifetimes that are highly sensitive to their immediate environment. Such dynamic behavior makes these parameters particularly well suited for investigating biomolecular interactions because it provides information on the state of a biomolecule.[8] For example, if a solvatochromic fluorophore is appended to a nucleic acid or the surface of a protein or a membrane at a site that is involved

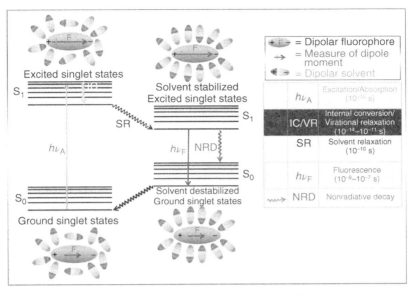

Figure 7.1   Origin of solvatochromic effects on fluorescence.

in a transient binding interaction or that undergoes a conformational change, then the probe will report such events provided that they are coupled to modifications in the local solvent sphere.

7.1.2   Origin of Solvatochromism

The effects of solvent polarity on the emission properties of a solvatochromic fluorophore are generalized in the Jablonski diagram, which depicts the energies of the different electronic states of the system (Fig. 7.1). As shown in the diagram, the dipolar fluorophore resides in the ground electronic state, S_0, surrounded by a sphere of polar solvent molecules. Upon absorption of light energy ($h\nu_A$), the fluorophore-solvent system is rapidly promoted to singlet excited state, S_1. During this event, the system adopts a new electronic configuration with an increased dipole moment that differs significantly from that of the ground state. This process of electronic excitation of solvent bound fluorophore occurs in a much faster rate than that of the motions of atomic nuclei (Frank–Condon principle).[9] Then, the solvent spheres reorient dipoles to accommodate the generated larger dipole of the fluorophore in the picoseconds timescale leading to the development of highly ordered arrangement in S_1 state. This step is known as solvent relaxation, which ultimately lowers the energy of the excited singlet state while simultaneously destabilizing the ground state. Therefore, the energy gap between the S_1 and S_0 state decrease significantly. The system finally returns to the ground state with the emission of a photon of a much longer wavelength (lower energy, $h\nu_F$; a fluorescence event) than that of the originally absorbed photon during excitation. Thus, the solvatochromic

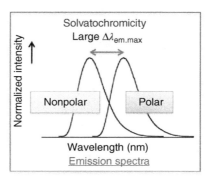

Figure 7.2   Schematic representation of solvatochromicity.

fluorophores emit at a much longer wavelength with increased solvent polarity as the degree of solvent relaxation increases (Fig. 7.2).

As the energy gap between the S_1 and S_0 states is reduced, there is a high possibility that solvatochromic fluorophores exhibit a marked increase in nonradiative decay (NRD). Thus, in many instances, the fluorophore returns spontaneously to the ground electronic state through a thermal nonradiative decay process (NRD, k_{nr}) that competes with fluorescence. This effect is much more prominent particularly in polar protic solvents such as water, methanol, and results in a decrease in the fluorescence quantum yield. The mechanisms for such processes include a range of events such as Intramolecular charge transfer (ICT), H-bonding, tautomerization, isomerization, and intersystem crossing to an excited triplet state (T_1). In such a competing scenario in polar solvents, perturbing the ordered solvent sphere, one can exploit the fluorophore to show sensitive switch-like emission property.

7.2   SOLVATOCHROMIC FLUORESCENT NUCLEOSIDE ANALOGS

The natural nucleobases only absorb light in the ultraviolet region and are not responsive to visible light nor can they be detected by normal fluorescence detection techniques as the intrinsic fluorescence of the naturally occurring nucleotide bases in DNA and RNA is extremely weak. These bases exhibit very short fluorescent decay times, generally in the range of a few picoseconds, and do not provide much structural information since signals are normally averaged over all bases in the oligonucleotide sequence. Thus, the lack of naturally occurring fluorescent bases has spurred the development of new fluorescent nucleosides that could be a probe molecule for DNA analysis. These nucleosides can be so designed as to get optimized fluorescence properties, that is, higher quantum yields and longer lifetimes.

7.2.1   Designing Criteria for Solvatochromic Fluorescent Nucleosides

To design an environmentally sensitive fluorophore, one needs to consider the following:

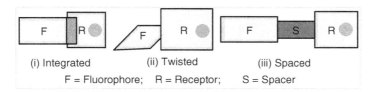

Figure 7.3   Three different ways of connecting fluorophores (F) and receptors (R).

(a) *Key Physical Parameters:* High molar extinction coefficient, high emission intensity, high-fluorescence brightness factor ($\varepsilon\Phi$), long excitation and emission wavelengths, high quantum yields, large Stokes shift, compatible size, hydrophobicity, hydrophilicity, and high stability.

(b) *Donor–Acceptor Units:* Environmentally sensitive fluorophores frequently possess an electron donor (**D**) and an electron acceptor (**A**) substituents attached to the same aromatic ring system, or linked via a linker or spacer. There are mainly three ways of connecting the donor and acceptor units as shown in Figure 7.3.

(c) *Possess ICT Character:* The electron donating and accepting groups on the aromatic ring confer a low-energy excited state with marked charge transfer character (ICT).

(d) *Solubility:* For biological application, a fluorophore should be water soluble.

For probing DNA structure, the ideal fluorescent nucleoside

(a) should have a bright fluorescence, which is sensitive to its local environment;

(b) should have a large Stokes shift;

(c) should be amenable to phosphoramidite chemistry for incorporation into oligonucleotides by solid-phase synthesis;

(d) should not disrupt duplex formation and should mimic one of the regular nucleosides. In general, maintaining Watson–Crick and Hoogsteen base-pairing hydrogen bonding is an important aspect in the design and synthesis of fluorescent nucleoside base analogs;

(e) should behave as a regular nucleoside in its interaction with proteins and enzymes;

(f) should be capable of being converted into the triphosphate and be incorporated into DNA with high efficiency by current commercial polymerases.

7.3   FLUORESCENTLY LABELED NUCLEOSIDES AND OLIGONUCLEOTIDE PROBES: COVALENT ATTACHMENT OF SOLVATOCHROMIC FLUOROPHORES ONTO THE NATURAL BASES

As discussed earlier, one of the most important strategies to install the fluorescence property into the natural nucleosides for biochemical application is to attach a fluorophore through a rigid acetylene linker. Such nucleosides are called fluorescently labeled nucleosides. Oligonucleotides containing such labeled nucleosides can widely be used in a variety of genomic assays including gene quantitation, allelic discrimination, expression analysis, and SNP typing.[10] Many assays exploit such fluorescently labeled oligonucleotide probes in which a biochemical event, such as DNA hybridization, causes an increase in fluorescence intensity. A rigid carbon linker is more advantageous compared to a flexible chain as the rigid linker has the ability to restrict the free motion of the attached fluorophore, leading to a finite disposition either inside the groove (groove-binding) or between base pairs (intercalation).

The lack of naturally occurring fluorescent bases has spurred the development of new nonnatural fluorescent nucleosides, which would be the probe for DNA analysis. These nucleosides can be designed to get optimized fluorescence properties, such as higher quantum yields and longer lifetimes, and can be incorporated into an oligonucleotide using standard automated synthetic methods. Typical examples of some covalently linked fluorescence nucleosides developed in our laboratory are shown in Figure 7.4.

7.3.1   Base-Discriminating Fluorescent Nucleosides (BDF)

Since the last decade we are engaged in designing covalently labeled nucleobases for SNPs typing in homogeneous solution. These bases are known as base-discriminating

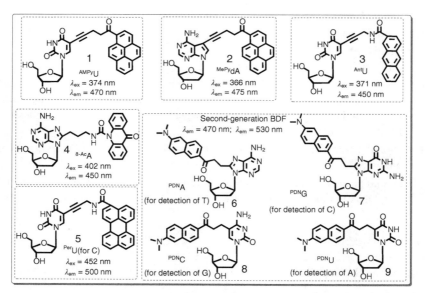

Figure 7.4   Examples of covalently linked fluorescent nucleosides developed in our laboratory.

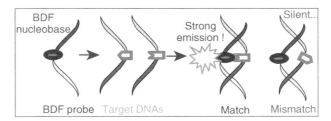

Figure 7.5   Schematic illustration for homogeneous SNP typing using BDF probe.

fluorescent (BDF) nucleosides,[11] which can be synthesized by attaching the fluorophores onto the natural nucleobases through an acetylenic linker via Sonogashira coupling protocol. The main aim to design BDF nucleosides was to dispose the fluorophoric unit of the BDF-labeled oligonucleotide probe either in the groove leading to groove-binding event or between a base pair leading to an intercalation in a DNA duplex. Exploiting these nucleosides, a number of conceptually new techniques for DNA detection have also been developed. The labeled nucleosides upon incorporation into an oligonucleotide are capable of detecting the change of microenvironment within the DNA and can report the presence of an opposite base in a target DNA via the generation of an enhanced fluorescence signal.

7.3.1.1   Design Concept of Base-discriminating Fluorescent Nucleosides: The
concept of BDF probe is based on the fluorescence change of the BDF base itself in response to an opposite base on a complementary target DNA strand (Fig. 7.5). When an oligonucleotide probe containing a BDF nucleoside hybridized to a fully matched complementary target DNA, the fluorophore has to reside outside the groove exposing itself to the more polar aqueous microenvironment resulting in a strong fluorescence signal. On the contrary, because of the lack of Watson–Crick base pairing in the mispaired position in the mismatched duplexes, the fluorophore remains inside the duplex facing highly hydrophobic microenvironment exhibiting a very weak fluorescence signal.[11h]

This concept would be very important and useful for devising new SNP typing methods. Due to the fluorescence change of BDF nucleobases, the bases on the complementary strands can be fluorimetrically read out without separation and washing steps. Therefore, the SNP typing method using BDF probes constitutes a very powerful homogeneous assay that does not require enzymes or time-consuming steps for fluorescence labeling during PCR and avoids hybridization errors.

7.3.1.2   Pyrene-Labeled BDF Probes
Fluorimetric Sensing by Four BDF Nucleobases: The pyrenecarbonyl derivatives, such as pyrene-1-carboxyaldehyde, show a strong solvatochromicity.[12] Thus, it was envisaged that if the pyrenecarbonyl fluorophore is attached to uracil at the C-5 position via a rigid acetylenic linker, the fluorophore has to extrude outside of the groove due to the Watson–Crick base pairing with the matched adenine base. Therefore, the

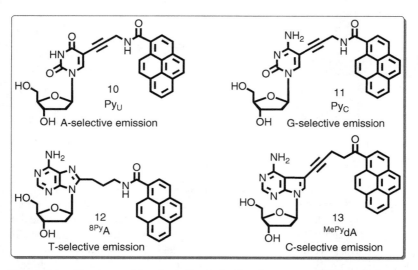

Figure 7.6   Structures of four base-discriminating fluorescent (BDF) nucleobases.

fluorophore has to face more polar aqueous microenvironment leading to a strong fluorescence emission that is selective for A by the pyrene-labeled BDF probe. On the other hand, when the pyrene fluorophore is engaged in intercalation due to the lack of base pairing in a mismatched DNA duplex, the BDF base would have a quenched emission because of the location of the fluorophore in a hydrophobic site.

On the basis of this concept, a new type of pyrene-labeled BDF nucleosides ^{Py}U and ^{Py}C were designed to detect opposite bases A and G, respectively (Fig. 7.6).[13] Thus, it was observed that these BDF bases, ^{Py}U and ^{Py}C, exhibited unique fluorescence properties depending on the nature of the base on the complementary strand and are capable of distinguishing A and G opposite to the BDF base, respectively, by a sharp change in fluorescence intensity (Fig. 7.7).

Such a sharp change in fluorescence intensity is mainly because of the difference in polarity of the microenvironments near the pyrenecarboxamide moieties of BDF bases, ^{Py}U and ^{Py}C. The difference in polarity was also supported by the energy-minimized structures for the duplexes containing a ^{Py}U/A or a ^{Py}U/G base pair (Fig. 7.8).

On the other hand, as was expected for the C analog, ^{Py}C exhibited a strong G-selective fluorescence emission. Thus, the fluorescence emission from the matched duplex, 5'-d(CGCAAC^{Py}CCAACGC)-3'/5'-d(GCGTTGNGTTGCG)-3' was found to be highly G-selective, which is unusual because of the quenching nature of G-base in general.[13]

BDF nucleosides, ^{8Py}A (**12**) and ^{MePy}A (**13**), were also designed, and they emit strong fluorescence when the opposite bases are T and C, respectively.[11h,14] The purine-type BDF probe containing ^{8Py}A selectively emits fluorescence only when the base opposite to BDF nucleoside is thymine and acts as effective reporter probes

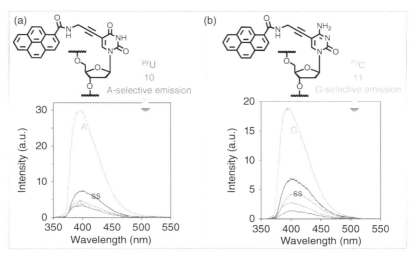

Figure 7.7 (a) The structures of ^{Py}U and the fluorescence spectra of the BDF probes hybridized with ODN with A, C, T, or G base opposite to BDF base. (b) Structures of ^{Py}C and the fluorescence spectra of the BDF probes hybridized with ODN with A, C, T, or G base opposite to the BDF base. "ss" denotes a single-stranded probe.

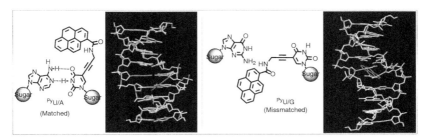

Figure 7.8 The energy-minimized structures of the duplexes containing a $^{Py}U/A$ or a $^{Py}U/G$ base pair. (*See color plate section for the color representation of this figure.*)

for homogeneous SNP typing (Fig. 7.9). The clear fluorescence change observed here is very useful for SNP typing.

These BDF probes enabled us to distinguish single-base alterations by simply mixing with a sample solution of target DNA and by reading the fluorescence signal. ^{Py}U-containing BDF probe was applied, for example, for discrimination of SNPs in human *c-Ha-ras* SNP sequence, which possesses a **C/A** SNP site.[13] On hybridization of the BDF probe of sequence, 5′-d(GGCGCCG^{Py}UCGGTGTG)-3′, with the target complementary sequences, the BDF probe showed an **A**-allele-specific fluorescence emission at 398 nm when excited at 344 nm. The fluorescence signal from the probe ODN was negligible upon hybridization with a target **C**-allele sequence opposite to the BDF base.

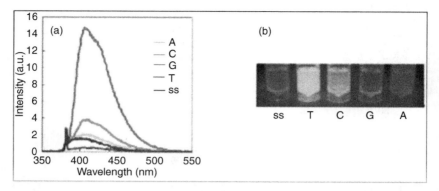

Figure 7.9 (a) Fluorescence spectra of ODN (^{8py}A)(2.5 µM) hybridized with ODN(N = A, G, T, or C, 2.5 µM), and single-stranded ODN (^{8py}A) (50 mM sodium phosphate, 0.1 M sodium chloride, pH 7.0., room temperature, λ_{ex} = 381 nm, "ss" = single-stranded BDF probe). (b) Fluorescence color change observed when illuminated with a 365-nm transilluminator. (*See color plate section for the color representation of this figure.*)

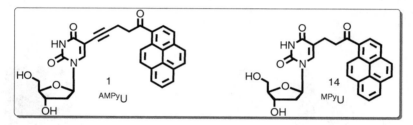

Figure 7.10 The structures of BDF nucleosides ^{AMPy}U and ^{MPy}U.

Alkanoyl Pyrene-Labeled Polarity-Sensitive BDF Probes: We have developed novel alkanoylpyrene-labeled BDF nucleosides, ^{AMPy}U (**1**) and ^{MPy}U (**14**) (Fig. 7.10). The nucleosides exhibit strong fluorescence emission at long wavelength that are highly sensitive to solvent polarity. BDF probes containing ^{AMPy}U selectively emit fluorescence when the base opposite to BDF nucleoside is adenine and act as effective reporter probes for homogeneous SNP typing.[11f] The homogeneous SNP typing method using ^{AMPy}U-containing BDF probes is a powerful alternative to conventional SNP typing as well as gene detection (Fig. 7.11).

Recently, we reported the design of labeled probe for SNPs detection via "Just-Mix and Read" strategy based on the matched/mismatched and fluorescence readout strategy. Photophysical properties of nucleoside $^{Oxo-Py}$U (**15**) revealed it as a highly polarity-sensitive molecule as shown in Figure 7.12.[11i]

Pyrene-Labeled BDF Probe: G-Specific Fluorescence Quenching: Guanine exhibits exceptionally high quenching efficiency toward many different types of fluorescent labels.[15] For the use of BDF probes, this quenching is considered as a drawback for

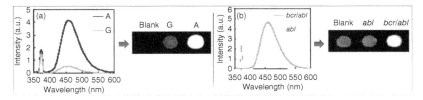

Figure 7.11 (a) Fluorescence spectra of ODN$_{ALDH2}$ (^{AMPy}U) hybridized with ODN$_{ALDH2}$ (A or G) and corresponding fluorescence image. (b) Fluorescence spectra of ODN$_{bcr/abl}$ (^{AMPy}U) hybridized with ODN$_{bcr/abl}$(A) or ODN$_{abl}$(G) and corresponding fluorescence image.

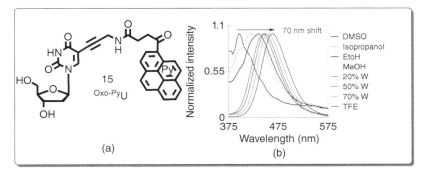

Figure 7.12 Structures of $^{Oxo-Py}$U and its solvatochromicity. (*See color plate section for the color representation of this figure.*)

SNP typing protocol. However, the G-specific quenching process can be used as an advantage in single base discrimination, if this quenching process could be switched on and off through base pairing with a BDF base. With this idea, we have developed a new strategy for the detection of single-base alterations through fluorescence quenching by guanine (G).[11g]

We devised a BDF oligonucleotide probe containing fluorescent nucleoside, 4'-pyrenecarboxamide-modified thymidine, $^{4'Py}$T (**16**, Fig. 7.13a), which exhibits intense fluorescence only when the $^{4'Py}$T of the probe is involved in a base pairing with A of a target complementary ODN.[11g] Pyrenecarboxamide was selected as the fluorophore due to its intercalative activity and its efficient quenching by G (Fig. 7.13).[16] Stable base pairing with A locates the fluorophore in the minor groove, where the fluorophore escapes efficient quenching by the flanking G bases. The pyrene group cannot intercalate into the DNA duplex in keeping with the base pairing, due to the short methylamide linker in the 4'-position. In contrast, when the complementary base of $^{4'Py}$T is mismatched (T, G, or C), then the hydrophobic pyrenyl group is likely to intercalate with the π-stacked DNA helix, breaking the weak hydrogen bonds. This intercalation enables the fluorophore to come into intimate contact with the flanking base pairs and results in the quenching of the fluorescence.[11g]

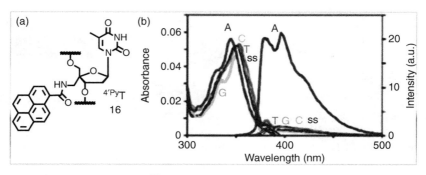

Figure 7.13 (a) Structure of $^{4'Py}$T. (b) Absorption and fluorescence spectra of $^{4'Py}$T-containing BDF probe hybridized with an ODN possessing A, C, T, or G base opposite to the $^{4'Py}$T base; "ss" denotes a single-stranded probe. From Ref. [16a–c].

Detection of Base Insertion by Pyrene Excimer Fluorescence: Insertion/deletion (indel) polymorphisms occupy approximately 10% of all the polymorphisms in the human genome[17] and lead to serious gene expression errors because they often cause a translational frame shift and create premature proteins. The development of pyrene-labeled ODNs suitable for the detection of extra bases would make it possible to judge the presence/absence of indel polymorphisms located at a specific site on a target DNA by simply hybridizing with target DNA.

Okamoto *et al.*, have designed a ^{Py2}**Lys**-containing (**17**) ODN, 5′-d(GTGTTAAG-CC^{Py2}**Lys**GCCAATATGT)-3′.[18] With excitation at 350 nm, the fluorescence of the single-stranded ODN was negligible. When ^{Py2}**Lys**-containing ODN was hybridized with ODN 5′-d(ACATATTGGCGGCTTAACAC)-3′, which does not possess the base opposite to ^{Py2}**Lys**, the fluorescence was still weak. In contrast, the fluorescence spectrum of the duplex with 5′-d(ACATATTGGCAGGCTTAACAC)-3′, where A is the base opposite to ^{Py2}**Lys**, had a strong fluorescence peak at 495 nm that is corresponding to the fluorescence wavelength from a pyrene excimer (Fig. 7.14).

The clear change in the fluorescence that depends on the presence or absence of the inserted base opposite to ^{Py2}**Lys** (A bulge) is useful for the detection of insertion polymorphisms (Fig. 7.14a). They tested the detection of an insertion mutation by hybridization of ^{Py2}**Lys**-containing probe using the coding sequence of the epithelial sodium channel *b* subunit (*b*ENaC) gene associated with Liddle's syndrome, which is an autosomal dominant form of hypertension with variable clinical expression.[19] Thus, the ^{Py2}**Lys**-containing probe, 5′-d(CTCACTGGGGTAGGGCCCAGT^{Py2}**Lys**GTTGGGGCT)-3′, was prepared and this probe was hybridized with *b*ENaC gene sequences, 5′-d(AGCCCCAAC(**G**)$_n$ ACTGGGCCCTACCCCAGTGAG)-3′ (wild type, $n = 0$; **G**-inserted mutant, $n = 1$). The fluorescence emission from the duplex with a G-inserted strand was very strong and clearly distinguishable from the poor fluorescence of the duplex with a wild-type strand (Fig. 7.14b). Therefore, the hybridization of ^{Py2}**Lys**-containing ODN with a target DNA facilitates the determination of the presence/absence of insertion

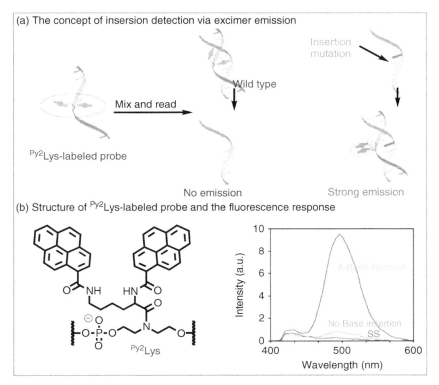

(a) The concept of insersion detection via excimer emission

Insertion mutation

Wild type

Mix and read

Py2Lys-labeled probe

No emission   Strong emission

(b) Structure of Py2Lys-labeled probe and the fluorescence response

Py2Lys

Figure 7.14 (a) Schematic representation of the concept. (b) Structure of **Py2Lys** (**17**) and the fluorescence spectra of **Py2Lys**-containing probe hybridized with an ODN with and without A bulge. "ss" denotes a single-stranded probe.

polymorphisms located at a specific site on the target DNA by a simple mixing strategy.[18]

7.3.1.3 Anthracene-Labeled BDF Nucleosides: Just like pyrene, anthracene is also known as an intercalator of DNA duplex. The fluorescence of 2-anthracenecarboxaldehyde and 2-acetylanthracene showed a strong dependence on solvent polarity.[20] As the solvent polarity increases, the emission intensity of these two anthracene derivatives increases and emit at a longer wavelength region. Thus, the anthracene-based fluorescent probe would also be attractive in sensing the microenvironment inside and outside the DNA duplexes.

Anthracene-containing BDF nucleoside, **2-Anth**U (**3**) was, therefore, prepared and the fluorescence behavior upon hybridization with their various target oligonucleotide sequences was examined.[21] The **2-Anth**U showed a strong fluorescence dependency on solvent polarity and is suitable as an effective probe for the DNA microenvironment. The BDF oligonucleotide probes selectively emit fluorescence only when the base opposite to BDF base is adenine (Fig. 7.15a and b).

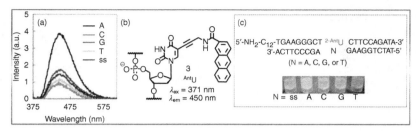

Figure 7.15 (a) Fluorescence spectra of ODN$_{abl}$ ($^{2\text{-Anth}}$U) hybridized with ODN$_{abl}$ (2.5 μM, **N = A**, full much), ODN$_{abl}$ (**N = T, G**, or **C**), and single-stranded ODN$_{abl}$ ($^{2\text{-Anth}}$U). Excitation wavelength was at 371 nm. (b) Structure of BDF base $^{2\text{-Anth}}$U. (c) Comparison of the fluorescence for the bases opposite to $^{2\text{-Anth}}$U (50 mM sodium phosphate, 0.1 M sodium chloride, pH 7.0, room temperature). "ss" denotes a single-stranded BDF probe. The sample solutions were illuminated with a 365-nm transilluminator. (*See color plate section for the color representation of this figure.*)

It should be mentioned that the fluorescence emission of $^{2\text{-Ant}}$U-containing BDF probes does not depend on the flanking base pairs. We incorporated BDF base $^{2\text{-Anth}}$U into a probe, ODN$_{abl}$ ($^{2\text{-Anth}}$U), [5′-NH$_2$-C$_{12}$-d(TGAAGGGCT$^{2\text{-Ant}}$UCTTCCAGA-TA)-3′] for the detection of the *abl* cancer gene. Thus, upon mixing of ODN$_{abl}$ ($^{2\text{-Anth}}$U), with the target matched ODN$_{abl}$ (**N = A**) [5′-d(TATCTGGAAGNAGCCC-TTCA)-3′], we observed a strong emission. For the mismatched duplexes, ODN$_{abl}$ ($^{2\text{-Anth}}$U)/ODN$_{abl}$ (**N = T, G**, or **C**), the emission was negligible (Fig. 7.15c). Thus, the $^{2\text{-Ant}}$U-containing BDF probe was shown to be effective for sensing adenosine on a target DNA by a fluorescence change at a longer wavelength than pyrenecarboxamide-labeled BDF probe.

7.3.1.4 Perylene-Labeled Nucleoside (^{Per}U):

We synthesized the perylene-containing fluorescent nucleoside ^{Per}U (**5**) and developed perylene-labeled oligonucleotide probes, d(5′-CGCAAC^{Per}UCAACGC-3′) and d(5′-CGCAAT^{Per}UTAAC-GC-3′), both of which are capable of detecting mismatched cytidine base (**C**) from the corresponding complementary target at a wavelength of 500 nm, having a long tail to 590 nm by a fluorescence enhancement (Fig. 7.16).[22]

The fluorescence enhancement was observed only when the base opposite to the ^{Per}U is cytidine irrespective of the sequences.[22] This observation is quite different from that of the pyrene-labeled BDF probes. It is assumed that the perylene moiety intercalates strongly when the complementary base is C. The optimized structures of the ^{Per}U containing homo- and hetero-duplexes showed that in fully matched duplexes the perylene carboxamide chromophore was extruded to the outside of the duplexes and thus exposed to a highly polar aqueous microenvironment. On the other hand, the duplexes containing ^{Per}U-C mismatched pair bound strongly via site-specific intercalative interaction to DNA bases and the aromatic rings of the perylene located in the major groove of the helix. The intercalated perylene itself resembles another base pair, being stacked between neighboring base pairs. Thus, in the case of mismatched duplexes, the perylene fluorophore is exposed to a

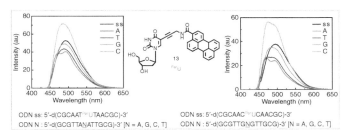

ODN ss: 5'-d(CGCAAT^PerUTAACGC)-3'   ODN ss: 5'-d(CGCAAC^PerUCAACGC)-3'
ODN N : 5'-d(GCGTTANATTGCG)-3' [N = A, G, C, T]   ODN N : 5'-d(GCGTTGNGTTGCG)-3' [N = A, G, C, T]

Figure 7.16   The structure of the perylene-labeled fluorescent nucleoside and the fluorescence behavior of a solution (2.5 μM) in a buffer (pH = 7.0) containing different ODNs; ss denotes single-stranded ODN and **A, T, G, C** denote double-stranded ODNs containing **A, T, G, C** bases opposite to ^{Per}U, respectively. Sequences are shown under the fluorescence spectra; (a) is for A/T pair flanking sequence, (b) structure of ^{Per}U, and (c) is for G/C pair flanking sequence. From Ref. [22]. Reprinted with permission of Elsevier. (*See color plate section for the color representation of this figure.*)

more hydrophobic microenvironment. It is clear that the microenvironment around the ^{Per}U is quite different in matched (^{Per}U-**A**) and mismatched (^{Per}U-**C**) cases irrespective of flanking base pairs. Discrimination based on mismatched base pair recognition offers an opportunity for selective targeting of mutated genes.

7.3.1.5   *DAN-Labeled Solvatochromic Fluorescent Nucleoside:*   Solvatochromic fluorophores such as 6-(dimethylamino)-2-acylnaphthalene (DAN) and its derivatives are among the most promising candidates for designing a dielectric-sensitive fluorescent probe for examining the change in the polarity of the DNA microenvironment and the DNA–protein interaction site.[23] In particular, when the fluorophore is tethered to a nucleobase in oligodeoxynucleotide (ODN) probes via a rigid linker, the change in the dielectric properties at a specific site on the target sequences would be easily monitored.

Okamoto *et al.* showed a method for mapping of the dielectric nature of the microenvironment around DNA by preparing a novel **DAN** family nucleoside, 6-(dimethylamino)-2-naphthalenecarboxamide (**DNC**)-tethered 2′-deoxyuridine (^{DNC}U, **18**, Fig. 7.17). This fluorescent nucleoside offers high sensitivity to solvent polarity. A series of fluorescence spectra for ^{DNC}U-containing ODN showed a large shift in wavelength, indicating a significant change in the dielectric properties of the microenvironment around ^{DNC}U. We also monitored the local dielectric properties of the DNA-binding region of DNA polymerase using the fluorescence change of ^{DNC}U-containing ODN probe.[24]

From the shift in fluorescence maxima (Fig. 7.17a) in solvent of different polarity, it is clear that solvatofluorochromic behavior was shown by ^{DNC}U in protic solvents, which were attributed to specific solute–solvent interactions such as hydrogen bonding. This solvatofluorochromic behavior was again clear from the plot of fluorescence maxima and the corresponding dielectric constant of the solvents and can be attributed to **DNC** chromophore-solvent interactions (Fig. 7.17b).[25]

Thus, ^{DNC}U nucleoside offers high-fluorescence sensitivity to the solvent polarity and is able to directly sense local dielectric properties by a change in fluorescence

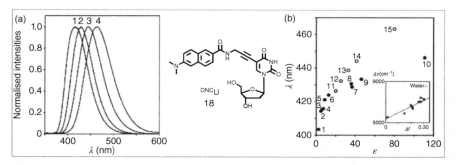

Figure 7.17 (a) Shift of fluorescence maxima of ^{DNC}U in different solvents; ethyl acetate (1); *N,N*-dimethylformamide (2); glycol (3); water (4). (b) A plot of emission maxima (λ_{fl}) of d^{DNC}U (2.5 μM) against dielectric constants (ε) of aprotic (filled circles) and protic solvents (open circles).

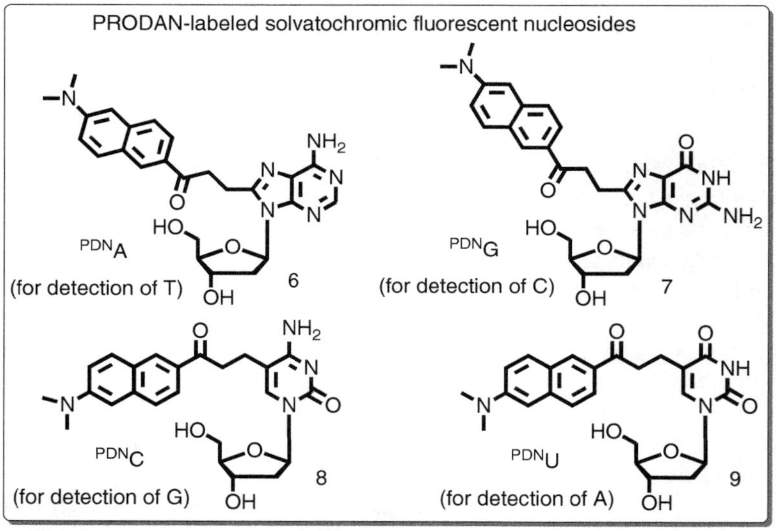

PRODAN-labeled solvatochromic fluorescent nucleosides

PDN$_A$
(for detection of T)   6

PDN$_G$
(for detection of C)   7

PDN$_C$
(for detection of G)   8

PDN$_U$
(for detection of A)   9

Figure 7.18 PRODAN-labeled fluorescent nucleosides.

signal. The ODN containing ^{DNC}U serves as an effective dielectric-sensitive fluorescent probe for monitoring the change of local dielectric properties on the interior of a DNA–protein complex and around the DNA surface.[24]

7.3.1.6 PRODAN-Labeled Fluorescent Nucleosides: Okamoto *et al.* have demonstrated a novel PRODAN-labeled fluorescent nucleobase, ^{PDN}U, (**9**, Fig. 7.18) for the mapping of the local dielectric properties around DNA.[26] The examination of photophysical properties of ^{PDN}U nucleoside in different solvents revealed that the fluorescence excitation spectra shifted depending on the nature of the solvent, which was attributed to specific interactions such as hydrogen bonding between ^{PDN}U and solvent molecules. It showed a large Stokes shifted emission with the change in solvent polarity. ^{PDN}U in protic solvents showed a larger solvatochromic shift, suggesting the existence of specific solute–solvent interactions, such as hydrogen bonding. Such solvatochromic behavior of ^{PDN}U made it possible to detect fluorimetrically the microstructural changes in DNA through the incorporation of ^{PDN}U into DNA and subsequent monitoring of the photochemical behavior.

Upon hybridization of the probe oligonucleotide containing ^{PDN}U with the perfectly matched target oligonucleotide, enhanced emission intensity was observed when excited at 450 nm, whereas the fluorescence emissions of the single-stranded probe and the duplexes containing single mismatched bases were suppressed. The fluorescence behavior of the probe ODN containing ^{PDN}U suggested that the effect of the nature of the neighboring base pair on the photochemical behavior of ^{PDN}U is not significant. Similarly, the ^{PDN}C-labeled oligonucleotide probe emit strongly when hybridized with its perfectly matched complementary strand, allowing the detection of a G base of a target opposite to ^{PDN}C (**8**, Fig. 7.18) of the probe ODN.[26]

7.4 NUCLEOSIDES WITH DUAL FLUORESCENCE FOR MONITORING DNA HYBRIDIZATION

Intramolecular charge transfer (ICT) often offers spectroscopists to detect a signal at a longer wavelength region and thereby to design a system for monitoring any perturbation in the local microenvironment around a biomolecule or to probe biomolecular events. It is also well known that a fluorophore-containing electron donor/acceptor unit such as 1-arylpyrene and 4-dialkylaminobenzonitrile derivatives shows a dual fluorescence, which originates from two relaxed singlet excited states, a locally excited (LE) state, and an intramolecular charge transfer (ICT) state.[27] Such fluorophores can be exploited in designing a fluorescent probe for monitoring the change in the microenvironment of biomolecules as they often sense the perturbation by a change in color.

To detect DNA hybridization by monitoring dual fluorescence, we tethered an N,N-dimethylaminophenyl-substituted pyrene derivative to 2'-deoxyuridine and incorporated it into a short oligonucleotide sequence. We observed that the dual fluorescence was effectively controlled into a duplex at ambient temperature, thus allowing us to probe DNA hybridization by a color change.[28] To assess the sensitivity of our designed probe, we have incorporated the nucleoside **19** (Fig. 7.19c) into a short oligonucleotide sequence ODN **1**, [5'-d(CGCAAT (**19**) TAACGC)-3'] and studied the photophysical behavior in the presence of target complementary strand ODN **2**, [5'-d(GCGTTAAATTGCG)-3']. For nucleoside **19**, a single fluorescence band was observed at 540 nm when excited at 380 nm (Fig. 7.19a). In contrast, for the single-stranded ODN **1** and the duplex ODN **1**/ODN **2**, another fluorescence band at a shorter wavelength (440 nm) but with highest intensity was observed in addition to the low intense band at 546 nm. This dual fluorescence was observed not only for a hybrid with a DNA strand but also for a DNA/RNA hybrid.[28]

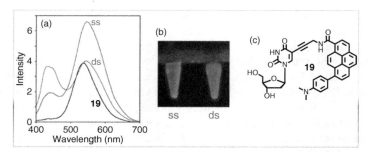

Figure 7.19   (a) Fluorescence spectra of **19** (7.3 μM), single-stranded ODN **1** (25 μM), and duplex ODN **1**/ODN **2** (25 μM) [50 mM sodium phosphate (pH 7.0); 0.1 M NaCl; 5 °C; $\lambda_{ex} = 380$ nm]. (b) Fluorescence image of the solutions of ODN **1** (25 μM, ss) and duplex ODN **1**/ODN **2** (25 μM, ds) illuminated with a 365 nm transilluminator. (c) Structure of *N,N*-dimethylaminophenyl-substituted pyrene derivative **19** with a dual fluorescence. (*See color plate section for the color representation of this figure.*)

The two fluorescence bands at 540 and 440 nm were assigned to the ICT and LE fluorescence bands, respectively, by investigating the photophysical behavior of the pyrene fluorophore, *N*-propargyl 8-(4′-(*N′,N′*-dimethylamino)phenyl)pyrene-1-carboxamide. Two structural states of ODN **1** showed different visible colors. An orange-colored fluorescence was observed for the single-strand ODN **1**, while duplex ODN **1**/ODN **2** emitted a pink fluorescence as a mixture of two colors of orange ICT and blue LE fluorescence (Fig. 7.19b). Thus, the change in the fluorescence color allowed us to monitor the change in the microenvironment around the fluorophore in the single-stranded state and in the duplex conformation, thus allowing the detection of target DNA. The conceptually new oligonucleotide probe containing dual fluorescent nucleoside, **19**, is capable of detecting the target DNA or RNA by a change in color originating from the LE and ICT state of the fluorophore.

7.5   APPROACH FOR DEVELOPING ENVIRONMENTALLY SENSITIVE FLUORESCENT (ESF) NUCLEOSIDES

7.5.1   Concept for Designing ESF Nucleosides

In contrast to previously described BDF nucleosides,[11] which show increasing or decreasing emission intensity at a fixed wavelength, ESF nucleosides are able to indicate the change in their local microenvironments by a drastic change in emission wavelength together with its intensity change. In order to devise such ESF nucleosides, molecular design of highly solvent-dependent fluorescent nucleobase is critically important. We describe below the design criteria for solvatofluorochromic nucleosides since ESF base should have high solvatochromicity.

1. ESF base should have high solvatofluorochromicity.
2. $\Delta\lambda_{em, max}$ of ESF base should be in the range of 50~120 nm from nonpolar to polar solvents. The schematic representation of solvatochromicity is shown in Figure 7.20.

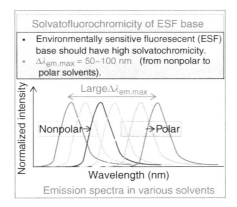

Figure 7.20   Schematic representation of solvatochromicity.

These criteria can be fulfilled by the following molecular design:

1. Extend the conjugation of nonemissive natural nucleobases with aromatic fluorophore via π bond.
2. Incorporate donor/acceptor system so as to induce ICT state (Fig. 7.21).
3. ICT state is strongly dependent on microenvironment such as polarity, hydrogen bonding, viscosity, and pH.
4. ICT state induced by donor/acceptor system is expected to produce large solvatochromicity.

Most of the efforts to impart useful photophysical properties upon nonemissive natural nucleobases involve the linking of natural nucleobases to fluorescent aromatic

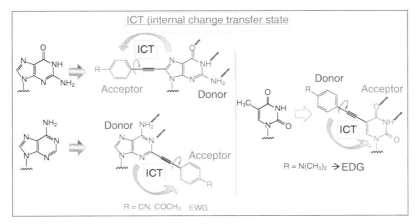

Figure 7.21   The schematic representation of donor- and acceptor-substituted nucleobases, which might show an efficient ICT emissive feature.

or heteroaromatic chromophores via an ethynyl linker[11h,29] or the direct attachment of fluorescent chromophores on the natural nucleobases.[5,30] By extensive studies on the substituent effects on 8-arylethynylated 2′-deoxyguanosine derivatives, we found that aromatics with donor/acceptor functional groups if linked via π-conjugation to the 8-position of guanosine and adenosine, 2-position of adenosine, and 5-position of deoxyuridines, can modulate the photophysical properties of the fluorescent nucleoside derivatives and might give rise to the design of a large library of solvatochromic fluorescent nucleosides.

7.5.2 Examples and Photophysical Properties of ESF Nucleosides

With the above design concept, several ESF nucleosides based on the combination of natural nucleobases and aromatic donor or acceptor chromophore units were synthesized and their solvatochromic emission properties were exploited. Environmentally sensitive donor/acceptor aromatics containing fluorescent deoxyuridine, deoxyadenosine, and deoxyguanosine derivatives were prepared mainly via palladium(0)-mediated Sonogashira cross-coupling reaction and their solvatochromicity was examined.

7.5.2.1 5-Substituted 2′-Deoxyuridine ESF Nucleoside Analogs: The synthesis of 5-arylethynylated-2′-deoxyuridines was achieved via Pd(0) mediated-Sonogashira coupling of 5-iodo-2′-deoxyuridine and the corresponding ethynylated aromatic derivatives (Scheme 7.1).[31] The photophysical study showed that 4-dimethylaminophenyl-substituted uridine (26) was solvatofluorochromic, whereas phenyl-substituted uridine (24) and 4-cyanophenyl-substituted uridine 25 did not show any solvatofluorochromicity (Fig. 7.22).

7.5.2.2 2′-Deoxyadenosine and Guanosine ESF Nucleoside Analogs
Synthesis of Purine ESF Nucleoside Analogs: The synthesis of 2-substituted adenosines is represented in Scheme 7.2. 3′, 5′-Dihydroxyl groups of guanosine (27) is protected with acetic anhydride and then 6-oxo-group was converted into chloro group by $POCl_3$ to afford 2-amino-6-chloro purine derivative (29). 2-Amino group of 2-amino-6-chloro purine derivative was converted into iodo derivative (30) by treating with amylnitrite and TMSI. 6-Chloro-2-iodo-purine was next converted into the amine by ammonia to afford 2-iodo-2′-deoxyadenosine, which was subsequently

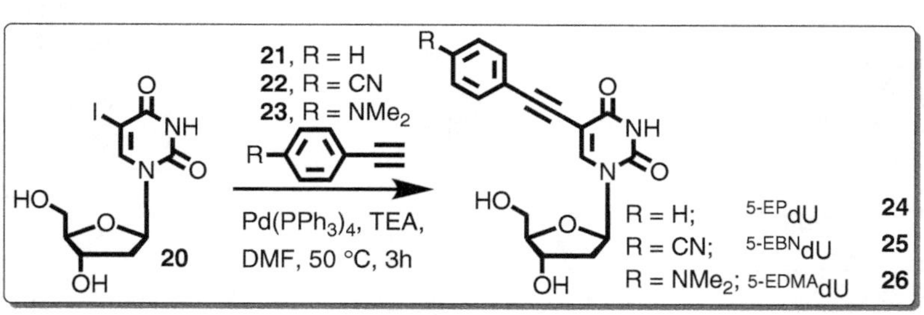

Scheme 7.1 Synthesis of 2′-deoxyuridine analogs of ESF nucleosides. From Ref. [31a,c].

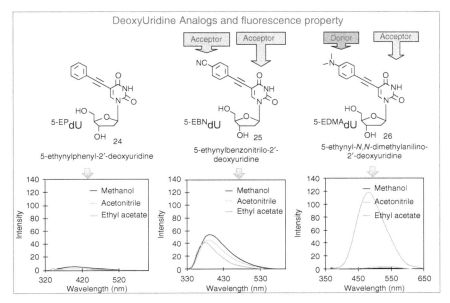

Figure 7.22 Donor- and acceptor-substituted 2′-deoxyuridine analogs of ESF nucleosides and their fluorescence spectra in different solvents of varying polarity.

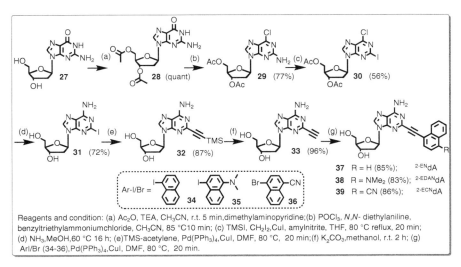

Scheme 7.2 Synthesis of 2-substituted 2′-deoxyadenosine ESF nucleoside analogs. From Ref. [32].

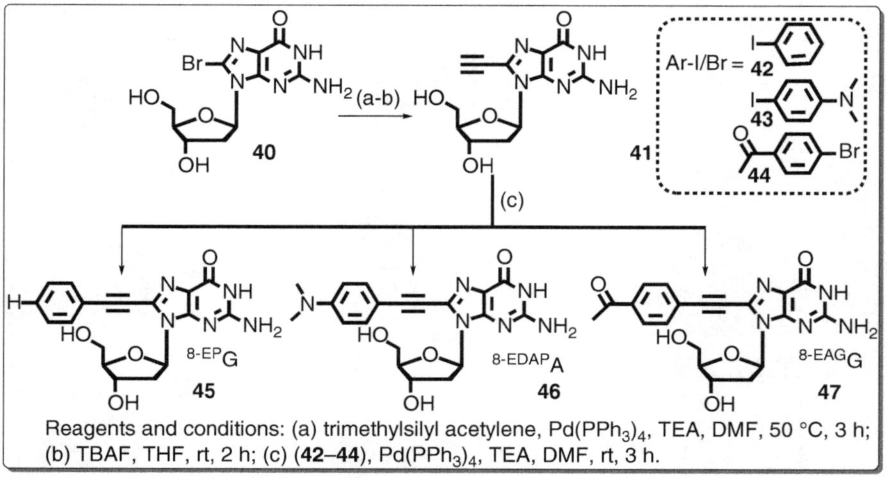

Scheme 7.3 Synthesis of 8-substituted 2′- deoxyguanosine ESF nucleoside analogs. From Refs. [33,34].

treated with TMS-acetylene under Sonogashira reaction conditions. TMS group of 2-trimethylsilylethynyl-2′-deoxyadenosine (**32**) was deprotected with K_2CO_3 and the alkynylated adenosines were treated separately with iodo- or bromo-substituted aromatics (**34–36**) under Sonogashira reaction conditions to afford 2-substituted 2′-deoxyadenosine derivatives (**37–39**) (Scheme 7.2).[32]

8-Arylethynylated-2′-deoxyadenosines or guanosines were prepared from the cross-coupling reaction of the corresponding 8-bromo-2′-deoxyadenosines or guanosines. By following Scheme 7.3, all the 8-arylethynylated 2′-deoxy guanosines or adenosines were prepared.[33,34]

8-Vinyl-substituted guanosine and adenosine derivatives were prepared by the route shown in Schemes 7.4 and 7.5, respectively. The amino group of 8-bromo-2′-deoxyguanosine **40** was protected by N,N-dimethylformamide diethylacetal to yield **48**, which was reacted with tetravinyltin (IV) under palladium (0)-mediated Stille coupling conditions to produce compound **49**. Compound **49** was coupled with 4-bromobenzene (**50**), 4-bromoacetophenone (**44**), and 4-cyanobromobenzene (**51**), respectively, under Heck reaction conditions, and the treatment with ammonia afforded acetylated (E)-8-styryl-deoxyguanosines **55–57** (Scheme 7.4).[34] The same strategy was utilized for the synthesis of 8-substituted styryl-2′-deoxyadenosine (**60**) (Scheme 7.5).

Looking at the high solvatochromic fluorescence property and wide usability of 6-propionyl-2-dimethylaminonaphthalene (**PRODAN**),[6,23,23e,35] we synthesized the pyrene analog of **PRODAN** with an expectation that it might have a stronger emission at longer wavelengths than **PRODAN** with reasonably high solvatochromicity.[36] We demonstrated solvatofluorochromic pyrene derivative **Apa** (**62**) possessing a carboxyl side chain, which was covalently attached to the C8-alkylamino-2′-deoxyguanosine to provide a highly solvatochromic fluorescent guanosine (**ApaG**, **65**) (Scheme 7.6).[36]

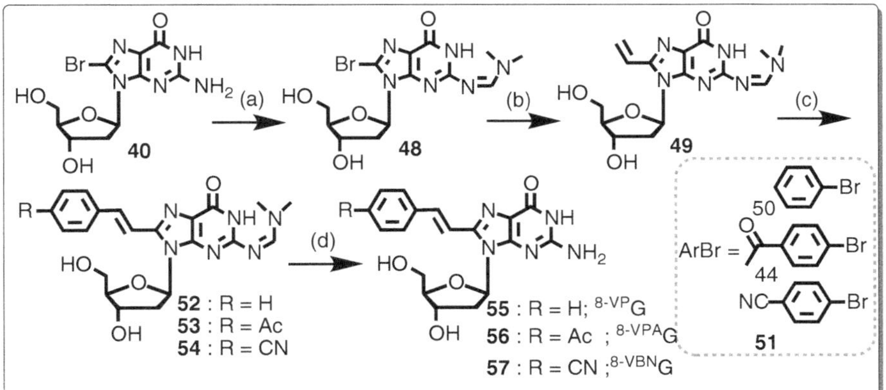

Reagents and condition: (a) DMFdiethylacetal,MeOH, 50 °C, 3 h; (b)Tetravinyltin(IV), Pd(PPh3)4, DMF, 120 °C, 3 h; (c) ArBr (**50**, **44**, **51**), Pd(PPh₃)₄, CH₃COONa, DMF, 100 °C, 8 h, (d) aq.NH4OH, MeOH, 50 °C, 5 h.

Scheme 7.4 Synthesis of 8-styrylguanosine derivatives. From Ref. [34].

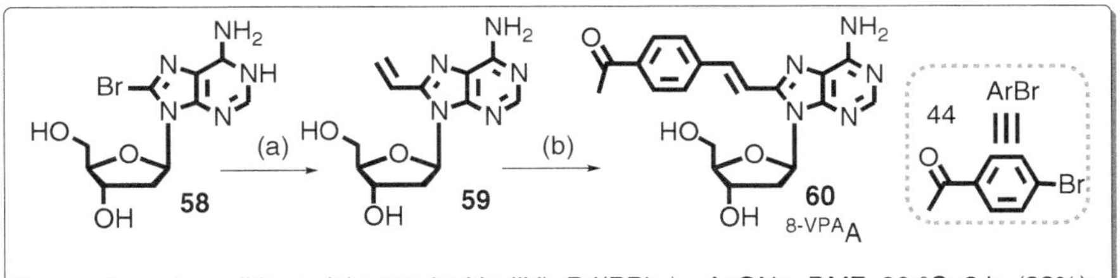

Reagents and conditions: (a) tetravinyltin (IV), Pd(PPh₃)₄, AcONa, DMF, 90 °C, 3 h, (82%); (b) ArBr (= 4-bromoacetophenone, **44**), Pd(PPh₃)₄, AcONa, DMF, 120 °C, 5 h, (36%).

Scheme 7.5 Synthesis of 8-styryladenosine derivatives.

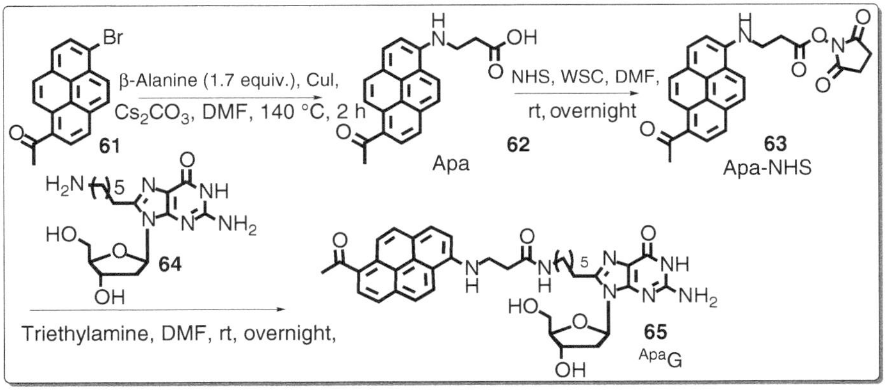

Scheme 7.6 Synthesis of solvatochromic pyrene tethered guanosine, ^{Apa}G. From Ref. [36].

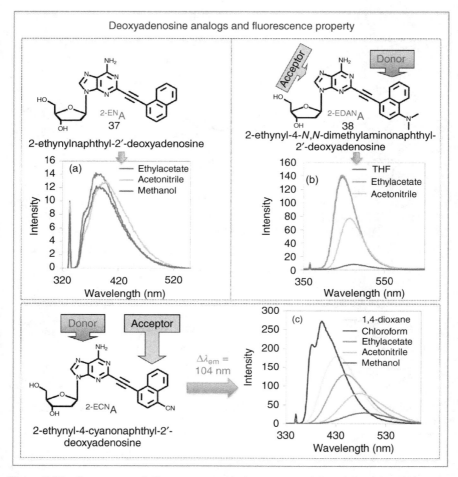

Figure 7.23 Structures and fluorescence emission spectra of 2-arylalkynylated 2′-deoxy-adenosine derivatives. From Ref. [32]. Reprinted with permission of Elsevier. (*See color plate section for the color representation of this figure.*)

Photophysical Properties of ESF Nucleosides: Fluorescence spectra of 2-arylethynylated-2′-deoxyadenosine (**37–39**) derivatives are shown in Figure 7.23.[32] While unsubstituted naphthalene is attached to the 2-position of adenosine via an ethynyl linker, solvatofluorochromicity was not observed (Fig. 7.23a). However, when, electron donor- or acceptor-containing naphthalenes are attached to adeno-sine, high solvatochromicities were observed when tested in various organic solvents of different dielectric properties. The solvatochromic nature of the ESF nucleosides is clearly evident from their steady state emission spectra depicted in Figure 7.23b and c. In general, more solvatofluorochromicities were exhibited by nucleosides containing donor or acceptor aromatic unit-linked adenosine derivatives.

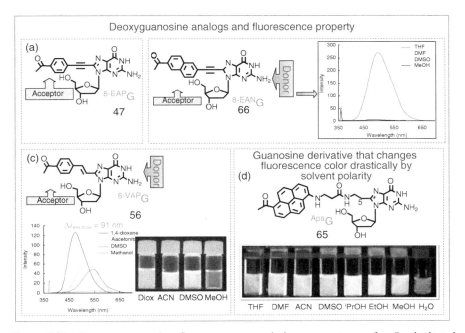

Figure 7.24 Structures and fluorescence emission spectra of 8-substituted 2′-deoxyguanosine derivatives. From Ref. [34]. Reprinted with permission of Elsevier. (*See color plate section for the color representation of this figure.*)

Particularly, 2-ethynylcyanonaphthyl-2′-deoxyadenosine (**39**) showed a remarkably large solvatochromicity ($\Delta\lambda_{em}$ = 104 nm, Fig. 7.23c).

Guanosine is the most electron-donating base among DNA bases. Thus, when an unsubstituted or electron-donor-substituted phenyl or naphthyl group was attached to C-8 position of guanosine via an ethynyl linker, a notable solvatochromicity was not observed. However, when an electron accepting group such as a cyano and an acetyl group was introduced on the phenyl or naphthyl ring, remarkably large solvatochromocity was observed (Fig. 7.24). For example, the fluorescence intensity of the acetyl analog **8-EAP**G (**47**) showed a strong solvent dependency, presumably because of the intermolecular charge transfer between the donor guanine moiety and the conjugated acceptor acetylphenyl unit.[33] This example showed that introduction of an acetyl group on the aromatic moiety could induce a strong solvatochromic fluorescence property to the donor guanine base (Fig. 7.24a). However, acetylnaphthalene-substituted guanosine **8-EAN**G (**66**) showed a strong fluorescence around 470 nm in ethyl acetate or THF but only very weak fluorescence at longer wavelengths in methanol.

Examination of the fluorescence behavior of 8-styryl-2′-deoxyguanosine derivatives (**8-VP**G, **55**) and 8-(4-acetylstyryl)-2′-deoxyguanosine (**8-VPA**G, **56**) revealed that nonsubstituted styryl conjugated nucleoside (**8-VP**G, **55**) exhibited negligible solvatochromic emission.[34] However, acetyl-substituted

TABLE 7.1   Photophysical Properties of Fluorescent Guanosine ^{Apa}G

	Solvent	$\lambda_{ab.max}$ (nm)	$\lambda_{fl.max}$ (nm)	Φ_{fl}	Δv (cm^{-1})	Δf
Aprotic	THF	451	523	0.62	2955	0.210
	DMF	452	543	0.63	3610	0.274
	Acetonitrile	440	546	0.72	4157	0.305
	DMSO	456	551	0.61	3637	0.263
Protic	2-Propanol	452	557	0.58	4155	0.276
	Ethanol	451	559	0.49	4186	0.289
	Methanol	447	565	0.44	4573	0.309
	Glycerol	464	567	0.38	3915	0.263
	Water	460	573	0.27	4478	0.320

8-(4-acetylstyryl)-2′-deoxyguanosine ($^{8-VPA}$G, **56**) behaved as a typical member of ESF nucleoside family as is supported by its fluorescence photophysical spectra shown in Figure 7.24c.[34] Thus, $^{8-VPA}$G exhibited a strong fluorescence emission in nonpolar solvents but very weak red shifted fluorescence emission in polar solvents with a solvatochromicity of $\Delta\lambda_{em}$ = 91 nm, indicating that it can be utilized as an effective ESF nucleoside for monitoring polarity around nucleic acid microenvironment (Fig. 7.24c).[34]

We also demonstrated an intriguing photophysical property of ^{Apa}G (**65**) nucleoside (Scheme 7.6 and Fig. 7.24d). ^{Apa}G showed a red shifted emission upon excitation at 450 nm as solvent polarity increased. For example, in changing the solvent from THF (λ_{em} = 523 nm) to water (λ_{em} = 573 nm), a wavelength shift of about 50 nm with decreasing fluorescence intensity was observed.[36] The fluorescence quantum yields markedly decreased with increasing solvent polarity (Table 7.1). The Stokes shift, which is the difference in wave numbers between the excitation maximum and the emission maximum, significantly changed depending upon solvent polarity. These observations indicated that ^{Apa}G has a reasonable solvatofluorochromicity ($\Delta\lambda_{fl.max}$ = 50 nm), which is also apparent from the fluorescence color change as shown in Figure 7.24d. It is notable that the fluorescence emission maximum of ^{Apa}G is more than 40–80 nm longer than that of the corresponding naphthalene analogs, **PRODAN** ($\lambda_{fl.max}$ = 531 nm in water)[6,23,35] and **Anap** ($\lambda_{fl.max}$ = 490 nm in water).[37]

From a plot of Stokes shifts (Δv) versus the orientational polarizability (Δf) of the solvent, it was observed that the change in the Stokes shift was roughly proportional to the orientational polarizability of the solvent. Larger solvatofluorochromic shifts of ^{Apa}G in protic solvents compared with aprotic solvents were attributed to specific solute–solvent interactions such as hydrogen bonding. The environmentally sensitive fluorophores **Apa** (**62**) and ^{Apa}G (**65**) can be easily incorporated into biopolymers such as proteins and nucleic acids by chemical modification. Most importantly, succinimidyl ester **Apa-NHS** (**63**) may be used as a chemical modifier that enables to introduce solvatochromic fluorophore into biomolecules containing the amino functionality.

7.6 BASE-SELECTIVE FLUORESCENT ESF PROBE

After evaluating the solvatochromic properties of the synthesized ESF nucleosides, their utility for DNA detection was examined. For such purposes, ESF nucleosides were incorporated into short ODN sequences via phosphoramidite chemistry using an automated DNA synthesizer. Oligonucleotide probes containing various ESF nucleosides, so called ESF-ODN probes, were utilized for hybridization with their complementary fully matched and single-base mismatched target oligonucleotides. The study of fluorescence emission properties showed that many of the ESF probes are efficient in detecting the presence of a base in a target ODN opposite to the ESF nucleoside of the probe ODN via the generation of a distinct fluorescence signal.

7.6.1 Cytosine Selective ESF Probe

As shown in Figure 7.25, it is clear from the fluorescence spectra that an ESF probe containing 4-*N,N*-dimethylaminophenyl-substituted ESF uridine ($^{5\text{-}EDMA}$U, **26**) was able to detect a cytosine base of the target ODN opposite to the ESF uridine of the probe ODN with a distinct and enhanced fluorescence signal generation. On the contrary, the probe containing a simple phenylethynyl-substituted 2′-deoxyuridine ($^{5\text{-}EP}$U, **24**) is unable to detect the opposite base of a target ODN with distinct emission, indicating the importance of donor/acceptor system of **26**. The probe containing ESF nucleoside $^{5\text{-}EDMA}$U (**26**) serves as a suitable reporter probe for the detection of C with a drastic enhancement of fluorescence intensity.

7.6.2 Thymine Selective Fluorescent ESF Probe

Highly solvatochromic fluorescent deoxyguanosine analog 8-(4-acetylphenylethynyl)-2′-deoxyguanosine, $^{8\text{-}EAP}$G (**47**), was incorporated into ODNs for a possible use in the detection of SNP.[33] Two ODNs containing $^{8\text{-}EAP}$G were, therefore, synthesized, one with flanking A/T base pair and the other with flanking G/C base pair, for testing whether SNP typing could be achieved universally or whether the detection ability of the probe would be sequence dependent.

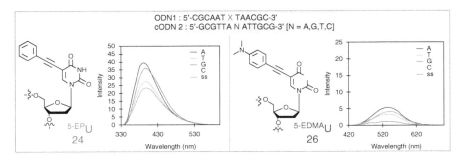

Figure 7.25 Detection of cytosine with $^{5\text{-}EDMA}$U-labeled ESF probe. (*See color plate section for the color representation of this figure.*)

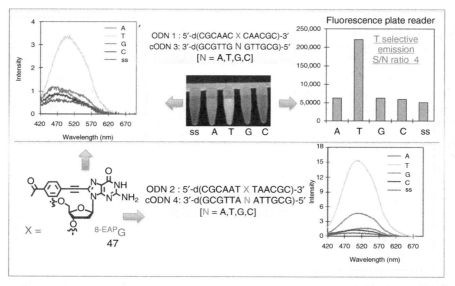

Figure 7.26 Detection of thymine with ESF probe containing [8-EAP]G. (*See color plate section for the color representation of this figure.*)

Fluorescence emission of ODN **1** and ODN **2** in the absence and presence of various complimentary target strands revealed the following observations. Fluorescence intensities of single-stranded ODN **1**, ODN **2**, and the duplexes formed by hybridization with the complimentary strands cODN **3** and cODN **4** (N = A, G, C), respectively, were found to be very weak. On the contrary, the fluorescence spectra of the duplexes ODN **1**/cODN **3** (N = T) and ODN **2**/cODN **4** (N = T) showed intense fluorescence emissions at around 510 nm (Fig. 7.26). Thus, the probe containing ESF nucleoside, [8-EAP]G (**47**), is highly efficient in sensing the presence of T base of target DNA opposite to the ESF nucleoside of probe DNA with an enhanced fluorescence signal that is again independent of the sequence or flanking bases. The fluorescence emission was visible with the naked eye only for T opposite to ESF base [8-EAP]G under illumination with 365 nm transilluminator. Acetyl-substituted fluorescent deoxyguanosine, [8-EAP]G is exceptionally useful as a T-specific ESF base regardless of the flanking sequences in a fluorescence hybridization assay. The reason for the T-specific fluorescence emission was explained on the basis of location of the chromophore at a hydrophobic site inside the groove.[33]

We have also designed ESF nucleoside, [8-EBN]G (**67**) containing oligonucleotide probe and exploited it for the detection of T base of a target DNA (Fig. 7.27). Thus, in a similar manner, we found that the probe ODN containing the ESF nucleoside [8-EBN]G is able to detect T base of target ODN in a sequence-independent manner opposite to the ESF base of probe ODN with an enhancement of fluorescence intensity.

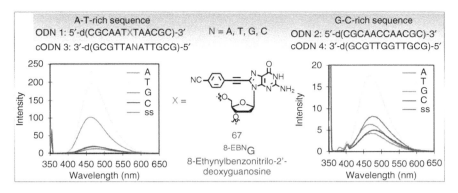

Figure 7.27   Detection of thymine with ESF probe containing $^{8\text{-EBN}}$G. (*See color plate section for the color representation of this figure.*)

7.6.3   Specific Detection of Adenine by Exciplex Formation with Donor-Substituted ESF Guanosine

Guanosine is the most electron-donating base but nonemissive. The attachment of aromatic chromophores such as phenyl or naphthyl group to the 8-position of guanosine through a triple bond induced a strong fluorescence to provide easily accessible fluorescent guanine derivatives.[29,38] Introduction of electron-donating groups such as a methoxy group on the aromatic ring induced an intense fluorescence at longer wavelengths as was exemplified by $^{8\text{-MPE}}$G (**68**) and $^{8\text{-MNE}}$G (**69**) (Fig. 7.28).[39] These methoxy-substituted fluorescent guanosine derivatives have an intriguing property, that is, they possess lower oxidation potentials than native guanosine and emit strong fluorescence at more than 380 nm. It was envisaged that stacking interaction between electron-donating guanosine, $^{8\text{-MPE}}$G and adenine base possessing a moderately electron-accepting property in a DNA duplex may give rise to the formation of an exciplex.

The fluorescence of single-stranded ODN **1** (X = $^{8\text{-MPE}}$G) was found to be relatively weak and appeared at around 380 nm. When the opposite base of a target complimentary strand is adenine (N = A) in a duplex ODN **1** (X = $^{8\text{-MPE}}$G, **68**) /ODN

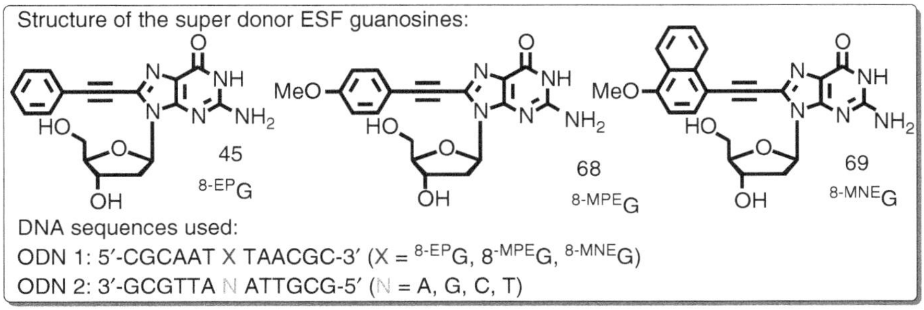

```
Structure of the super donor ESF guanosines:

        45                      68                      69
      8-EPG                   8-MPEG                   8-MNEG

DNA sequences used:
ODN 1: 5'-CGCAAT X TAACGC-3' (X = 8-EPG, 8-MPEG, 8-MNEG)
ODN 2: 3'-GCGTTA N ATTGCG-5' (N = A, G, C, T)
```

Figure 7.28   Structures of donor-substituted 2′-deoxyguanosine analogs of ESF base and the sequences of ESF ODN probe for the detection of adenosine. From Ref. [39].

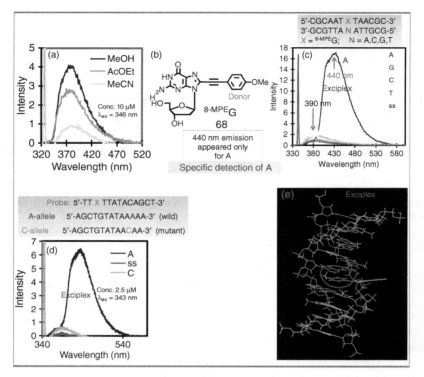

Figure 7.29 (a) Fluorescence emission of $^{8\text{-}MPE}$G in different solvents. (b) Structure of donor-substituted ESF base, $^{8\text{-}MPE}$G. (c) Emission of the ESF probe in the presence of target DNAs, showing the detection of A of target DNA with enhanced fluorescence emission. (d) Detection of wild-type A-allele. (e) Amber* conformation of the duplex, showing $\pi - \pi$ interaction between A of target ODN and the aromatic moiety of $^{8\text{-}MPE}$G. (*See color plate section for the color representation of this figure.*)

2 (N = A), a new and strong emission was observed at a longer wavelength at around 440 nm (Fig. 7.29). The emission at 435 nm is not observed when the opposite bases of complimentary strands are G, C, and T, suggesting that the new emission is an exciplex emission from a complex formed between $^{8\text{-}MPE}$G and adenine (A). However, when ODN (X = $^{8\text{-}EP}$G, **45**) containing unsubstituted phenyl group was used, such exciplex emission has not been observed, indicating a requirement of the electron donor substituent for the exciplex formation.[39]

A more prominent example of exciplex formation is the emission from methoxy-substituted naphthalene derivative $^{8\text{-}MNE}$G (**69**). Thus, a characteristic fluorescence from the naphthalene chromophore appeared at 390–430 nm for single-stranded ODN **1** (X = $^{8\text{-}MNE}$G). However, in a duplex [ODN **1** (**X** = $^{8\text{-}MNE}$G) /ODN **2** (N = A)], a new exciplex emission at 470 nm emerged together with the appearance of original naphthalene fluorescence, suggesting a partial formation of excited state complex (Fig. 7.30). The formation of ground state complex between

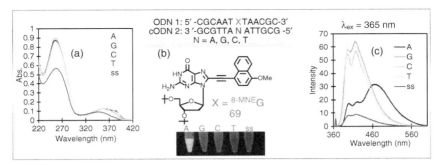

Figure 7.30 (a) UV-visible spectra of ESF probe in the presence of various target DNAs, showing the formation ground state complex between the A of target DNA opposite to [8-MNE]G and [8-MNE]G. (b) Structure of [8-MNE]G, ODN sequences used, the fluorescence image of various duplexes. (c) Emission spectra of the ESF probe in the presence of various target DNAs, allowing the detection of A of target DNA via exciplex emission. (*See color plate section for the color representation of this figure.*)

the naphthalene chromophore and A was indicated by the appearance of a new absorption band in the UV-visible spectrum of the duplex ODN **1** (X = [8-MNE]G) /ODN **2** (N = A) (Fig. 7.30a). The fluorescence color change was also observable by the naked eye upon illumination with a 365-nm transilluminator. The study of fluorescence lifetime indicated the existence of two species that are responsible for the fluorescence emission and suggest that longer lifetime component with a strong emission at 470 nm is due to the formation of an exciplex between the [8-MNE]G and adenosine (A).[39]

7.7 MOLECULAR BEACON (MB) AND ESF NUCLEOSIDES

7.7.1 Ends-Free and Self-Quenched MB

After popularization of the idea of "molecular beacon" technology by Tyagi *et al.*, in 1996, it became a powerful tool for genetic analysis and is widely used in chemistry and biology due to easiness of synthesis, unique functionality, molecular specificity, and structural tolerance to various modifications. [40] Molecular beacons (MBs) are hairpin-shaped oligonucleotide probes containing stem and loop structures with a quencher and a fluorophore at both ends, at which fluorescence is restored when bound to its target oligonucleotide sequence.

It is known that the interaction between guanosine (**G**) base and neighboring electron-withdrawing aromatic fluorophores normally results in a strongly quenched fluorescence of the fluorophore. We envisaged that it is worthwhile if we exploit the intrinsic quenching property of **G**-base and design a molecular beacon having no quencher at the 3′-end of the stem.[41] Our efforts in designing practically useful fluorescent probe has led to a rational design of self-quenched MB with both 3′- and 5′-ends free that require no terminal **G** and possess fluorophore covalently

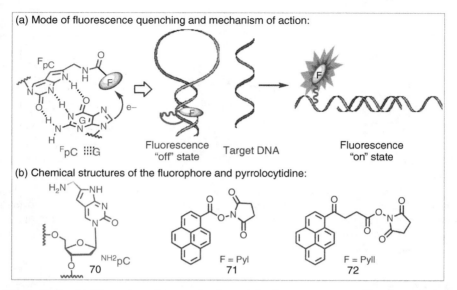

Figure 7.31 (a) Illustration of the mechanism of the fluorescence quenching for ends-free self-quenched MB. (b) Structures of pyrrolocytidine (**pC**) and fluorophores (**PyI, PyII**). From Ref. [42a,b].

attached to pyrrolocytidine (**pC**) (**70**) placed in the middle of the stem as shown in Figure 7.31.[42] In this design, guanine base opposite to **pC** serves as a quencher of the intercalated pyrene fluorophore **PyI** (**71**) or **PyII** (**72**). By using our conceptually designed novel quencher-free MBs, we were able to detect the target DNA irrespective of sequences and the length of the stem with high efficiency. The designed ends-free MBs produced excellent "on"/"off" signals that can thus be widely used as a sensitive probe.

7.7.2   Single-Stranded Molecular Beacon Using ESF Nucleoside in a Bulge Structure

While MBs have been used successfully in DNA assays, their utility in a cell is limited mainly because the stem structure of MB is destroyed by nonspecific binding in a cell. Therefore, single-stranded MB is highly desirable for DNA detection in a cell. In an effort to explore the possible use of designed ESF nucleosides, we exploited solvatochromic fluorescent deoxyadenosine, 8-(4-*N,N*-dimethylaminophenylethynyl)-2′-deoxyadenosine, [8-EDA]A (**73**). [8-EDA]A was synthesized by following the procedure as that was adopted for the synthesis of 8-substituted guanosines (Scheme 7.3). We envisaged that ODN containing highly solvatochromic ESF base [8-EDA]A can serve as a single-stranded MB, because in a fully matched structure the bulky [8-EDA]A resides outside the groove, whereas [8-EDA]A moiety is located inside groove when [8-EDA]A forms a bulged

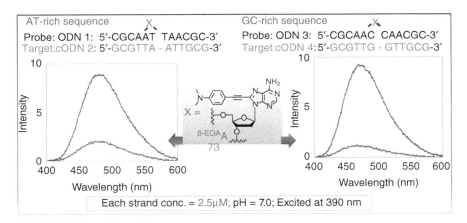

Figure 7.32 Fluorimetric detection of target DNA by single-stranded MB containing [8-EDA]A. Y. Saito *et al.*, Unpublished Results.

structure upon hybridization with a fully matched complementary target DNA. The fluorescence spectrum of probe ODN **1** containing [8-EDA]A at the middle was measured in the absence and presence of complementary target 15-mer cODN **2** (5′-GCGTTA-ATTGCG-3′) forming a bulge structure at [8-EDA]A (Fig. 7.32). It was clear that the fluorescence intensity of single-stranded ODN **1** was very weak (λ_{em} = 480 nm). However, the fluorescence intensity of the duplex between ODN **1** and target cODN **2** containing bulge at [8-EDA]A was drastically enhanced. Hybridization with mismatched target DNAs did not produce such fluorescence enhancement. It was also confirmed that the fluorescence change of probe ODN does not depend on the flanking bases of [8-EDA]A (Fig. 7.32) (Y. Saito, Unpublished Results).

7.8 SUMMARY AND FUTURE OUTLOOK

The lack of significant fluorescence in naturally occurring nucleobases has prompted the design of extrinsic synthetic fluorescent nucleobase analogs with improved photophysical properties for nucleic acids research. Thus, either the inspiration from nature or a rational approach for rational designing of emissive analogs has been employed for the generation of signaling nucleobases via the conjugation of existing solvatochromic fluorophores to the natural nucleobases. This chapter has highlighted mostly our own work for the design concept, synthesis, photophysical properties, and the importance of solvatochromic fluorescent nucleosides and various methods in which ESF oligonucleotide probes (ESF-ODN probes) were used as a molecular signaling device in DNA detection.

Fluorescent probes containing BDF nucleoside have been used by many research groups. The clear and drastic change in fluorescence intensity of BDF nucleosides when the BDF-labeled probe is hybridized with a target sequence is very useful for SNP genotyping. The BDF probes described here are now applied in chip-based SNP detection technologies.

We explored numerous ESF nucleosides including 8-substituted purine nucleosides. However, 8-substituted adenosines and guanosine derivatives are in anti-conformation as to the glycoside bond due to the bulky substituents at purine C-8 position. Incorporation of such 8-substituted adenosine and guanosine bases into DNA oligomers induced destabilization of the duplex due to the lack of base pairing, whereas 5-substituted uridines, 2-substituted adenosines, and 7-substituted 7-deazapurine ESF nucleosides do not always destabilize DNA duplex structures. The concept for designing ESF nucleosides and the application of such a fluorescent probe containing ESF bases might find special attention in the future of gene detection technology. The field of designing fluorescent nucleosides of high solvatofluorochromicity is also flourishing, and many more advancements toward this end are expected in the near future. It is expected that tuning the chromophore's structure to achieve predictable photophysical properties might allow us to create fluorescent nucleoside analogs to be applicable for bioimaging as well as for gene detection technology.

ACKNOWLEDGMENTS

We thank all coworkers involved in this project. Particularly, we thank Prof. Akimitsu Okamoto, The University of Tokyo, for collaboration throughout this work. We also thank Associate Prof. Yoshio Saito, Nihon University for continuous collaboration. Dr. Bag is thankful to his family members, and his daughter (Shreyasi) and Son (Saswata), for their never-ending lovingly inspirations and encouragements during the course of this write-up. Financial supports from Grant-in-Aid for Scientific Research of MEXT, Japanese Government to Prof. Saito and Department of Science and Technology (DST), New Delhi, Govt. of India (Grant No. SR/S1/OC-69/2008) to Dr. S. S. Bag are also gratefully acknowledged.

REFERENCES

1. (a) Sameiro, M.; Goncalves, T. *Chem. Rev.* **2009**, *109*, 190. (b) Lakowicz, J. R. Principles of Fluorescence Spectroscopy, 3rd ed.; Springer: New York, **2006**. (c) Sinkeldam, R. W.; Tor, Y. *Org. Biomol. Chem.* **2007**, *5*, 2523. (d) Azzi, A. *Q. Rev. Biophys.* **1975**, *8*, 237.
2. (a) Fakhrai-Rad, H.; Pourmand, N.; Ronaghi, M. *Hum. Mutat.* **2002**, *19*, 479. (b) Nazarenko, I.; Pires, R.; Lowe, B.; Obaidy, M.; Rashtchian, A. *Nucleic Acids Res.* **2002**, *30*, 2089. (c) Ward, D. C.; Reich, E.; Stryer, L. *J. Biol. Chem.* **1969**, *244*, 1228. (d) Menger, M.; Tuschl, T.; Eckstein, F.; Porschke, D. *Biochemistry* **1996**, *35*, 14710. (e) Lacourciere, K. A.; Stivers, J. T.; Marino, J. P. *Biochemistry* **2000**, *39*, 5630. (f) Secrist III, J. A.; Barrio, J. R.; Leonard, N. J. *Science* **1972**, *175*, 646. (g) Holmen, A.; Albinsson, B.; Norden, B. *J. Phys. Chem.* **1994**, *98*, 13460. (h) Seela, F.; Zulauf, M.; Sauer, M.; Deimel, M. *Helv. Chim. Acta* **2000**, *83*, 910.
3. (a) Hurley, D. J.; Seaman, S. E.; Mazura, J. C.; Tor, Y. *Org. Lett.* **2002**, *4*, 2305. (b) Fauth, C.; Speicher, M. R.; *Cytogen. Cell Gen.* **2001**, *93*, 1. (c) Pernthaler, J.; Glockner, F. O.; Schonhuber, W.; Amann, R.; *Meth. Microbiol.* **2001**, *30*, 207. (d) Amann, R.; Fuchs, B. M.; Behrens, S. *Curr. Opin. Biotech.* **2001**, *12*, 231. (e) Sieben, V. J.; Debes-Marun, C. S.;

Pilarski, P. M.; Kaigala, G. V.; Pilarski, L. M.; Backhouse, C.; *IET Nanobiotechnol.* **2007**, *1*, 27.

4. (a) Janzen, W.P. High Throughput Screening: Methods and Protocols*;* Humana Press, **2002**. (b) Pernthaler, A.; Pernthaler, J.; Amann, R. *Appl. Environ. Microbiol.* **2002**, *68*, 3094. (c) Wagner, M.; Horny, M.; Daimsz, H.; *Curr. Opin. Microbiol.* **2003**, *6*, 302. (d) Hertzberg, R. P.; Pope, A. J. *Curr. Opin. Chem. Biol.* **2000**, *4*, 445.

5. For reviews, (a) Hawkins, M. E. *Cell Biochem. Biophys.* **2001**, *34*, 257; (b) Rist, M. J. ; Marino, J. P. *Curr. Org. Chem.* **2002**, *6*, 775; (c) Ranasinghe, R. T.; Brown, T. *Chem. Commun.* **2005**, 5487; (d) Tor, Y. Ed., Tetrahedron symposium in print number 128, **2007**, *63*(17), 3421–3614; (e) Venkatesan, N. Y.; Seo, J.; Kim, B. H. *Chem. Soc. Rev.* **2008**, *37*, 648; (f) Sinkeldam, R. W.; Greco, N. J.; Tor, Y. *Chem. Rev.* **2010**, *110*, 2579.

6. (a) Haugland, R. P. The Handbook: A Guide to Fluorescent Probes and Labeling Technologies; Molecular Probes, **2005**. (b) Weber, G.; Farris, F. J. *Biochemistry* **1979**, *18*, 3075. (c) Saroja, G.; Ramachandram, B.; Saha, S.; Samanta, A. *J. Phys. Chem. B* **1999**, *103*, 2906. (d) Grabchev, I.; Chovelon, J. M.; Qian, X. *J. Photochem. Photobiol., A* **2003**, *158*, 37.

7. (a) Reichardt, C. Solvents and Solvent Effects in Organic Chemistry. VCH: Weinheim, **1988**. (b) Bayliss, N. *J. Chem. Phys.***1950**, *18*, 292.

8. (a) Loving, G. S.; Imperiali, B. *Bioconjug. Chem.* **2009**, *20*, 2133. (b) Loving, G. S.; Imperiali, B. *J. Am. Chem. Soc.* **2008**, *130*, 13630. (c) Loving, G. S.; Sainlos, M.; Imperiali, B. *Trends Biotechnol.* **2009**, *28*, 73.

9. Condon, E.U. *Phys. Rev.* **1928**, *32*, 858.

10. (a) International Human Genome Sequencing Consortium *Nature* **2001**, *409*, 860. (b) Whitcombe, D.; Theaker, J.; Guy, S. P.; Brown, T.; Little, S. *Nat. Biotechnol.* **1999**, *17*, 804. (c) Blackburn, G. M.; Gait, M. J. Nucleic Acids in Chemistry and Biology. 2nd ed.; Oxford University Press: Oxford, England, New York, **1996**. (d) Bloomfield, V. A.; Crothers, D. M.; Tinoco, I. Nucleic Acids: Structures, Properties, and Functions. University Science Books: Sausalito, CA, **2000**.

11. (a) Okamoto, A.; Tainaka, K.; Saito, I. *J. Am. Chem. Soc.* **2003**, *125*, 4972. (b) Okamoto, A.; Tainaka, K.; Saito, I. *Chem. Lett.* **2003**, *32*, 684. (c) Okamoto, A.; Tainaka, K.; Fukuta, T.; Saito, I. *J. Am. Chem. Soc.* **2003**, *125*, 9296. (d) Okamoto, A.; Tainaka, K.; Saito, I. *Tetrahedron Lett.* **2003**, *44*, 6871. (e) Saito, Y.; Miyauchi, Y.; Okamoto, A.; Saito, I. *Chem. Commun.* **2004**, 1704. (f) Saito, Y.; Miyauchi, Y.; Okamoto, A.; Saito, I. *Tetrahedron Lett.* **2004**, *45*, 7827. (g) Dohno, C.; Saito, I. *ChemBioChem* **2005**, *6*, 1075. (h) For a review; Okamoto, A.; Saito, Y.; Saito, I. *J. Photochem. Photobiol., C***2005**, *6*, 108. (i) Bag, S. S.; Kundu, R.; Matsumoto, K.; Saito, Y.; Saito, I. *Bioorg. Med. Chem. Lett.* **2010**, *20*, 3227.

12. (a) Kalyanasundaram, K.; Thomas, J. K. *J. Phys. Chem.* **1977**, *81*, 2176. (b) de Silva, A. P.; Gunaratne, H. Q. N.; Gunnlaugsson, T.; Huxley, A. J. M.; McCoy, C. P.; Rademacher, J. T.; Rice, T. E. *Chem. Rev.* **1997**, *97*, 1515.

13. Okamoto, A.; Kanatani, K.; Saito, I. *J. Am. Chem. Soc.* **2004**, *126*, 4820.

14. (a) Saito, Y.; Kanatani, K., Ochi, Y.; Okamoto, A.; Saito, I. *Nucl. Acids Res. Sym. Series* **2004**, *48*, 243. (b) Saito, Y.; Kanatani, K.; Ochi, Y.; Okamoto, A.; Saito, I. *Nucl. Acids Res. Sym. Series* **2005**, *49*, 153.

15. (a) Seidel, C. A. M.; Schulz, A.; Dauer, M. H. M. *J. Phys. Chem.* **1996**, *100*, 5541. (b) Marras, S. A. E.; Kramer, F. R.; Tyagi S. *Nucleic Acids Res.* **2002**, *30*, e122. (c) Manoharan, M.; Tivel, K. L.; Zhao, M.; Nafisi, K.; Netzel, T. L.; *J. Phys. Chem.* **1995**, *99*, 17461. (d) Jean, J. M.; Hall. K. B. *Proc. Natl. Acad. Sci. U. S. A.* **2001**, *98*, 37.

16. (a) Lewis, F. D.; Zhang, Y.; Letsinger, R. L. *J. Am. Chem. Soc.* **1997**, *119*, 5451. (b) Paris, P. L.; Langenhan, J. M.; Kool, E. T. *Nucleic Acids Res.* **1998**, *26*, 3789. (c) Bryld, T.; Højland, T.; Wengel, J. *Chem. Commun.* **2004**, 1064.

17. (a) Haga, H.; Yamada, Y.; Ohnishi, Y.; Nakamura, Y.; Tanaka, T. *J. Hum. Genet.* **2002**, *47*, 605. (b) Tabara, Y.; Kohara, K.; Miki, T. et al. *Hypertens. Res.*, **2012**, doi: 10.1038/hr.2012.41. (c) Thompson, E. R.; Gorringe, K. L.; Choong, D. Y. H.; Eccles, D. M.; ConFab, K.; Mitchell, G.; Campbell, I. G. *Breast Cancer Res. Treat.* **2012**, doi: 10.1007/s10549-012-2088-3.

18. (a) Okamoto, A.; Ichiba, T.; Saito, I. *J. Am. Chem. Soc.* **2004**, *126*, 8364. (b) Okamoto, A.; Ochi, Y.; Saito, I. *Bioorg. Med. Chem. Lett.* **2005**, *15*, 4279.

19. (a) Hansson, J. H.; Schild, L.; Lu, Y.; Wilson, T. A.; Gautschi, I.; Shimkets, R.; Nelson-Williams, C.; Rossier, B. C.; Lifton, R. P. *Proc. Natl. Acad. Sci. U. S. A.* **1995**, *92*, 11495. (b) Hiltunen, T. P.; Hannila-Handelberg, T.; Petajaniemi, N.; Kantola, I.; Tikkanen, I.; Virtamo, J.; Gautschi, I.; Schild, L.; Kontula, K. *J. Hypertens.* **2002**, *20*, 2383.

20. (a) Modukuru, N. K.; Snow, K. J.; Perrin, B. S. Jr.; Thota, J.; Kumar, C. V. *J. Phys. Chem. B* **2005**, *109*, 11810. (b) Tan, W. B.; Bhambhani, A.; Duff, M. R.; Rodger, A.; Kumar, C. V. *Photochem. Photobiol.* **2006**, *82*, 20.

21. Saito, Y.; Motegi, K.; Bag, S. S.; Saito, I. *Bioorg. Med. Chem.* **2008**, *16*, 107.

22. Bag, S. S.; Saito, Y.; Hanawa, K.; Kodate, S.; Suzuka, I.; Saito, I. *Bioorg. Med. Chem. Lett.* **2006**, *16*, 6338.

23. (a) Macgregor, R. B.; Weber, G. *Nature* **1986**, *319*, 70. (b) Balter, A.; Nowak, W.; Pawelkiewicz, W.; Kowalczyk, A. *Chem. Phys. Lett.* **1988**, *143*, 565. (c) Pierce, D. W. ; Boxer, S. G. *J. Phys. Chem.* **1992**, *96*, 5560. (d) Samanta, A.; Fessenden, R. W. *J. Phys. Chem. A* **2000**, *104*, 8972. (e) Cohen, B. E.; McAnanney, T. B.; Park, E. S.; Jan, Y. N.; Boxer, S. G.; Jan, Y. J. *Science*, **2002**, *296*, 1700.

24. Okamoto, A.; Tainaka, K.; Saito, I. *Bioconjug. Chem.* **2005**, *16*, 1105.

25. Rachofsky, E. L.; Osman, R.; Ross, J. B. A. *Biochemistry* **2001**, *40*, 946.

26. (a) Tainaka, K.; Ikeda, S.; Tanaka, K.; Nishiza, K.-i.; Fujiwara, Y.; Okamoto, A.; Saito, I. *Nucleic Acids Symp. Ser.* **2006**, *50* (1), 133. (b) Tainaka, K.; Tanaka, K.; Ikeda, S.; Nishiza, K.-i.; Unzai, T.; Fujiwara, Y.; Saito, I.; Okamoto, A. *J. Am. Chem. Soc.* **2007**, *129*, 4776.

27. (a) Lippert, E.; Luder, W.; Boss, H. In Advances in Molecular Spectroscopy; Marngini, A., Ed.; Pergamon Press: Oxford, **1962**; p 443. (b) Wiessner, A.; Huttmann, G.; Kuhnle, W.; Staeck, H. *J. Phys. Chem.* **1995**, *99*, 14923. (c) Weigel, W.; Rettig, W.; Dekhtyar, M.; Modrakowski, C.; Beinhoff, M.; Schluter, A. D. *J. Phys. Chem. A* **2003**, *107*, 5941. (d) Rettig, W. *Angew. Chem. Int. Ed. Engl.* **1986**, *25*, 971. (e) Grabowski, Z. R.; Rotkiewicz, K.; Rettig, W. *Chem. Rev.* **2003**, *103*, 3899.

28. Okamoto, A.; Tainaka, K.; Nishiza, K.-i.; Saito, I. *J. Am. Chem. Soc.* **2005**, *127*, 13128.

29. (a) Volpini, S.; Costanzi, S.; Lambertucci, C.; Vittori, S.; Klotz, K.-N.; Lorenzen, A.; Cristalli, G. *Bioorg. Med. Chem. Lett.* **2001**, *11*, 1931. (b) Costanzi, S.; Lambertucci, C.; Vittori, S.; Volpini, R.; Cristalli, G. *J. Mol. Graph. Model.* **2003**, *21*, 253. (c) Abou-Elkhair, R. A. I.; Netzel, T. L. *Nucl. Nucleot. Nucl. Acids* **2005**, *24*, 85. (d) Firth, A. G.; Fairlamb, I. J. S.; Darley, K.; Baumann, C. G. *Tetrahedron Lett.* **2006**, *47*, 3529. (e) Seo, Y. J.; Rhee, H.; Joo, T.; Byeang H. Kim, *J. Am. Chem. Soc.* **2007**, *129*, 5244. (f) O'Mahony, G.; Ehrman, E.; Grøtli, M. *Tetrahedron* **2008**, *64*, 7151. (g) Abou-Elkhair, R. A. I.; Dixon, D. W.; Netzel, T. L. *J. Org. Chem.* **2009**, *74*, 4712.

30. (a) Liu, C. H.; Martin, C. T. *J. Mol. Biol.* **2001**, *308*, 465. (b) Wilhelmsson, L.M.; Holmén, A.; Lincoln, P.; Nielsen, P.E.; Nordén, B. *J. Am. Chem. Soc.* **2001**, *123*, 2434. (c) Engman, K. C.; Sandin, P.; Osborne, S.; Brown, T.; Billeter, M.; Lincoln, P.; Nordén, B.; Albinsson, B.; Wilhelmsson, L. M. *Nucleic Acids Res.* **2004**, *32*, 5087. (d) Tor, Y.; Del Valle, S.; Jaramillo, D.; Srivatsan, S. G.; Rios, A.; Weizman, H. *Tetrahedron* **2007**, *63*, 3608. (e) Liu, H.; Gao, J.; Lynch, S.; Maynard, L.; Saito, D.; Kool, E. T. *Science* **2003**, *302*, 868. (f) Krueger, A. T.; Kool, E. T. *J. Am. Chem. Soc.* **2008**, *130*, 3989. (g) Lee, A. H. F.; Kool, E. T. *J. Am. Chem. Soc.* **2006**, *128*, 9219. (h) Greco, N. J.; Tor, Y. *J. Am. Chem. Soc.* **2005**, *127*, 10784. (i) Riedl, J.; Pohl, R.; Rulíšek, L.; Hocek, M. *J. Org. Chem.* **2012**, *77*, 1026. (j) Shin, D.; Sinkeldam, R. W.; Tor, Y. *J. Am. Chem. Soc.* **2011**, *133*, 14912. (k) Sato, K.; Sasaki, A.; Matsuda, A. *ChemBioChem* **2011**, *12*, 2341. (l) Hirose, W.; Sato, K.; Matsuda, A. *Angew. Chem. Int. Ed.* **2010**, *49*, 8392.

31. (a) Sonogashira, K.; Tohda, Y.; Hagihara, N. *Tetrahedron Lett.* **1975**, *16*, 4467. (b) Sonogashira, K. *J. Organomet. Chem.* **2002**, *653*, 46. (c) Basak, A.; Mandal, S.; Bag, S. S. *Chem. Rev.* **2003**, *103*, 4077. (d) Negishi, E.; Anastasia, L. *Chem. Rev.* **2003**, *103*, 1979. (e) Chinchilla, R.; Najera, C.; *Chem. Rev.* **2007**, *107*, 874.

32. Tanaka, M.; Kozakai, R.; Saito, Y.; Saito, I. *Bioorg. Med. Chem. Lett.* **2011**, *21*, 7021.

33. Shinohara, Y.; Matsumoto, K.; Kugenuma, K.; Morii, T.; Saito, Y.; Saito, I. *Bioorg. Med. Chem. Lett.* **2010**, *20*, 2817.

34. Matsumoto, K.; Takahashi, N.; Suzuki, A.; Morii, T.; Saito, Y.; Saito, I. *Bioorg. Med. Chem. Lett.* **2011**, *21*, 1275.

35. MacGregor, R. B.; Weber, G. *Ann. N.Y. Acad. Sci.* **1981**, *366*, 140.

36. Saito, Y.; Shinohara, Y.; Ishioroshi, S.; Suzuki, A.; Tanaka, M.; Saito, I. *Tetrahedron Lett.* **2011**, *52*, 2359.

37. Lee, H. S.; Guo, J.; Lemke, E. A.; Dimia, R. D.; Schultz, P. G. *J. Am. Chem. Soc.* **2009**, *131*, 12921.

38. (a) Okamoto, A.; Kanatani, K.; Ochi, Y.; Saito, Y.; Saito, I. *Tetrahedron Lett.* **2004**, *45*, 6059. (b) Okamoto, A.; Ochi, Y.; Saito, I. *Chem. Commun.* **2005**, 1128.

39. Saito, Y.; Kugenuma, K.; Tanaka, M.; Suzuki, A.; Saito, I. *Bioorg. Med. Chem. Lett.* **2012**, *22*, 3723.

40. (a) Tyagi, S.; Kramer, F. R. *Nat. Biotechnol.* **1996**, *14*, 303. (b) Broude, N. E.; *Trends Biotechnol.* **2002**, *20*, 249. (c) Steemers, F. J.; Ferguson, J. A.; Walt, D. R. *Nat. Biotechnol.* **2000**, *18*, 91. (d) Li, J.; Fang, X. H.; Schuster, S. M.; Tan, W. *Angew. Chem. Int. Ed.* **2000**, *39*, 1049. (e) Silverman, A. P.; Kool, E. T. *Trends Biotechnol.* **2005**, *23*, 225. (f) Yang, C. J.; Medley, C. D.; Tan, W. *Curr. Pharm. Biotechnol.* **2005**, *6*, 445. (g) Wu, Y.; Yang, C. J.; Moroz, L. L.; Tan, W. *Anal. Chem.* **2008**, *80*, 3025. (h) Marras, S. A. E. *Methods Mol. Biol.* **2006**, *335*, 3. (i) Marras, S. A. E.; Tyagi, S.; Kramer, F. R. *Clin. Chim. Acta* **2006**, *363*, 48. (j) Kim, Y.; Sohn, D.; Tan, W. *Int. J. Clin. Exp. Pathol.* **2008**, *1*, 105. (k) Li, Y.; Zhou, X.; Ye, D. *Biochem. Biophys. Res. Commun.* **2008**, *373*, 457. (l) Hwang, G. T.; Seo, Y. J.; Kim, B. H. *J. Am. Chem. Soc.* **2004**, *126*, 6528. (m) Seo, Y. J.; Ryu, J. H.; Kim, B. H. *Org. Lett.* **2005**, *7*, 4931. (n) Friedrich, A.; Hoheisel, J. D.; Marme, N.; Knemeyer, J.-P. *FEBS Lett.* **2007**, *581*, 1644.

41. (a) Saito, Y.; Mizuno, E.; Bag, S. S.; Suzuka, I.; Saito, I. *Chem. Commun.* **2007**, *43*, 4492. (b) Matsumoto, K.; Shinohara, Y.; Bag, S. S.; Takeuchi, Y.; Morii, T.; Saito, Y.; Saito, I. *Bioorg. Med. Chem. Lett.* **2009**, *19*, 6392.

42. (a) Saito, Y.; Shinohara, Y.; Bag, S. S.; Takeuchi, Y.; Matsumoto, K.; Saito, I. *Nucleic Acids Symp. Ser.* **2008**, (52), 361. (b) Saito, Y.; Shinohara, Y.; Bag, S. S.; Takeuchi, Y.; Matsumoto, K.; Saito, I. *Tetrahedron* **2009**, *65*, 934.

8

EXPANDING THE NUCLEIC ACID CHEMIST'S TOOLBOX: FLUORESCENT CYTIDINE ANALOGS

KIRBY CHICAS AND ROBERT H.E. HUDSON

Department of Chemistry, The University of Western Ontario, London, Canada

8.1 INTRODUCTION

The important and complex nature of nucleic acids has generated a great demand for tools and techniques to examine their environmental properties for diagnostic and structural purposes. Of the many tools available, none may find the ease of use and potential for broad application as readily as fluorescent nucleoside analogs. Possessing no analytically exploitable fluorescence,[1] natural nucleosides are readily endowed with fluorescent properties through judicious choice of chemistry, which enable them to be used in combination with fluorescence spectroscopy. This combination enables them to fulfill a number of requirements needed for an ideal detection device or assay.

Ideal detection devices and assays must offer great selectivity, excellent sensitivity, ease of use, and low cost of production.[2] Nucleic acids have long been admired for their genetic coding properties but have recently emerged as important materials for molecular diagnostic technologies. Nucleic acids satisfy ideal detection device and assay requirements by a number of assets such as specific Watson–Crick base pairing, high stability, low cost of synthesis, and excellent adaptability to modifications.[2] Nucleic acids are often used in combination with fluorescence spectroscopy for a number of reasons: (1) a large selection of fluorophores for nucleic acid conjugation exists; (2) there are minimal health risks associated with

Fluorescent Analogs of Biomolecular Building Blocks: Design and Applications, First Edition.
Edited by Marcus Wilhelmsson and Yitzhak Tor.
© 2016 John Wiley & Sons, Inc. Published 2016 by John Wiley & Sons, Inc.

fluorophore handling; (3) there is instrumentation capable of detection at ultralow concentrations; (4) portability of instrumentation for onsite detection; and (5) the relatively long shelf life of fluorophores.[2]

Fluorescent nucleoside analogs have shown utility in a wide range of applications including but not limited to single nucleotide polymorphism (SNP) detection,[3–7] nucleic acid structure and function, and microenvironmental studies. Structure and function experiments have allowed for the resolution of hybridization events,[8] folding,[9] conformational change,[10] and enzyme action.[11] Microenvironmental probing studies with fluorescent nucleosides have shown nucleobase damage,[12] depurination/depyrimidation,[13] and base flipping.[9] To this day, new uses for fluorescent nucleoside analogs continue to emerge broadening the utility of the technique and driving development in the field.

Numerous modifications have been explored to introduce favorable fluorescent properties to nucleic acids. Classical fluorophores, such as fluorescein, rhodamine, and their congeners such as the Alexa dyes, have been appended to oligonucleotides by linkers to the sugar phosphate backbone or have been tethered to nucleobases. Modifications have also been conjugated to the base or utilize the base as the fluorophore itself. Due to the extensive number of modifications and the ongoing possibilities for modifications, numerous classifications and terms have been assigned to the fluorescent nucleoside field. Tor has divided the field into five categories: (1) chromophoric base analogs, (2) pteridines, (3) expanded nucleobases, (4) extended nucleobases, and (5) isomorphic bases.[14] Wilhelmsson has segmented the field in terms of internal (base) and external (sugar/phosphate) modifications.[15] Further classifications have been offered by Asseline[16] and terms such as base-discriminating fluorophore (BDF) have been introduced by Saito and coworkers.[17]

This review focuses upon modifications of *cytosine* for the fluorescent probing of nucleic acids. Typical modifications fall into two fields capable of *canonical* base pairing: base analogs possessing pendant fluorophores (extrinsic) and intrinsically fluorescent nucleoside analogs,[18] a term that has been in sporadic use and our group has attempted to popularize because of its clarity of description.

The pendant class often has an advantage in that higher overall brightness (defined as $\Phi \times \varepsilon$) is obtained by the attachment of a well-characterized fluorophore. The high brightness is usually due to the combination of efficient luminescence (high quantum yield, Φ) and large molar absorptivity coefficients (ε). Despite the usually favorable photophysical properties, this class suffers from a number of drawbacks. The use of a fluorophore covalently bound by a tether to a nucleobase can allow for the independent movement of the fluorophore and subsequent interpretation of these results can become complex, especially in FRET measurements. Also, since the fluorophore is remote from the base pairing moiety, it does not directly report on the environment; thus, it does not reflect hybridization at a specific base, protonation, or other electronic changes at the site of interest.[18]

Intrinsically fluorescent base analogs are those in which fluorescence is observed from the nucleobase itself and not from an appended (extrinsic) moiety. Intrinsically fluorescent nucleosides are attractive as the base itself communicates (micro)environmental changes, thus enabling one to elucidate events occurring in close proximity to the nucleobase. Modest chemical modifications can also yield stunning photophysical properties allowing them to be competitive with traditional fluorescent probes. Intrinsically fluorescent base analogs have historically been able to communicate hybridization change save a handful of exceptions.

8.2 DESIGN AND CHARACTERIZATION OF FLUORESCENT C ANALOGS

Described by Kossel and Steudel in 1903,[19] the pyrimidine cytosine enjoys a three hydrogen bond Watson–Crick face for complementary pairing with guanine. Cytosine exhibits some flexibility in its binding modes due to the basic N^1 position. Unprotonated cytosine forms the normal W/C bond and in this arrangement may also be found in the underlying duplex portion of a triple helix. In the purine motif triplex, guanosine of the underlying duplex is recognized by G, whereas in the pyrimidine motif a protonated C is engaged in Hoogsteen bonding to the G of the underlying triplex[20] (Fig. 8.1). Hemi-protonated C-rich sequences are also able to form interesting intercalated tetraplex structures.[21]

Cytosine is readily halogenated at the 5-position, providing a handle for well-established carbon–carbon bond forming chemistry. Modification of the exocyclic amine also provides entry into fluorescent C analogs. Modifications of the C scaffold either by manipulation at the 5-position or of the amine generally produces a C analog, where additional substituents or structures are found in the major groove

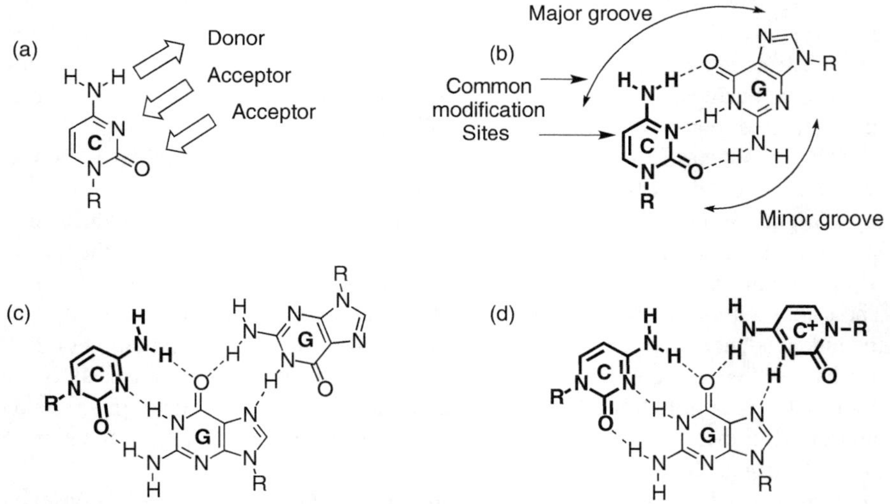

Figure 8.1 Cytosine-based hydrogen bonding motifs. (a) Hydrogen bond donor and acceptor site of cytosine nucleobase, (b) cytosine–guanine Watson/Crick base pair, (c) purine motif triplet, and (d) pyrimidine motif triplet.

(Fig. 8.1). These modifications tend to be nonperturbing in nucleic acids and in many cases increase duplex stability by favorable base stacking or additional hydrogen bond engagement of G. Modification at the 6-position is generally regarded to be detrimental due to the steric interaction of the sugar moiety in the anti-glycosidic conformer.[22]

8.2.1 1,3-Diaza-2-Oxophenothiazine (tC)

R = ribose, deoxyribose,
N-(2-aminoethyl)-glycine

The tricyclic cytosine (tC) analog was designed as a helix stabilizing modification for antisense applications in 1995 by Matteucci and coworkers. Original studies showed melt temperature increases with respect to 5-methylcytidine and good discrimination between guanine and adenine.[23] From 1995 to 2001, tC saw modest development with patent applications filed in the year of its first publication. In 2001, tC underwent a revival as a fluorescent nucleobase analog through the work of Wilhelmsson et al.[24] and has since seen thorough characterization and application. In their work, Wilhelmsson et al. showed that tC could be selectively excited over DNA (λ_{abs} = 260 nm) at a wavelength of 375 nm with corresponding emission at 505 nm.[24] CD and NMR studies showed tC readily adopting a position in B form DNA irrespective of neighboring base combinations. As was observed by Matteucci in 1995, a melt temperature increase was observed by the incorporation of tC into the synthetic oligodeoxynucleotides (ODNs).[25] Quantum yield determinations by Albinsson and coworkers showed moderate quantum yield values for the free nucleoside (Φ = 0.13) in water and a slightly higher quantum yield for the acetic acid derivative of tC (Φ = 0.16).[26] Incorporation of tC into ODNs produced slightly greater quantum yields than that of the free nucleoside. Values ranged from 0.17 to 0.24 in the single-stranded form and 0.16–0.21 in the double-stranded form. Very little change in quantum yield was observed with respect to flanking bases or incorporation into dsDNA making it a relatively insensitive analog to microenvironment. It was found that tC displayed single fluorescence lifetimes in oligonucleotides and on average over 10 sequences obtained lifetime values of τ_f (single-stranded) = 5.7 ns and τ_f (double-stranded) = 6.3 ns.[15] The monoexponential decay further bolstered support provided by NMR studies that tC occupied a well-defined position and geometry within a DNA helix.[26]

The photophysical characteristics defining tC have made it a unique exception to the otherwise hybridization communicating class of nucleosides. Its relative insensitivity to the environment and combination of firm/predictable stacking, low base flipping rate, and single fluorescence lifetime were thought to be advantageous for the observation of rotational dynamics of up to ~50-mer single-stranded and double-stranded oligonucleotide and oligonucleotide/protein complexes.[26] tC has seen application in FRET,[27] in DNA[28]/RNA[29] polymerase experiments in their respective ribo- and deoxy-forms, and in DNA/protein interaction experiments.[30]

8.2.2   1,3-Diaza-2-Oxophenoxazine (tC^O)

R = ribose, deoxyribose,
N-(2-aminoethyl)-glycine

The oxo variant of tC, referred to as tC^O, displays photophysical properties different from those of tC. Also initially synthesized by Matteucci and coworkers for the same purposes as tC, tC^O saw revival years later by Wilhelmsson and coworkers for its fluorescence properties. It is excited at a wavelength (~360 nm) well resolved from that of DNA with emission observed at ~465 nm. Acting similarly to tC, tC^O achieves firm stacking and acceptance into B form DNA with minimal perturbations to its environment. On average, a melt temperature increase of ~3 °C could be observed for tC^O incorporated ODNs. Pyrimidines 5′ to tC^O caused a +5 °C change in melt temperature whereas purines 5′ to tC^O were observed to not cause a significant change at all. Albinsson proposed that careful selection of the 5′ neighbor could allow for potential duplex stability control. Somewhat surprisingly, tC^O exhibited a greater sensitivity to its environment in comparison to its thio congener. The free nucleoside obtained a quantum yield of 0.30 but when incorporated into a single-stranded ODN, tC^O quantum yields varied from 0.14 to 0.41. A strong quenching effect was observed by sequences containing G at the 5′ side of the fluorescent C analog affecting one-quarter of the possible nearest neighbor combinations. The other nearest neighbor combinations obtained quantum yields from 0.29 to 0.41, where some of the brightest variations retained a 3′-G. Fluorescent decays proved more complex in the tC^O case and were fitted against two fluorescent lifetimes.[31]

The photophysical properties of tC^O became less variable upon incorporation into dsDNA. Quantum yields averaged 0.22 ± 0.05 irrespective of base sequence and complement. Furthermore, all emission decay curves were fitted to a single fluorescence lifetime. While structurally similar to tC, the tC^O analog exhibited microenvironment sensitivity and much greater brightness than that of tC.[31] tC^O remains one of the brightest intrinsically fluorescent nucleoside analogs in a duplex. tC^O has been used in the monitoring of melting processes of complex nucleic acids,[32] DNA polymerase experiments,[30] PCR product labeling[33] and in the development of the first nucleobase

FRET pair.[34] 1,3-diaza-2-oxophenoxazine has further been incorporated into peptide nucleic acid (PNA) and has been observed to exhibit unusual sequence-dependent binding affinities.[35]

8.2.3 7-Nitro-1,3-Diaza-2-Oxophenothiazine (tC$_{nitro}$)

In 2009, the Wilhelmsson group reported the development of the first nucleobase FRET pair.[34] The FRET pair utilized the previously described tC^O as a donor with the then newly developed tC analog – tC$_{nitro}$ as the acceptor. It was thought that the characteristics of the tC analogs (*vide supra*) made it a viable candidate to obtain the highest possible control over donor/acceptor orientation in nucleic acids systems.[15] This control would then allow for the mitigation of extraneous probe movement or nucleic acid perturbations caused by traditional molecular beacon technologies. Thermal denaturation experiments yielded stabilization on par with those observed for tC and tC^O with ready incorporation into B form DNA.[34] tC$_{nitro}$ was characterized as having an absorption band from 375 to 525 nm neatly overlapping the emission band of tC^O.[36] It was found that distances of 2–13 bases could be monitored reliably, where the FRET efficiency was dependent upon both distance and orientation.[34]

8.2.4 G-Clamp and 8-oxoG-Clamp

"G-clamp"   "8-oxo G-clamp"   R = ribose, deoxyribose, N-(2-aminoethyl)-glycine

Variations of the tC scaffold were explored by Matteucci yielding the "G-clamp", designed to engage guanine on both the W/C and Hoogsteen faces. Stability experiments yielded an +18 and +16 °C increase in melt temperature for complementary

DNA and RNA targets compared to the 5-methylC control.[37] Matteucci and coworkers further showed the utility of the C analog as a potent antisense modification in phosphorothioate-modified oligonucleotides.[38] The G-clamp has also enjoyed incorporation into PNA[39, 40] with similar results to those observed in DNA,[37] and RNA albeit exhibiting sequence-dependent binding affinities.[35]

It was not until 2007 when Sasaki and coworkers developed a probe for 8-oxoG did the G-clamps' potential as a fluorescent C analog come to light. Three amine-protected G-clamp variations (Cbz, Bz, and Ac) were synthesized in the hopes of engaging additional hydrogen bond interactions with the 7-NH group of 8-oxoG or repulsive interactions with G. All the G-clamp variants exhibited fluorescence (λ_{excit} = 365 nm, λ_{emiss} = 450 nm) allowing for binding studies through fluorescence quenching. Titration experiments of the free nucleosides showed that in buffered chloroform, G-clamp became quenched upon the addition of 8-oxo-dG and moderately quenched by the addition of dG.[41] The protected variants exhibited quenching only upon the addition of 8-oxo-dG. Quenching was not observed upon addition of the other nucleosides.[41] Less impressive selectivity for 8-oxoG over G was observed in unbuffered chloroform and preliminary attempts at screening in aqueous media showed good potential but required the components to be solubilized by detergent (Triton X-100) exceeding the critical micellar concentration.

Studies on the Cbz variant of 8-oxoG-clamp surprisingly showed slight destabilization compared to dC (when located in the center of an ODN) and no change was observed when the C analog was terminally located. It was hypothesized that H-bond formation between the 7-NH functionality of 8-oxoG and the G-clamp was not successful due to steric repulsion of the phenyl ring and the sugar phosphate backbone of the complementary strand. It was further hypothesized that locating the 8-oxoG-clamp at the terminal position proved fruitless due to solvation effects of the aqueous environment disturbing H-bond formation. Fluorescence characterization of the ODNs showed that selective quenching of the fluorescent C analog by 8-oxoG was retained, but little selectivity (as determined by T_m) between dG and the oxidation product was observed.[42]

The Sasaki group initiated further studies to improve upon the selectivity of the 8-oxoG-clamp family and determined that improved binding properties were obtained by the use of carbamate-protected amines as opposed to the tested acyl derivatives.[43] From this, it was assumed that the sp^3 oxygen of the carbamate was better located for hydrogen bonding with 8-oxoG (Fig. 8.2) and therefore responsible for the selectivity of 8-oxoG-clamp. Of the numerous clamps tested, an oxoG-clamp with a 2-pyren-1-yl-ethoxycarbonyl moiety (Fig. 8.2) proved to be the best for recognizing the oxidation product of G. Increased base stacking by the pyrenyl moiety was further proposed (not shown). Incorporation of this probe into an ODN has yet to be reported.[43]

In an attempt to address the quenching effect of dG upon G-clamp, Sasaki and coworkers investigated 7-substitutions upon the 1,3-diazaphenoxazine scaffold. It was thought that manipulation of the S$_0$ state by 7-substitutions would mitigate the hypothesized PET quenching to the native G species and improve sensitivity toward the reactive oxygen product 8-oxoG.[44] Of the derivatives tested,

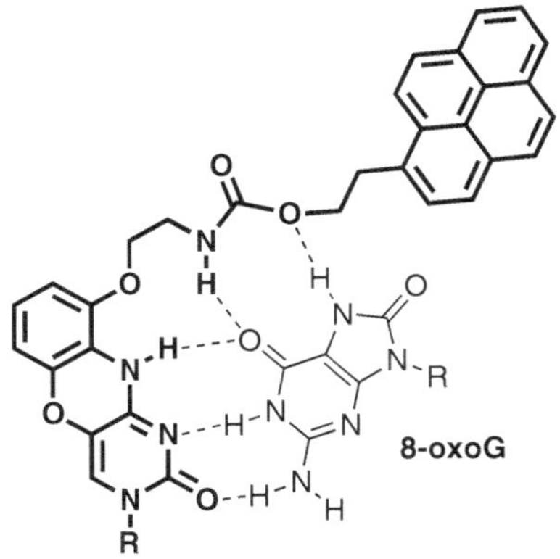

Figure 8.2 Sasaki's modified G-clamp and its proposed binding mode to 8-oxoG.

a 7-phenyl-substituted 8-oxoGclamp was found to be insensitive to dG while maintaining high selectivity for 8-oxo-dG.[44] Lately, Sasaki's group has recognized that 8-oxoG would likely adopt a syn-glycosidic conformer with in a duplex, resulting in a different H-bonding face being presented to the "G-clamp" and new designs have addressed this issue.[45]

8.2.5 Ç and Çᶠ

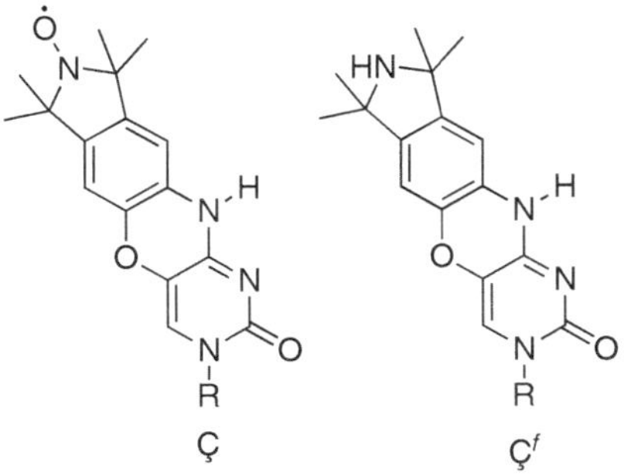

R = deoxyribose

Desiring a rigid spin label for EPR studies on distances and molecular motions in nucleic acids, Sigurdsson and coworkers looked toward the well-characterized tC base for their purposes.[46] En route to the target molecule (Ç) an intermediate (Çᶠ) was observed to be highly fluorescent. Mild reduction of the spin-labeled nucleoside by dithiothreitol or sodium dithionite yielded other fluorescent nucleoside variants (a hydroxylamine and a sulfurous ester). The hydroxylamine was quickly reoxidized upon exposure to oxygen, whereas the sulfurous ester did not. Similar to other tC analogs, their excitations were observed at ~360 nm with emissions at ~450 nm.[46] The quantum yield of the sulfurous ester was found to be considerably less (Φss = 0.09, Φds = 0.03) than those observed for other tC analogs.[46] Later studies found that reduction of the spin label with sodium sulfide in water produced the corresponding amine (C^f) with a quantum yield of 0.31. This fluorescent C analog was then used to discriminate between the four natural bases for the purposes of SNP detection[47] while C^f has also been utilized as a dual EPR/fluorescence probe for aptamer studies.[48]

8.2.6   Benzopyridopyrimidine (BPP)

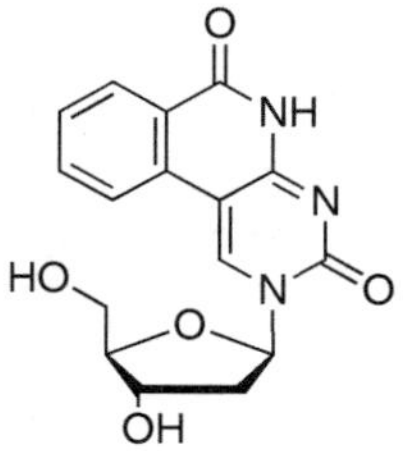

In 2003, Saito, then at Kyoto University, coined the term base-discriminating fluorophore (BDF) to describe his fluorescent nucleobase analog BPP. Fluorescent probes had been commonplace up to this time, yet the field lacked a probe with the ability to provide a clear distinction of the opposite base by fluorescence change. Saito's BPP was the first such example of a base analog that could provide clear distinction of the purine bases on the complementary strand by fluorescence. Saito proposed using this base as an on/off reporter for A/G SNP detection. It was found that BPP was W/C pairing competent for G and was further able to pair with A via a wobble mode (Fig. 8.3).

Similar to the other C analogs described herein, BPP exhibited excitation ($\lambda_{\text{excit}} = 347$ nm in a single-stranded ODN) resolved from that of DNA. Further photophysical characterization revealed that a BPP-containing ODN complementarily matched to a target ODN became quenched ($\Phi = 0.0018$), while hybridization with an A mismatch caused fluorescence emission at 390 nm ($\Phi = 0.035$).[49] Despite this promising on/off response making naked eye judgment of SNP detection possible, the BPP base suffered from a number of shortcomings. BPP showed sensitivity to nearest neighbor combinations, where quenching was observed for flanking G/C bases.[17] When BPP was held constant at the 5′-terminus and the cross strand 3′ base was varied, a fluorescence decrease of 81% comparative to that of the free nucleoside

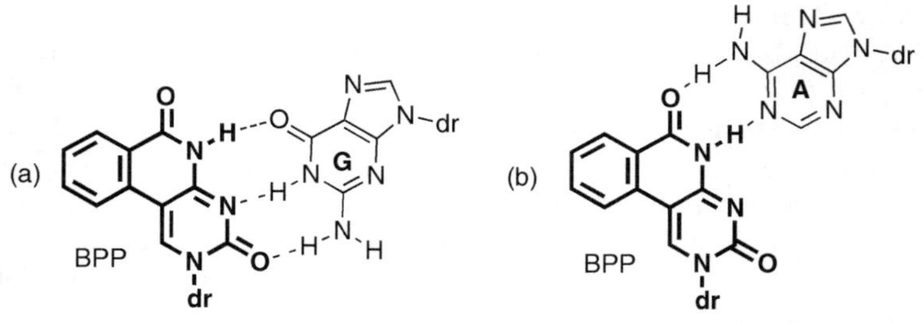

Figure 8.3   Proposed binding of BPP to guanine and wobble pairing to adenine.

was observed when G was present. Fluorescence response remained constant in the presence of the other base combinations.[17] Despite the difference between on/off signaling (20-fold), the quantum yield values obtained were undesirably low. Further complicating matters, the single-stranded state was observed to be more fluorescent than the A mismatch. These observations prompted the development of another fluorescent C analog NPP (*vide infra*). BPP was successfully used in A/G SNP typing in both DNA and RNA[3, 49] as the deoxynucleoside.

8.2.7 Naphthopyridopyrimidine (NPP)

The Saito group continued their work by the development of the fluorescent C analog naphthopyridopyrimidine (NPP). It was observed that unlike BPP, NPP in the single-stranded state was less emissive than that of the wobble pair with A. It was hypothesized that some quenching was observed in the single-stranded state through hydrophobic aggregation of the nucleobases in aqueous media.[50] The fluorescence response was greatly improved over that of BPP, where the greatest response was obtained for A ($\Phi = 0.096$) and quenching was again observed for G ($\Phi = 0.007$).[17, 50] No data on neighboring base effects have been reported.

8.2.8 dC^{hpp}

Polycyclic C analogs have also been investigated by Sekine and coworkers at the Tokyo Institute of Technology. Their C analog dC^{hpp} was determined to have a quantum yield of 0.12 for the free nucleoside in 10 mM sodium phosphate with an observed excitation at 300 nm and emission at 360 nm. Thermal denaturation showed a slight duplex stability increase of 1.1 °C for the complementary sequence. While a modest change was observed for the match case, a much more dramatic change was observed for the A mismatch in which a 10 °C increase was observed. It was recognized that the analog could form a base pair with adenine by way of a reverse Watson–Crick base pair (like BPP/NPP contributing to stabilization). It was also postulated that the analog might switch tautomeric form between the cytosine amine and imino form. ODN studies showed fluorescence quenching for the proper W/C base pair and no fluorescence change for the A mismatch.[51] No quantum yield was reported for the dC^{hpp} double-stranded and single-stranded states.

8.2.9   dC^{hpd}, dC^{mpp}, dC^{tpp}, dC^{ppp}

dC^{hpd} dC^{mpp} dC^{tpp} dC^{ppp}

R = deoxyribose

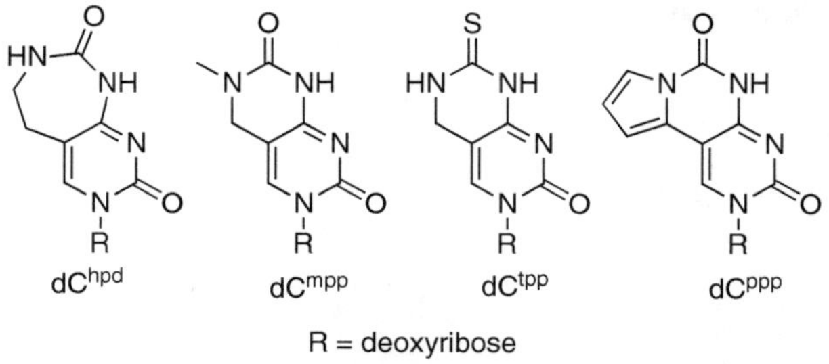

Interested in understanding the properties of dC^{hpp}, four analogs were synthesized and studied by the Sekine group in 2006. A study of the dC^{hpd} showed that planarity of the bicyclic structure was key to achieving favorable fluorimetric properties. If coplanarity was not achieved, significantly reduced fluorescence was observed. N-methylation of the carbamoyl (dC^{mpp}) also produced fluorescence quenching effects with respect to the dC^{hpp} scaffold. The importance of maintaining the carbonyl moiety also became apparent when zero fluorescence was observed for the dC^{tpp} variant. The most interesting fluorescence properties were those attributed to dC^{ppp}. Introduction of the pyrrolo ring induced a marked redshift in comparison to the original dC^{hpp} nucleoside. Excitation was reported at 360 nm (60 nm red to NPP) with emission at 490 nm affording a large Stokes shift of 130 nm. The new fluorescence emission was dramatically redshifted from that of the dC^{hpp} analog by 121 nm. The quantum yield of dC^{ppp} was determined to be 0.11 in 10 mM sodium phosphate (pH 7)[52] but has yet to be incorporated into oligonucleotides for study.

8.2.10   dC^{PPI}

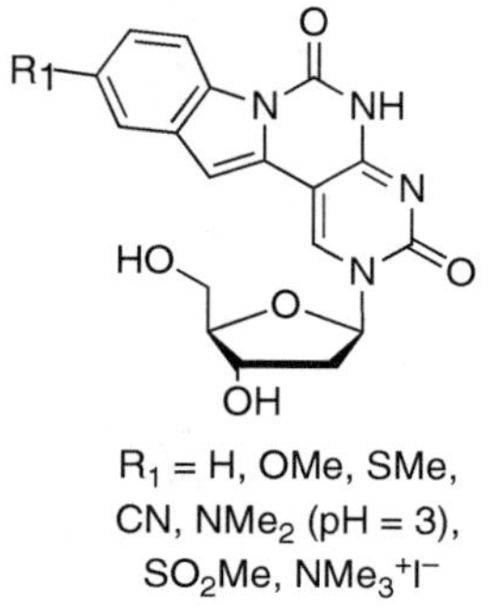

R$_1$ = H, OMe, SMe,
CN, NMe$_2$ (pH = 3),
SO$_2$Me, NMe$_3$$^+I^-$

The Sekine group continued their C analog studies by way of their analog – dC^{PPI}. A range of substituents were tested for their effect on fluorescence in a number of solvents of varied polarity.[53] In a buffered aqueous environment (pH = 7.4) the unmodified dC^{PPI} nucleoside showed considerable quenching ($\Phi = 0.006$) compared to the previous dC^{hpp} analog ($\Phi = 0.11$). Electron-donating substituents under these conditions also lent themselves to fluorescence quenching while the electron-withdrawing groups increased fluorescence intensity considerably. The greatest quantum yield obtained for the analog possessing the strongly electron-withdrawing substituent NMe_3^+, which was determined to have a quantum yield of 0.096 while the dimethylamine variant returned a yield of <0.001. Stern–Volmer titrations of the nucleosides revealed considerable sensitivity of the R_1 = CN variant toward dGMP, whereas dGMP least affected the fluorescence emission of the methoxy derivative. Oligonucleotides incorporating dC^{PPI} derivatives R_1 = H, OMe, and CN were observed to become quenched upon base pairing with G in the complementary strand. This quenching was attributed to a PET mechanism. Quenching due to G was also observed in the single strand thereby exhibiting sensitivity to neighboring bases. The fluorescence of oligonucleotides incorporating dC^{PPI} derivatives increased upon pairing to the counter strands except in the complementary G case.[54]

8.2.11   dxC

Benzo-expanded nucleoside analogs were first reported by Leonard *et al.* in the mid 1970s.[55] He would utilize a benzo-expanded A in the investigation of ATP-dependent enzymes[56] and later report a deoxybenzoA congener.[57] He would further explore size-expanded nucleosides with an expanded GTP variant.[58]

The Kool group of Stanford drew inspiration from Leonard's work and produced the complete set of expanded DNA mimics of all four natural nucleobases and incorporated them into oligomers to study their hybridization properties.[59] Synthesized and described by the Kool group in 2004, xDNA has been the subject of detailed studies for base pairing and helix forming properties.[60–64] In addition to increased stability, the benzo homologation has imparted extended conjugation endowing the xDNA bases with fluorescent properties.

It was found that xDNA could be excited at either the natural wavelength of 260 nm or at longer wavelengths (320–333 nm) for fluorescence emission. ssODNs containing successive terminal dxC bases (1–4) changed their emission properties with an increase in overall quantum yield with respect to number of dxC bases. Furthermore, interesting and somewhat peculiar quenching or fluorescence

enhancement properties were observed when the nature and number of the complementary base was varied. Seemingly, the most interesting was a severe quenching
effect observed for dxC-containing ODNs opposite complementary poly-G strands,
where up to a 95% quenching effect was observed.[59] dxC has recently been used in
template-independent polymerase studies.[65]

8.2.12   rxC

The Kool group further explored expanded nucleoside analogs and ported their
design to RNA and developed a full genetic set of size-expanded ribonucleosides. The
rxC analog unlike the dxC variant lacked a 6-methyl substituent reflecting a required
change in synthesis. The ribonucleoside was synthesized in an overall 19.2% yield
over eight steps.[66] The photophysical characteristics of rxC in methanol were found
to be similar to those of the dxC congener. The xrC nucleoside has yet to be reported
in ORNs.

8.2.13   Methylpyrrolo-dC (MepdC)

6-Substituted pyrrolo-dC has had a relatively long history as a fluorescent C
analog dating back to the late 1980s.[67] Despite a handful of reports during the
1990s, methylpyrrolo-dC went relatively exploited until the turn of the century. In
the early 2000s, methylpyrrolo-dC saw a resurgence[68] in use and has since become
one of the more popular fluorescent C analogs, probably due in large part to its
commercial availability.[69] Methylpyrrolo-dC has been shown to act similarly to C in
terms of hybridization selectivity and stability.[70] While the pC base itself possesses
a quantum yield of ~0.038 (pH 7 phosphate buffer), the ribo- and deoxy-congeners
undergo an approximate 39% decrease in fluorescence.[71] Quantum yields reported
for single-stranded pyrrolo-dC-containing ODNs were found to be approximately
0.03.[71] Preliminary studies of neighboring base effects show increased sensitivity
toward a G neighbor but relative insensitivity to the other natural bases. Some uses
of methylpyrrolo-dC include the characterization of the transcription bubble of T7

RNA polymerase,[68] the kinetics of DNA repair by a human alkyl transferase,[72] investigations of the HIV-1-polypurine tract,[73] and the structure and dynamics of single strands able to form DNA hairpins.[74] Incorporation of ribo-MepC into oligomers has been achieved, and its ability to signal structural change is preserved.[8, 10]

8.2.14   5-(Fur-2-yl)-2′-Deoxycytidine (C^{FU})

The Tor group at UCSD investigated a number of conjugated five-member ring systems based upon 2-phenylfuran. 2-Phenylfuran is a known fluorophore ($\lambda_{excit} = 280\,\text{nm}$; $\lambda_{emiss} = 320\,\text{nm}$) with a desirably high molar extinction coefficient of $20{,}000\,\text{M}^{-1}\,\text{cm}^{-1}$ and good quantum yield ($\Phi = 0.4$).[14] By conjugation of a furan moiety to the six-member aromatic system of cytosine, Tor hoped to emulate the favorable properties of the parent molecule. Following a previously reported synthesis,[75] Tor produced the furan-labeled C analog from 5-iodo-2-deoxyuridine. Unfortunately, the quantum yield of the C^{FU} nucleoside proved to be less than ideal with a value of 0.02 in water. Although possessing a low efficiency, the C^{FU} base could still be selectively excited over DNA at ~310 nm with emission at ~440 nm. The Tor group proceeded to investigate the fluorescent C analog as a potential candidate for the detection of 8-oxoG. C^{FU} was found to be a good candidate as a nonperturbing DNA base as determined by thermal denaturation. Stern–Volmer titrations proved the sensitivity of C^{FU} to 8-oxo-G and relative insensitivity to unmodified G[12]. Anticipating transverse mutation, the fluorescence response with respect to T was also measured and determined to provide the greatest fluorescence intensity. Moreover, the emission wavelength was observed to change with complementary base. It was proposed that C^{FU} could facilitate rapid and nondestructive real-time fluorescence-based methods for the *in vitro* monitoring of oxidative stress.[12]

8.2.15   Thiophen-2-yl pC

R = H, OH

Despite the success of the heterocycle-conjugated and fused pyrimidines, the Tor group has sought ways to increase the efficiency of their substitutions. In looking to improve quantum yields of their furan- and thiophene-substituted pyrimidines, the group turned their attention to the pC core, which had been shown to possess favorable fluorescence characteristics in other systems. The group synthesized both the deoxy and ribose variants of the thiophen-2-yl pC base. Utilizing an acetylation protection procedure followed by Sonogashira cross coupling, they were able to synthesize a thiophene cross-coupled C analog. Foregoing acylation of the exocyclic amino group, it was observed that the intermediate nucleoside resisted annulation by copper and therefore formation of the pyrrolo ring. The intermediate nucleoside was then screened against other metal catalysts to induce cyclization. It was found that sodium tetrachloroaurate(III) dihydrate produced the desired cyclized product in low-to-moderate yield. Deprotection of the 3' and 5' hydroxyls was carried out by ammonia treatment to yield the free nucleosides.[76]

Photophysical characterization of the deoxy and ribose thiophen-2-yl pC nucleosides yielded little to no difference in their photophysical properties despite their different sugar moieties. It was found that they underwent excitation at ~370 nm and emission at ~471 nm in water with an efficiency of ~0.42. A greater efficiency was observed in dioxane (~0.48)[76] with redshifted absorption and blueshifted emission corresponding to decreased Stokes shifts with respect to those in water. The thiophene substitution on pC led to greater molar extinction coefficients, especially compared to the fused thiophene version (see Section 8.2.16). The extinction coefficients when taken into consideration with quantum yield led to brightness factors that were approximately 14–24 times brighter than those of the MepC or MepdC.

8.2.16   Thiophene Fused pC

R = H, OH

Studies on solvent viscosity and fluorescence showed that heterocycle conjugated pyrimidines displayed increased fluorescence in more viscous environments.[77] It was posited that the increased fluorescence was due to a reduction in free rotation of the heterocycle moiety with respect to the pyrimidine. In accordance with this observation, it was proposed that the removal of rotatable bonds by fusion of the thiophene ring with that of the pC base would yield increased efficiencies.[76]

The thiophene fused pC was synthesized by Stille coupling of the thiophene heterocycle to unprotected or O-acetyl protected 5-iodouridine or 5-iodo-2-deoxyuridine.

Bromination of the thiophene ring followed by atom exchange and cyclization under basic conditions afforded to thiophene fused pC analog in overall ~20% yield.[76]

Similar to that of the thiophene pC analog described previously (*vide supra*), the ribose and deoxyribose variants of the thiophene fused system displayed nearly identical photophysical properties to one another. Excitation at 357 nm and emission at ~476 nm was observed with a low quantum yield of 0.01 in water. Quantum efficiency in dioxane, however, was determined to be excellent at approximately 0.73 with redshifted absorption and blueshifted emission with respect to those values in water. Despite the great gain in quantum yield, the molar extinction coefficients of the thiophene fused systems were determined to be only about 1/3–1/5 those of the thiophen-2-yl pC systems. This decrease in extinction coefficient led to decreased brightness factors that were only a fraction of those of the thiophene pCs. Despite the decreased brightness factors, a much greater fluorescence change was observed between that of the thiophene fused pC in water and that in dioxane, where an approximate 100-fold difference was observed.

8.2.17 Thieno[3,4-d]-Cytidine (thC)

Further expanding their understanding of thiophene-modified nucleosides; the Tor group completed a synthesis of an emissive RNA alphabet based upon a thieno[3,4-*d*]-pyrimidine core. The thC analog was accessed by treatment of thU with 2,3,4-triisopropylbenzenesulfonyl chloride followed by methanolic ammonolysis. Utilizing this method, thC was accessed in 37% yield over two steps.[78]

Crystal structure analysis of the thC ribonucleoside showed what appeared to be an intermediate sugar conformation resembling that of a C2′-endo pucker. The crystal structure further revealed that the ribonucleoside adopted an anti-orientation despite its modification.[78]

Characterization of the fluorescence properties for the thC analog was undertaken in water and dioxane. Little difference was observed in water and dioxane for the excitation and emission maxima, where excitation was observed to fall at ~324 nm with emission at ~425 nm. While the wavelengths of excitation and emission remained relatively unchanged with respect to solvent, a drastic difference in quantum yield was observed in water ($\Phi = 0.41$) in comparison to that of dioxane ($\Phi = 0.01$).[78] The thC analog was further found to obtain a low molar extinction coefficient at λ_{em} in both solvents thereby providing low brightness factors. Correlation of Stokes shifts with respect to changing solvent composition (dioxane/water) provided the lowest polarity sensitivity with respect to the other members of the emissive alphabet.

8.2.18 Triazole Appended

R = deoxyribose

Triazole appended cytidine analogs were investigated by the Hudson group in the hopes of developing a method for the rapid derivatization of the C nucleobase. This rapid derivatization would provide facile entry into BDFs. The research would further judge the amenability of click chemistry to produce modifications on the C base. It was also postulated that careful selection of the alkyne substrate could also allow for entry into polymer chemistry or electrochemistry. Yields of the triazole appended nucleosides varied from poor (15% for the dithiophene) to good (78% for the phenyl). The fluorescence efficiency for the triazolyl nucleosides were disappointingly poor although on par with many fluorescent bases in use today (Φ ranged from 0.01 to 0.08 in EtOH). The triazolyl nucleosides displayed pronounced solvatochromicity and in aqueous conditions underwent on average a 10-fold decrease in fluorescence emission. The solvatochromism, especially for the fluorenone-labeled base, was observed to be nonlinear with respect to solvent composition change (EtOH $\rightarrow$ water) and it was hypothesized that multiple excited states could be attributed to the solvatochromism. The triazolyl nucleosides have yet to be incorporated into oligonucleotides for study,[79] primarily due to their weak fluorescence.

8.3 IMPLEMENTATION

Fluorescent C analogs have been successfully incorporated into synthetic DNA, RNA, and PNA oligonucleotides, and there is a wide variety of applications, which have demonstrated their potential usefulness. Once again the reader is referred to recent reviews on the area for a comprehensive overview,[9, 15, 17, 18, 60] herein, we describe the synthesis, incorporation, and subsequent studies of substituted pC analogs that have been carried out at The University of Western Ontario.

The general features for the synthesis of pC analogs are illustrated in what follows. The chemistry utilized depends on the scaffold of the oligonucleotide analog. For PNA, the starting point most conveniently has utilized cytosine[80, 81] or uracil[82] nucleobases. For nucleic acid backbone structures, the deoxy- or ribonucleosides are used.[70, 83] The appropriate substrates are prepared for derivatization by halogenation at the C5 position, most often iodination, in preparation for Sonogashira/Castro–Stephens cross-coupling chemistry. The desired, fused bicyclic structure is achieved by the intramolecular cyclization (5-endo dig annulation) of the 5-alkynyl pyrimidine. The annulation reaction (Scheme 8.1. Approach 1)

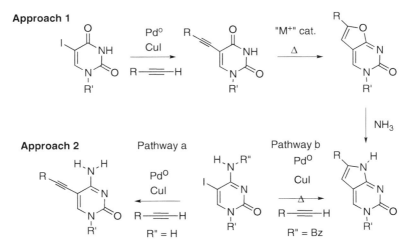

Scheme 8.1   Synthetic approaches to pyrrolcytidines via furanouracil/uridine (approach 1) or by starting with cytidine/cytosine (approach 2).

usually occurs under mild conditions for the uracil derivative, although there is some substance dependence. For instance, electron-rich alkynes cyclize more rapidly than those that are electron deficient. The cyclization reaction is metal,[84–89] base,[90, 91] or electrophile catalyzed[92] and is facilitated by conventional heating or microwave irradiation.[93] Lewis acidic, alkynophilic metals have proved to be effective for this type of reaction in general ($M^+ = Cu(I)$,[67] $Zn(II)$,[86] $Au(III)$,[76, 84] $Hg(II)$,[87] $Pd(0)$[89]). Utilizing approach one, the final transformation is the atom exchange reaction to convert the fluorouracil nucleobase to pyrrolocytosine by treatment with ammonia.[94] This step is critical, of course, because the cyclization of uracil to furanouracil results in a change of the hydrogen bond donor (HBD) and hydrogen bond acceptor (HBA) sites of the base such that no longer pairs with any of the natural nucleobases. Complementarity to guanine is manifested once the furanouracil is converted to the pyrrolocytosine (Fig. 8.4).

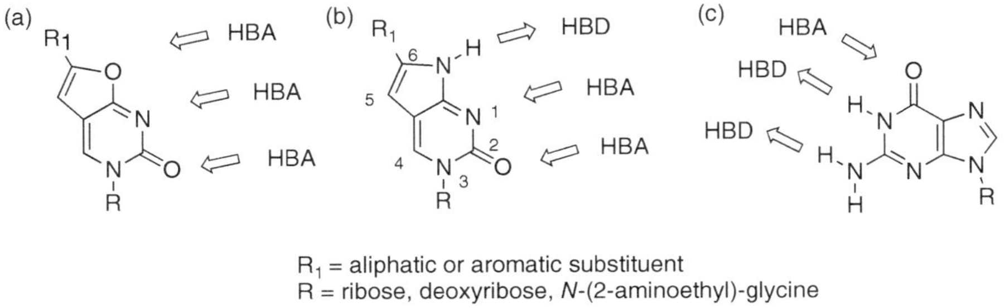

(a)   (b)   (c)

R_1 = aliphatic or aromatic substituent
R = ribose, deoxyribose, N-(2-aminoethyl)-glycine

Figure 8.4   Comparison of hydrogen bond donor (HBD) and hydrogen bond acceptor (HBA) faces of (a) furanouracil, (b) pyrrolocytosine, and (c) guanine.

The second approach starts with cytosine/cytidine and the stage is set once again for the cross-coupling reaction by iodination of the base. In approach 2 (Scheme 8.1), partitioning of the products between the simple cross-coupled 5-alkynylcytosine and the annulated pyrrolocytosine is conveniently controlled by the substrate. If the nucleobase is unprotected (pathway a), and the conditions are not forcing (room temperature, for instance), then the 5-alkynylcytosine derivatives are achieved in good yield.[95] For the intramolecular cyclization to occur (pathway b), acylation of the N^4 must be undertaken prior to Sonogashira chemistry. We preferably use the benzoyl derivative. Using this substrate and elevated temperatures (60–80 °C), a domino reaction sequence of cross coupling and cyclization occurs during which there is partial loss of the benzoyl group. In our experience, the N-benzoylpyrrolocytosine is not fluorescent and also not entirely stable to silica gel column chromatography conditions. Thus, the benzoyl group is removed to facilitate isolation and purification of the desired pyrrolocytosine product. Forgoing benzoylation of the exocyclic amine, cyclization has been reported using a gold catalyst providing low-to-moderate yields.[76] The *best* synthetic route toward a pC analog is case specific and must be chosen appropriately as each modification to the pC scaffold will generate its own synthetic challenges.

Utilizing the above-outlined syntheses, the Hudson Group at The University of Western Ontario has explored modifications of the pC scaffold and their consequent fluorescence properties in PNA, DNA, and RNA.

8.3.1 PNA

Initially, with the intent of synthesizing 5-alkynylpyrimidines to determine their effect on the biophysical properties of PNA oligomers,[96] Hudson and coworkers "rediscovered" the cyclization of N^4-acyl-protected cytosine to pyrrolocytosine – an observation first made by Ohtsuka that had been largely underappreciated.[67] During these studies, the conditions were defined for synthesis of the simple cross-coupled products versus the annulated pyrrolocytosine (*vide supra*). It was found that structurally simple 5-alkynylcytosine derivatives were luminescent,[80] much as is the case for uracil derivatives[82]; however, the pCs were found to be better fluorophores.[97] Pyrrolocytosines that possess aromatic substitution are remarkably good fluorophores, better than those with aliphatic substitution (Fig. 8.5). All of the pCs shown are blue fluorophores ($\lambda_{emiss} \sim 450$ nm) except for the *para*-(N,N-dimethylamino)phenyl that shows a bathochromic shift ($\lambda_{emiss} \sim 500$ nm) and the *para*-nitrophenyl that displays weak orange fluorescence ($\lambda_{emiss} \sim 575$ nm), and an absorbance band that overlaps the emission of 6-phenylpyrrolocytosine (PhpC). Being the structurally simplest aromatic substituent of the series shown and having good brightness ($\Phi_{(EtOH)} = 0.63$, $\varepsilon_{(\lambda max)} = 6200$), this modification has been the most studied. Furthermore, good luminescence is maintained in water ($\Phi \sim 0.35$), giving approximately 10-fold greater the luminescence MepC.

The 6-phenyl-substituted pC derivative, PhpC acetate, was further converted into the Fmoc PNA monomer for PNA synthesis using standard protocols (Scheme 8.2). Notably, neither monomer synthesis nor oligomerization chemistry required pyrrole nitrogen protection. The hybridization behavior for the PNA toward DNA was

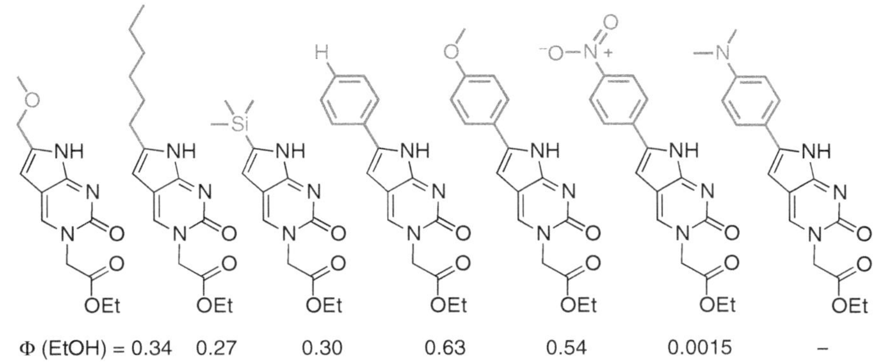

Figure 8.5 Structural variation of pyrrolocytosine derivatives and corresponding fluorescence quantum yields as measured in ethanol.

examined by temperature-dependent UV–vis spectrophotometry. Under standard ionic strength (100 mM NaCl, 10 mM Na_2HPO_4, 0.1 mM EDTA, and pH 7.1) the PhpC-containing PNA bound complementary DNA with a stability on par with the corresponding unmodified PNA displaying the general trend that an internal sequence modification was mildly stabilizing (0–2.5°C/insert), whereas a terminal modification was neutral or slightly destabilizing. In all the cases examined thus far, PhpC-containing PNA oligomers display excellent sequence discrimination for a complementary G versus mismatch that rival or best natural C.

In order to obtain an environmentally sensitive C analog with tighter binding properties, similar to that of Mateucci's G-clamp, the Hudson group further elaborated the PhpC architecture by the introduction of *ortho*-aminoethoxy substituents. The choice of substituent proved to be dual purpose: (1) increase solubility of the PhpC and (2) putatively engage the Hoogsteen face of G in an additional H-bond. These analogs were termed "G-clamps"[98] (Fig. 8.6). On the basis of molecular modeling, O^6 is more proximate to the pendant amine although the possibility of a bifurcated H-bond to O^6, N^7 cannot be dismissed.

The syntheses of the boPhpC and moPhpC G-clamp ethyl esters and eventual PNA monomers were accomplished by a domino Sonogashira cross-coupling/annulation reaction between the appropriately synthesized alkyne and ethyl (N^4-benzoyl-5-iodocytosin-1-yl)acetate (Scheme 8.2). The ethyl esters were then prepared for PNA synthesis by conversion into the Fmoc monomer.[98]

The alkynes were synthesized by displacement reactions of 2-iodophenol or 2,6-diiodophenol with Boc protected 2-bromoethylamine to afford the aryl Boc-protected aminoethoxy linker iodides. Further modification was achieved by Sonogashira cross coupling of TMS acetylene followed by deprotection to yield the appropriate alkynes[98] (Scheme 8.3).

Both moPhpC and boPhpC were incorporated into PNA oligomers and tested against RNA and DNA targets. Binding affinity for the DNA target increased by ~10°C for both moPhpC and boPhpC in the match case, which was ascribed to

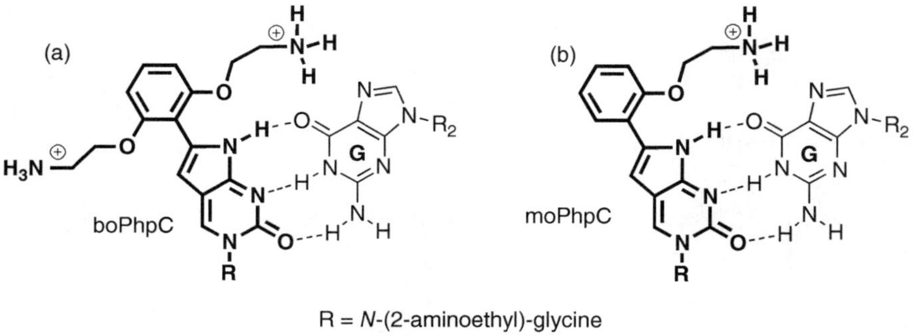

R = N-(2-aminoethyl)-glycine

Figure 8.6   Structures of the pyrrolocytosine G-clamp analog.

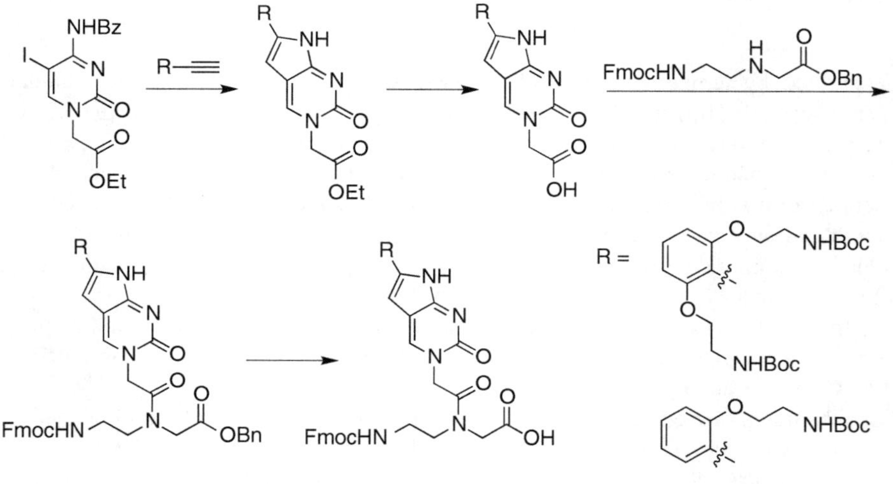

Scheme 8.2   Synthetic route to pyrrolocytosine-based PNA monomer.

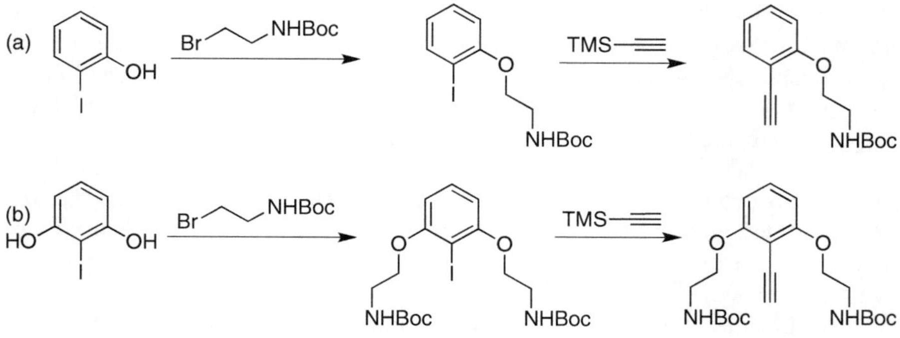

Scheme 8.3   Preparation of alkynes required for pyrrolocytosine G-clamp synthesis.

additional hydrogen bond interactions. A decrease of 10 °C in melt temperature was observed for the DNA mismatch cases. Binding affinity for the RNA target was only moderately improved. Quantum yield determination of the boPhpC ethyl ester provided a value of 0.32 in aqueous buffer with dramatic solvatochromatic properties. It was observed that quantum efficiency increased in lower polarity solvents (up to 0.75) and became less efficient in more polar media. Incorporation of boPhpC into ssPNA increased fluorescence efficiency to greater than that of the nucleobase in aqueous solution suggesting a less polar environment for the boPhpC in PNA due to hydrophobic collapse. Fluorescence of the boPhpC PNA was observed to become considerably quenched upon hybridization with the DNA target (~50%).[98]

Around the same time period, the Corey Group at UT Dallas Southwestern Medical School was using PNA oligomers for the selective inhibition of mutant HTT protein. Mutant HTT is responsible for Huntington's disease (HD), an incurable neurological disorder. The Corey group reasoned that silencing of the mutant HTT protein to be a useful strategy for the treatment of Huntington's disease. Furthermore, they reasoned that a successful strategy would require selective inhibition of the mutant allele while preserving expression of the wild-type allele as the wild-type protein is required for survival. Their approach was to use short PNAs complementary to the expanded triplet repeat portion of the mRNA encoding the protein huntingtin (HTT). For the clinical development of anti-HTT PNAs, it was required that potency and allele-selectivity be optimized. It was rationalized that PNAs modified with the boPhpC moiety might be beneficial for the aforementioned optimization as they provided the potential for oligomer affinity tailoring.[99]

When oligos containing the boPhpC moiety were added at 1 µM concentration, selective HTT inhibition was observed. The introduction of one or two boPhpC substitutions did not greatly increase the potency of inhibition of mutant HTT or improve selectivity. Alternatively, introduction of three or four boPhpC bases significantly eroded selectivity or potency.[99] This demonstrated that PNAs bearing the boPhpC modification behaved like unmodified PNAs in the antisense application but were not the optimal modification for this particular target. However, the boPhpC insert had the advantage of intrinsic fluorescence that permitted visualization of intracellular localization without the need for an appended fluorophore. Confocal fluorescence microscopy showed punctate intracellular PNA distribution consistent with known uptake/distribution mechanisms for PNA. In another study, fluorescent boPhpC PNAs were used to track the intracellular trafficking of anti-miRNAs.[100]

With the success of the PhpC G-clamp family upon binding to DNA, studies were undertaken to address its lower binding affinity toward RNA. Building on the PhpC scaffold, structural variants in which the position (*ortho-* or *meta-*), length of tether (ethyl or propyl), and nature of terminal group (amino or guanidino) were investigated. These studies identified led to *meta*-guanidinoethoxy-substituted PhpC (mmGuaPhpC). The monomer was synthesized in a manner similar to that of boPhpC and moPhpC with further modification by guanylation to provide the mmGuaPhpC PNA base (Fig. 8.7). Designed to engage G (Fig. 8.7) in a binding mode similar to that of Manoharan's guanidino G-clamp,[101] it was observed to increase binding affinity toward RNA by 9 °C, on average.[102]

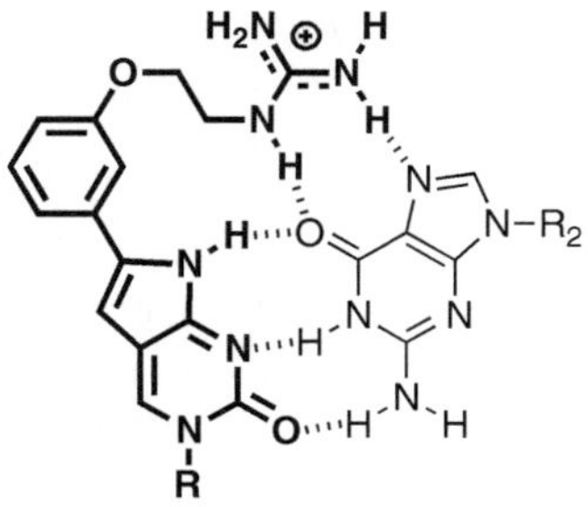

R = *N*-(2-aminoethyl)-glycine
R_2 = ribose, deoxyribose

Figure 8.7 Proposed structural model for the binding of the pyrrolocytosine-based guanidino G-clamp to complementary guanine.

In addition to increasing binding affinity of the PNA probe to RNA, it maintained its previously documented ability to communicate hybridization change.

8.3.2 DNA

The first of our pC studies in DNA reported the synthesis and incorporated into ODNs of the 6-methoxymethyl derivative (MmepdC) (Fig. 8.8) for the purpose of selective fluorimetric detection of guanosine-containing sequences.[103] The synthesis of the modified nucleoside allowed for pyrrolo-dC structure evaluation by comparison to the MepdC structural congener. The synthesis of the MOM-substituted pC utilized dimethoxytritylation protection of 5-iodo-2′-deoxyuridine followed by Sonogashira cross coupling and annulation with methylpropargyl ether. Smooth atom exchange and phosphitylation was reported to yield the phosphoramidite derivative in overall 42% yield from the nucleoside.[70, 103] The amidite was then subjected to DNA synthesis by the phosphoramidite method. Coupling efficiency was reported to be on par with those of the commercial amidites.[103]

A duplex stabilization of $+1.3\,^\circ$C (with respect to C) was observed when the MmepdC nucleoside occupied a central location in the ODN of study, whereas the

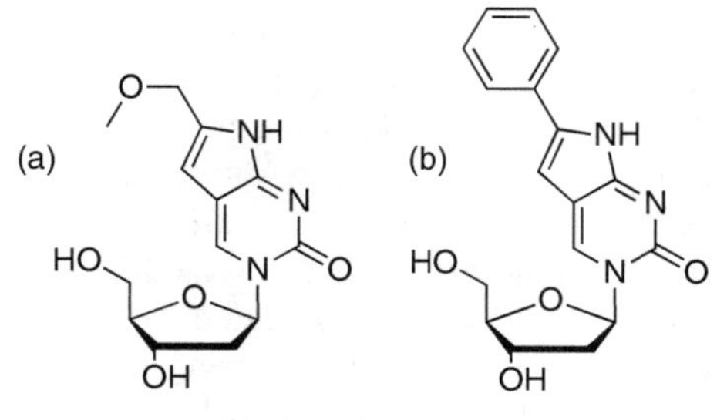

Figure 8.8 Structures of (a) ^{Mme}p-dC and (b) ^{Ph}pdC.

MepdC analog showed a destabilization of -4.7°C. Fluorescence responses for the C analogs were almost identical with a threefold fluorescence decrease (central position) upon hybridization to a complementary strand.[103]

Prior to the synthesis of the MmepdC analog, it was well known from PNA systems that aromatic substitutions provided the most dramatic fluorescence response and that interesting characteristics would be observed from a PhpdC analog (Fig. 8.8). The analog was investigated by the means of the deoxyuridine route (*vide supra*). It was found that the PhpdC phosphoramidite could be synthesized in three steps with good overall yield.[104]

Thermal denaturation experiments of a centrally located PhpdC yielded a moderate increase in duplex stability (+3.3 °C)[104] relative to dC while maintaining excellent mismatch discrimination that was equal to or better than that of pyrrolo-dC. Furthermore, emission from the PhpdC ODN was 18 times greater than that of pyrrolo-dC.[104] Additional fluorescence studies showed the ability of PhpdC to communicate the identity of the complementary base. Similar to other C analogs, PhpdC showed sequence-dependent quenching and has yet to undergo comprehensive neighboring base studies.

The effect of substituent variation on pyrrolocytosine ring has been investigated with respect to heterocycle variation, ring fusion, and aromatic ring extension.[105] When the phenyl substituent of PhpdC is replaced by either an indole or benzofuran, Fig. 8.9, the fluorophore brightness increases as a result of improvement of the quantum yield and extinction coefficient for the longest wavelength absorption band in organic solvents. Although the wavelength of emission for the benzofuryl- and indolyl-substituted pyrrolocytidines is redshifted compared to PhpdC, the effect is not dramatic (≤35 nm in water). However, despite the similarity absorption and emission wavelengths, the polarity sensitivity of these fluorophores is quite different. While the PhpdC and the benzofuryl-derivative share a similar responsiveness to the change

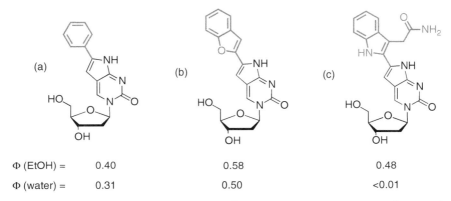

	(a)	(b)	(c)
Φ (EtOH) =	0.40	0.58	0.48
Φ (water) =	0.31	0.50	<0.01

Figure 8.9 Structures of phenylpyrrolo-2′-deoxycytidine (a) compared to the 2-benzofuryl-substituted analog (b) and the 2-substituted 3-methylenecarboxamide indolyl analog (c).

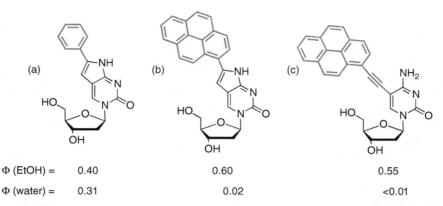

Φ (EtOH) =   0.40   0.60   0.55

Φ (water) =   0.31   0.02   <0.01

Figure 8.10   Structures of the parent PhpdC (a) compared to derivatives possessing the much larger pyrene aromatic group on the pdC scaffold (b) or the related, uncyclized 5-ethynylpyrene dC (c).

in environment from EtOH to water, the indole-substituted analog's behavior is quite different displaying pronounced quenching effect in water.

We were interested to determine how large an aromatic substituent attached to the pdC moiety could be while maintaining the properties of an intrinsic nucleobase fluorophore, that is, the nucleobase and substituent together constitute the fluorophore and the substituent does not act as an independent label. To this end, we compared the PhpdC analog to the pdC that was derivatized with a 1-pyrenyl- and 5-(1-pyrenylethynyl)-2'-deoxycytidine, Fig. 8.10. The pyrenyl-conjugate of pdC exhibited photophysical properties associated with an intrinsic nucleobase fluorophore – the broad, featureless emission similar to PhpdC, whereas the emission of the pyrenylethynyl conjugate of dC displayed vibronic structure, more similar to pyrene itself. Moreover, DFT calculations of the electronic structure of the pyrene-conjugated pC (Fig. 8.10b) indicate that the HOMO and LUMO are both distributed over the entire chromophore, whereas in the case of the pyrenylethynyl dC (Fig. 8.10c), the HOMO and LUMO are centered on the pyrene moiety. This result lends credence to the notion that the pyrenylpC is an intrinsic fluorophore and pyrenylethynyl dC behaves as an appended fluorophore.

Earlier, we established that PhpdC is an intrinsic nucleobase fluorophore; however, it possesses a rotatable bond as part of the chromophore. Since it is possible that the fluorescence efficiency is affected by the rotatable bond, possibly providing avenues for nonradiative decay of the excited state, we investigated the effect of ring fusion, that is, transforming PhpdC into the benzo-fused pdC derivative, Figure 8.11b. Such a derivative was described by Matteucci and von Krosigk as a carbazole nucleoside analog some time ago,[106] yet photophysical characterization was not reported. We have found that fusing the benzo group to the pC nucleus has a profound effect on the quantum yield and environmental responsiveness. This fluorescent base has been incorporated into PNA by our group, yet failed to provide a fluorescence-based response to hybridization.[107]

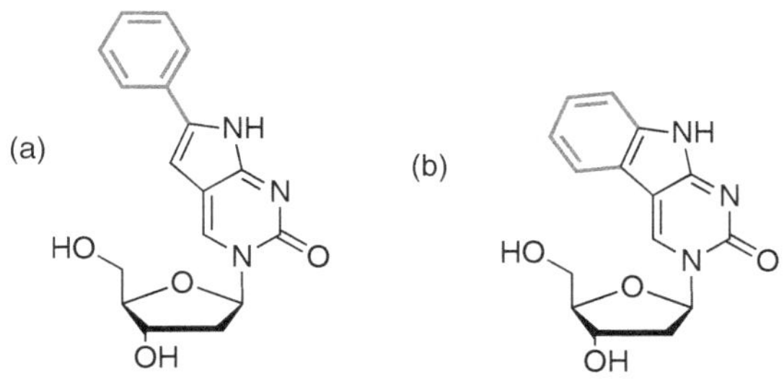

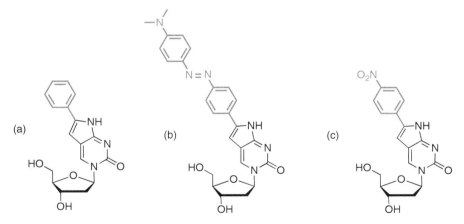

Φ (EtOH) =	0.40	>0.99
Φ (water) =	0.31	<0.01

Figure 8.11 Structural comparison of the structures and quantum yields for PhpdC (a) and benzopdC (b).

While PhpdC (see Fig. 8.12a) and related derivatives provide a fluorescence change during hybridization, some analogs do not. In order to construct a molecular beacon, or to perform studies on nucleic acid structure by fluorescence resonance energy transfer (FRET),[108] a fluorescence acceptor (or dark fluorophore, i.e., a quencher) is required. Maintaining the theme base-pairing competence, the well-known dark fluorophore *N,N*-dimethyl-4-(phenyldiazenyl)aniline was grafted onto the pC core to produce a quencher with properties similar to the universal quencher DABCYL (4-((4-(dimethylamino)phenyl)azo)benzoic acid).[109, 110] Inspired by Wilhelmsson's report of the quenching ability of tC$_{nitro}$ (see Section 8.2.3), we investigated the ability of ^{nitroPh}pC base acetic acid (Fig. 8.5) to act as a quencher, which does but less well than the azo-based quencher (Fig. 8.12b), nonetheless this has prompted us to also study the nucleoside (Fig. 8.12c).

Figure 8.12 From fluorophore (a) to quencher: based on DABCYL (b) or incorporating the simple nitro-substituent (c).

8.3.3 RNA

The promising work of pC analogs in the realms of PNA and DNA soon followed into the world of RNA. PhpC and MmepC ribonucleosides were synthesized for study and implementation as nucleic acid/enzyme probes.

Oligoribonucleotides (ORN) containing a single insert of either PhpC or MmepC in an 18-mer was prepared. It was determined that the modified ribonucleosides did not impair binding to DNA or RNA targets compared to the RNA control. Modifications located in a central position improved binding more than those placed near the 3'-end. Fluorescence characteristics of the ribonucleosides proved to be similar to those of the PNA and DNA analogs with considerable fluorescence observed in the single-stranded state and quenching in the double-stranded state.

In collaboration with the Damha group of McGill University, they set out to design and implement a simplified molecular beacon technology for HIV-1 RT RNase H assays as RNase H had been identified as a potential target for antiretroviral therapy. In developing simplified fluorescence assays, it was thought that high-throughput screening for the testing of anti-RNase H agents could be better facilitated. In addition to a more simple design and synthesis, it was found that the use of the DABCYL quencher for the RNA/DNA heteroduplex could be avoided, which is important since it has been seen to hinder RNaseH catalytic efficiency in this system.[11]

The most common RNase H fluorescence assay design was a molecular beacon[111] possessing a fluorescent moiety (fluorescein or rhodamine) at a RNA 3'-terminus across from a quencher (DABCYL) at the 5'-DNA terminus (Fig. 8.13). Quenched in the heteroduplex, this probe would become emissive upon RNase H action.

The proposed design utilized a single PhpC ribonucleoside insert at either a terminus or central location in an oligoribonucleotide. These inserts were imagined to behave in a quenched manner in the duplex and become fluorescent upon enzyme action (Fig. 8.14). Synthesis of the heteroduplex yielded expected results with respect to PhpC fluorescence, where an approximate 75% reduction (comparative to the ssORN) in fluorescence was observed.

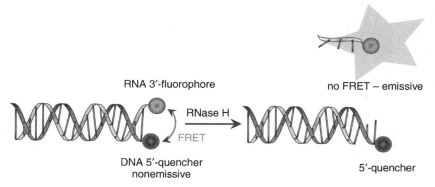

Figure 8.13 A dual-chromophore molecular beacon design RNase H sensor. The donor fluorophore is commonly fluorescein and the acceptor fluorophore DABCYL.

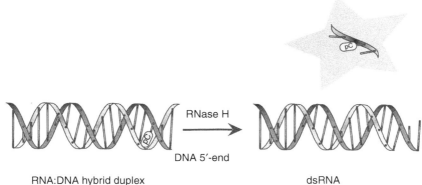

RNase H

DNA 5'-end

RNA:DNA hybrid duplex
quenched state

dsRNA
emissive state

Figure 8.14 Design of a turn-on RNase H sensor based on a single PhprC embedded in a RNA:DNA heteroduplex.

Quantum yield studies of neighboring base effects showed fluorescence independence with respect to the phosphate but dependence upon the neighboring bases for trinucleotide sequences. The greatest quantum yield was observed for a U-PhpC-U trinucleotide ($\Phi = 0.41$), whereas the lowest quantum yield was found to be 0.11 for the G-PhpC-U trinucleotide. The dramatic decrease in fluorescence for the neighboring G sequence was attributed to base-stacking interactions affecting the oscillator strength for the fluorescence transition. The high quantum yield value for the flanking U PhpC sequence was thought to be due to shielding of the PhpC moiety from solvent while avoiding other nonradiative deactivation pathways.[11]

Standard 5'-^{32}P-label/PAGE assay allowed for the determination of the compatibility of the PhpC-labeled substrates with the enzyme. It was found that unlike the classic MB technology, the PhpC systems did not compromise enzymatic activity. The change in enzyme reaction time in comparison to the classical MB was considerable as the optimized PhpC system reached completion in approximately 20 min, whereas the traditional MB design (RNA 3'-Fluorescein/DNA 5'-DABCYL) was not complete for 130 min. Not only did PhpC not perturb enzyme action, it provided a 14-fold fluorescence change detectable by the naked eye, whereas the classic MB design only provided an 8 times change from the quenched state to the emissive state. Further utility was provided by the ability to monitor RNase H activity on the PhpC containing heteroduplex system by fluorescence polarization. To test the ability of the PhpC MB as a high-throughput screening device, it was tested in a 96-well plate spectrofluorimeter by fluorescence intensity and fluorescence polarization. It was found that RNase H activity could be detected down to a 10 nM substrate concentration (comparable to [32]PAGE assays) or 60 nM in fluorescence polarization mode.[11]

The intrinsic fluorescence of PhpC in RNA has also been exploited in investigation of the intracellular distribution of siRNAs.[112] The PhpC inserts were found to be compatible with silencing of gene expression in HeLaX1/5 cells that incorporated the luciferase reporter. Modifications were tolerated in either the guide or template strand and both, even next to the scissile phosphodiester bond, but less tolerated at the 5′-end or near the 3′-end of the antisense strand. The PhpC-labeled siRNAs allowed the visualization of Lipofectamine™ delivery siRNA to HeLa cells.

8.4   CONCLUSIONS

In this chapter, we have briefly described the fluorescent properties and characteristics of a generous selection of fluorescent C analogs focusing on those nucleo-side/nucleobase modifications that adhere to the principles of minimal structural modification and noninterference with canonical base pairing properties. There are many more examples of fluorescent C analogs, especially of the extrinsic category that have not been included in this short review and readers are referred to recent reviews.[9, 15]

We have illustrated, through presentation of examples taken from the literature and from our own work, a variety of C analogs that possess widely varying characteristics ranging from the environmentally sensitive to the environmentally insensitive that may be prepared through the relatively modest modification of cytosine. We have also observed that fluorescence efficiencies can range from poor to extraordinarily good, yet all of these fluorescent C analogs have potential utility as demonstrated in the early reports of their practical application. From this, it is clear that nucleoside analogs that combine fluorescence with its innate molecular recognition properties are useful aids in the study of nucleic acids and allied fields.

8.5   PROSPECTS AND OUTLOOK

The design, synthesis, and application of intrinsically fluorescent C analogs are currently well over a decade old, and other areas of fluorescent nucleoside chemistry are steeped in a longer history, by as yet this area of endeavor is by no means completely mature. New demands placed by the ever-increasing desire to explore biological functions at high resolution, lower detection limits, or in real time within complex systems will continue to motivate developments in this field. For C analogs, in particular, there remains the need for bright analogs to compete with traditional fluorophores. Although some gains can be made in quantum efficiencies, there is still need for molecules with very large molar extinction coefficients. Fluorescent base analogs that have tunable fluorescence, especially with respect to wavelength, would be advantageous, as most currently are blue emitters. New analogs that are predictably responsive to microsequence/structural effects or which are completely immune to them would be useful. Further, analogs that are accurate reporters of environmental conditions beyond base pairing, such as local pH, or oxidative/reductive environments could serve as the foundation of new probes. Finally, analogs that are compatible with enzymatically driven processes such as transcription, reverse transcription, and translation would not only provide a reporter group for studying these processes but may find use in technological applications such as SELEX for the development of functional molecules. Achievements in this field will likely result from the fruitful collaboration between spectroscopists, computational and synthetic chemists.

ACKNOWLEDGMENTS

This work was made possible by continuous funding from the Natural Science and Engineering Research Council of Canada and a talented cohort of undergraduate and graduate students over the course of the past decade. Mr. Adam Elmehriki is thanked for photophysical characterization of some of the pyrrolocytosine compounds described in this chapter. We also gratefully acknowledge our collaborators, Prof. Masad Damha (McGill University, Canada), Prof. David Corey (UT Southwestern Medical School, USA), and their coworkers for their contributions.

REFERENCES

1. Vaya, I.; Gustavsson, T.; Miannay, F. A.; Douki, T.; Markovitsi, D. *J. Am. Chem. Soc.* **2010**, *132*, 11834.
2. Lam, J. C. F.; Aguirre, S.; Li, Y. Nucleic Acids as Detection Tools: The Chemical Biology of Nucleic Acids; Mayer, G., Ed.; Chichester, West Sussex, John Wiley & Sons Ltd.: United Kingdom, **2010**, pp. 401–431.
3. Okamoto, A.; Tainaka, K.; Saito, I. *Chem. Lett.* **2003**, *32*, 684.
4. Okamoto, A.; Tanaka, K.; Fukuta, T.; Saito, I. *J. Am. Chem. Soc.* **2003**, *125*, 9296.
5. Hattori, M.; Ohki, T.; Yanase, E.; Ueno, Y. *Bioorg. Med. Chem. Lett.* **2012**, *22*, 253.
6. Yoshida, Y.; Niwa, K.; Yamada, K.; Tokeshi, M.; Baba, Y.; Saito, Y.; Okamoto, A.; Saito, I. *Chem. Lett.* **2010**, *39*, 116.
7. Saito, Y.; Motegi, K.; Bag, S. S.; Saito, I. *Bioorg. Med. Chem.* **2008**, *16*, 107.
8. Tinsley, R. A.; Walter, N. G. *RNA* **2006**, *12*, 522.
9. Sinkeldam, R. W.; Greco, N. J.; Tor, Y. *Chem. Rev.* **2010**, *110*, 2579.
10. Zhang, C. M.; Liu, C. P.; Christian, T.; Gamper, H.; Rozenski, J.; Pan, D. L.; Randolph, J. B.; Wickstrom, E.; Cooperman, B. S.; Hou, Y. M. *RNA* **2008**, *14*, 2245.
11. Wahba, A. S.; Esmaeili, A.; Damha, M. J.; Hudson, R. H. E. *Nucleic Acids Res.* **2010**, *38*, 1048.
12. Greco, N. J.; Sinkeldam, R. W.; Tor, Y. *Org. Lett.* **2009**, *11*, 1115.
13. Tanpure, A. A.; Srivatsan, S. G. *Chemistry* **2011**, *17*, 12820.
14. Greco, N. J.; Tor, Y. *Tetrahedron* **2007**, *63*, 3515.
15. Wilhelmsson, L. M. *Q. Rev. Biophys.* **2010**, *43*, 159.
16. Asseline, U. *Curr. Org. Chem.* **2006**, *10*, 491.

17. Okamoto, A.; Saito, Y.; Saito, I. *J. Photochem. Photobiol. C* **2005**, *6*, 108.

18. Dodd, D. W.; Hudson, R. H. E. *Mini Rev. Org. Chem.* **2009**, *6*, 378.

19. Kossel, A. Steudel, H. Z. *Physiol. Chem.* **1903**, *38*, 49.

20. Lipsett, M. N. *J. Biol. Chem.* **1964**, *239*, 1256.

21. Gehring, K.; Leroy, J. L.; Gueron, M. *Nature* **1993**, *363*, 561.

22. Schweize, M. P.; Kreishma, G. P. *J. Magn. Reson.* **1973**, *9*, 334.

23. Lin, K. Y.; Jones, R. J.; Matteucci, M. *J. Am. Chem. Soc.* **1995**, *117*, 3873.

24. Wilhelmsson, L. M.; Holmen, A.; Lincoln, P.; Nielson, P. E.; Norden, B. *J. Am. Chem. Soc.* **2001**, *123*, 2434.

25. Engman, K. C.; Sandin, P.; Osborne, S.; Brown, T.; Billeter, M.; Lincoln, P.; Norden, B.; Albinsson, B.; Wilhelmsson, L. M. *Nucleic Acids Res.* **2004**, *32*, 5087.

26. Sandin, P.; Wilhelmsson, L. M.; Lincoln, P.; Powers, V. E. C.; Brown, T.; Albinsson, B. *Nucleic Acids Res.* **2005**, *33*, 5019.

27. Tahmassebi, D. C.; Millar, D. P. *Biochem. Biophys. Res. Commun.* **2009**, *380*, 277.

28. Stengel, G.; Gill, J. P.; Sandin, P.; Wilhelmsson, L. M.; Albinsson, B.; Norden, B.; Millar, D. *Biochemistry* **2007**, *46*, 12289.

29. Stengel, G.; Urban, M.; Purse, B. W.; Kuchta, R. D. *Anal. Chem.* **2010**, *82*, 1082.

30. Stengel, G.; Purse, B. W.; Wilhelmsson, L. M.; Urban, M.; Kuchta, R. D. *Biochemistry* **2009**, *48*, 7547.

31. Sandin, P.; Borjesson, K.; Li, H.; Martensson, J.; Brown, T.; Wilhelmsson, L. M.; Albinsson, B. *Nucleic Acids Res.* **2008**, *36*, 157.

32. Borjesson, K.; Sandin, P.; Wilhelmsson, L. M. *Biophys. Chem.* **2009**, *139*, 24.

33. Stengel, G.; Urban, M.; Purse, B. W.; Kuchta, R. D. *Anal. Chem.* **2009**, *81*, 9079.

34. Borjesson, K.; Preus, S.; El-Sagheer, A. H.; Brown, T.; Albinsson, B.; Wilhelmsson, L. M. *J. Am. Chem. Soc.* **2009**, *131*, 4288.

35. Ortega, J.-A.; Blas, J. R.; Orozco, M.; Grandas, A.; Pedroso, E.; Robles, J. *Org. Lett.* **2007**, *9*, 4503.

36. Preus, S.; Borjesson, K.; Kilsa, K.; Albinsson, B.; Wilhelmsson, L. M. *J. Phys. Chem. B* **2010**, *114*, 1050.

37. Lin, K. Y.; Matteucci, M. D. *J. Am. Chem. Soc.* **1998**, *120*, 8531.

38. Flanagan, W. M.; Wolf, J. J.; Olson, P.; Grant, D.; Lin, K. Y.; Wagner, R. W.; Matteucci, M. D. *Proc. Natl. Acad. Sci. U. S. A.* **1999**, *96*, 3513.

39. Ausin, C.; Ortega, J. A.; Robles, J.; Grandas, A.; Pedroso, E. *Org. Lett.* **2002**, *4*, 4073.

40. Rajeev, K. G.; Maier, M. A.; Lesnik, E. A.; Manoharan, M. *Org. Lett.* **2002**, *4*, 4395.

41. Nakagawa, O.; Ono, S.; Li, Z.; Tsujimoto, A.; Sasaki, S. *Angew. Chem. Int. Ed.* **2007**, *46*, 4500.

42. Nasr, T.; Li, Z.; Nakagawa, O.; Taniguchi, Y.; Ono, S.; Sasaki, S. *Bioorg. Med. Chem. Lett.* **2009**, *19*, 727.

43. Li, Z. C.; Nakagawa, O.; Koga, Y.; Taniguchi, Y.; Sasaki, S. *Bioorg. Med. Chem.* **2010**, *18*, 3992.

44. Koga, Y.; Fuchi, Y.; Nakagawa, O.; Sasaki, S. *Tetrahedron* **2011**, *67*, 6746.

45. Taniguchi, Y.; Koga, Y.; Fukabori, K.; Kawaguchi, R.; Sasaki, S. *Bioorg. Med. Chem. Lett.* **2012**, *22*, 543.

46. Barhate, N.; Cekan, P.; Massey, A. P.; Sigurdsson, S. T. *Angew. Chem. Int. Ed.* **2007**, *46*, 2655.

47. Cekan, P.; Sigurdsson, S. T. *Chem. Commun.* **2008**, *1*, 3393.

48. Cekan, P.; Jonsson, E. O.; Sigurdsson, S. T. *Nucleic Acids Res.* **2009**, *37*, 3990.

49. Okamoto, A.; Tainaka, K.; Saito, I. *J. Am. Chem. Soc.* **2003**, *125*, 4972.

50. Okamoto, A.; Tainaka, K.; Saito, I. *Tetrahedron Lett.* **2003**, *44*, 6871.

51. Miyata, K.; Tamamushi, R.; Ohkubo, A.; Taguchi, H.; Seio, K.; Santa, T.; Sekine, M. *Org. Lett.* **2006**, *8*, 1545.

52. Miyata, K.; Mineo, R.; Tamamushi, R.; Mizuta, M.; Ohkubo, A.; Taguchi, H.; Seio, K.; Santa, T.; Sekine, M. *J. Org. Chem.* **2007**, *72*, 102.

53. Mizuta, M.; Seio, K.; Miyata, K.; Sekine, M. *J. Org. Chem.* **2007**, *72*, 5046.

54. Mizuta, M.; Seio, K.; Ohkubo, A.; Sekine, M. *J. Phys. Chem. B* **2009**, *113*, 9562.

55. Leonard, N. J.; Sprecker, M. A.; Morrice, A. G. *J. Am. Chem. Soc.* **1976**, *98*, 3987.

56. Leonard, N. J. *Acc. Chem. Res.* **1982**, *15*, 128.

57. Lessor, R. A.; Gibson, K. J.; Leonard, N. J. *Biochemistry* **1984**, *23*, 3868.

58. Leonard, N. J.; Keyser, G. E. *Proc. Natl. Acad. Sci. U. S. A.* **1979**, *76*, 4262.

59. Krueger, A. T.; Kool, E. T. *J. Am. Chem. Soc.* **2008**, *130*, 3989.

60. Wilson, J. N.; Kool, E. T. *Org. Biomol. Chem.* **2006**, *4*, 4265.

61. Liu, H. B.; Gao, J. M.; Kool, E. T. *J. Am. Chem. Soc.* **2005**, *127*, 1396.

62. Lynch, S. R.; Liu, H. B.; Gao, J. M.; Kool, E. T. *J. Am. Chem. Soc.* **2006**, *128*, 14704.

63. Gao, J. M.; Liu, H. B.; Kool, E. T. *J. Am. Chem. Soc.* **2004**, *126*, 11826.

64. Liu, H. B.; Lynch, S. R.; Kool, E. T. *J. Am. Chem. Soc.* **2004**, *126*, 6900.

65. Jarchow-Choy, S. K.; Krueger, A. T.; Liu, H. B.; Gao, J. M.; Kool, E. T. *Nucleic Acids Res.* **2011**, *39*, 1586.

66. Hernandez, A. R.; Kool, E. T. *Org. Lett.* **2011**, *13*, 676.

67. Inoue, H.; Imura, A.; Ohtsuka, E. *Nippon Kagaku Kaishi* **1987**, *7*, 1214.

68. Liu, C. H.; Martin, C. T. *J. Mol. Biol.* **2001**, *308*, 465.

69. Glen Research Catalog **2012**. http://www.glenresearch.com/Catalog/structural.html#p64 (accessed 28/08/12).

70. Berry, D. A.; Jung, K. Y.; Wise, D. S.; Sercel, A. D.; Pearson, W. H.; Mackie, H.; Randolph, J. B.; Somers, R. L. *Tetrahedron Lett.* **2004**, *45*, 2457.

71. Hardman, S. J. O.; Botchway, S. W.; Thompson, K. C. *Photochem. Photobiol.* **2008**, *84*, 1473.

72. Zang, Z.; Fang, Q. M.; Pegg, A. E.; Guengerich, F. P. *J. Biol. Chem.* **2005**, *280*, 30873.

73. Dash, C.; Rausch, J. W.; Le Grice, S. F. J. *Nucleic Acids Res.* **2004**, *32*, 1539.

74. Zhang, X.; Wadkins, R. M. *Biophys. J.* **2009**, *96*, 1884.

75. Wigerinck, P.; Pannecouque, C.; Snoeck, R.; Claes, P.; Declercq, E.; Herdewijn, P. *J. Med. Chem.* **1991**, *34*, 2383.

76. Noe, M. S.; Rios, A. C.; Tor, Y. *Org. Lett.* **2012**, *14*, 3150.

77. Sinkeldam, R. W.; Wheat, A. J.; Boyaci, H.; Tor, Y. *ChemPhysChem.* **2011**, *12*, 567.

78. Shin, D.; Sinkeldam, R. W.; Tor, Y. *J. Am. Chem. Soc.* **2011**, *133*, 14912.

79. Dodd, D. W.; Swanick, K. N.; Price, J. T.; Brazeau, A. L.; Ferguson, M. J.; Jones, N. D.; Hudson, R. H. E. *Org. Biomol. Chem.* **2010**, *8*, 663.

80. Hudson, R. H. E.; Dambenieks, A. K.; Viirre, R. D. *Synlett* **2004**, *13*, 2400.

81. Hudson, R. H. E.; Viirre, R. D.; McCourt, N.; Tse, J. *Nucleos. Nucleot. Nucl.* **2003**, *22*, 1029.

82. Hudson, P. H. E.; Moszynski, J. M. *Synlett* **2006**, *18*, 2997.

83. Robins, M. J.; Barr, P. J. *J. Org. Chem.* **1983**, *48*, 1854.

84. Iritani, K.; Matsubara, S.; Utimoto, K. *Tetrahedron Lett.* **1988**, *29*, 1799.

85. Arcadi, A.; Bianchi, G.; Marinelli, F. *Synthesis* **2004**, *4*, 610.

86. Sniady, A.; Durham, A.; Morreale, M. S.; Marcinek, A.; Szafert, S.; Lis, T.; Brzezinska, K. R.; Iwasaki, T.; Ohshima, T.; Mashima, K.; Dembinski, R. *J. Org. Chem.* **2008**, *73*, 5881.

87. Kurisaki, T.; Naniwa, T.; Yamamoto, H.; Imagawa, H.; Nishizawa, M. *Tetrahedron Lett.* **2007**, *48*, 1871.

88. Trost, B. M.; McClory, A. *Angew. Chem. Int. Ed.* **2007**, *46*, **2074**.

89. Larock, R. C. *J. Organomet. Chem.* **1999**, *576*, 111.

90. Rodriguez, A. L.; Koradin, C.; Dohle, W.; Knochel, P. *Angew. Chem. Int. Ed.* **2000**, *39*, 2488.

91. Yu, C. J.; Yowanto, H.; Wan, Y. J.; Meade, T. J.; Chong, Y.; Strong, M.; Donilon, L. H.; Kayyem, J. F.; Gozin, M.; Blackburn, G. F. *J. Am. Chem. Soc.* **2000**, *122*, 6767.

92. Rao, M. S.; Esho, N.; Sergeant, C.; Dembinski, R. *J. Org. Chem.* **2003**, *68*, 6788.

93. Carpita, A.; Ribecai, A. *Tetrahedron Lett.* **2009**, *50*, 6877.

94. Woo, J. S.; Meyer, R. B.; Gamper, H. B. *Nucleic Acids Res.* **1996**, *24*, 2470.

95. Hudson, R. H. E.; Viirre, R. D.; Liu, Y. H.; Wojciechowski, F.; Dambenieks, A. K. *Pure Appl. Chem.* **2004**, *76*, 1591.

96. Hudson, R. H. E.; Li, G.; Tse, J. *Tetrahedron Lett.* **2002**, *43*, 1381.

97. Hudson, R. H. E.; Dambenieks, A. K.; Moszynski, J. M. *Proc. SPIE Int. Soc. Opt. Eng.* **2005**, *5969*, 59690J/1–59690J/10.

98. Wojciechowski, F.; Hudson, R. H. E. *J. Am. Chem. Soc.* **2008**, *130*, 12574.

99. Hua, J. X.; Dodd, D. W.; Hudson, R. H. E.; Corey, D. R. *Bioorg. Med. Chem. Lett.* **2009**, *19*, 6181.

100. Torres, A. G.; Fabani, M. M.; Vigorito, E.; Williams, D.; Al-Obaidi, N.; Wojciechowski, F.; Hudson, R. H. E.; Seitz, O.; Gait, M. J. *Nucleic Acids Res.* **2012**, *40*, 2152.

101. Wilds, C. J.; Maier, M. A.; Tereshko, V.; Manoharan, M.; Egli, M. *Angew. Chem. Int. Ed.* **2002**, *41*, 115.

102. Wojciechowski, F.; Hudson, R. H. E. *Org. Lett.* **2009**, *11*, 4878.

103. Hudson, R. H. E.; Choghamarani, A. G. *Nucleos. Nucleot. Nucl.* **2007**, *26*, 533.

104. Hudson, R. H. E.; Ghorbani-Choghamarani, A. *Synlett* **2007**, *6*, 870.

105. Elmehriki, A. A. H.; Suchy, M.; Chicas, K. J.; Wojciechowski, F.; Hudson, R. H. E. *Artif. DNA: PNA XNA* **2014**, *5*(2), e29174-1–e29174-15.

106. Matteucci, M. D.; von Krosigk, U. *Tetrahedron Lett.* **1996**, *37*, 5057.

107. Suchy, M.; Hudson, R. H. E. *J. Org. Chem.* **2014**, *79*, 3336.

108. Forster, T. *Ann. Physik.* **1948**, *2*, 55.

109. Charles, M. Synthesis and Spectroscopic Studies of Substituted Pyrrolocytidines. *University of Western Ontario – Electronic Thesis and Dissertation Repository*, Paper 1125, http://ir.lib.uwo.ca/etd/1125, **2013**.

110. Ettles, C. M. Progress Toward Synthesis of Molecular Beacons Incorporating DABCYL Analog Quenchers. *University of Western Ontario – Electronic Thesis and Dissertation Repository,* Paper 1790, http://ir.lib.uwo.ca/etd/1790, **2013**.

111. Tyagi, S.; Kramer, F. R. *Nat. Biotechnol.* **1996**, *14*, 303.

112. Wahba, A. S.; Azizi, F.; Deleavey, G. F.; Brown, C.; Robert, F.; Carrier, M.; Kalota, A.; Gewirtz, A. M.; Pelletier, J.; Hudson, R. H. E.; Damha, M. J. *ACS Chem. Biol.* **2011**, *6*, 912.

9

SYNTHESIS AND FLUORESCENCE PROPERTIES OF NUCLEOSIDES WITH PYRIMIDOPYRIMIDINE-TYPE BASE MOIETIES

KOHJI SEIO

Department of Life Science, Tokyo Institute of Technology, Yokohama, Japan

TAKASHI KANAMORI

Education Academy of Computational Life Sciences, Tokyo Institute of Technology, Yokohama, Japan

AKIHIRO OHKUBO AND MITSUO SEKINE

Department of Life Science, Tokyo Institute of Technology, Yokohama, Japan

A family of fluorescent nucleosides having a pyrimidopyrimidine motif was developed. The pyrimidopyrimidoindole nucleoside dC^{PPI} exhibits enhanced fluorescence intensity and longer emission wavelength in a hydrophobic environment compared with an aqueous environment. Because of its large Stokes shift, dC^{PPI} can be used in combination with 2-aminopurine for doubly labeled oligonucleotides. In addition, a DNA triplex system in which the Watson–Crick G-dC^{PPI} pair is recognized by triplex-forming oligonucleotides containing abasic sites such as a C3 linker was developed. The unique fluorescence and hybridization properties of dC^{PPI} have great potential for use in structural analyses as well as the detection of biologically important nucleic acids.

Fluorescent Analogs of Biomolecular Building Blocks: Design and Applications, First Edition.
Edited by Marcus Wilhelmsson and Yitzhak Tor.
© 2016 John Wiley & Sons, Inc. Published 2016 by John Wiley & Sons, Inc.

9.1   INTRODUCTION

Fluorescent nucleic acids are important tools for the visualization of DNA and RNA both *in vitro* and *in vivo*. Nucleic acids have various functional groups such as hydroxyl groups, base moieties, and internucleotidic phosphate groups, which can be labeled with various fluorescent dyes.[1] Highly bright fluorescent dyes such as fluorescein, Cy3, and Cy5 can be conjugated to oligonucleotides for producing fluorescent oligonucleotides that can be used as highly sensitive hybridization probes for DNA microarray[2] and PCR analyses.[3]

There exists another class of fluorescent oligonucleotides that have modified nucleotide residues bearing a fluorescent heterocyclic group instead of a natural nucleobase.[4] Because these fluorescent base moieties are small, their fluorescence intensities are weaker than those of the abovementioned general fluorescent dyes. On the other hand, the fluorescent nucleobases can be designed in a manner such that they form Watson–Crick base pairs in DNA and/or RNA duplexes. In addition, in duplexes, these nucleobases can be stacked on upstream or downstream natural nucleobases. Therefore, these fluorescent oligonucleotides can be used as fluorescent indicators for nucleic acid hybridization, higher order structures, and nucleic acid–protein interactions, which are generally sensitive to base pair formation and base stacking.

Fluorescent nucleobases can be structurally categorized into several groups. One recently developed group is the pyrimidine derivatives with additional ring systems. Some of the fluorescent pyrimidine bases include various 5-arylpyrimidine derivatives such as 5-(thien-2-yl) and 5-(furan-2-yl)uracil,[5] 5-(furan-2-yl)cytosine,[5h] 5-(pyridin-2-yl)uracil,[6a] and 5-biaryluracil.[6b] In addition, quinazoline derivatives[7] or thieno[3,2-*d*]pyrimidine and thieno[3,4-*d*]pyrimidine nucleosides,[8] containing an aromatic ring fused at the C5 and C6 positions, have been reported as fluorescent pyrimidine nucleosides. Furthermore, nucleobases with aromatic rings fused at the C4 and C5 positions such as 1,3-diaza-2-oxo-phenoxazine (tC^O),[9a–c] 1,3-diaza-2-oxo-phenothiazine (tC),[9b,10] 3*H*-pyrrolo[2,3-*d*]pyrimidin-2-one (pyrrolo-C) derivatives,[11] and benzo-[12] or naphthopyridopyrimidine[12] (base-discriminating fluorescent nucleoside) have been reported. Recently, we developed a series of nucleosides having pyrimidopyrimidine derivatives as fluorescent nucleobases categorized as C5- and C4-fused derivatives,[13–15] including dihydropyrimidopyrimidine nucleoside (dC^{hpp}), pyrimidopyrimidine nucleoside (dC^{PPP}), pyrimidopyrimidoindole nucleoside (dC^{PPI}), and dC^{PPI} derivatives with different ring structures and substituents (Fig. 9.1). These derivatives exhibit interesting fluorescence and base recognition properties with the advantage that they can be used in nucleic acid duplex and triplex systems.

9.2   DISCOVERY, DESIGN, AND SYNTHESIS OF PYRIMIDOPYRIMIDINE NUCLEOSIDES

9.2.1   Synthesis and Fluorescence Properties of dC^{hpp}

The discovery of dC^{hpp} was a coincidence. Initially, we studied the synthesis and hybridization properties of the oligonucleotide incorporating 4-*N*-carbamoyl-

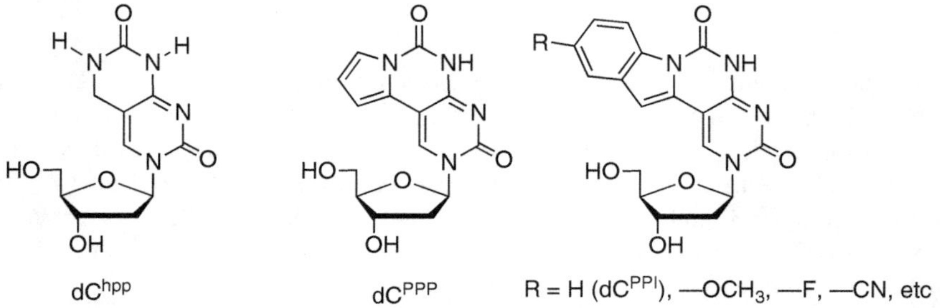

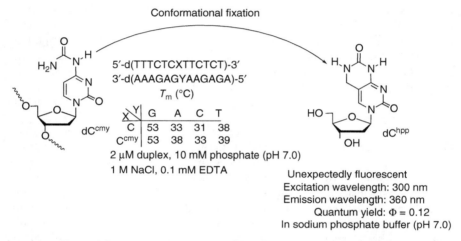

Figure 9.1 Pyrimidopyrimidine derivatives dC^hpp, dC^PPP, and dC^PPI.

Figure 9.2 *N*-Carbamoylcytidine dC^cmy and its cyclic derivative dC^hpp.

deoxycytidine dC^cmy (Fig. 9.2). The UV-melting experiments using the duplexes 5'-d(TTTCTCXTTCTCT)-3'/3'-(AAAGAGYAAGAGA)-5', where X = dC or dC^cmy and Y = dG, dA, dC, or T, showed that oligonucleotides containing unmodified deoxycytidine (dC) and dC^cmy hybridize to the complementary or single-base mismatched strands with comparable stabilities giving the same T_m values of 53 °C.[16] This result suggested that the carbamoyl modification of the cytosine amino group could be used for the development of hybridization probes for biological applications. Therefore, to improve hybridization affinity and base selectivity, we designed dC^hpp with an additional six-membered ring, which is a fixed conformational analog of dC^cmy.

The dC^hpp nucleoside can be synthesized according to the scheme shown in Figure 9.3, starting from 3',5'-*O*-bis(TBDMS)thymidine via six steps.[13] Briefly, the methyl group of 3',5'-*O*-bis(TBDMS)thymidine was brominated under radical conditions. Subsequently, the bromomethyl group was converted into an azidomethyl group. Then, the amino group was introduced to position 4 of the thymine base and the azido group was reduced to an amino group. Subsequent cyclization and deprotection gave the desired nucleoside dC^hpp. The nucleoside dC^cmy having a

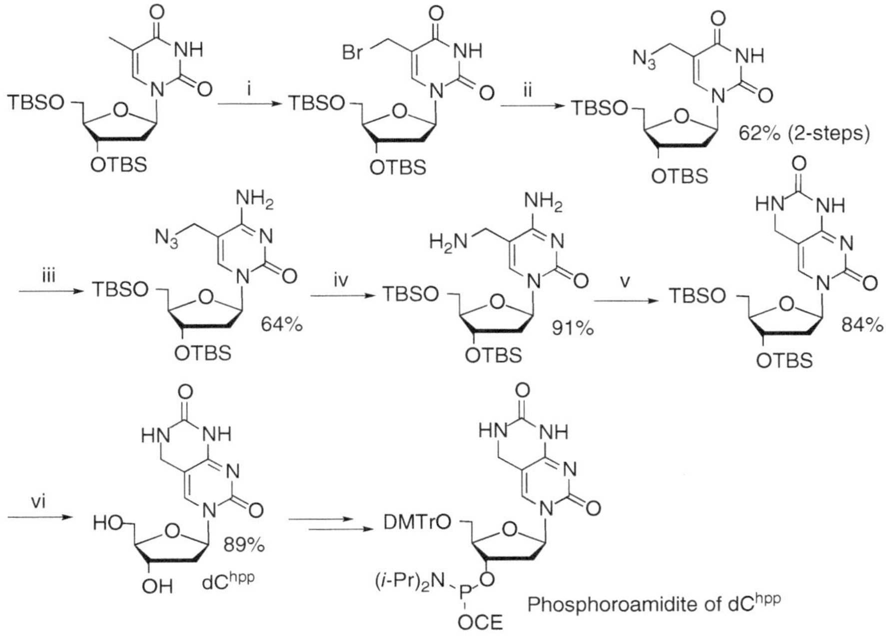

Figure 9.3   Synthesis of dC^{hpp} nucleoside: (i) *N*-bromosuccinimide, AIBN, CCl$_4$, reflux, 2 h; (ii) NaN$_3$, DMF, 1 h; (iii) (a) triisopropylbenzenesulfonyl chloride, NBu$_4$Br (cat) CH$_2$Cl$_2$–aq. Na$_2$CO$_3$, 2 h; (b) NH$_3$, 1,4-dioxane, 6 h; (iv) H$_2$, Pd–C, 2 h; (v) carbonyldiimidazole, DMF, 2 h; (vi) NBu$_4$F, THF, 1 h.

noncyclized carbamoyl group showed an absorption maximum at around 280 nm, which is redshifted compared to that of dC due to the presence of the carbamoyl group. Interestingly, dC^{hpp}, which has a cyclized carbamoyl group, exhibited a more redshifted absorption maximum at around 300 nm. In addition, dC^{hpp} proved to be a fluorescent nucleoside. Upon being irradiated at 300 nm, dC^{hpp} exhibited fluorescence at 360 nm with a quantum yield of 0.12. It was previously reported that dC^{cmy} and its 5-methyl derivative did not show any fluorescence. Thus, the fluorescence properties of dC^{hpp} suggested that the conformational fixation by incorporating a ring structure is essential for its fluorescence. The properties of oligodeoxynucleotides containing dC^{hpp}, which were synthesized by using the phosphoramidite unit of dC^{hpp}, as shown in Figure 9.3, were also studied.[13] Both oligonucleotide 5'-d(CGCAAT[C^{hpp}]TAACGC)-3'-containing dC^{hpp} and its duplex with 3'-d(GCGTTAAATTGCG)-5' exhibited fluorescence at 360 nm with an intensity of 40% of that of dC^{hpp}. In contrast, the fluorescence of the duplex of the abovementioned oligonucleotide with 3'-d(GCGTTA<u>G</u>ATTGCG)-5' (which has a dG–dC^{hpp} base pair at the underlined position) is quenched to about 5% of that of dC^{hpp}. Since the thermal stability of the 13mer duplex containing a G–C^{hpp} base pair with a T_m value of 57°C[13] is almost the same as that of the duplex containing an unmodified G–C base pair in the same position with a T_m value of 58°C,[13] this

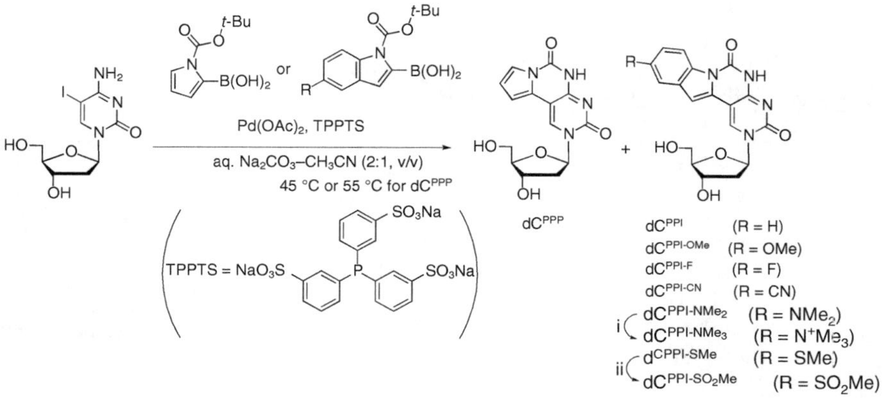

Figure 9.4 Synthesis of dC^{PPP}, dC^{PPI}, and dC^{PPI} derivatives: (i) CH$_3$I/DMF (ii) OXONE®/H$_2$O–MeOH (1:1, v/v). Yields: dC^{PPP} (55%); dC^{PPI} (75%); dC$^{PPI-OMe}$ (72%); dC^{PPI-F} (82%); dC^{PPI-CN} (46%); dC$^{PPI-NMe_2}$ (84%); dC$^{PPI-NMe_3}$ (quant.); dC$^{PPI-SMe}$ (79%); dC$^{PPI-SO_2Me}$ (93%).

quenching is attributed to the quenching by the nearby guanine moiety and not to conformational perturbation.

9.2.2 Design, Synthesis, and Fluorescence Properties of dC^{PPP}, dC^{PPI}, and dC^{PPI} Derivatives

We designed dC^{PPP}, dC^{PPI}, and derivatives of dC^{PPI} to examine the effects of attaching an aromatic ring system to dC^{hpp}. With this ring fusion, we also expected increased redshifts in the absorption and emission wavelengths. These nucleosides were synthesized according to the procedures shown in Figure 9.4.[14,15] Briefly, 5-iodocytidine was condensed with various pyrrole- or substituted indole boronic acids whose N–H groups were protected with a t-butyloxycarbonyl group in the presence of palladium diacetate and triphenylphosphine trisulfonic acid trisodium salt (TPPTS). Using this scheme, dC^{PPP}, dC^{PPI}, and dC^{PPI} derivatives having substituents such as methoxy, fluoro, cyano, dimethylamino, and methylthio on the indole ring could be synthesized. In addition, dC$^{PPI-NMe_2}$ having a dimethylamino group and dC$^{PPI-SMe}$ having a methylthio group were converted into dC$^{PPI-NMe_3}$ having a trimethylammonium group by methylation and dC$^{PPI-SO_2Me}$ having a methylsulfonyl group using OXON®, respectively.

The photophysical properties of dC^{PPP}, dC^{PPI}, and dC^{PPI} derivatives are shown in Table 9.1. Here, we focus on the relationship between structure and brightness, which is the product of the quantum yield (Φ) and absorption coefficient (ε_{max}). Comparing dC^{PPP} with dC^{PPI}, we found that brightness reduced from 500 to 25 M^{-1} cm^{-1} by replacing the pyrrole ring in dC^{PPP} with an indole ring in dC^{PPI}, mainly due to the decrease in Φ from 0.11 for dC^{PPP} to 0.006 for dC^{PPI}. The incorporation of an electron-donating group such as OMe (dC$^{PPI-OMe}$) or SMe (dC$^{PPI-SMe}$) did not increase the brightness. Moreover, the dC$^{PPI-NMe_2}$ derivative having more

TABLE 9.1 **Photophysical Properties of dC^{PPP}, dC^{PPI}, and dC^{PPI} Derivatives in Aqueous Buffer**

	λ_{max}^{abs} (nm)	ε_{max} (M^{-1} cm^{-1})	λ_{max}^{flu} (nm)	Φ	Brightness $\varepsilon_{max} \times \Phi$ (M^{-1} cm^{-1})
dC^{PPP}	369	4760	490	0.105	500
dC^{PPI}	374	4157	513	0.006	25
dC$^{PPI\text{-}OMe}$	375	5321	511	0.005	27
dC$^{PPI\text{-}SMe}$	372	4110	511	0.002	8
dC$^{PPI\text{-}F}$	371	6284	505	0.019	119
dC$^{PPI\text{-}CN}$	369	6998	487	0.061	427
dC$^{PPI\text{-}SMe}$	372	4110	511	0.002	8
dC$^{PPI\text{-}NMe_3}$	368	7380	483	0.096	813
dC$^{PPI\text{-}SO_2Me}$	367	8468	484	0.078	576

Reprinted with permission from Mizuta et al.[15]. Copyright 2007 American Chemical Society.

TABLE 9.2 **Solvatochromic Data of dC^{PPI} (R = H)**

Solvent	λ_{max}^{abs} (nm)	ε_{max} (M^{-1} cm^{-1})	λ_{max}^{flu} (nm)	Φ	Brightness $\varepsilon_{max} \times \Phi$
Toluene ($\varepsilon = 2.4$)	372	5230	467	0.236	1234
CH$_2$Cl$_2$ ($\varepsilon = 8.9$)	372	5370	477	0.168	902
Methanol ($\varepsilon = 32.7$)	381	6930	488	0.114	790
CH$_3$CN ($\varepsilon = 36.7$)	373	6970	470	0.221	1540

Reprinted with permission from Mizuta et al.[15]. Copyright 2007 American Chemical Society.

electron-donating dimethylamino groups exhibited negligible fluorescence (data not shown).

In contrast, the incorporation of electron-withdrawing substituents such as fluoro (dC$^{PPI\text{-}F}$), cyano (dC$^{PPI\text{-}CN}$), trimethylammonium (dC$^{PPI\text{-}NMe_3}$), and methylsulfonyl (dCP$^{PPI\text{-}SO_2Me}$) significantly increased the brightness of dC^{PPI} from 25 M^{-1} cm^{-1} to 119, 427, 813, and 576 M^{-1} cm^{-1}, respectively. Particularly, the trimethylammonium derivative dC$^{PPI\text{-}NMe_3}$ exhibited the highest brightness value of 813 M^{-1} cm^{-1}, due to both high ε_{max} and Φ.

Table 9.2 shows the solvatochromic data of dC^{PPI} obtained from the fluorescence measurements in various solvents. These data suggest that the brightness of dC^{PPI}'s fluorescence increases in organic solvents in comparison with that in aqueous buffer as shown in Table 9.1. For example, the brightness is 1234 M^{-1} cm^{-1} in toluene, which is 50 times higher than that in an aqueous solution. Similarly, in all cases, the fluorescence intensity of dC^{PPI} increased in organic solvents. These solvatochromic data indicated that the fluorescence of dC^{PPI} increases after the incorporation into single- and double-stranded DNAs because of the less polar environment in oligonucleotides.

9.2.3 Fluorescence Properties of the Oligonucleotides Containing dC^{PPI}

A dC^{PPI} residue can be incorporated into oligonucleotides using the corresponding phosphoramidite unit and standard phosphoramidite chemistry.[17] The sequences and properties of the oligonucleotides and duplex DNA incorporating dC^{PPI} residues are shown in Table 9.3. T_m values obtained from UV-melting experiments of the 5'-d(CGCAAT[C^{PPI}]TAACGC)-3'/3'-d(GCGTTA[X]ATTGCG)-5' duplexes, where X = A, T, G, or C suggested that the dC^{PPI} residue had a unique base pairing property. The duplex with base X = G showed the highest T_m (58°C), probably due to the formation of Watson–Crick-type base pairs (Fig. 9.5a). The duplex with base X = A showed a comparable T_m (54°C), which could be explained by the formation of a reverse-wobble-type dC^{PPI}-A base pair (Fig. 9.5b). When the nucleobases at the X positions were T and C, the T_m values were 47 and 48 °C, respectively, which are much lower than that obtained when X = G.

The single-strand data show that the fluorescence brightness increased from 25 to 175 M^{-1} cm^{-1}, which is the abovementioned value obtained for the nucleoside dC^{PPI}, by incorporating dC^{PPI} into the 5'-d(CGCAAT[C^{PPI}]TAACGC)-3' sequence. Moreover, the formation of duplexes from 5'-d(CGCAAT[C^{PPI}]TAACGC)-3' and complementary strand 3'-d(GCGTTA[X]ATTGCG)-5' further increased the brightness to 293, 605, and 245 M^{-1} cm^{-1} from that of the single strand when the base at X (opposite C^{PPI}) was A, C, and T, respectively. In contrast, when the base at X was G, the brightness decreased to 59 M^{-1} cm^{-1}, which was lower than that in the single-strand state. One of the possible reasons of this fluorescence quenching might be the photo-induced electron transfer from guanine to the excited dC^{PPI} residue. This was supported by comparing the energies of the

TABLE 9.3 Photophysical Data of the dC^{PPI}-Modified ODN: 5'-d(CGCAAT[C^{PPI}]TAA CGC)-3' and the Duplexes with 3'-d(GCGTTA[X]ATTGCG)-5'

	λ_{max}^{abs} (nm)	ε_{366} (M^{-1} cm^{-1})	λ_{max}^{flu} (nm)	Φ	Brightness $\varepsilon_{366} \times \Phi$
Single strand	383	5822	505	0.030	175
X = A (T_m = 54°C)	385	4718	500	0.062	293
X = C (T_m = 48°C)	379	5708	489	0.106	605
X = T (T_m = 47°C)	385	4718	500	0.052	245
X = G (T_m = 58°C)	390	3957	511	0.015	59

Reprinted with permission from Mizuta et al. [17]. Copyright 2009 American Chemical Society.

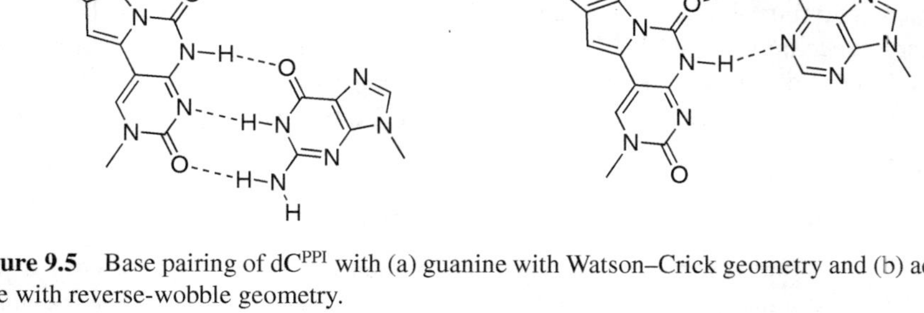

Figure 9.5 Base pairing of dC^{PPI} with (a) guanine with Watson–Crick geometry and (b) adenine with reverse-wobble geometry.

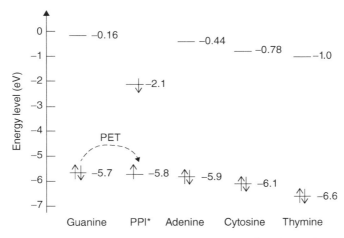

Figure 9.6   Energy diagram of HOMOs and LUMOs of nucleobases.

frontier molecular orbitals of nucleobases such as guanine, adenine, thymine, cytosine, and PPI, which is the aglycone part of the dC^{PPI} residue. Based on density functional theory (DFT) calculations of the ground states at the B3LYP/6-31G** level, HOMO and LUMO levels were found to be as follows: HOMO of guanine (-5.7 eV) > PPI (-5.8 eV) > adenine (-5.9 eV) > cytosine (-6.1 eV) > thymine (-6.6 eV); and LUMO of guanine (-0.16 eV) > adenine (-0.44 eV) > cytosine (-0.78 eV) > thymine (-1.0 eV) > PPI (-2.1 eV). The HOMO and LUMO energy diagram is shown in Figure 9.6. This diagram suggests that the electron transfer from the HOMO of guanine (-5.7 eV) to the singly occupied lower energy orbital of the excited state of the dC^{PPI} residue is an energetically favorable process. This electron transfer may be one of the processes that change the fluorescence intensity of the dC^{PPI} residue in the duplex with base X = G. However, the diagram does not explain the fluorescence intensities of the other duplexes. It is unclear why the fluorescence intensity is higher when X = A compared to when X = T and why the highest fluorescence intensity occurs when X = C. Therefore, other factors such as the structural differences and dipole–dipole interactions between the nucleobases in the duplex should be considered. For example, in the case where X = A or G, the difference of the base pairing geometry shown in Figure 9.5 may change the stacking geometry of dC^{PPI} with the upper and lower bases and change the microenvironment around the dC^{PPI} residue.

9.3   IMPLEMENTATION

9.3.1   Application to a DNA Triplex System

We tried to expand the application of dC^{PPI} to a DNA triplex system. A DNA triplex is a supramolecular structure of DNA formed by the recognition of a DNA duplex by a single-stranded DNA called "triplex forming oligonucleotide" (TFO).[18]

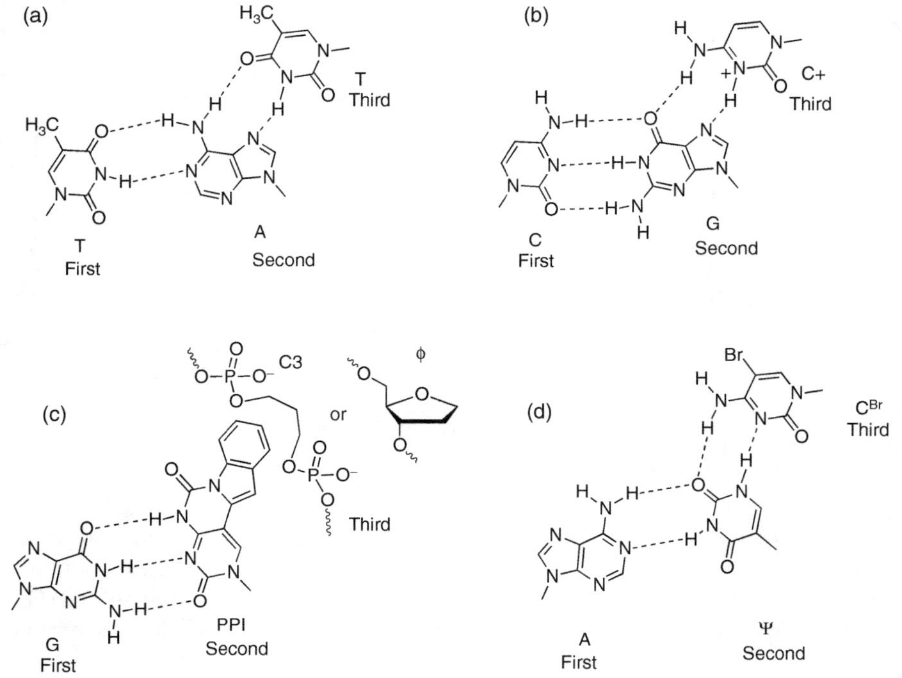

Figure 9.7   Base triads of (a) T–A·T, (b) C–G·C+, (c) G–PPI·C3 (or G–PPI·φ), and (d) A–ψ·C^Br.

A DNA triplex structure is stabilized by two types of base pairs, that is, the canonical Watson–Crick base pairs (such as T–A and C–G), and Hoogsteen base pairs (shown in Fig. 9.7a and b). In these figures, T in TFO recognizes A in a T–A base pair from the major groove and protonated C (C+) recognizes G in a C–G base pair. As described earlier, the base moiety of dC^PPI can form a Watson–Crick base pair with G. Therefore, we attempted to develop a new TFO incorporating an artificial nucleotide residue that could selectively recognize the base moiety of dC^PPI in the duplex. In this chapter and later, the base moiety of dC^PPI is abbreviated as PPI. Because PPI has a large aromatic ring that protrudes into the major groove, we designed an abasic nucleoside surrogate such as a propylene linker (C3) or tetrahydrofuran residue (φ) (Fig. 9.7c).[19] In these G–PPI·C3 or G–PPI·φ triads, the large aromatic PPI ring fits into the space formed by the sugar–phosphate backbone of the abasic site and the 5′-upstream and 3′-downstream nucleobases in TFO (Fig. 9.8).

It should be noted that when canonical nucleobases are used in TFO, the nucleobases of the first strand of the target DNA should always be T and/or C because of the Hoogsteen base pairing pattern requirement (Fig. 9.7a and b). In contrast, the combination of the canonical Hoogsteen base pairs and G–PPI·C3 or G–PPI·φ triads expands the triplex sequence variation such that the first strand can involve T,

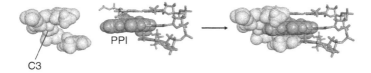

Figure 9.8   Molecular model of the recognition of PPI in a duplex by C3.

3'-GTTTTTTCT X TCTTTG^T T
5'-CAAAAAAGA Y AGAAAC$_T$ T

HP-ta: XY = T-A HP-gppi: XY = G-PPI
HP-cg: XY = C-G HP-aΨ: XY = A-Ψ

5'-TTTTTTCT Z TCTTT-3'

TFO-T: Z = T TFO-C3: Z = C3
TFO-C: Z = C TFO-ϕ: Z = ϕ
TFO-A: Z = A TFO-Br: Z = C$_{Br}$
TFO-G: Z = G

Figure 9.9   Oligonucleotides used for preparing triplexes.

C, and/or G. In addition, we designed another A–ψ·C^{Br} triad, where ψ is the base moiety of pseudouridine, and C^{Br} is 5-bromocytosine (see Fig. 9.7d).[20] The Br group was introduced to reduce the basicity of the N3 position of the cytosine ring so that C^{Br} recognizes ψ rather than G. Using these four base triads, we developed a DNA triplex system consisting of T–A·T, C–G·C+, G–PPI·C3 (or G–PPI·ϕ), and A–ψ·C^{Br}, in which the first base of the triads could be T, C, G, or A.

To clarify the stability and selectivity of the G–PPI·C3 and G–PPI·ϕ triads, various hairpin oligodeoxynucleotides (HPs) and TFOs were synthesized for the measurement of T_m values of the triplexes (Fig. 9.9).

First, the stability and selectivity of the triplex containing PPI as the second base were evaluated using HP-gppi, which is a hairpin duplex incorporating a G–PPI base pair, and TFO-T, -C, -A, -G, -C3, and -ϕ, which are TFOs incorporating thymine, cytosine, adenine, guanine, C3 linker, and tetrahydrofuran residue at the counter position of PPI, respectively. We found that the hairpin duplexes containing G–PPI bound strongly to the TFOs with abasic sites C3 and ϕ at position X, giving T_m values of 56°C and 52°C, respectively. In contrast, the TFOs incorporating the canonical bases at position X showed lower T_m values between 45 and 49 °C. When C3 and ϕ are compared, it is proved that C3 has a higher affinity for PPI than ϕ. Thus, the new base triad G–PPI·C3 could be used as an artificial base pair for incorporating a fluorescent moiety into a DNA triplex.[8]

Next, the stability and selectivity of all the base triads in Figure 9.7 were studied using various combinations of HPs and TFOs (Fig. 9.9). The results are shown in Table 9.4. When each row in the table was examined, it is clear that each HP (-cg, -ta, -gppi, and aψ) bound most tightly to the base that matched the TFOs (TFO-C, -T, C3, and C^{Br}, respectively). Similarly, when each column was examined, it is clear that TFO-C, -T, and -C3 bound most tightly to the base that matched the HPs (-cg, HP-ta, and HP-gppi, respectively). The only exception was TFO-C^{Br}, which bound to both HP-ppig and HP-aψ with comparable stability because of the generally high stability of the triplex formed by HP-gppi and TFOs.

TABLE 9.4 T_m (°C) Values of the Triplex Incorporating Canonical and Artificial Triads

	TFO-C	TFO-T	TFO-C3	TFO-C^{Br}
HP-cg	<u>53</u>	21	12	45
HP-ta	18	<u>48</u>	14	21
HP-gppi	49	45	<u>56</u>	49
HP-aψ	42	36	21	<u>48</u>

Conditions: 2.0 μM oligonucleotides, 10 mM cacodylate (pH 5.4), 50 mM MgCl$_2$, and 500 mM NaCl. The underlined cells indicate the matched combination of HPs and TFOs.

Fluorescence spectra of the DNA duplex, HP-gppi, and the triplex with TFO-C3 are shown in Figure 9.10a. The fluorescence intensity was higher upon the formation of the G–PPI·C3 pair compared to the duplex state. This fluorescence enhancement is explained by the structure shown in Figure 9.8, in which the PPI moiety is covered by the upper and lower hydrophobic bases of the TFO. This model is supported

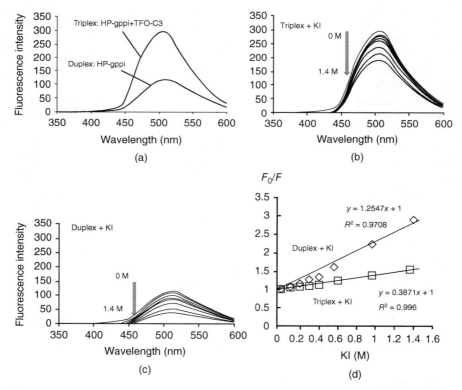

Figure 9.10   (a) Fluorescence of the duplex and triplex incorporating dC^{PPI} measured at 10 °C; (b) triplex fluorescence quenching by KI; (c) duplex fluorescence quenching by KI; (d) Stern–Volmer steady-state fluorescence plot for fluorescence quenching by KI. Conditions: 1.0 μM oligonucleotides in 10 mM sodium cacodylate, 50 mM MgCl$_2$, and 500 mM NaCl (pH 5.4).

by quenching experiments[21] on the triplex (Fig. 9.10b) and duplex (Fig. 9.10c). As shown in Figure 9.10b, the fluorescence intensity of the triplex decreased with the addition of potassium iodide (KI). Similarly, the fluorescence of the duplex also decreased as shown in Figure 9.10c, but the concentration dependence of fluorescence was much greater in the duplex than in the triplex. The dependence of the duplex and triplex fluorescence intensities on KI concentration was quantitatively compared using the Stern–Volmer steady-state fluorescence plot (Fig. 9.10d). The Stern–Volmer equation is as follows. In the following equation, F_o and F are the fluorescence intensities in the absence and presence of KI, respectively. $[Q]$ is the concentration of KI. K_{SV} is the Stern–Volmer constant calculated from the slope of the F_o/F versus $[Q]$ plot.

$$\frac{F_o}{F} = 1 + K_{SV}[Q]$$

This clearly shows that the fluorescence of the duplex was sharply quenched by KI giving a K_{SV} value of 1.25, which is higher than that of 0.39 for the triplex. These data suggest that the PPI moiety is buried into the pocket formed by the abasic site of C3 and flanked by the upper and lower TFO bases, which effectively hide the approach of the fluorescence quencher, I^-.

9.3.2   Double Labeling of an Oligonucleotide with dC^{PPI} and 2-Aminopurine

As shown in Table 9.1, the dC^{PPI} residue was excited at 374 nm and emitted at 513 nm. Thus, the Stokes shift of dC^{PPI} is calculated to be 139 nm, which is higher than that of most other fluorescent nucleobases. For example, the Stokes shift of 2-aminopurine (2-AP, the most commonly used fluorescent nucleobase) is only 70 nm ($\lambda_{excitation} = 300$ nm and $\lambda_{emission} = 370$ nm). In addition, because dC^{PPI} has a large absorption peak at around 250–300 nm, this fluorescent nucleoside can be simultaneously excited with 2-AP by irradiation at 300 nm to give two well-separated fluorescence signals if they were incorporated into a single nucleic acid molecule. Several examples of the incorporation of two different fluorescent nucleosides into DNA and RNA structures have been reported, but the separation of the fluorescence signals were not clear. In one example, pyrrolo-dC and 2-AP are incorporated into a hairpin-type molecular beacon.[22] In another example, 5-(furan-2-yl)uridine and 2-AP are incorporated into an RNA duplex.[5f] Therefore, it might be useful if nucleic acid molecules are doubly labeled with dC^{PPI} and 2-AP, giving two well-separated fluorescence signals at 370 and 513 nm. To demonstrate this possibility, several oligonucleotides incorporating dC^{PPI} and 2-AP were synthesized, and their fluorescence was measured simultaneously.[23] Figure 9.11a shows the fluorescence spectrum of a duplex consisting of 24 base pairs, two of which are dC^{PPI}-A and 2-AP-T base pairs. Clearly, the emissions from 2-AP and dC^{PPI} were observed at around 370 and 510 nm, respectively, as separated peaks. In this sequence, the position of dC^{PPI} was designed so that it forms a base pair with A instead of G to prevent quenching by G. Although the signal from 2-AP is significantly smaller than that from dC^{PPI} in this case, the intensity can become comparable to that of dC^{PPI}

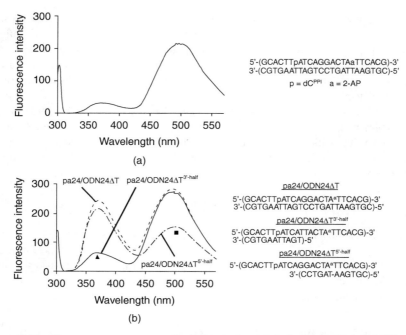

Figure 9.11 Fluorescence spectrum of a DNA duplex labeled with 2-AP and dC^{PPI}. Conditions: 0.5 µM oligonucleotides in 50 mM sodium phosphate and 0.1 M NaCl (pH 7.0). The closed triangle and square indicate the peak heights of 2-AP and dC^{PPI}, respectively, in the single-strand state. (Spectra reproduced from Seio *et al.* [23]. Fluorescence properties of oligonucleotides doubly modified with an indole-fused cytosine analog and 2-aminopurine, with permission from Elsevier).

by designing the sequence so that the 2-AP residue bulges out from the duplexes (Fig. 9.11b). In addition, the fluorescence signals of 2-AP and dC^{PPI} can be modified almost independently by hybridization with different oligonucleotides. For example, hybridization with an oligonucleotide that only covered the dC^{PPI} residue reduced the intensity of the 2-AP emission keeping that of dC^{PPI} unchanged. On the contrary, the hybridization of the 2-AP region reduced the intensity of dC^{PPI} selectively. These results clearly suggest that the combination of 2-AP and dC^{PPI} is suitable for the fluorescent double labeling of oligonucleotides for the simultaneous monitoring of structural changes such as duplex formation.

9.4 CONCLUSIONS

We developed fluorescent nucleosides with a pyrimidopyrimidine structure. The pyrimidopyrimidoindole nucleoside dC^{PPI} exhibited enhanced fluorescence intensity in a hydrophobic environment compared with an aqueous environment and emission at a longer wavelength such as 510 nm with a large Stokes shift. Because the pyrimidopyrimidoindole skeleton of dC^{PPI} is chemically stable, various substituents can be introduced to this fluorophore to modify its fluorescence properties. Because of the large Stokes shift, dC^{PPI} can be used for the double fluorescent labeling of oligonucleotides in combination with 2-AP. In addition, we developed a DNA triplex system in which the Watson–Crick G-dC^{PPI} pair is recognized by TFO containing abasic sites such as the C3 linker. When the triplex is formed, the fluorescence of the dC^{PPI} increases because of the more hydrophobic environment around dC^{PPI}.

9.5 PROSPECTS AND OUTLOOK

There are several possible applications of dC^{PPI}. First, because the oligonucleotides doubly modified with 2-AP and dC^{PPI} derivatives show two well-separated fluorescence signals (Fig. 9.11), the double labeling of oligonucleotides is possible. By using this combination, one may monitor the structural alterations of distal positions of various functional nucleic acids such as DNA aptamers. Thus, the independence of fluorescence derived from 2-AP and dC^{PPI} first has to be established. Because the dC^{PPI} absorption overlaps with the emission of 2-AP at around 370 nm, photophysical interactions may occur. For application to functional RNAs, ribonucleoside derivatives of dC^{PPI} need to be developed.

In combination with DNA triplex systems, dC^{PPI} could also be applied to nucleic acid detection. As described here, dC^{PPI} is recognized by abasic sites such as C3, and it is incorporated into DNA triplexes stabilized by base triads such as T–A·T, C–G·C+, G–PPI·C3, and A–ψ·C^{Br}. It should be noted that, no matter what the first strand sequence is, one can design second and third strands that can recognize dC^{PPI} derivatives. Moreover, as shown in Figure 9.10, the fluorescence of dC^{PPI} increases when a triplex is formed. Thus, a triplex system can be used for the detection of target DNA or RNA by capturing the DNA or RNA as the first strand. Unfortunately, the fluorescence of the triplex is merely two times as intense as that of the duplex. Therefore, more sensitive and brighter dC^{PPI} derivatives need to be developed for this application. These improvements could be achieved by designing new fluorescence properties on the basis of the knowledge obtained from the study of dC^{PPI} derivatives.

REFERENCES

1. Davies, M. J.; Shah, A.; Bruce, I.J. *Chem. Soc. Rev.* **2000**, *29*, 97.

2. (a) Schena, M.; Shalon, D.; Davis, R. W.; Brown, P. O. *Science*, **1995**, *270*, 467; (b) DeRisi, J.; Penland, L.; Brown, P. O.; Bittner, M. L.; Meltzer, P. S.; Ray, M.; Chen, Y. D.; Su, Y. A.; Trent, J. M. *Nat. Genet.* **1996**, *14*, 457; (c) Brenner, S.; Johnson, M.; Bridgham, J.; Golda, G.; Lloyd, D. H.; Johnson, D.; Luo, S. J.; McCurdy, S.; Foy, M.; Ewan, M.; Roth, R.; George, D.; Eletr, S.; Albrecht, G.; Vermaas, E.; Williams, S. R.; Moon, K.; Burcham, T.; Pallas, M.; DuBridge, R. B.; Kirchner, J.; Fearon, K.; Mao, J.; Corcoran, K. *Nat. Biotechnol.* **2000**, *18*, 630.

3. (a) Livak, K. J.; Flood, S. J. A.; Marmaro, J.; Giusti, W.; Deetz, K. *PCR Methods Appl.* **1995**, *6*, 357; (b) Whitcombe, D.; Theaker, J.; Guy, S. P.; Brown, T.; Little, S. *Nat. Biotechnol.* **1999**, *17*, 804; (c) Ranasinghe, R. T.; Brown, T. *Chem. Commun.* **2005**, 5487.

4. (a) Rist, M. J.; Marino, J. P. *Curr. Org. Chem.* **2002**, *6*, 775; (b) Sinkeldam, R. W.; Greco, N. J.; Tor, Y. *Chem. Rev.* **2010**, *110*, 2579; (c) Wilhelmsson, L. M. *Q. Rev. Biophys.* **2010**, *43*, 159.

5. (a) Greco, N. J.; Tor, Y. *J. Am. Chem. Soc.* **2005**, *127*, 10784; (b) Greco, N. J.; Tor, Y. *Tetrahedron* **2007**, *63*, 3515; (c) Srivatsan, S. G.; Tor, Y. *Nat. Protoc.* **2007**, *2*, 1547; (d) Greco, N. J.; Tor, Y. *Nat. Protoc.* **2007**, *2*, 305; (e) Srivatsan, S. G.; Tor, Y. *Tetrahedron* **2007**, *63*, 3601; (f) Srivatsan, S. G.; Tor, Y. *J. Am. Chem. Soc.* **2007**, *129*, 2044; (g) Pesnot, T.; Wagner, G. K. *Org. Biomol. Chem.* **2008**, *6*, 2884; (h) Greco, N. J.; Sinkeldam, R. W.; Tor, Y. *Org. Lett.* **2009**, *11*, 1115; (i) Sinkeldam, R. W.; Wheat, A. J.; Boyaci, H.; Tor, Y. *ChemPhysChem* **2011**, *12*, 567; (j) Sinkeldam, R. W.; Hopkins, P. A.; Tor, Y. *ChemPhysChem* **2012**, *13*, 3350.

6. (a) Sinkeldam, R. W.; Marcus, P.; Uchenik, D.; Tor, Y. *ChemPhysChem* **2011**, *12*, 2260; (b) Riedl, J.; Pohl, R.; Rulíšek, L.; Hocek, M. *J. Org. Chem.* **2012**, *77*, 1026.

7. (a) Godde, F.; Toulmé, J. J.; Moreau, S. *Biochemistry* **1998**, *37*, 13765; (b) Godde, F.; Toulmé, J. J.; Moreau, S. *Nucleic Acids Res.* **2000**, *28*, 2977; (c) Xie, Y.; Dix, A. V.; Tor, Y. *J. Am. Chem. Soc.* **2009**, *131*, 17605; (d) Xie, Y.; Maxson, T.; Tor, Y. *Org. Biomol. Chem.* **2010**, *8*, 5053; (e) Xie, Y.; Maxson, T.; Tor, Y. *J. Am. Chem. Soc.* **2010**, *132*, 11896.

8. (a) Tor, Y.; Del Valle, S.; Jaramillo, D.; Srivatsan, S. G.; Rios, A.; Weizman, H. *Tetrahedron* **2007**, *63*, 3608; (b) Srivatsan, S. G.; Weizman, H.; Tor, Y. *Org. Biomol. Chem.* **2008**, *6*, 1334; (c) Srivatsan, S. G.; Greco, N. J.; Tor, Y. *Angew. Chem. Int. Ed.* **2008**, *47*, 6661; (d) Srivatsan, S. G.; Tor, Y. *Chem. Asian J.* **2009**, *4*, 419; (e) Shin, D.; Sinkeldam, R. W.; Tor, Y. *J. Am. Chem. Soc.* **2011**, *133*, 14912.

9. (a) Nakagawa, O.; Ono, S.; Li, Z.; Tsujimoto, A.; Sasaki, S. *Angew. Chem. Int. Ed.* **2007**, *46*, 4500; (b) Sandin, P.; Stengel, G.; Ljungdahl, T.; Börjesson, K.; Macao, B. Wilhelmsson, L. M. *Nucleic Acids Res.* **2009**, *37*, 3924; (c) Nasr, T.; Li, Z.; Nakagawa, O.; Taniguchi, Y.; Ono, S.; Sasaki, S. *Bioorg. Med. Chem. Lett.* **2009**, *19*, 727.

10. (a) Wilhelmsson, L. M.; Holmén, A.; Lincoln, P.; Nielsen, P. E.; Nordén, B. *J. Am. Chem. Soc.* **2001**, *123*, 2434; (b) Wilhelmsson, L. M.; Sandin, P.; Holmén, A.; Albinsson, B.; Lincoln, P.; Nordén, B. *J. Phys. Chem. B* **2003**, *107*, 9094; (c) Sandin, P.; Wilhelmsson, L. M.; Lincoln, P.; Powers, V. E. C.; Brown, T.; Albinsson, B. *Nucleic Acids Res.* **2005**, *33*, 5019; (d) Tahmassebi, D. C.; Millar, D. P. *Biochem. Biophys. Res. Commun.* **2009**, *380*, 277; (e) Porterfield, W.; Tahmassebi, D. C. *Bioorg. Med. Chem. Lett.* **2009**, *19*, 111.

11. (a) Liu, C.; Martin, C. T. *J. Mol. Biol.* **2001**, *308*, 465; (b) Berry, D. A.; Jung, K. Y.; Wise, D. S.; Sercel, A. D.; Pearson, W. H.; Mackie, H.; Randolph, J. B.; Somers, R. L. *Tetrahedron Lett.* **2004**, *45*, 2457; (c) Tinsley, R. A.; Walter, N. G. *RNA* **2006**, *12*, 522; (d) Seela, F.; Sirivolu, V. R. *Org. Biomol. Chem.* **2008**, *6*, 1674; (e) Noé, M. S.; Ríos, A. C.; Tor, Y. *Org. Lett.* **2012**, *14*, 3150; (f) Hudson, R. H. E.; Ghorbani-Choghamarani, A. *Synlett* **2007**, 870.

12. (a) Okamoto, A.; Tainaka, K.; Saito, I. *J. Am. Chem. Soc.* **2003**, *125*, 4972; (b) Okamoto, A.; Tainaka, K.; Saito, I. *Tetrahedron Lett.* **2003**, *44*, 6871.

13. Miyata, K.; Tamamushi, R.; Ohkubo, A.; Taguchi, H.; Seio, K.; Santa, T.; Sekine, M. *Org. Lett.* **2006**, *8*, 1545.

14. Miyata, K.; Mineo, R.; Tamamushi, R.; Mizuta, M.; Ohkubo, A.; Taguchi, H.; Seio, K.; Santa, T.; Sekine, M. *J. Org. Chem.* **2007**, *72*, 102.

15. Mizuta, M.; Seio, K.; Miyata, K.; Sekine, M. *J. Org. Chem.* **2007**, *72*, 5046.

16. Miyata, K.; Kobori, A.; Tamamushi, R.; Ohkubo, A.; Taguchi, H.; Seio, K.; Sekine, M. *Eur. J. Org. Chem.* **2006**, *67*, 3626.

17. Mizuta, M.; Seio, K.; Ohkubo, A.; Sekine, M. *J. Phys. Chem. B* **2009**, *113*, 9562.

18. (a) Duca, M.; Vekhoff, P.; Oussedik, K.; Halby, L.; Arimondo, P. B. *Nucleic Acids Res.* **2008**, *36*, 5123; (b) Arya, D. P. *Acc. Chem. Res.* **2011**, *44*, 134; (c) Malnuit, V.; Duca, M.; Benhida, R. *Org. Biomol. Chem.* **2011**, *9*, 326.

19. Mizuta, M.; Banba, J.; Kanamori, T.; Tawarada, R.; Ohkubo, A.; Sekine, M.; Seio, K. *J. Am. Chem. Soc.* **2008**, *130*, 9622.

20. (a) Trapane, T. L.; Ts'o, P. O. P. *J. Am. Chem. Soc.* **1994**, *116*, 10437; (b) Trapane, T. L.; Christopherson, M. S.; Roby, C. D.; Ts'o, P. O. P. *J. Am. Chem. Soc.* **1994**, *116*, 8412; (c) Kanamori, T.; Masaki, Y.; Mizuta, M.; Tsunoda, H.; Ohkubo, A.; Sekine, M.; Seio, K. *Org. Biomol. Chem.* **2012**, *10*, 1007.

21. (a) Lehrer, S. S. *Biochemistry* **1971**, *10*, 3254; (b) Lakowicz, J. R. Principles of Fluorescence Spectroscopy. 3rd ed., Springer Science+Business Media, LLC: New York, USA, **2006**.

22. Mari, A. A.; Jockusch, As.; Li, Z.; Ju, J.; Turro, N. J. *Nucleic Acids Res.* **2006**, *34*, e50.

23. Seio, K.; Kanamori, T.; Tokugawa, M.; Ohzeki, H.; Masaki, Y.; Tsunoda, H.; Ohkubo, A.; Sekine, M. *Bioorg. Med. Chem.* **2013**, *21*, 3197.

10

FÖRSTER RESONANCE ENERGY TRANSFER (FRET) BETWEEN NUCLEOBASE ANALOGS – A TOOL FOR DETAILED STRUCTURE AND DYNAMICS INVESTIGATIONS

L. Marcus Wilhelmsson

Chemistry and Chemical Engineering/Chemistry and Biochemistry, Chalmers University of Technology, Gothenburg, Sweden

10.1 INTRODUCTION

The three-dimensional structure and the conformational dynamics of nucleic acids and their conformational changes upon interaction with other biomolecules are fundamental to the understanding of these key players of life and their role in the central dogma of molecular biology as well as RNAi-mediated gene activity moderation. Traditionally, atomic resolution structure information of biomolecules is achieved using X-ray crystallography[1] and NMR spectroscopy[2]. There is no doubt that these methods are vital in structure (crystallography and NMR) and dynamics (NMR) investigations. However, they also suffer from significant drawbacks like the need for crystals and, thus, the deficiency of solvent molecules in crystallography, and the large sample amounts as well as upper molecular weight limit for NMR. Lower structure resolution techniques such as small-angle scattering[3], cryo-electron microscopy[4], site-directed spin labeling[5] as well as metal nanoparticles used as "plasmon rulers"[6] and in small-angle X-ray scattering interference[7] can also be

Fluorescent Analogs of Biomolecular Building Blocks: Design and Applications, First Edition.
Edited by Marcus Wilhelmsson and Yitzhak Tor.
© 2016 John Wiley & Sons, Inc. Published 2016 by John Wiley & Sons, Inc.

successfully used for nucleic acid investigations. However, several of these methods are expensive and less straightforward to perform. Moreover, in spin labeling there is a lack of commercially available probes, whereas in the nanoparticle-based methods the size and heterogeneity of the particles usually complicates measurements. Thus, none of these lower resolution structure or ruler techniques are as ubiquitously exploited in biosciences as Förster resonance energy transfer (FRET)[8]. FRET is inexpensive, easy, rapid, and allows for measurements of large (bio)molecular complexes with just a few nmol of sample. Moreover, FRET can be performed at the single-molecule level. The major drawbacks of the method as a high-structure resolution technique are the dependency on the photophysical properties of the fluorophores (Φ_D and J in Eq. 10.2) and the uncertainty in probe position and orientation[9] (κ^2 in Eq. 10.2) typically associated with the linker used to connect the fluorophore to the biomolecule under investigation (Fig. 10.1).

The FRET efficiency, E, (Eq. 10.1) can be measured using either steady-state or time-resolved techniques and in combination with the characteristic Förster distance, R_0, (Eq. 10.2; distance where half of the excited energy is lost via transfer to the acceptor) give distance, R_{DA} (<10 nm), as well as orientation, κ, thereby providing three-dimensional structural information.

$$E = 1 - \frac{I_{DA}}{I_D} = 1 - \frac{\tau_{DA}}{\tau_D} = \frac{R_0^6}{R_0^6 + R_{DA}^6} \qquad (10.1)$$

$$R_0 = 9.7 \cdot 10^3 (J\kappa^2 n^{-4}\Phi_D)^{\frac{1}{6}} \qquad (10.2)$$

In Equation 10.2, n refers to the refractive index of the medium between the fluorophores and Φ_D to the quantum yield of the donor. Both parameters are easily accessible. The J-integral, J, which is a measure of the energy overlap between the donor emission and the acceptor absorption, is also a reasonably easily accessible parameter. On the other hand, as was briefly mentioned above, due to the flexibility of the linkers commonly used to connect fluorophores to nucleic acids, the orientation factor κ could be more challenging to estimate accurately. The orientation factor[9] is calculated using Equation 10.3 and is determined by the

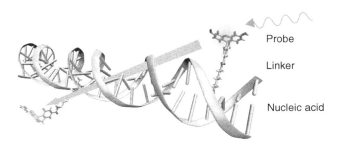

Figure 10.1 Schematic representation of typical external FRET probes attached covalently to DNA via a flexible linker. Reprinted with permission from Wiley.

relative orientation of the transition dipole moments of the donor, e_D, and acceptor, e_A, and the direction of the vector interconnecting them, e_{DA}, where e refers to unit vectors.

$$\kappa = e_D \cdot e_A - 3(e_D \cdot e_{DA})(e_{DA} \cdot e_A) \qquad (10.3)$$

For most nucleic acid FRET systems that contain fluorophores covalently attached through flexible linkers, a κ^2 of 2/3 is commonly utilized. This refers to a totally randomized distribution during the time-lapse of the energy transfer process and has been used both correctly and incorrectly over the years. To increase the accuracy of the κ^2 value, the distribution of fluorophore orientation during the energy transfer process is traditionally described by a simple wobbling-in-cone model, with semiangles obtained from fluorescence anisotropy measurements allowing for estimation of the limits of $\langle \kappa^2 \rangle$.[10] Even more accurate is to have a practically complete control of κ, which can be accomplished by knowing the direction of the transition dipole moments within the fluorophores (e_D and e_A) and having them virtually fixed during the process of energy transfer. As will be clear from the description of the development of the first nucleic acid base analog FRET pair below, this condition is essentially fulfilled for that system.

By combining Equations 10.1 and 10.2, it is obvious that in order to obtain high-resolution structure information from easily accessible steady-state or time-resolved measurements, a fluorophore is required with stable photophysical properties (gives constant J and Φ_D) as well as an accurately determined orientation factor. This means that the fluorophore should have a spectral envelope and Φ that are essentially insensitive to changes in its microenvironment, which may occur when new species are added to the sample, as well as having an easily defined position within the nucleic acid system under investigation. As has been presented in previous chapters in this book, the vast majority of fluorophores, including available fluorescent base analogs, are highly sensitive to their immediate surroundings.[11] Although a requirement for many applications, this is, however, not a desired property in a FRET experiment. To enable accurate FRET in nucleic acid–containing systems, my group has therefore developed a family of base analogs, the tricyclic cytosines, which meet most of the criteria that are important to become a member of an excellent FRET pair. This chapter discusses the development and characterization of these cytosine analogs, their use as the first nucleic acid base analog FRET pair, and some recent applications.

10.2   THE TRICYCLIC CYTOSINE FAMILY

The tricyclic cytosines, tC, tC^O, and tC$_{nitro}$ (Fig. 10.2), are derivatives of phenothiazine, which is a well-known pharmacophore often used in antipsychotic and antihistaminic drugs. Two of these cytosine analogs, tC and tC^O, were first synthesized by Lin *et al.* and introduced as base analogs that, with their increased base-stacking interactions, could have a potential use in antisense technology.[12]

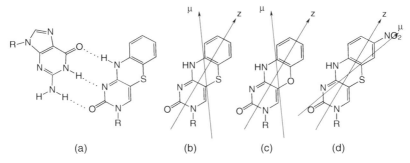

(a)   (b)   (c)   (d)

Figure 10.2 (a) Structure of G-tC base pair (R = rest of DNA). (b) Structure of tC with its molecular long axis (z) and the direction of the transition dipole moment (μ) of the lowest energy transition of tC. (c) As (b) but for tC^O. (d) As (b) but for tC$_{nitro}$.

A couple of years later, while involved in a PNA antisense-related project, we came across a PNA-oligonucleotide containing a tC-monomer.

We recognized that tC is a derivative of phenothiazine, in itself a fluorophore, that becomes a fluorophore with excellent properties upon modifying its core to methylene blue. Since no studies of such properties were previously reported, we investigated the fluorescence properties of this tC-PNA in great detail. The study confirmed that tC has interesting and unique properties as a fluorescent base analog (*vide infra*).[13]

Today the phosphoramidites of the tC-family (tC, tC^O, and tC$_{nitro}$) are all commercially available ready for solid-phase DNA synthesis. This was obviously not the case when Lin *et al.* first synthesized the DMT-activated H-phosphonate of tC and tC^O for further solid-phase DNA chemistry.[12] They started from the 5-iodo-2′-deoxyuridine and 5-bromo-2′-deoxyuridine for tC and tC^O, respectively. The three-ring system was built using 2-aminothiophenol and 2-aminophenol and cyclization was performed using potassium *tert*-butoxide and potassium fluoride, respectively. Later, we made some minor modifications to the synthetic route for tC^O, though, keeping the overall strategy of Lin *et al.*. On the contrary, for tC we changed the synthetic route substantially.[14] Instead of building from a nucleoside such as 5-iodo-2′-deoxyuridine, we used 5-bromouracil as a starting material and prepared the tC nucleobase and the sugar moiety separately. After glycosylation, we activated the nucleoside for further DNA solid-phase synthesis. This new route, similar to the one by Lin *et al.*, gives a relatively low overall yield but is significantly more straightforward with fewer purification steps.[14] Moreover, established synthetic protocols for the synthesis of the tC nucleobase[15] and the sugar moiety allow for scale up to give access to plenty of nucleoside. It is noteworthy to mention that this synthetic approach, coupling a nucleobase to a sugar moiety, is more general compared to starting from commercially available nucleosides. When synthesizing the FRET-acceptor tC$_{nitro}$, we used a similar approach to the one developed by us for tC. First 5-nitro-2-amino-thiophenol[16] was coupled to 5-bromouracil whereafter a cyclization afforded the nucleobase.[17] Incorporation of the tC-family base analogs into DNA via solid-phase synthesis and subsequent purification of oligonucleotides are then performed according to standard protocols with the exception that the coupling time is increased compared to the standard phosphoramidites.[14]

10.2.1   Structural Aspects, Dynamics, and Ability to Serve as Cytosine Analogs

A nucleobase analog should be able to serve as a faithful structural mimic of one of the nucleobases in naturally occurring nucleosides (A, C, G, T, or U). To report correctly on properties of the natural systems under investigation, it is important that the modified nucleoside leaves the native nucleic acid structure as unperturbed as possible. This subchapter is divided into two parts: (1) the nucleosides and (2) inside nucleic acid systems. It will become apparent that knowledge of the structural properties of the base analog itself and its effect on the structure of nucleic acids after incorporation are important for a full understanding of its use in fluorescence-based assays and in FRET experiments in specific.

10.2.1.1   Nucleosides: The structure of the parent compound of tC and tC_{nitro}, phenothiazine, was early on investigated using X-ray crystallography and reported to have a folded conformation over the sulfur–nitrogen axis of the middle ring.[18] In contrast, X-ray crystal structures have shown that the tricyclic core of tC^O is planar.[19]

We have used density functional theory (DFT) and time-dependent DFT (TDDFT) with the B3LYP functional to examine both the ground- and excited-state levels of tC, tC_{nitro}, and tC^O.[20] In agreement with the crystal structures of the parent compound, we find that tC and tC_{nitro} indeed have a folded structure in the ground state (Fig. 10.3). The potential energy surfaces reveal two local energy minima for both molecules corresponding to two geometries that are mirror images and folded $\sim25°$ along the sulfur–nitrogen axis. The energy barrier between these minima was found to be low ($\sim0.05\,eV$) for both tC and tC_{nitro}, making interconversion between the two states possible. In NMR-studies of tC incorporated into a dodecamer of DNA, only one set of chemical shifts was observed, giving further evidence of a fast interconversion between the two conformations.[21]

The structure of the first excited state of these two phenothiazine derivatives, on the other hand, was calculated to be planar with steep increases in energy even at low angles ($\sim10°$).[20] Corresponding calculations for the ground state of tC^O show, in agreement with the crystal structures, that it is planar (Fig. 10.3).[20] This is also the case for the first excited state of tC^O.

In pH titrations of the tricyclic cytosines, we have shown that their neutral forms are predominant at and around physiological pH.[22,23] Upon significant increase in the pH, the deprotonated forms appear with spectral and emissive features distinct from the neutral forms. Calculations for tC show its deprotonated form to be planar, which is further supported by more structured spectral features.[23a] In AM1 calculations of the protonated form of tC, we obtained evidence supporting a folded structure for the phenothiazine derivatives under acidic conditions as well.[23a]

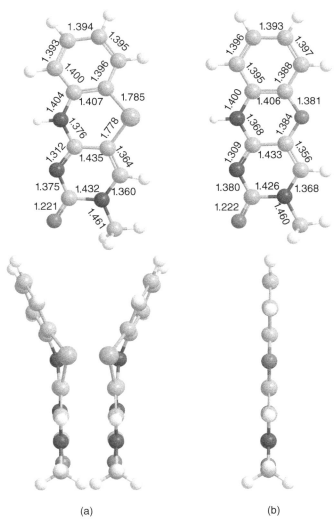

(a)   (b)

Figure 10.3 B3LYP/6-31G(d,p) optimized ground-state structures of (a) tC and (b) tC^O. Top: front view. Bottom: side view (both local minima are shown for tC). Bond lengths in Ångström. Reproduced from Ref. [20] with permission from the PCCP Owner Societies.

For proper base pairing with guanine the tricyclic cytosines need to be in their amino tautomer (Fig. 10.4). Early studies reported that tC and tC^O discriminate well between G and A targets suggesting that the amino tautomer is indeed the dominating form.[12] In an NMR investigation, we have also shown evidence of the amino form being present when tC is placed opposite of G.[21] However, later it has been demonstrated that during, for example, the action of DNA polymerases tC and tC^O in the template strand can code both for G and A as well as tC- and tC^O-nucleotides

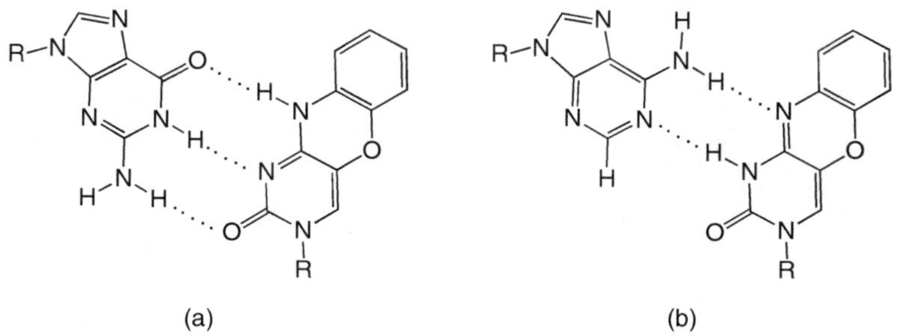

(a)   (b)

Figure 10.4   (a) Amino-tautomer of tC^O base-paired with guanine. (b) Imino-tautomer of tC^O base-paired with adenine.

be incorporated across G or A templates.[24] This behavior was explained by a possible stabilization of the imino-form of the tricyclic cytosines in the presence of A (Fig. 10.4) in the polymerase site.[24b] Moreover, in a recent X-ray crystallography study of an oligonucleotide-RB69 polymerase complex by the Konigsberg laboratory, it was shown that two hydrogen bonds between A and tC^O could indeed be formed in addition to the common three hydrogen bonds between G and tC^O.[25]

10.2.1.2   Inside Nucleic Acid Systems: To be a good probe molecule, the fluorescent base analog should minimally perturb the structure and dynamics of the nucleic acid. To understand how this family of base analogs influences nucleic acids, we have performed extensive structure and dynamics characterization of them incorporated in DNA. As previously mentioned, it is also essential for the detailed FRET studies we describe below, to have a good knowledge regarding the position, orientation, and dynamics of the involved donor and acceptor.

For all three tricyclic cytosines, tC, tC^O, and tC_{nitro}, we find that they, on average, increase the stability with a couple of degrees centigrade of 10–20-mer oligonucleotides.[17,22,26] A plausible explanation for this is the increased $\pi-\pi$ overlap between the extended tricyclic ring systems and the neighboring bases. For certain base environments, exchanging a C with one of the tricyclic cytosine results in the same melting temperature, allowing a base replacement without affecting the overall duplex stability.[17,22,26] Moreover, we have used circular dichroism (CD) to investigate the overall secondary structure of DNA after exchanging one cytosine in a short oligonucleotide with a tC, tC^O, or tC_{nitro}, and find that the DNA remains in its natural B-form.[17,22,26] For tC we have also performed an NMR study in which we replaced one cytosine in a self-complementary dodecamer.[21] All NMR data in this investigation, that is, scalar couplings and assigned NOEs, are consistent with an overall B-form DNA structure. Locally, NMR supports that tC exists in its anti conformation and the expected base pairing pattern with G. However, data suggest that tC is slightly pulled toward the middle of the base-stack compared to a normal C. This is not unexpected considering the favorable increased base-stacking interactions this results in between the outer rings of tC and its adjacent bases. Furthermore, in a recent investigation using DFT and TDDFT, we find evidence suggesting that tC, and also the other phenothiazine derivative tC_{nitro}, positioned in a B-DNA environment prefers to be in only one of its possible bent conformations.[20] The steric hindrance caused by the surrounding bases makes one of the two local minima mentioned above higher in energy, thus forcing the outer ring to be directed away from the 5'-end. Finally, it has been shown by other laboratories using X-ray crystallography that the phenoxazine tC^O and also its structurally similar nitroxide derivative[27] both form proper base pairs with guanine. However, in an RB69 polymerase mutant, the crystal structure also showed possible base pairing with dATP through two hydrogen bonds.[25]

The dynamics of the tricyclic cytosines is important to understand for a variety of applications and vital when estimating an accurate orientation factor, κ^2, to be used during a FRET experiment. In the previously mentioned NMR study of a tC-containing dodecamer, we found evidence from the NOESY cross peaks to the water signal that suggest no increased opening dynamics in the tC-G base pair compared to a normal one.[21] Further indications that suggest a rigid stacking of the tricyclic cytosine analogs are (1) the more structured UV–Vis spectra upon incorporation into DNA and especially in a duplex context, (2) the average increase in melting temperature compared to unmodified duplexes, and (3) the single emission lifetime observed for tC and tC^O for virtually all combinations of base surroundings in DNA duplexes.[17,22,26] Maybe the most important evidence of low mobility of tC and tC^O inside DNA in the timescale of the fluorescence event is what we have found in anisotropy investigations. Comparing calculated with experimental values, we found that tC and tC^O indeed report merely on the overall rotation and tumbling of the DNA molecule and that the mobility of the probes relative to the DNA is of very small amplitude.[22] Moreover, we have demonstrated in a highly viscous solution, where the rotation and tumbling of the nucleic acid is totally hindered, that the anisotropy of the tC/tC^O probe inside the nucleic acid approaches that of the limiting anisotropy giving further evidence to a very low rate of dynamics compared to the timescale of the emission (Preus and Wilhelmsson, unpublished results). Later in this chapter, where the tC^O–tC_{nitro} FRET pair will be discussed, it is also obvious from the sinusoidal shape of the FRET curve, resulting from placing the donor and acceptor at various positions inside a DNA duplex, that the probes are in fact rigidly stacked. A FRET pair where the donor and acceptor have high base-flipping rates would have resulted in a significantly less distinct sinusoidal shape.

10.2.2 Photophysical Properties

Before going into details, it should be mentioned that the emissive members of the tC-family, tC and tC^O, are not like the vast majority of fluorescent base analogs when it comes to their photophysical properties. As has been described in the previous chapters, the emission intensity (to some extent also the position and shape of the spectral envelope) of fluorescent base analogs normally changes dramatically from single- to double-stranded environments and is dependent on neighboring bases.[11]

From numerous examples in this book, it is apparent that this environmental sensitivity of fluorescence properties can be extremely useful. On the other hand, for mere detection purposes, an environmentally insensitive fluorescence quantum yield in combination with a large extinction coefficient, resulting in a high brightness, is more important.[11c] Environment-insensitive fluorescent base analogs also have significant advantages in FRET and anisotropy experiments.[11c] In these techniques, properties such as a high and virtually unaffected quantum yield as well as a single lifetime simplify accurate measurements significantly. Both tC (in all contexts) and tC^O (in duplexes) are environment-insensitive fluorescent base analogs and have these desirable properties.

The emission maximum of tC is situated at ~505 nm both in its monomeric form[23a] and in single- and double-stranded systems[13,26] (Fig. 10.5). The quantum yield of the emission is virtually constant in single- and double-stranded systems with an average of 20% (Table 10.1).[26] Moreover, in both environments, the emission displays single-exponential fluorescence decays with an approximately lifetime of 6 ns.[26] This excited state can be reached by exciting tC to S_1 by 400 nm radiation. This pure state gives an absorption band from 350 to 450 nm and we have determined the transition dipole moment to be directed approximately 35° counterclockwise from the long axis of the tC three-ring system (Fig. 10.2),[23a] an important parameter for accurate calculation of orientation factors in FRET experiments. Finally, the absorption and emission spectral envelopes of tC are essentially the same and irrespective of the surrounding environment under physiological conditions.[23a,26]

The oxo-analog of tC, tC^O, on the other hand, has an emission maximum around 460 nm that can be reached by exciting to the maximum of the S_1-state at 365 nm and, like in tC, the absorption and emission spectral envelopes are unaffected by the microenvironment like in tC (Fig. 10.5).[22] The transition dipole moment corresponding to the S_1-state of tC^O is directed 33° counterclockwise from the

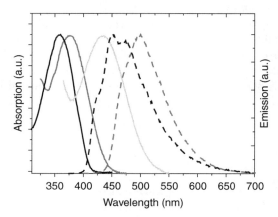

Figure 10.5 Normalized absorption (full lines) and emission (dashed lines) spectra of tC^O (black), tC (dark gray) and tC_{nitro} (light gray) in duplex context. No emission spectrum for the virtually nonfluorescent tC_{nitro} is presented. (*See color plate section for the color representation of this figure.*)

TABLE 10.1 Fluorescence Quantum Yields of tC and tC^O in Single- and Double-Stranded DNA

Neighboring Bases[a]	Φ_D, tC, ss	Φ_D, tC, ds	Φ_D, tC^O, ss	Φ_D, tC^O, ds
GA	0.17	0.18	0.14	0.17
GG	0.18	0.18	0.14	0.18
GC	0.19	0.16	0.17	0.20
GT	[b]	[b]	0.15	0.17
AA	0.21	0.18	0.38	0.23
AG	[b]	[b]	0.41	0.24
AC	[b]	[b]	0.33	0.27
AT	[b]	[b]	0.36	0.27
TA	0.21	0.20	0.34	0.21
TG	0.24	0.21	0.36	0.19
TC	[b]	[b]	0.29	0.23
TT	0.22	0.20	0.30	0.22
CA	0.20	0.20	0.34	0.18
CG	[b]	[b]	0.41	0.19
CC	0.22	0.19	0.31	0.21
CT	0.21	0.20	0.30	0.23

[a]Full sequence is 5'-CGCAXMYTCG-3', where M is tC or tC^O and X and Y the neighboring bases in the table.
[b]Not established for tC.

long axis of the tC^O three-ring system (Fig. 10.2) and is more than twice as intense ($\varepsilon = 9000\,M^{-1}\,cm^{-1}$) as for tC.[22] As for tC a single fluorescence lifetime ($\sim$4 ns) and a stable quantum yield with an average of 22% is found for tC^O in dsDNA (Table 10.1). In single-stranded systems, tC^O has a quantum yield that changes with surrounding bases (Table 10.1).[22] The values range from 41% to almost three times lower, which is still a moderate change considering the quantum yield variations seen for environment-sensitive fluorescent base analogs like 2-aminopurine.[28] Moreover, for most cases, the emission decay needs two time components to be fitted well and the average lifetime $\langle\tau\rangle$ varies from 2.5 to 5.7 ns.[22] When tC^O has an adenine as the opposite base, and therefore exists in its imino-tautomer (*vide supra*), the quantum yield stays essentially the same as when base-paired with a G. However, the emission spectrum for this form has a lot of vibrational fine structure.[24a]

The third member of the three-cyclic cytosine family, tC$_{nitro}$, is essentially nonfluorescent at room temperature.[17] Upon decreasing the temperature to 150 K in a rigid propylene glycol (PG) glass, we have shown that tC$_{nitro}$ reaches a quantum yield of almost 20% (increases in quantum yield in rigid glasses have been observed also for tC and tC^O).[20] We suggest that the deactivation process resulting in the nonemissive nature of tC$_{nitro}$ is through rotation or vibration of its nitro group and that this motion is hindered in the rigid PG glass. The lowest energy absorption band of tC$_{nitro}$ from 375 to 525 nm (Fig. 10.5), the envelope of which is stable under physiological conditions, has shown to be the result of a single in-plane polarized electronic transition with a dipole moment oriented 27° from the long axis toward the nitro group (Fig. 10.2; note that this is the opposite direction compared to tC/tC^O).[23b] This transition has its maximum around 440 nm and $\varepsilon_{max} = 5400\,M^{-1}\,cm^{-1}$. This means that it has a good spectral overlap with the emission of tC ($\lambda_{em,max} = 505$ nm) and an excellent overlap with tC^O ($\lambda_{em,max} = 460$ nm) enabling the development of a nucleic base analog FRET pair as described in Section 10.3.[23b]

10.3   DEVELOPMENT OF THE FIRST NUCLEIC ACID BASE ANALOG FRET PAIR

With accurately determined directions of transition dipole moments and a high control of probe molecule position due to the firm base-pairing and -stacking, virtually constant quantum yields in duplex systems as well as stable spectral envelopes, the tricyclic cytosines offer a unique possibility to get highly accurate values for the orientation factor (κ^2), donor emission (Φ_D), and the J-integral (see Eq. 10.2 above), respectively. This in turn results in an accurately determined Förster distance (R_0 in Eqs. 10.2 and 10.1), thus enabling high precision in distance measurements (R_{DA} in Eq. 10.1) and high structural resolution. As a consequence of a better energy overlap (Fig. 10.6) and thus a calculated R_0 that is 16% longer than for tC–tC$_{nitro}$[17,23b] we choose to develop tC^O–tC$_{nitro}$ to be the first nucleic base analog FRET pair. However, it should be noted that tC–tC$_{nitro}$ also would work well as a FRET pair and that the even less sensitive quantum yield of tC, which was discussed above, could be an additional advantage in this context.

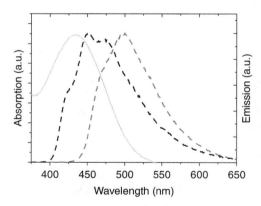

Figure 10.6   Absorption spectrum of tC$_{nitro}$ (light gray line) and representative, normalized emission spectra of tC^O (black dashed line) and tC (dark gray dashed line) in duplex. The extensive energy overlap between the donor tC^O emission and the acceptor tC$_{nitro}$ absorption can be seen under the light gray line and black dashed line. The slightly smaller overlap between tC emission and tC$_{nitro}$ absorption can be seen under the light gray line and dark gray dashed line.

10.3.1   The Donor–Acceptor Pair tC^O–tC$_{nitro}$

The calculated Förster distance for the tC^O–tC$_{nitro}$ donor–acceptor pair is 27 Å if using a $\kappa^2 = 2/3$ (merely to facilitate comparison with common FRET pairs).[17] This R_0 is approximately half of Förster distances for commonly used external FRET pairs and therefore enables accurate measurements of conformations in duplexes with the tC^O and tC$_{nitro}$ being separated up to 1.5–2 turns of a DNA helix.[17] In the initial study of this base analog FRET pair, we used seven 33-mers oligonucleotides, of which three contained the tC^O donor at different positions and four complementary ones contained the tC$_{nitro}$ acceptor at various positions.[17] In combining these strands in all possible ways, we, thus, got duplexes in which the donor and acceptor were separated by between 2 and 13 bases. Moreover, in order to get as similar Φ_D as possible and in doing so facilitate evaluation of the measurements, we choose the same surrounding bases around tC^O in all donor sequences. We measured the change in tC^O–tC$_{nitro}$ FRET efficiency using the donor emission (tC$_{nitro}$ is virtually nonfluorescent) in both steady-state and time-resolved fluorescence experiments (Fig. 10.7).[17]

The experimental results show a FRET efficiency that is highly dependent on both distance and orientation as the separation and therefore also the direction of the transition dipoles of tC^O and tC$_{nitro}$ are altered in a stepwise manner. Qualitatively, the data come out exactly as expected for a rigidly positioned FRET pair situated at different positions within the DNA duplex: the efficiency (E in Eq. 10.1) decreases sharply with distance while oscillating between local maxima and minima as the transition dipole moments of tC^O and tC$_{nitro}$ change between more parallel and more perpendicular configurations. Similar studies of semirigid/rigid FRET pairs in a DNA context

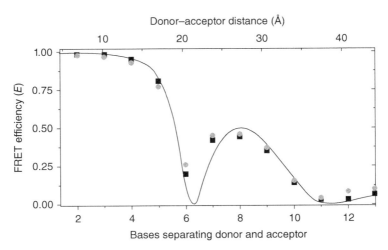

Figure 10.7   Efficiency of energy transfer for the base analog FRET pair tC^O–tC$_{nitro}$ estimated using decreases in tC^O, donor, emission (circles) and tC^O average emission lifetimes (squares) as the two analogs are separated by 2–13 bases in a DNA duplex. Curve fitting using equations in Börjesson *et al.* with α and J_{DA} as fit parameters is shown as solid line. Reprinted with permission from the American Chemical Society.

have been performed before by, for example, the groups of Lewis,[29] Lilley,[30] and Tor[31]. In all these interesting studies, the FRET pair had more dynamics within the time frame of the energy transfer and thus the local maxima and minima of the FRET decay curve are less pronounced.

To analyze the measured efficiencies in a more quantitative way, we used a model that take into consideration vector distances and accurate orientations between chromophores and they become increasingly similar to the rough model in which the chromophores are placed on top of each other along the DNA helix, with increasing base separation (for details see Börjesson et al.)[17]. The excellent fit, where the J-integral and phase angle (α) are varied, to the experimental data (Fig. 10.7) and the distinct changes between maxima and minima not only confirm that the donor quantum yield and the spectral envelopes are virtually constant, but also strongly indicate that these C-analogs have practically no dynamics (though very fast dynamics cannot be ruled out) on the timescale of the fluorescence. Hence, this data suggest that we have indeed successfully designed an excellent nucleic acid base analog FRET pair with the characteristics that we desired. The results imply that a set of strands with the tC^O–tC_{nitro} FRET pair at strategically chosen positions, that is, the steep slopes in Figure 10.7, will make it possible to accurately distinguish distance- from orientation-changes using FRET and therefore provide detailed 3-D structure information of nucleic acid–containing systems.

The model used for a more quantitative fit of the data from the FRET data in Figure 10.7 is still a very crude one where, for example, no variations in rise or rise angle for different base pairs in the duplex are taken into account. To take advantage of the high-resolution structure tool that our FRET pair constitutes, we have also developed a general and much more accurate freeware methodology, FRETmatrix, to simulate and analyze base–base FRET in any kind of nucleic acid–containing system.[32] This MATLAB-based freeware provides a general framework for the simulation of FRET in nucleic acids that is particularly useful in systems involving constrained probes including not only base analogs but also rigidly bound probes.[30] The simulation part of the methodology has main purposes to enable a fast and rough evaluation of the FRET measured for a system under investigation, and in particular for simulating expected FRET efficiencies given any previously determined structure to enable an optimal design of sequences and donor and acceptor positions for the system you are about to study. Here an all-atom, 3-D nucleic acid model is constructed using, for example, standard A/B-DNA parameters (shift, slide, rise, tilt, roll, and twist)[33] or pdb structures, followed by a FRET simulation between static transition dipoles in the donor and acceptor. In the analysis part, which makes up the second part of the freeware, we extend the static model to a more information-rich methodology that uses directional probability distributions to simulate the nanosecond dynamics of the donor–acceptor transition dipoles involved. This is accompanied by an analysis routine performing a global fit of multiple time-resolved donor intensity decays from FRET measurements comprising various donor–acceptor separations. This analysis methodology will hence enable (1) a more *accurate structure determination* than the simulation part and (2) suggest the *dynamics* of involved probe molecules and the system they are a part of. In investigations using our freeware FRETmatrix,[32] we have gained information that is otherwise only obtainable hypothetically by means of much more complex and time-consuming NMR or single-molecule methods. Hence, our methodology shows great promise for the future.

10.3.2 Applications of Tricyclic Cytosines in FRET Measurements

Besides having been used in numerous biochemical investigations, especially regarding DNA/RNA polymerases,[24b,25,34] and also in a nanotechnological study,[35] the tricyclic cytosine has found use in several FRET investigations. In the first report of the fluorescence properties of tC in nucleic acids, it was also demonstrated that tC works well as a FRET donor of an external rhodamine acceptor on the opposite strand.[13] In this donor–acceptor pair, tC was assumed to be rigidly positioned and rhodamine to be randomly oriented giving a κ^2 of 1/3. The resulting Förster distance was calculated to be 40 Å and the measured FRET efficiency resulted in a distance that was within 10% of that estimated using a computer model of the nucleic acid duplex. Later tC was used as a donor in a DNA polymerase study in which an Alexa-555 served as an acceptor on the Klenow fragment polymerase.[36] In this investigation, the tC-Alexa-555 FRET pair was utilized to monitor conformational changes of a polymerase-DNA complex during selection and binding of nucleotide substrates (Fig. 10.8).

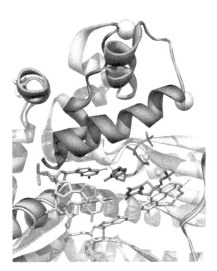

Figure 10.8 Close-up of the catalytic center of KF DNA polymerase. Shown are overlaid crystal structures of the open binary (light gray ribbon) and closed ternary (dark green ribbon) complexes of the polymerase. The O-helix is shown in red, highlighting the different positions of this helix in the open (upper) and closed (lower) conformations. Yellow balls correspond to positions of Alexa-555. The bound DNA is displayed in stick representation. The tC base is the tricyclic yellow base in the front on the right, modeled into the structure in place of normal cytosine. Reprinted with permission from the American Chemical Society. (*See color plate section for the color representation of this figure.*)

The FRET experiments indicated that the rate-limiting step in the overall cycle of nucleotide incorporation for the Klenow fragment-DNA system is the nucleotide-induced conformational transition of the KF finger domain. Both these studies serve as good examples of investigations in which the emissive tricyclic cytosines, tC and tC^O, can be part of a FRET pair including an external, commercially available, and bright standard fluorophore.

We have recently started to utilize the base analog FRET pair tC^O–tC$_{nitro}$ in biochemical investigations. In an ongoing study, we monitor the conformational change of a DNA upon binding of the P2C repressor and use the FRET change to estimate binding constant and stoichiometry (ms in preparation). Moreover, we investigate the requirement for the mitochondrial transcription factor A (TFAM) for transcription.[37] In this study, the tC^O–tC$_{nitro}$ FRET pair has been incorporated into various positions and reports both on significant structural changes to the DNA upon TFAM binding its requirement for mitochondrial RNA polymerase and TFB2M to bind to the DNA.

10.4   CONCLUSIONS

In this chapter, the advantages of rigidly positioned environment-insensitive fluorescent base analogs in nucleic acid system FRET measurements have been discussed. The environment insensitivity, resulting in a virtually constant quantum yield as well as absorption and emission spectral envelopes, gives unparalleled control of the Förster distance whereas the rigid stacking of the base analog inside the nucleic acid allows for an accurate estimation of the orientation factor. Consequently, FRET measurements performed with such probe molecules give the opportunity to realize high-resolution 3-D structures of biomolecular systems involving nucleic acids. The first nucleic acid base analog FRET pair, tC^O–tC$_{nitro}$, has all the features necessary for such probe molecules as well as being excellent analogs of cytosine. Using a set of strands containing our FRET pair at strategically chosen positions and evaluating the energy transfer, which is essentially like using a large set of NOEs in a nucleic acid NMR experiment, thus enables high structural resolution. In combination with the freeware FRETmatrix, developed by my group, that takes advantage of a global fit of all time-resolved data in such a set of tC^O–tC$_{nitro}$ FRET experiments, we obtain an even higher structural resolution in addition to the possibility to present data for the nanosecond dynamics of individual base pairs and also of the surrounding overall structure. This kind of information is highly important for the understanding of a myriad of processes fundamental to life and very difficult and/or cumbersome to obtain using other techniques like NMR and single-molecule methods for large, complex nucleic acid–containing systems.

10.5   PROSPECTS AND OUTLOOK

Methods to investigate the three-dimensional structure of biomolecules are and will continue to be vital for an increased understanding of nature and life. Techniques such as NMR and X-ray crystallography will obviously also be key players in the future. However, the need for crystals in X-ray crystallography and the complexity, upper molecular weight limit and large sample amount needed in NMR constitute drawbacks of these methods. Other lower resolution techniques such as small-angle scattering, cryo-electron microscopy, site-directed spin labeling, and nanoparticle plasmon rulers are also important to biomolecular structure determination but are less ubiquitous in the biosciences than FRET.

FRET can be performed routinely and fast on the ensemble scale in most laboratories and also on the single-molecule scale in a large number of laboratories. Today most FRET experiments on biomolecules utilize external, commercially available fluorophores such as fluoresceins, rhodamines, Cy-, Alexa-, or ATTO-dyes. They are all bright molecules giving long Förster distances thus enabling monitoring of long distances. However, their typically external attachment on nucleic acids makes it difficult to predict their orientation and dynamics relative to the nucleic acid, resulting in less accurate estimates of the orientation factor and therefore lower structure resolution. In contrast, the FRET methodology presented in this chapter, using nucleobase analogs as a FRET pair, facilitates both high structural resolution and dynamics investigations. With the methodology to evaluate data from this FRET pair (or fluorophores with similar properties) that has recently been developed in my laboratory, we foresee that base–base FRET will be a significant technique in the near future. However, there are still issues that need to be addressed. One of the main challenges is to increase the brightness, photostability, spectral coverage of the nucleobase analogs, and to develop such analogs for all the natural bases. With improved photophysical characteristics, this method could be used even at single-molecule level, which potentially will give very interesting information about nucleic acid structure and dynamics. Moreover, the improved brightness would increase the distances possible to screen in the nucleic acid system under investigation. In conclusion, we envision quantitative FRET-methods, like the one presented in this chapter, to have an increased impact on biosciences in the coming decades and to be one of the key methodologies to unravel new knowledge regarding elementary processes of large biomolecular complexes in living organisms.

ACKNOWLEDGMENTS

I wish to thank all coauthors of my articles concerning the tricyclic cytosine analogs and special thanks go to the persons involved in the laboratory work concerning the synthesis and characterization of the first nucleobase FRET pair Dr Karl Börjesson, Dr Søren Preus, Dr Peter Sandin as well as members of the group of Professor Tom Brown at the University of Southampton, UK. Finally, I would like to thank my previous student Dr Anke Dierckx for the ongoing investigations, where the tC^O-tC_{nitro} FRET pair is utilized and for proofreading this chapter.

REFERENCES

1. Holbrook, S. R. *Annu. Rev. Biophys.* **2008**, *37*, 445.
2. Foster, M. P.; McElroy, C. A.; Amero, C. D. *Biochemistry* **2007**, *46*, 331.

3. Rambo, R. P.; Tainer, J. A. *Curr. Opin. Struct. Biol.* **2010**, *20*, 128.

4. Frank, J. *Q. Rev. Biophys.* **2009**, *42*, 139.

5. (a) Marko, A.; Denysenkov, V.; Margraft, D.; Cekan, P.; Schiemann, O.; Sigurdsson, S. T.; Prisneet, T. F. *J. Am. Chem. Soc.* **2011**, *133*, 13375; (b) Schiemann, O.; Prisner, T. F. *Q. Rev. Biophys.* **2007**, *40*, 1.

6. (a) Reinhard, B. M.; Siu, M.; Agarwal, H.; Alivisatos, A. P.; Liphardt, J. *Nano Lett.* **2005**, *5*, 2246; (b) Reinhard, B. M.; Sheikholeslami, S.; Mastroianni, A.; Alivisatos, A. P.; Liphardt, J. *Proc. Natl. Acad. Sci. U. S. A.* **2007**, *104*, 2667; (c) Sonnichsen, C.; Reinhard, B. M.; Liphardt, J.; Alivisatos, A. P. *Nat. Biotechnol.* **2005**, *23*, 741.

7. (a) Mastroianni, A. J.; Sivak, D. A.; Geissler, P. L.; Alivisatos, A. P. *Biophys. J.* **2009**, *97*, 1408; (b) Mathew-Fenn, R. S.; Das, R.; Silverman, J. A.; Walker, P. A.; Harbury, P. A. B. *PLoS One* **2008**, *3*, e3229.

8. (a) Förster, T. *Ann. Phys.* **1948**, *2*, 55; (b) Stryer, L.; Haugland, R. P. *Proc. Natl. Acad. Sci. U. S. A.* **1967**, *58*, 719; (c) Stryer, L. *Annu. Rev. Biochem.* **1978**, *47*, 819; (d) Preus, S.; Wilhelmsson, L. M. *ChemBioChem* **2012**, *13*, 1990.

9. Dale, R. E.; Eisinger, J. *Biopolymers* **1974**, *13*, 1573.

10. (a) Kinosita, K.; Kawato, S.; Ikegami, A. *Biophys. J.* **1977**, *20*, 289; (b) Dale, R. E.; Eisinger, J.; Blumberg, W. E. *Biophys. J.* **1979**, *26*, 161; (c) Lipari, G.; Szabo, A. *Biophys. J.* **1980**, *30*, 489; (d) Parkhurst, L. J.; Parkhurst, K. M.; Powell, R.; Wu, J.; Williams, S. *Biopolymers* **2001**, *61*, 180; (e) Ivanov, V.; Li, M.; Mizuuchi, K. *Biophys. J.* **2009**, *97*, 922.

11. (a) Dodd, D. W.; Hudson, R. H. E. *Mini-Rev. Org. Chem.* **2009**, *6*, 378; (b) Sinkeldam, R. W.; Greco, N. J.; Tor, Y. *Chem. Rev.* **2010**, *110*, 2579; (c) Wilhelmsson, L. M. *Q. Rev. Biophys.* **2010**, *43*, 159.

12. Lin, K.; Jones, R. J.; Matteucci, M. *J. Am. Chem. Soc.* **1995**, *117*, 3873.

13. Wilhelmsson, L. M.; Holmén, A.; Lincoln, P.; Nielsen, P. E.; Nordén, B. *J. Am. Chem. Soc.* **2001**, *123*, 2434.

14. Sandin, P.; Lincoln, P.; Brown, T.; Wilhelmsson, L. M. *Nat. Prot.* **2007**, *2*, 615.

15. (a) Roth, B.; Hitchings, G. H. *J. Org. Chem.* **1961**, *26*, 2770; (b) Roth, B.; Schloemer, L. A. *J. Org. Chem.* **1963**, *28*, 2659.

16. Chedekel, M. R.; Sharp, D. E.; Jeffery, G. A. *Synth. Commun.* **1980**, *10*, 167.

17. Börjesson, K.; Preus, S.; El-Sagheer, A. H.; Brown, T.; Albinsson, B.; Wilhelmsson, L. M. *J. Am. Chem. Soc.* **2009**, *131*, 4288.

18. (a) Bell, J. D.; Blount, J. F.; Briscoe, O. V.; Freeman, H. C. *Chem. Commun.* **1968**, 1656; (b) McDowell, J. J. H. *Acta Crystallogr., Sect. B: Struct. Sci.* **1976**, *32*, 5.

19. (a) Wilds, C. J.; Maier, M. A.; Tereshko, V.; Manoharan, M.; Egli, M. *Angew. Chem. Int. Ed.* **2002**, *41*, 115; (b) Wilds, C. J.; Maier, M. A.; Manoharan, M.; Egli, M. *Helv. Chim. Acta* **2003**, *86*, 966.

20. Preus, S.; Kilså, K.; Wilhelmsson, L. M.; Albinsson, B. *Phys. Chem. Chem. Phys.* **2010**, *12*, 8881.

21. Engman, K. C.; Sandin, P.; Osborne, S.; Brown, T.; Billeter, M.; Lincoln, P.; Nordén, B.; Albinsson, B.; Wilhelmsson, L. M. *Nucleic Acids Res.* **2004**, *32*, 5087.

22. Sandin, P.; Börjesson, K.; Li, H.; Mårtensson, J.; Brown, T.; Wilhelmsson, L. M.; Albinsson, B. *Nucleic Acids Res.* **2008**, *36*, 157.

23. (a) Wilhelmsson, L. M.; Sandin, P.; Holmén, A.; Albinsson, B.; Lincoln, P.; Nordén, B. *J. Phys. Chem. B* **2003**, *107*, 9094; (b) Preus, S.; Börjesson, K.; Kilså, K.; Albinsson, B.; Wilhelmsson, L. M. *J. Phys. Chem. B* **2010**, *114*, 1050.

24. (a) Börjesson, K.; Sandin, P.; Wilhelmsson, L. M. *Biophys. Chem.* **2009**, *139*, 24; (b) Stengel, G.; Purse, B. W.; Wilhelmsson, L. M.; Urban, M.; Kuchta, R. D. *Biochemistry* **2009**, *48*, 7547.

25. Xia, S. L.; Beckman, J.; Wang, J. M.; Konigsberg, W. H. *Biochemistry* **2012**, *51*, 4609.

26. Sandin, P.; Wilhelmsson, L. M.; Lincoln, P.; Powers, V. E. C.; Brown, T.; Albinsson, B. *Nucleic Acids Res.* **2005**, *33*, 5019.

27. Edwards, T. E.; Cekan, P.; Reginsson, G. W.; Shelke, S. A.; Ferre-D'Amare, A. R.; Schiemann, O.; Sigurdsson, S. T. *Nucleic Acids Res.* **2011**, *39*, 4419.

28. Ward, D. C.; Reich, E.; Stryer, L. *J. Biol. Chem.* **1969**, *244*, 1228.

29. Lewis, F. D.; Zhang, L. G.; Zuo, X. B. *J. Am. Chem. Soc.* **2005**, *127*, 10002.

30. (a) Iqbal, A.; Wang, L.; Thompson, K. C.; Lilley, D. M. J.; Norman, D. G. *Biochemistry* **2008**, *47*, 7857; (b) Ouellet, J.; Schorr, S.; Iqbal, A.; Wilson, T. J.; Lilley, D. M. J. *Biophys. J.* **2011**, *101*, 1148.

31. Hurley, D. J.; Tor, Y. *J. Am. Chem. Soc.* **2002**, *124*, 13231.

32. Preus, S.; Kilsa, K.; Miannay, F. A.; Albinsson, B.; Wilhelmsson, L. M. *Nucleic Acids Res.* **2013**, *41*, e18.

33. (a) Olson, W. K.; Bansal, M.; Burley, S. K.; Dickerson, R. E.; Gerstein, M.; Harvey, S. C.; Heinemann, U.; Lu, X. J.; Neidle, S.; Shakked, Z.; Sklenar, H.; Suzuki, M.; Tung, C. S.; Westhof, E.; Wolberger, C.; Berman, H. M. *J. Mol. Biol.* **2001**, *313*, 229; (b) Lu, X. J.; Olson, W. K. *Nucleic Acids Res.* **2003**, *31*, 5108; (c) Zheng, G. H.; Lu, X. J.; Olson, W. K. *Nucleic Acids Res.* **2009**, *37*, W240.

34. (a) Sandin, P.; Stengel, G.; Ljungdahl, T.; Börjesson, K.; Macao, B.; Wilhelmsson, L. M. *Nucleic Acids Res.* **2009**, *37*, 3924; (b) Stengel, G.; Urban, M.; Purse, B. W.; Kuchta, R. D. *Anal. Chem.* **2009**, *81*, 9079; (c) Stengel, G.; Urban, M.; Purse, B. W.; Kuchta, R. D. *Anal. Chem.* **2010**, *82*, 1082; (d) Xia, S. L.; Christian, T. D.; Wang, J. M.; Konigsberg, W. H. *Biochemistry* **2012**, *51*, 4343; (e) Walsh, J. M.; Bouamaied, I.; Brown, T.; Wilhelmsson, L. M.; Beuning, P. J. *J. Mol. Biol.* **2011**, *409*, 89.

35. Sandin, P.; Tumpane, J.; Börjesson, K.; Wilhelmsson, L. M.; Brown, T.; Nordén, B.; Albinsson, B.; Lincoln, P. *J. Phys. Chem. C* **2009**, *113*, 5941.

36. Stengel, G.; Gill, J. P.; Sandin, P.; Wilhelmsson, L. M.; Albinsson, B.; Nordén, B.; Millar, D. P. *Biochemistry* **2007**, *46*, 12289.

37. Shi, Y.; Dierckx, A.; Wanrooij, P. H.; Wanrooij, S.; Larsson, N.-G.; Wilhelmsson, L. M.; Falkenberg, M.; Gustafsson, C. M. *Proc. Natl. Acad. Sci. U. S. A.* **2012**, *109*, 16510.

11

FLUORESCENT PURINE ANALOGS THAT SHED LIGHT ON DNA STRUCTURE AND FUNCTION

ANAËLLE DUMAS, GUILLAUME MATA, AND NATHAN W. LUEDTKE

Department of Chemistry, University of Zurich, Zurich, Switzerland

11.1 INTRODUCTION

Linear sequences of nucleobases contain the information necessary for the construction and function of every living organism on Earth. The five common nucleobases (A, G, C, T, U) form specific base pairs that are stabilized by hydrogen bonding and base stacking in *Watson–Crick* base pairing – where guanosine bonds with cytidine and adenosine pairs with thymidine in duplex DNA and RNA (Fig. 11.1a). These nucleobases can also participate in a wide variety of noncanonical pairing interactions that stabilize diverse DNA and RNA secondary structures such as hairpin, triplex, G-quadruplex, and i-motif (Figs. 11.1b and 11.2).[1,2] Recent evidence suggests that certain nonduplex DNA structures may play a direct role in regulating gene expression, mediating recombination, and stabilizing chromosomal ends.[3–5] Analogously, in some respects, RNA molecules contain specific binding sites for small molecules that can modulate specific cellular events.[6–9] Many of these mechanisms, however, remain poorly understood – especially in the context of living systems. The elucidation of structure–function relationships for nucleic acids *in vivo* has therefore become an important area of scientific investigation. Fluorescence-based assays can provide powerful and direct means for studying the folding and localization of biological macromolecules,[10] but relatively few imaging techniques are available for nucleic

Fluorescent Analogs of Biomolecular Building Blocks: Design and Applications, First Edition.
Edited by Marcus Wilhelmsson and Yitzhak Tor.

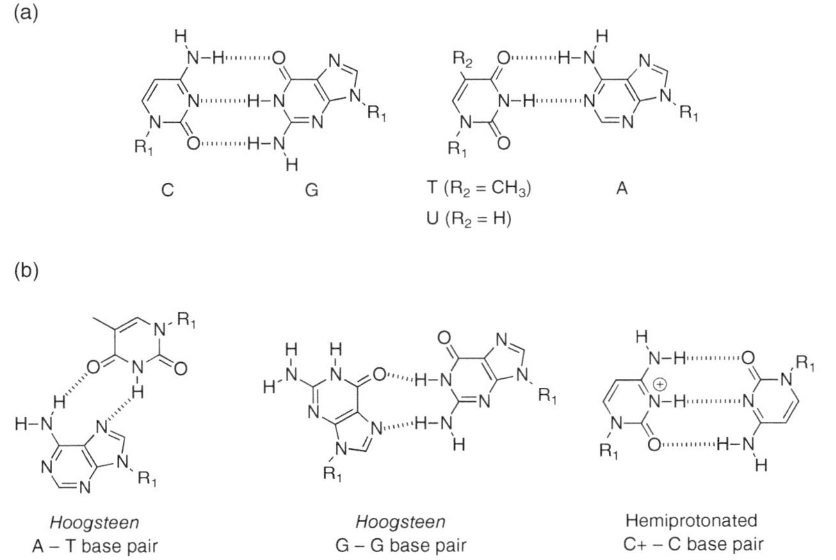

(a)

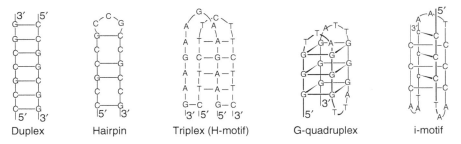

C   G   T (R₂ = CH₃)   A
              U (R₂ = H)

(b)

Hoogsteen   Hoogsteen   Hemiprotonated
A – T base pair   G – G base pair   C+ – C base pair

Figure 11.1 (a) Canonical *Watson–Crick* base pairs. (b) Selected examples of noncanonical base pairs. $R_1 = 2'$-D-deoxyribose (DNA) or D-ribose (RNA).

Duplex   Hairpin   Triplex (H-motif)   G-quadruplex   i-motif

Figure 11.2 Schematic representation of selected DNA secondary structures.

acids – most of which are limited by their low sensitivity to alternatively folded structures, large perturbations to native systems, and/or inability to be applied in whole cells and organisms.

A wide variety of biophysical techniques has been used to characterize nucleic acids structure and function *in vitro*. These include NMR,[11–14] circular dichroism,[15] UV absorption,[16,17] X-ray crystallography,[18] Förster resonance energy transfer (FRET),[19–22] and immunostaining.[23] Most of these methods require pure samples and are incompatible with conformational analyses in living cells. Fluorescent probes that resemble biomolecular building blocks can enable powerful approaches for probing the folding and activities of biological macromolecules.[10] Fluorescence spectroscopy benefits from excellent sensitivity and versatility in a wide variety of

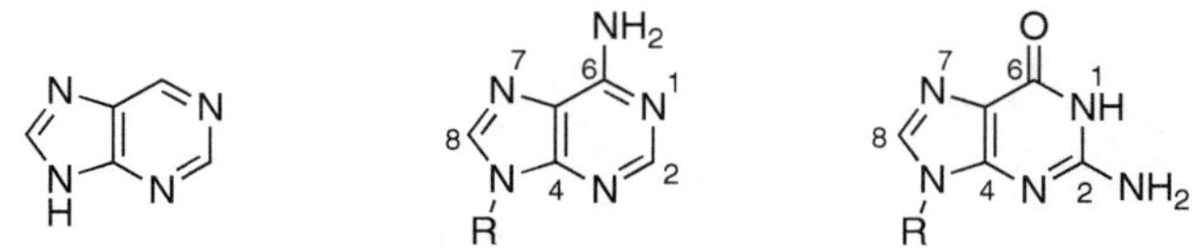

Purine	Adenosine	Guanosine
λ_{abs} = 259 nm	λ_{abs} = 259 nm	λ_{abs} = 252 nm
λ_{em} = 375 nm	λ_{em} = 312 nm	λ_{em} = 340 nm
$\Phi < 2 \times 10^{-3}$	$\Phi = 5 \times 10^{-5}$	$\Phi = 8 \times 10^{-5}$

Figure 11.3 Structures of purine, adenosine, and guanosine and their photophysical properties.[24,25] R = 2'-deoxyribose or ribose.

techniques including steady-state and time-resolved intensity measurements, FRET, and fluorescence anisotropy. These methods can be applied on single molecules and in whole biological systems to provide information about the environment, dynamics, folding, location, and ligand binding of nucleic acids.

The five canonical nucleobases exhibit very low quantum yields of fluorescence ($\Phi = (0.5-1) \times 10^{-4}$, Fig. 11.3). This can be viewed as a major challenge for the biophysical analysis of native nucleic acids, or alternatively, as providing a low background environment for implementing highly emissive probes. Due to their small size and predictable location, "internal" fluorescent probes that mimic endogenous residues offer some advantages over external (noncovalent) and end-conjugated probes. If properly designed, internal fluorescent probes can maintain the structural and electronic characteristics of natural nucleobases, and therefore have minimal impacts on the folding and stability of nucleic acids. In many cases, the photophysical properties of such probes are highly sensitive to their local environment and can thereby act as reporters for binding interactions and conformational changes with single nucleotide resolution.[26–38]

Internal fluorescent probes can be derived from both purine and pyrimidine residues by introducing small chemical modifications to the natural structures.[39] Due to their larger surface areas, purine derivatives offer some advantages over pyrimidine derivatives including redshifted excitation/emission wavelengths. Purines can also act as energy/electron donors,[40,41] and they exhibit stronger base-stacking interactions than pyrimidines.[42–44] Emissive purine analogs can therefore probe DNA and RNA structures by reporting changes in local base-stacking interactions that are characteristic of specific secondary and tertiary structures.[40]

11.2 DESIGN, PHOTOPHYSICAL PROPERTIES, AND APPLICATIONS OF PURINE MIMICS

An ideal fluorescent analog of guanine or adenine should fulfill the following criteria: (1) report changes in local environment via marked differences in photophysical properties, (2) retain high quantum efficiency in the context of folded nucleic acids, (3) exhibit redshifted wavelengths of excitation and emission, and (4) introduce little or no perturbation to the folding or stability of nucleic acids containing it. This last point requires the preservation of structural similarity with the natural nucleobases, including retention of the base pairing faces and glycosidic bond conformations. As a complicating factor, the very same modifications that can improve the photophysical properties of nucleobases (extended π-systems, increased symmetry, etc.) can lead to substantial structural perturbations. The design of new fluorescent base analogs, therefore, requires a delicate balance of minimizing structural impact while maximizing quantum yields, wavelengths of excitation/emission, and environmental sensitivity.

11.2.1   Early Examples of Fluorescent Purine Mimics

Purine is the heterocyclic core of guanine and adenine residues (Fig. 11.3). All three heterocycles exhibit weak fluorescence in water, with quantum yields (Φ) ranging from about 10^{-5} to 10^{-3}. Early studies by Ward *et al.* in 1969 reported emissive adenosine analogs (Fig. 11.4): formycin (**P.1**), 2-aminopurine (**2AP, P.2**), and 2,6-diaminopurine (**P.3**).[45] These compounds have redshifted absorption (280–303 nm) and emission (340–370 nm) maxima compared to adenine. **P.1** and **P.3** are only weakly emissive, but **2AP** exhibits a high quantum yield ($\Phi = 0.68$) in water.[45] **2AP** can form base pairs with thymidine and uracil without significant perturbation of double helical structures.[46] **2AP** can also form a wobble base pair with cytosine and is, therefore, considered to be both an adenine and guanine mimic.[47–49] The fluorescence properties of **2AP** are highly sensitive to its local environment. Numerous assays involving **2AP** as an internal fluorescent probe have exploited these characteristics for monitoring DNA–DNA hybridization,[46,50,51] DNA–RNA polymerase binding,[52–54] RNA–aminoglycoside recognition,[55–58] enzymatic activities,[35,46,51] DNA/RNA folding,[33,59,60] and base-pairing interactions/dynamics.[48,49,61–66] **2AP** has also been used as an energy donor in the construction of FRET-based molecular beacons containing pyrrolo-dC as an energy acceptor.[41,67–72] Despite its widespread use, **2AP** exhibits a number of important limitations that include fluorescence quenching upon its incorporation into nucleic acids, overlapping absorption with natural DNA bases, and imperfect *Watson–Crick* pairing interactions. Another early example of a highly fluorescent

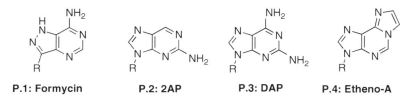

P.1: Formycin   **P.2: 2AP**   **P.3: DAP**   **P.4: Etheno-A**

Figure 11.4   Early examples of fluorescent purine analogs.

TABLE 11.1 Spectroscopic Properties of Early Fluorescent Purine Analogs

Number	Name	Condition	λ_{abs} (nm)	λ_{em} (nm)	Φ	References
P.1	Formycin	Water	295	340	0.06	[45]
P.2	2AP	Water	303	370	0.68	[45]
	2AP	ssDNA	310	370	0.01–0.03	[67–72,76]
	2AP	dsDNA	310	370	0.002–0.05	[67–72,76]
P.3	DAP	Water	280	350	0.01	[45]
P.4	Etheno-A	Water	275	415	0.56	[73–75]

adenosine analog is 1,N^6-ethenoadenosine (**Etheno-A, P.4,** $\Phi = 0.56$ in water).[73,74] **Etheno-A** nucleotide triphosphate can mimic ATP for certain enzymes,[73,74] and **Etheno-A**-modified DNA can be used to detect protein-binding interactions.[75] The *Watson–Crick* face of **Etheno-A** is masked by the etheno group, thereby preventing proper base-pairing interactions with thymidine. The resulting perturbation of duplex stability is similar in magnitude to that of a base pair mismatch, thereby posing serious limitations to the application of **Etheno-A** in nucleic acids. Although imperfect, these early examples of fluorescent purine analogs demonstrated the power of internal fluorescent probes and motivated the design and synthesis of new, minimally disruptive nucleobase analogs for DNA and RNA (Table 11.1).

11.2.2 Chromophore-Conjugated Purine Analogs

Purine residues provide attachment sites for the conjugation of fluorescent metal complexes and organic fluorophores (Fig. 11.5). By using flexible, saturated linkers, nucleobase analogs with photophysical characteristics similar to those of the parent fluorophores are obtained. In contrast, the use of conjugated, unsaturated linkers between the fluorophore and the purine generates chromophores with unique photophysical properties (Table 11.2). Many examples of chromophoric purine analogs utilize a pyrene unit (Fig. 11.5). The fluorescence of pyrene is solvatochromic, and it is usually quenched by stacking interactions with natural nucleobases. Numerous sensing applications have exploited these properties to detect conformational changes. For example, the guanosine derivative **P.5** was used to detect DNA hybridization according to its redshifted emissions. This nucleobase analog is also a photoinducible electron donor used in electron transfer studies in duplex DNA and modified peptides.[77,78] The identities of the neighboring residues, base pair mismatches, and global DNA structure all contribute to local stacking interactions and thereby impact the fluorescence properties of the pyrene unit. Accordingly, **P.6,**[79] **P.7,**[79,80] and **P.8**[82,83] were shown to be capable of detecting base pair mismatches. **P.8** was also used for the detection of B $\rightarrow$ Z transitions in duplex DNA,[84] as well as alternative G-quadruplex folds in single-stranded DNA.[85,86] **P.10** was prepared by postsynthetic modification of a C8 alkylamino-substituted 2'-deoxyguanosine unit in DNA, and was used to detect strand hybridization.[92]

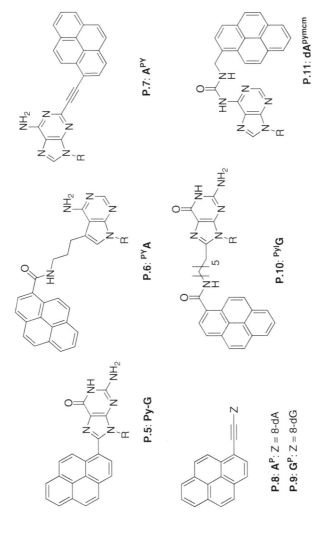

Figure 11.5 Chromophore-conjugated purine analogs ($R = 2'$-deoxyribose or ribose).

P.7: A^{PY}

P.11: dA^{pymcm}

P.6: ^{PY}A

P.10: ^{Pyl}G

P.5: Py-G

P.8: A^P: Z = 8-dA
P.9: G^P: Z = 8-dG

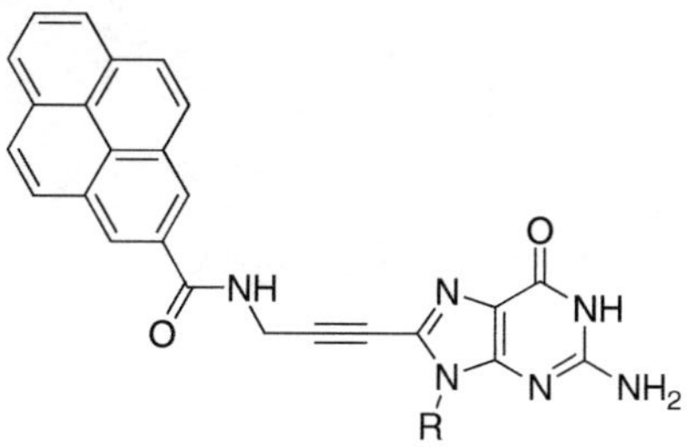

P.15: ^{dan}**G**

P.19

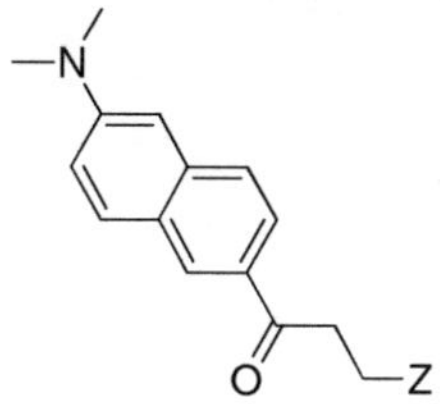

P.13: ^{PDN}**G**; Z = 8-dG
P.14: ^{PDN}**A**: Z = 8-dA

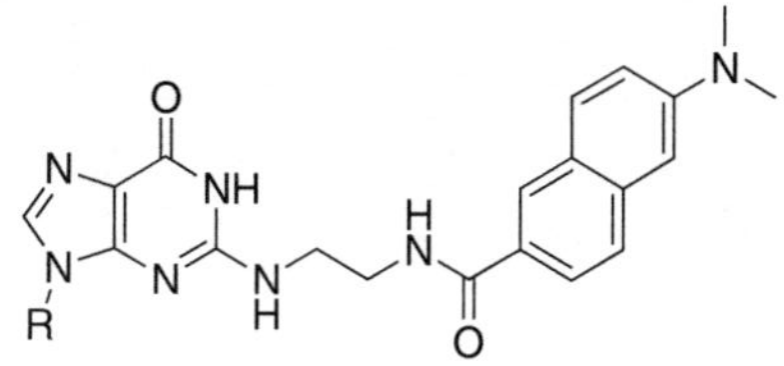

P.17: ^{8FV}**G**

P.12: ^{8Py}**G**

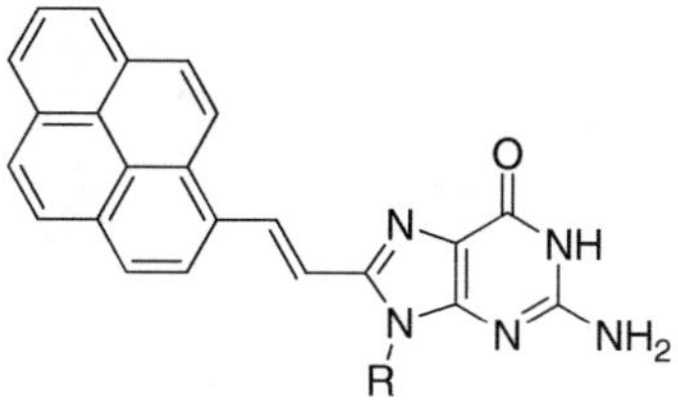

P.16: ^{VPy}**G**

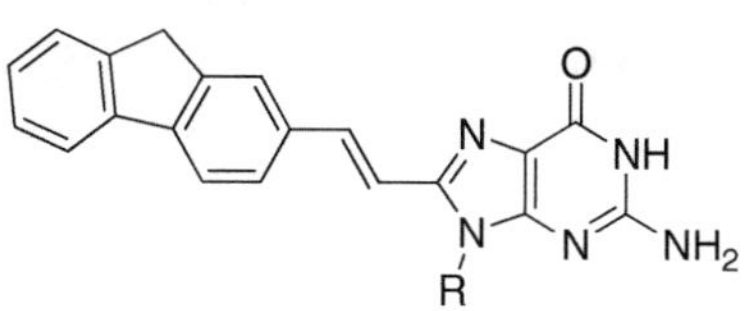

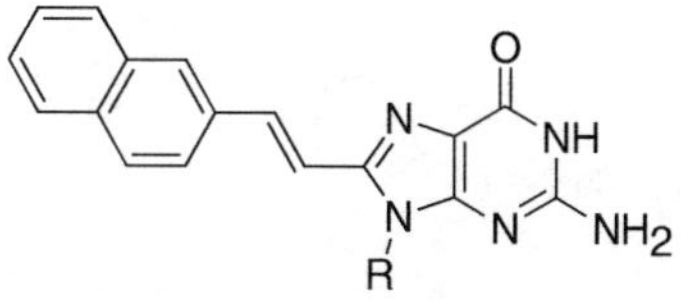

P.18: ^{8NV}**G**

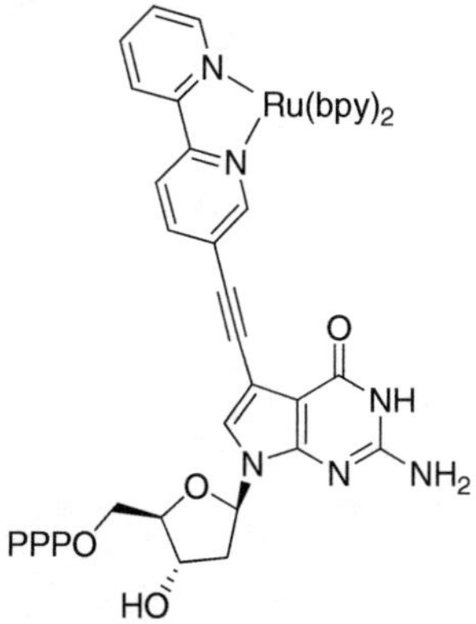

Figure 11.5 *(Continued)*

TABLE 11.2   Spectroscopic Properties of Chromophore-Conjugated Purine Analogs

Number	Name	Condition	λ_{abs} (nm)	λ_{em} (nm)	Φ	References
P.5	Py-G	ssDNA	350	445	0.08–0.11	77,78
	Py-G	dsDNA	370	455	0.25–0.34	77,78
P.6	^{PY}A	ssDNA	347	388	0.064	79
P.6	^{PY}A	dsDNA	350–353	390–397	0.006–0.1	79
P.7	A^{PY}	dsDNA	380–405	450–480	0.11–0.88	80,81
P.8	A^{P}	MeOH	420	434	0.73	82–89
	A^{P}	DNA	420	451–480	n.d.	82–89
P.9	G^{P}	DNA	423	438–507	n.d.	36,90,91
P.10	^{PyI}G	DNA	350	400–535	0.01–0.26	92
P.11	dA^{pymcm}	DNA	350	400	n.d.	93
P.12	^{8Py}G	Methanol	342	405–428	0.10	94
	^{8Py}G	DNA	372	405–458	0.007–0.116	94
P.13	^{PGN}G	DNA	373–406	515–524	0.07–0.13	95
P.14	^{PGN}A	DNA	382–405	517–526	0.07–0.20	95
P.15	^{dan}G	DNA	325	443	n.d.	96,97
P.16	^{VPy}G	Methanol	405	489	0.43	98
P.17	^{8NV}G	DNA	373	470	n.d.	99
P.18	^{8FV}G	DNA	370	470	n.d.	99,100
P.19	dG^{Ru}TP	CH$_3$CN	450	635	0.01	101

The close proximity of two pyrene units can lead to excimers with redshifted emission at ca. 480 nm compared to monomeric pyrene (~390 nm). Nucleic acids containing two pyrene units can therefore report conformational changes that result in a close proximity of the two probes. Using this strategy, nucleosides **P.7**,[80,81] **P.8**,[87–89] **P.9**,[90] and **P.11**[93] were used to discriminate duplex versus hairpin structures. **P.8** and **P.9** were also used in combination with 5-(1-ethenylpyrene)-cytidine and 5-(1-ethenylpyrene)-uridine to evaluate B → Z conformational changes in DNA.[91] Saito and coworkers developed a pyrene-based system involving ^{8Py}**G** (**P.12**) that was capable of distinguishing the folded state of a DNA oligonucleotide as being single-stranded, duplex, or G-quadruplex.[94]

Chromophoric purine analogs were prepared by attaching PRODAN (PGN) to the C8-position of deoxyguanosine (**P.13**), deoxyadenosine (**P.14**), or the related DAN fluorophore to N2 of deoxyguanosine (**P.15**). The excitation wavelengths and quantum yields of **P.13**–**P.15** were highly sensitive to hydration and the polarity of the local environment. These characteristics were used to probe DNA hybridization and single-base mismatches using **P.13** and **P.14**,[95] as well as A → B and B → Z transitions in duplex DNA using **P.15**.[96,97] The conjugation of pyrene, fluorene, or naphthalene via an alkene linker to the C8-position of guanosine yielded nucleobases **P.16**–**P.18** that exhibited reasonably high quantum yields when in the *trans* configuration. Upon irradiation with UV light, these compounds photoisomerize to relatively nonemissive *cis* isomers.[98–100] In the context of oligonucleotides, purines **P.17** and **P.18** could act as a photoswitch of DNA conformation, converting duplex into G-quadruplex structures upon photoisomerization.[99,100]

The fluorescence properties of 7-deazapurine analogs conjugated to bipyri-
dine, phenanthroline, terpyridine, and their corresponding ruthenium-containing
complexes were investigated by Hocek and coworkers.[101,102] Notably, the triphosphate **P.19** was incorporated into modified oligonucleotides by certain DNA
polymerases.[101] As a result of template-specific incorporation, **P.19** could be used
for the detection of nucleotide polymorphisms using its luminescence and/or redox
properties.[101]

The specific properties and applications of chromophoric nucleobase analogs are
summarized in Table 11.8. In almost all cases, the incorporation of fluorophores into
DNA/RNA caused fluorescent quenching of the probe by photoinduced electron
transfer.[134] The impact of these nucleobase analogs on the global structure and
stability of modified nucleic acids is strongly dependent on the shape, size, and
hydrogen-bonding capabilities of each probe. In general, conjugation sites that
utilized position C8 of purines or C7 of 7-deaza purines had minimal impact on
structure and stability.

11.2.3 Pteridines

Pteridines are naturally occurring fluorescent purine mimics. Optimized fluorophores
from this family include **3-MI (P.20)**, **6-MI (P.21)**, **6-MAP (P.22)**, and **DMAP (P.23)**
(Fig. 11.6). As nucleosides, these fluorophores display strong visible fluorescence
($\lambda_{em} = \sim$430 nm; $\Phi = 0.39$–0.88), but their emission is significantly quenched upon
incorporation into oligonucleotides ($\Phi = 0.004$–0.29), where neighboring purine
residues are efficient quenchers (Table 11.3).[135,136] Since they contain methyl
groups on the *Watson–Crick* face, **3-MI** and **DMAP** are highly disruptive to duplex
DNA, resulting in destabilizations similar to base pair mismatches. Despite this,
3-MI-modified oligonucleotides have been used in numerous applications such as
protein-binding assays and the study of intracellular transport of oligonucleotides
(Table 11.8).[103–107] Compared to **3-MI**, **6-MAP** and **6-MI** introduce much less
duplex destabilization and show evidence for hydrogen bonding with cytosine and
thymidine.[103,137]

The fluorescence intensities of pteridine nucleotides depend on their interactions with neighboring bases. This provides a means to assay conformational
changes induced by the cleavage, unwinding or bending of DNA upon protein
binding.[31,32,75,105] With similar underlying principles, alkyl transferase activity,

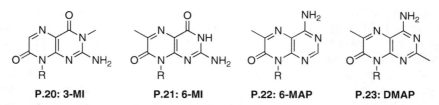

Figure 11.6 Examples of pteridines (R = 2′-deoxyribose or ribose).

TABLE 11.3   Spectroscopic Properties of Pteridines Fluorophores

Number	Name	Condition	λ_{abs} (nm)	λ_{em} (nm)	Φ	References
P.20	**3-MI**	MeOH	348	430	0.88	[32,103]
	3-MI	ssDNA			0.005–0.29	[135,137]
	3-MI	dsDNA			0.004–0.05	[135,137]
P.21	**6-MI**	MeOH	340	431	0.70	[32,103]
	6-MI	ssDNA			0.03–0.30	[137]
	6-MI	dsDNA			0.04–0.25	[137]
P.22	**6-MAP**	MeOH	320	430	0.39	[32,136]
	6-MAP	ssDNA			0.01–0.04	[136]
	6-MAP	dsDNA			0.005–0.01	[136]
P.23	**DMAP**	MeOH	310	430	0.48	[32,136]
	DMAP	ssDNA			<0.01–0.11	[136]

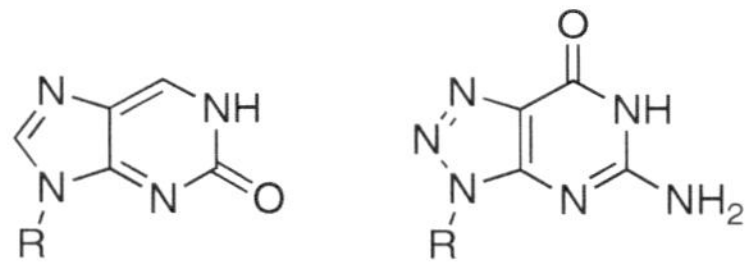

P.24: Isoinosine   P.25: 8-AzadG

Figure 11.7   Examples of isomorphic nucleobases (R = 2'-deoxyribose or ribose).

[104] DNA/RNA polymerase interactions,[103] antibiotic binding,[106] and DNA structures[108] have been assayed using **6-MAP**. **6-MAP** also has a reasonably good two photon cross sections and is therefore a potential probe for whole-animal imaging studies.[109] In addition to fluorescence intensity changes resulting from variable base-stacking interactions, the fluorescence anisotropy of **3-MI** has been used to detect protein binding.[105] **3-MI** is also sufficiently bright for single-molecule detection[107] and for the study of intracellular transport of oligonucleotides.[103]

11.2.4   Isomorphic Purine Analogs

Subtle modification of natural purine structures can provide isomorphic nucleobase analogs of guanosine and adenosine that exhibit useful fluorescence properties. The high structural similarity of the resulting analogs to native nucleobases can result in little or no perturbation of nucleic acid structure or stability. The most famous example of this family is **2AP** (Fig. 11.4), previously described in Section 2.1. Additional examples of isomorphic purine analogs are shown in Figure 11.7 and their spectroscopic characteristics are summarized in Table 11.4.

The isomorphic purine analog 2'-deoxyisoinosine (**P.24**) is a fluorescent isomer of the naturally occurring inosine. However, its incorporation into DNA resulted in little base-pairing selectivity and significantly destabilized duplexes.[138] 8-Aza-2'-deoxyguanosine (**8-AzadG**, **P.25**) is a purine isomorph that exhibits weak

TABLE 11.4   Spectroscopic Properties of Isomorphic Purines

Number	Name	Condition	λ_{abs} (nm)	λ_{em} (nm)	Φ	References
P.24	**Isoinosine**	Water	320	382	n.d.	[138]
	Isoinosine	DNA	318	383	n.d.	[138]
P.25	**8-AzadG**	Buffer	256	347	0.01–0.55	[110–112,139]
	8-AzadG	RNA	285	377	0.005–0.06	[110–112,139]

fluorescence emissions at neutral pH ($\Phi = 0.01$) and strong fluorescence upon ionization under alkaline conditions ($\Phi = 0.55$).[111,112,139] Upon incorporation into oligonucleotides, **8-AzadG** fluorescence is significantly quenched, although not as much so as **2AP**.[111] The presence of an electronegative nitrogen atom at position 8 decreases the pK_a of N1H by one unit ($pK_a = 8.3$) compared to 2′-deoxyguanosine ($pK_a = 9.3$). **8-AzadG** was therefore used as a reporter of purine ionization in RNAs[110] and for base pair mismatches in DNA. **8-AzadG** induces almost no destabilization of duplexes[111,140] but was found to be incompatible with G-quadruplex formation.[140]

11.2.5   Fused-Ring Purine Analogs

Tricyclic, size-expanded purines such as **Etheno-A** (Fig. 11.4) can be highly emissive chromophores (Fig. 11.8, Table 11.5). Tricyclic systems generally exhibit red-shifted excitation/emission spectra and larger extinction coefficients as compared to their bicyclic counterparts. **dxA** (**P.26**) and **dxG** (**P.27**) are size-expanded adenosine and guanosine analogs prepared by Kool and coworkers that contain a benzene ring inserted between the imidazole and pyrimidine rings. The quantum yields for fluorescence of these nucleosides ($\lambda_{em} = 393$–413 nm, $\Phi = 0.40$–0.44) is reduced upon their incorporation into oligonucleotides ($\Phi = 0.02$–0.08). Due to their size, duplexes containing a single **dxA** or **dxG** unit are significantly destabilized as compared to unmodified oligonucleotides.[114] In contrast, when **dxA** or **dxG** are used at all positions of a double helix, a highly stable, expanded DNA duplex is obtained.[114] Certain DNA polymerases can even read the chemical information stored in these ring-expanded genetic codes, and **dxA**-containing single-stranded DNA can encode for dT by the replication machinery of *Escherichia coli*.[115]

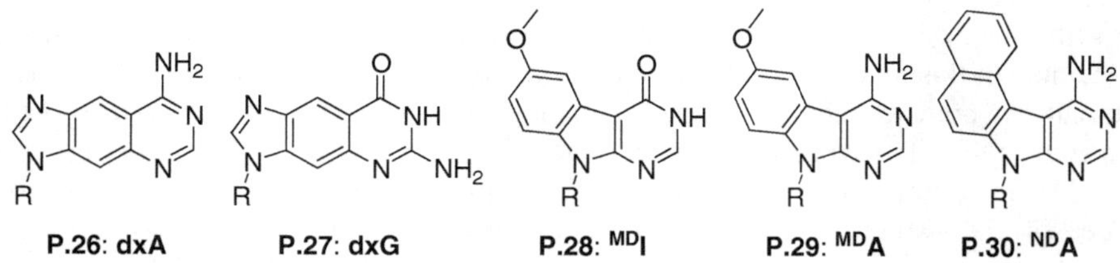

P.26: dxA   P.27: dxG   P.28: MD| P.29: MDA   P.30: NDA

Figure 11.8   Examples of ring-fused purine analogs (R = 2′-deoxyribose or ribose).

TABLE 11.5   Spectroscopic Properties of Size-Expanded Purine Analogs

Number	Name	Condition	λ_{abs} (nm)	λ_{em} (nm)	Φ	References
P.26	dxA	MeOH	333	393	0.44	113–115,141–143
	dxA	ssDNA	333	393	0.02–0.04	113
P.27	dxG	MeOH	320	413	0.40	113–115,141,143,144
	dxG	ssDNA	320	390	0.06–0.08	113
P.28	MDI	Buffer	315	442	0.12	116
	MDI	DNA	315	424	0.002–0.01	116
P.29	MDA	Buffer	327	427	0.12	116
	MDA	DNA	327	424	0.001–0.08	116
P.30	NDA	DNA	352–354	383	0.005–0.027	117

The addition of a fused aromatic ring to positions 7 and 8 of 7-deaza purines can generate new fluorescent nucleobases. For example, the inosine and adenosine mimics ^{MD}I (**P.28**) and ^{MD}A (**P.29**) exhibit good quantum yields as nucleosides ($\Phi = 0.12$) but are quenched in the context of oligonucleotides. ^{MD}I and ^{MD}A have been used as base-discriminating fluorescent probes that exhibit brighter fluorescence when paired with T and C, respectively.[116] The naphthodeazaadenine derivative NDA (**P.30**) has been utilized in a FRET-based system with fluorescein for discriminating base-pairing interactions.[117]

11.2.6   Substituted Purine Derivatives

Fused tricyclic systems are typically too large and rigid to be compatible with the folding of native duplex, triplex, and quadruplex structures. To overcome this limitation, natural nucleobases can be substituted with small, nonfluorescent moieties to generate substituted purine derivatives as fluorescent probes.[29] The addition of a simple vinyl or aryl group to adenosine or guanosine is often sufficient for generating a fluorescent product (Fig. 11.9). These purine derivatives generally show redshifted emission maxima compared to isomorphic nucleobase, and they are typically less disruptive to DNA/RNA folding and stability as compared to the fused nucleobase analogs.

Castellano and coworkers reported a series of "push–pull" purine fluorophores (**P.31–P.39**, Fig. 11.9) having electron-donating groups such as amino, methylamino, dimethylamino, or benzylether at position 2 or 6. These groups were complemented by an electron-accepting group such as cyano or methyl ester moieties at position 8. Nucleobases bearing these complementary groups (**P.34–P.39**) exhibit favorable photophysical characteristics such as redshifted absorption and emission maxima, enhanced quantum yields, and greater solvatochromism as compared to their acceptor-free analogs (**P.31–P.33**, Table 11.6). This strategy provides a means for tunable purines across the UV-to-blue spectrum.[30] All of these derivatives, however, contain disrupted hydrogen-bonding faces and are therefore incompatible with native base-pairing interactions in DNA/RNA molecules.

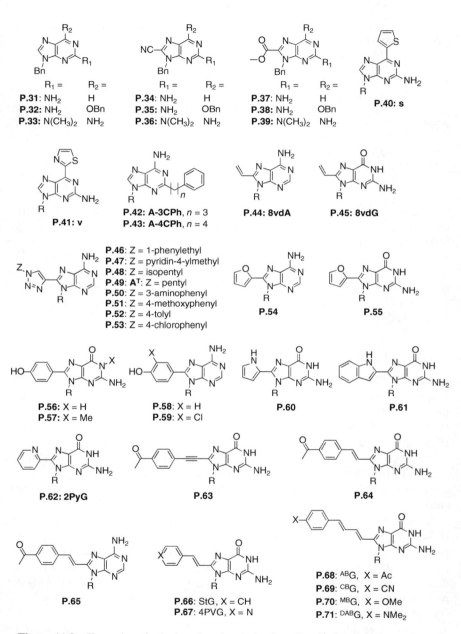

Figure 11.9 Examples of substituted purine derivatives (R = 2′-deoxyribose or ribose).

TABLE 11.6   Spectroscopic Properties of Purine Analogs Developed by *Castellano and coworkers* **in** $CH_2Cl_2{}^{30}$

Number	λ_{abs} (nm)	λ_{em} (nm)	Φ	Number	λ_{abs} (nm)	λ_{em} (nm)	Φ
P.31	304	357	0.20	**P.36**	336	387	>0.95
P.32	281	360	0.03	**P.37**	328	379	0.42
P.33	299	360	0.12	**P.38**	315	371	>0.95
P.34	326	371	0.20	**P.39**	338	409	>0.95
P.35	311	355	0.81				

Data from Ref. 30.

The introduction of thiophene or thiazole at position 6 of **2-AP** furnishes 2-amino-6-(2-thienyl)purine (**s, P.40**) and 2-amino-6-(2-thiazolyl)purine (**v, P.41**), respectively. Compared to **2-AP** (**P.2**), these compounds exhibit redshifted absorption ($\lambda_{abs} = 348$–$359\,nm$) and emission maxima ($\lambda_{em} = 434$–$461\,nm$), as well as high quantum efficiencies ($\Phi = 0.41$ and 0.46 in aqueous solution). Unlike **2-AP**, these derivatives retain good quantum yields ($\Phi = 0.02$–0.20) upon their incorporation into nucleic acids, but they do not exhibit proper base-pairing interactions with natural pyrimidines. The fluorescence properties of **P.40** and **P.41** are highly sensitive to local environment and base-stacking interactions.[118] **P.40** and **P.41** can form stable, unnatural base pairs with 2-oxo-($1H$)pyridine, allowing for its site-specific enzymatic incorporation into RNA molecules for aptamer labeling.[119] **P.40** can also form a noncanonical base pair with pyrrole-2-carbaldehyde and was thereby used to analyze local conformational changes of RNA molecules.[120,121]

While extension of the purinic π-system is a common and highly successful strategy to provide fluorescent derivatives, the introduction of nonconjugated phenylalkyl groups at position 2 of adenosine can also yield compounds with remarkable photophysical properties. As free nucleosides in solution, 2-(3-phenylpropyl)adenosine (**A-3CPh, P.42**) and 2-(4-phenylbutyl)adenosine (**A-4CPh, P.43**) exhibit relatively low quantum yields ($\Phi = 0.01$ and 0.007, respectively). Remarkably, upon their incorporation into RNA hairpins, these values are significantly enhanced ($\Phi = 0.08$–0.57, Table 11.7). The origin of this effect is unknown.[122]

The 8 position of adenine and guanine is a highly attractive site for modification because it is synthetically accessible and it is not involved in hydrogen-bonding interactions along either the *Watson–Crick* or *Hoogsteen* faces. The addition of very small substituents to this position is sufficient for endowing fluorescent properties to the purine derivative. For example, 8-vinyl-2′-deoxyadenosine (**8vdA, P.44**) exhibits a high quantum yield of fluorescence in water ($\Phi = 0.66$). This nucleobase was shown to be less disruptive and less quenched than **2AP** in the context of duplex DNA.[145] Time-resolved fluorescence studies demonstrated that **8vdA** can be used to differentiate the opposing base.[123] More recently, this approach was extended to 8-vinyl-2′-deoxyguanosine (**8vdG, P.45**), which was found to exhibit conformation-sensitive emission properties in the context of polymorphic G-quadruplex structures.[124]

TABLE 11.7   Spectroscopic Properties of Substituted Purine Derivatives

Number	Name	Condition	λ_{abs} (nm)	λ_{em} (nm)	Φ	References
P.40	**S**	Buffer	348	434	0.41	[118]
	S	DNA	352	432	0.07–0.20	[118]
P.41	**V**	Buffer	359	461	0.46	[118]
	V	DNA	363	460	0.02–0.17	[118]
P.42	**A-3CPh**	Water	292	385	0.01	[122]
	A-3CPh	RNA	292	399–422	0.17–0.32	[122]
P.43	**A-4CPh**	Water	278	396	0.007	[122]
	A-4CPh	RNA	278	395–416	0.08–0.57	[122]
P.44	**8vdA**	Buffer	290	382	0.66	[145]
	8vdA	ssDNA	290	387–389	0.014–0.037	[145]
	8vdA	dsDNA	290	382–384	0.005–0.016	[145]
P.45	**8vdG**	Buffer	277	400	0.72	[124]
P.46	—	THF	289	342	0.63	[146]
P.47	—	THF	289	346	0.49	[146]
P.48	—	THF	289	343	0.62	[146]
P.49	**A^T**	THF	289	342	0.62	[146]
	A^T	ssDNA	282	353–358	0.005–0.21	[125]
	A^T	dsDNA	282	349–354	0.003–0.011	[125]
P.50	—	THF	296	402	0.38	[146]
P.51	—	THF	294	370	0.03	[146]
P.52	—	THF	294	368	0.05	[146]
P.53	—	THF	294	396	0.05	[146]
P.54	—	Water	304	374	0.69	[29]
P.55	—	Water	294	378	0.57	[29]
P.56	—	Buffer	278	390	0.47	[126]
P.57	—	Buffer	279	378	0.25	[126]
P.58	—	Buffer	282	391	0.56	[126]
P.59	—	Buffer	290	383	0.22	[126]
P.60	—	Buffer	292	379	0.10	[127]
P.61	—	Buffer	321	390	0.78	[127]
P.62	**2PyG**	Water	300	415	0.02	[131,147]
	2PyG	CH$_3$CN	316	395	0.71	[131,147]
	2PyG	DNA	305–320	415	0.01–0.10	[131,147]
P.64	—	THF	370	499	0.49	[130]
P.65	—	THF	350	470	0.37	[130]
P.66	**StG**	Water	340	450	0.49	[131,132]
	StG	DNA	360	450	0.20–0.45	[131]
P.67	**4PVG**	Water	355	490	0.16	[131]
	4PVG	CH$_3$CN	355	490	0.57	[131]
	4PVG	DNA	375	475	0.02–0.07	[131]
P.68	**^{AB}G**	CH$_3$CN	392	547	0.57	[133]
P.69	**^{CB}G**	CH$_3$CN	387	526	0.47	[133]
P.70	**^{MB}G**	CH$_3$CN	352	441	0.68	[133]
P.71	**^{DAB}G**	CH$_3$CN	349	443	0.75	[133]

Azide–alkyne "click" cycloaddition reactions were used to prepare a small library of 8-substituted adenosine derivatives containing a variably substituted triazole ring (**P.46–P.53**).[146] Variation of the substituent on the triazole caused changes in emission wavelengths but not absorbance maxima. Aliphatic substituents yielded compounds with high quantum yields ($\Phi = 0.49$–0.64), and the inclusion of aromatic moieties resulted in redshifting of absorbance maxima by 25–55 nm. With the notable exception of **P.50** ($\Phi = 0.38$), derivatives containing aromatic substituents were only moderately emissive ($\Phi = 0.03$–0.05).[146] The pentyl derivative **P.49** (**A^T**) was incorporated into DNA using phosphoramidite chemistry. Remarkably, this adenine analog causes only minor destabilization of duplexes, where it exhibits quantum yields ranging from 0.003 to 0.20, depending on the neighboring bases and hybridization state of the oligonucleotide.[125]

Arylation of the 8 position of purines with furan (**P.54** and **P.55**),[29] phenol derivatives (**P.56–P.59**), pyrrole (**P.60**), indole (**P.61**), or pyridine rings (**2PyG**, **P.62**) gives a series of fluorescent guanosine and adenosine analogs with redshifted absorption ($\lambda_{abs} = 278$–321 nm) and emission maxima ($\lambda_{em} = 378$–415 nm) as compared to guanosine itself. These compounds exhibit a wide range of quantum yields in water ($\Phi = 0.02$–0.78). The emission intensities of the phenolic derivatives **P.57–P.59** are highly pH-sensitive.[126] The pyrrole and indole derivatives (**P.60** and **P.61**) can serve as fluorescent reporters of hydrogen bonding that are quenched upon addition of dC, but enhanced upon addition of G.[127]

Among the C8-arylated purine analogs reported to date, 8-(2-pyridyl)-2′-deoxyguanosine (**2PyG**, **P.62**) exhibits one of the highest emission wavelengths ($\lambda_{em} = 415$ nm). Due to its "push–pull" character, this fluorophore exhibits strong solvatochromism.[128] Due to excited-state proton transfer reactions with bulk solvent, **2PyG** is quenched by water ($\Phi = 0.02$) but is highly emissive in aprotic solvents such as acetonitrile ($\Phi = 0.71$). Together, these features result in highly environmentally-sensitive fluorescence properties. Remarkably, upon its incorporation into oligonucleotides, **2PyG** exhibits enhanced quantum yields ($\Phi = 0.03$–0.15) as compared to the free nucleoside in water, possibly due to the desolvation of the pyridyl nitrogen in the context of folded nucleic acids.[147] **2PyG** is minimally disruptive to both duplex and G-quadruplex structures, and it has enabled the quantification of nucleobase-to-probe energy transfer reactions within G-quadruplex structures. In addition, the (2-pyridyl)-imidazole moiety of **2PyG** facilitates the binding and localization of transition metal ions to arbitrary N7 sites in nucleic acids.[128]

Consistent with "the particle in a box" description of molecular orbitals, enlarging the size of conjugated π-systems is one straightforward strategy to purine analogs with redshifted excitation and emission maxima. Following this approach, conjugated linkers have been used to link aryl groups with the 8 position of purines.[148] The addition of an acetylene bridge between guanosine and phenyl generates compounds with redshifted emission maxima ($\lambda_{em} = 475$ nm) (**P.63**) as compared to the corresponding 8-phenyl derivative. The fluorescence properties of **P.63** were shown to be strongly solvent dependent. In the context of oligonucleotides, **P.63** acts as a T-specific base-discriminating fluorophore.[129]

Acetyl-substituted 8-styryl deoxyguanosine and deoxyadenosine derivatives (**P.64** and **P.65**) exhibit relatively low quantum yields but strong solvatochromism and emission maxima ranging from 477 to 558 nm (1,4-dioxane → methanol).[130] The corresponding 8-styryl-2′-deoxyguanosine derivative (**StG**, **P.66**) exhibits relatively little environmental sensitivity, but it retains a high quantum yield for emission over a broad range of solvents including water and acetonitrile ($\Phi = 0.49$ and 0.74). **StG** was shown to photoisomerize upon irradiation with UV light.[132] The presence of a nitrogen atom in the para position of the phenyl ring in 8-[2-(pyrid-4-yl)-ethenyl]-2′-deoxyguanosine (**4PVG**, **P.67**) largely eliminated the light-induced photoisomerization observed for **StG**. **4PVG** exhibits environmentally sensitive quantum yields ($\Phi = 0.16$ in water and $\Phi = 0.57$ in acetonitrile), as well as redshifted emissions ($\lambda_{em} = 490$ nm) as compared to **StG** ($\lambda_{em} = 450$ nm). **StG** and **4PVG** exhibit very little quenching upon their incorporation into oligonucleotides ($\Phi = 0.13$–0.45 and 0.03–0.10, respectively). In analogy to **2PyG**, **StG**, and **4PVG** can also act as energy acceptors from unmodified DNA bases, allowing the utilization of energy transfer measurements for discriminating unfolded, G-quadruplex, and duplex DNA structures.[131] For **StG,** the combination of high energy transfer efficiencies, high probe quantum yield, and high oligonucleotide molar extinction coefficient provide a highly sensitive and reliable readout of G-quadruplex formation even in sample solutions diluted below 1 nM.[131] Further elaboration of these types of derivatives includes 8-arylbutadienyl-2′-deoxyguanosine derivatives (**P.68–P.71**). The inclusion of electron-donating groups, such as methoxy (**^{MB}G**, **P.70**) or dimethylamine (**^{DAB}G**, **P.71**), provides compounds with high quantum yields ($\Phi = 0.56$–0.75) that are relatively insensitive to environment. These compounds also exhibit similar absorption and emission wavelengths ($\lambda_{abs} = 343$–354 nm and $\lambda_{em} = 429$–443 nm) as the styryl derivatives **StG** and **4PVG**. In contrast, the addition of electron-withdrawing groups such as acetyl (**^{AB}G**, **P.68**) or cyano (**^{CB}G**, **P.69**) gives fluorophores with "push–pull" characteristics. This is accompanied by redshifted excitation and emission ($\lambda_{abs} = 387$–399 nm and $\lambda_{em} = 486$–547 nm, respectively) as well as environmentally sensitive quantum yields ($\Phi = 0.02$–0.59 and 0.32–0.75 for **^{AB}G** and **^{CB}G**, respectively).[133]

11.3   IMPLEMENTATION

Modification of position 8 of guanosine can generate highly fluorescent products, with emission properties that are highly sensitive to DNA folding. While the addition of bulky groups to the C8 position of guanosine can shift the conformational equilibrium of the glycosidic bond from *anti* to *syn*, [145,149–154] DNA folding can force 8-modified guanosines to adopt *anti* conformations with relatively small energetic penalties to DNA folding ($\Delta\Delta G < 1$ kcal/mol).[14,40,124,131,147,150,151,154] In this section, the potential applications of 8-substituted guanosine will be further illustrated by describing the scope of 8-(2-pyridyl)-2′-deoxyguanosine (**2PyG**). **2PyG** has played a key role in a number of *in vitro* experiments for the study of G-quadruplex folding, energy transfer, and metal binding. The remaining challenges faced by internal fluorescent probes for their application in cell-based studies will be presented at the end of the section.

11.3.1   Probing G-Quadruplex Structures with 2PyG

The **2PyG** nucleoside (**P.62**) exhibits environment-sensitive photophysical characteristics including an excellent correlation between *Stokes* shift and solvent polarity (E_T^N).[155,156] This is consistent with solvent-mediated stabilization of a charge-separated emissive state having a larger dipole moment than the ground state.[30,157] The quantum yield of **2PyG** strongly correlates to solvent acidity (SA),[158] consistent with an excited state proton transfer as a dominant nonradiative decay pathway of **2PyG** in aqueous and organic solvents. Since the solvent exposure of guanine residues is lower in G-quadruplex versus other structures, these results suggested that **2PyG** could be used as a conformation-specific fluorescent probe that exhibits enhanced fluorescence upon G-quadruplex folding.

G-quadruplexes are intriguing nucleic acids nanostructures possessing potential biological relevance,[3–5,159–167] as well as interesting photophysical and material properties.[168–180] Although G-quadruplexes were considered for many years as structural curiosities, evidence for their presence in genomic DNA and RNA is rapidly accumulating. Computational analyses of bacterial and human genomes for guanine-rich tracts revealed an underrepresentation of G-quadruplex-type forming sequences in open reading frames, but an overabundance of such sequences in promoter regions and 3′ telomeric ends of chromosomes.[159,181] While the exact biological relevance of these structures remains an open question, DNA sequences with the ability to fold into G-quadruplex structures have been implicated in regulating gene transcription,[160] recombination,[3] chromosome stability,[5,182–184] and programmed cell death.[185] In addition, energy and electron transfer reactions involving G-quadruplexes can mediate DNA damage and repair.[175–177]

G-quadruplexes are highly demanding structures in terms of their compatibility with internal fluorescent probes. For example, the common fluorescent purine analogs **2AP** and **3-MI** are incompatible with the folding of native G-tetrads and also exhibit fluorescence quenching by neighboring guanine residues.[47,67,68,135–137,147] **2PyG**, in contrast, can be directly incorporated into the G-tetrads of natively folded G-quadruplexes and the Watson–Crick base pairs of duplex DNA (Fig. 11.10), with relatively little impact on global structure or stability.[40,131,147] **2PyG** was found to be minimally disruptive to both duplex $(\Delta T_m = -3.5\,^\circ C$ on average) and G-quadruplex structures $(\Delta T_m = -1.0\,^\circ C$ on average) as determined by temperature-dependent circular dichroism measurements.[147]

Upon their incorporation into nucleic acids, fluorescent guanine mimics are typically quenched by base-stacking interactions with purine residues. The quantum yield (Φ) of **2AP**, for example, is inversely proportional to the extent of base-stacking interactions with neighboring purine residues, decreasing approximately 200-fold upon its incorporation into G-rich DNA $(\Phi = 0.002–0.004)$.[47,67,68,147] To date, there are only a small number of reported internal fluorescent probes that have similar

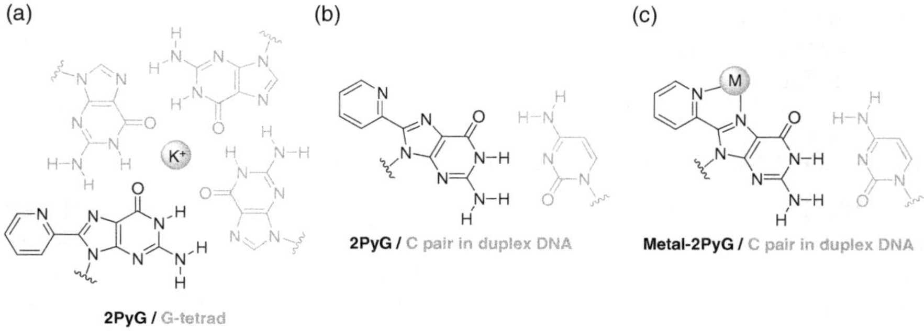

Figure 11.10 (a) **2PyG** (black) incorporated into a G-tetrad (gray). (b) *Watson–Crick* base pair between **2PyG** and cytidine (gray). (c) Base pair between metal-coordinated **2PyG** residue (black) and cytidine (gray).

or higher quantum yields in the context of folded nucleic acids. Examples include 1,3-diaza-2-oxophenothiazine,[186] 5-(fur-2-yl)-2′-deoxyuridine[29] and the adenine derivatives **A-3CPh** (**P.42**) and **A-4CPh** (**P.43**).[122,187] None of these compounds, however, are effective mimics of guanine residues.

2PyG (**P.62**) exhibits a higher quantum yield upon incorporation into folded oligonucleotides ($\Phi = 0.03$–0.15) as compared to the free nucleoside in water ($\Phi = 0.02$).[40,131,147] Using phosphoramidite chemistry, **2PyG** was incorporated into six different DNA sequences capable of duplex–quadruplex polymorphism, and the purified oligonucleotides were folded into four different structures by changing the ions present in the buffer. For example, **2PyG** was incorporated into position 9 of the human telomeric sequence (hTeloG9) and the purified oligonucleotide was folded into polymorphic G-quadruplexes in the presence of K^+ or Na^+; or prepared as unfolded single strands in Li^+; or hybridized with its complementary strand in the presence of Na^+ ions to generate a duplex. The resulting emission spectra are shown in Figure 11.11a. The quantum yield of the **2PyG** nucleoside ($\Phi = 0.02$) increased 50–100% in the context of single-stranded and duplex DNA ($\Phi = 0.03$–0.04), and by 400–500% in G-quadruplex structures ($\Phi = 0.08$–0.10, Fig. 11.11b). Similar results were obtained when **2PyG** was placed at positions G17 and G23 of the same "hTelo" DNA sequence, as well as positions G10 and G15 of a G-quadruplex-forming sequence derived from the ckit(2) promoter.[40,128,131,147] Consistent with the presence of **2PyG**-G base-stacking interactions, the excitation maximum of the **2PyG** nucleoside monomer ($\lambda_{ex} = 300$ nm) was redshifted by approximately 30 nm in duplex and G-quadruplex structures.[188] The modest quantum yields of **2PyG** are compensated by its relatively large molar extinction coefficient ($\varepsilon_{280nm} \approx 20,000$ cm^{-1} M^{-1}) to give good brightness. **2AP**, in comparison, is approximately 100 times less bright than **2PyG** due to its low molar extinction coefficient ($\varepsilon_{305nm} = 5600$ cm^{-1} M^{-1}) and low quantum yield ($\Phi \approx 0.003$) when base-stacked with guanine residues. These results suggested that **2PyG** is resistant to G-mediated quenching. Indeed, Stern–Volmer plots revealed dramatic differences between **2AP** and **2PyG** nucleosides in terms of their susceptibility to fluorescence quenching by guanosine monophosphate (GMP).

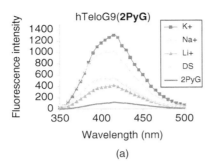

Name	Φ (330 nm)
2PyG monomer	0.02
Double stranded	0.03
Single stranded	0.04
G-quadruplex	0.09

(a)   (b)

Figure 11.11   (a) Fluorescence spectra ($\lambda_{ex} = 330$ nm) of the **2PyG** nucleoside compared to "hTeloG9(**2PyG**)" DNA that was prepared as unfolded single strands (Li$^+$), G-quadruplex structures (K$^+$ and Na$^+$), or double-stranded DNA (DS). (b) Quantum yields calculated for the hTeloG9(**2PyG**) oligonucleotide prepared in various folds. Sequence of hTeloG9(**2PyG**): TTGGG TTA(**2PyG**)G GTTAG GGTTA GGGA. All samples were prepared in 10 mM cacodylate buffer (pH 7.4) containing 100 mM of LiCl, NaCl, or KCl.

In both cases, the Stern–Volmer plots were nonlinear. The F_o/F plots for **2AP** exhibited upward curvature (increasing quenching) with increasing GMP concentrations, whereas F_o/F plots of **2PyG** reached a plateau at high GMP concentrations.[147] As GMP is known to self-assemble into G-tetrads at high concentrations, these results suggested that **2PyG** was incorporated into GMP tetrads. Similar photophysical changes, including a 30-nm redshift in fluorescence emission, were observed at high GMP concentrations as for the incorporation of **2PyG** into G-quadruplex-forming oligonucleotides. These results demonstrated that the self-assembly of GMP at high concentrations can mimic the local nucleobase environments of G-quadruplex structures.[147] This may provide a simplified means to screen new purine derivatives for identifying candidates that exhibit enhanced fluorescence properties upon their incorporation into nucleic acids.

11.3.2   Energy Transfer Quantification

The enhanced fluorescence intensity of **2PyG** in the context of G-quadruplex versus other DNA conformations is notable but the effect is not large enough for an unambiguous determination of structure. The quantification of energy transfer efficiencies, however, can provide a robust readout of conformation. In this approach, a fluorescent nucleobase analog serves as an emissive energy acceptor, whereas the proximal ensemble of unmodified residues serves as an energy donor.[41,67–69,189] The resulting differences in energy transfer efficiencies can readily be interpreted in terms of the folded state of the strand.[40,68,131]

Previous studies have suggested that G–G base stacking, hydrogen bonding, and/or restricted motions within G-quadruplexes can enhance the photo-excited lifetimes and energy transfer properties of native guanine residues.[190–193] However, the poor quantum yield ($\Phi \approx 10^{-4}$), short-lived excited state ($\tau \approx 1$ ps), and overlapping emission wavelengths of guanine residues with other bases prevents the quantification of energy transfer efficiencies (η_t) in unmodified DNA.[190–193] Due to their good spectral overlap, **2PyG** ($\lambda_{ex} \approx 330$ nm in DNA) can act as a FRET energy acceptor for unmodified guanine residues ($\lambda_{em} \approx 330$ nm in DNA). Since its fluorescence emissions are not quenched by neighboring residues, **2PyG** can be used to quantify energy transfer efficiencies (η_t) involving unmodified residues in G-quadruplex, single-stranded, and duplex DNA. Due to the low quantum yields of unmodified nucleobases, only those in close proximity should be capable of serving as energy donors to **2PyG**.

The excitation spectra of oligonucleotides containing **2PyG** exhibit two maxima corresponding to direct excitation of the probe ($\lambda_{ex} = 330$ nm), and its indirect excitation via energy transfer from unmodified nucleobases ($\lambda_{ex} = 260$ nm, Fig. 11.12a). The ratio of these excitation peaks weighted by the absorbance properties of the DNA and the quantum yield of **2PyG** were used to calculate energy transfer efficiency (η_t) values.[40,41] η_t is defined as the number of photons transferred from all possible energy donors to **2PyG**, divided by the total number of photons absorbed by all nucleobases at 260 nm. Interestingly, energy transfer efficiencies from unmodified bases to **2PyG** were three- to fourfold higher in G-quadruplexes ($\eta_t = 0.11$–0.41) compared to the same oligonucleotides in duplex DNA ($\eta_t = 0.01$–0.07). These values were independent of the exact structure and the stability of the G-quadruplex used. Unfolded single-stranded DNAs in Li$^+$ solutions gave intermediate efficiencies ($\eta_t = 0.05$–0.11). The combination of strong DNA molar absorptivity ($\varepsilon_{260nm} \approx 250{,}000$ cm^{-1} M^{-1}) together with high η_t values resulted in fluorescence enhancements that were 10- to 30-fold higher for **2PyG**-containing G-quadruplexes as compared to the same oligonucleotides prepared as single-stranded or duplex structures (Fig. 11.12b). While calculating energy transfer efficiencies requires multiple measurements and data processing, a simple comparison of emission intensities resulting from probe excitation at 330 nm versus DNA excitation at 260 nm can

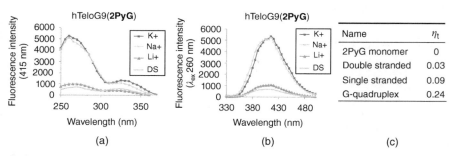

Figure 11.12 (a) Excitation ($\lambda_{em} = 415$ nm) and (b) emission ($\lambda_{ex} = 260$ nm) spectra of "hTeloG9(**2PyG**)" prepared as unfolded single strands (Li$^+$), G-quadruplex structures (K$^+$ and Na$^+$), or double-stranded DNA (DS). (c) Energy transfer efficiencies (η_t) measured for **2PyG** in the hTeloG9(**2PyG**) oligonucleotide prepared in various folds. Similar results were obtained when **2PyG** was placed at positions G17 and G23 of "hTelo", as well as positions G10 and G15 of a G-quadruplex-forming sequence derived from the ckit(2) promoter. All samples were prepared in 10 mM cacodylate buffer (pH 7.4) containing 100 mM of LiCl, NaCl, or KCl.

differentiate duplex from G-quadruplex structures according to a robust fourfold difference in emission intensities.[131] Oligonucleotides prepared as duplexes, in contrast, gave emission ratios close to 1.0. Given the simplicity and sensitivity of this approach, it should be compatible with conformational analyses using fluorescence microscopy and even single-molecule spectroscopy.

To evaluate if the exceptionally high energy transfer efficiencies in G-quadruplex structures were a result of the unusual photophysical properties of G-tetrads themselves, two alternative 8-substituted-2′-deo-xyguanosines, 8-(2-phenylethenyl)-2′-deoxyguanosine (**StG**), and 8-[2-(pyrid-4-yl)-ethenyl]-2′-deoxyguanosine (**4PVG**) were used to measure energy transfer efficiencies within G-quadruplex and duplex structures. Due to their preference for adopting an *anti* glycosidic bond conformation, **StG** and **4PVG** exhibit less potential for perturbing DNA structure/stability as compared to **2PyG**.[131] **4PVG** and **StG** also exhibit redshifted excitation and emission maxima, and higher quantum yields ($\Phi = 0.03–0.44$) than **2PyG** in the context of folded DNA. Similar to **2PyG**, the excitation spectra of the **4PVG**- and **StG**-modified oligonucleotides exhibit two maxima corresponding to direct excitation ($\lambda_{ex} = 350–400\,nm$) and indirect excitation via unmodified nucleobases ($\lambda_{ex} = 260\,nm$). Once again, the calculated energy transfer efficiencies (η_t) were highly indicative of structure, with η_t values for G-quadruplex ($\eta_t = 0.12–0.36$) > unfolded ($\eta_t = 0.03–0.17$) > duplex ($\eta_t = 0.02–0.05$). For oligonucleotides containing **StG**, the combination of high quantum yield ($\Phi = 0.20–0.45$), high energy transfer efficiencies, and high molar extinction coefficients of the oligonucleotides provided a highly sensitive and reliable reporter of G-quadruplex conformation in sample solutions diluted below 250 pM.[131]

One of the key electronic differences between G-quadruplex versus duplex structures is the coordination of alkali metal ions to the O6-position of guanine residues in G-tetrads (Fig. 11.10). To evaluate the potential role of O6-metal ion coordination in mediating the exceptionally high energy transfer efficiencies in G-quadruplexes, oligonucleotides containing **2PyG** were folded into G-quadruplex structures under salt-deficient conditions by partial dehydration in 40% polyethylene glycol (PEG) 200.[194–197] Under these conditions, Li^+ cations from the buffer are absent or only weakly bound inside the G-quadruplex structure. Interestingly, almost no energy transfer ($\eta_t = 0.02–0.07$) was observed when G-quadruplexes were folded under these conditions.[40] Quadruplexes prepared in the presence of both PEG and 100 mM NaCl, in contrast, exhibited energy transfer efficiencies similar to those prepared in 100 mM NaCl and water (Fig. 11.13). These results suggested that the G-tetrad core by itself was not sufficient for promoting efficient energy transfer in G-quadruplexes, and that the ions contained within the structure are required for high energy transfer efficiencies (Fig. 11.14). To evaluate the exact role played by these ions, experiments were conducted using unmodified G-quadruplexes. These experiments revealed that cation coordination to O6 results in enhanced quantum yields of the guanine residues in G-quadruplex structures, therefore making them more efficient energy donors than the guanine residues in single-stranded or duplex DNA.[40]

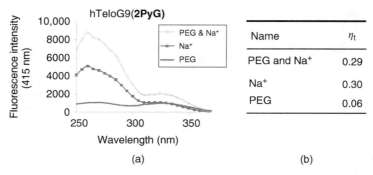

Figure 11.13 Excitation spectra (a) and energy transfer efficiencies (b) measured for hTeloG9(**2PyG**) folded in the presence of 10 mM lithium-cacodylate buffer (pH 7.4) containing either 40% PEG, 100 mM NaCl, or 40% PEG and 100 mM NaCl.

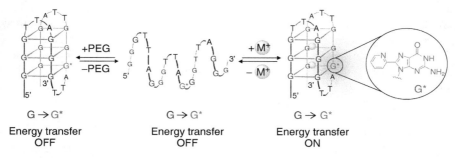

Figure 11.14 Cartoon summarizing the changes in guanine-to-**2PyG** energy transfer in G-quadruplexes folded under cation-deficient (a) or cation-rich (b) conditions, where "PEG" = 40% polyethylene glycol 200 and "M$^+$" = Na$^+$, K$^+$, Rb$^+$, or NH$_4{}^+$.

11.3.3 Metal-Ion Localization to N7

Metal ions coordinated to nucleobases play critical roles in mediating the catalytic activities of natural ribozymes,[198–201] as well as synthetic DNA-based catalysts.[202,203] Complex mixtures of products are normally observed when transition metals are added to unmodified nucleic acids.[204–210] The ability to direct metal ions to specific and arbitrary sites will provide new opportunities for catalyst preparation, as well as a powerful tool for studying thermodynamics of metal–DNA interactions[205–209] and metal-dependent energy/charge transfer processes.[211–216] 8-(2-Pyridyl)-2′-deoxyguanosine (**2PyG**) exhibits selective binding of Cu(II), Ni(II), Cd(II), and Zn(II) via a bidentate effect provided by the N7 atom of guanine and 2-pyridine group.[128] By monitoring **2PyG** fluorescence, metal-binding reactions were directly monitored. Upon addition of certain transition metals, the double absorbance maximum of the **2PyG** nucleoside (280–300 nm) was redshifted to 325–375 nm. This is consistent with an enhancement of the "push–pull" character of the **2PyG** fluorophore upon metal ion binding. Upon saturation with Cd(II) or Zn(II) the fluorescence intensity of **2PyG** increased approximately 40-fold in buffered water (pH 7.4), resulting in quantum yields $(\Phi) \approx 0.49$. Cu(II) and Ni(II), in contrast, quenched **2PyG**.[128] In the context of folded DNA, **2PyG** selectively bound to Cu(II) and Ni(II) with equilibrium dissociation constants (K_d) ranging from 25 to 850 nM, depending on the folded state of the oligonucleotide (duplex > G-quadruplex) as well as the identity of the metal ion (Cu > Ni ≫ Cd). These binding affinities were approximately 10- to 1000-fold higher than unmodified metal-binding sites in DNA. **2PyG** therefore provides a means for the site-specific control of transition metal binding on nucleic acids.

2PyG-modified oligonucleotides have been used to evaluate the competition between N7 hydrogen bonding and metal ion binding at guanine residues involved in the G-tetrads of G-quadruplex structures.[128] Consistent with the involvement of N7 in both metal binding and G-tetrad formation (Fig. 11.10), Cu(II), Ni(II), and Cd(II) all exhibited higher affinities for **2PyG**-containing duplexes as compared to the corresponding G-quadruplex structures. Taken together, these studies have demonstrated how the addition of a simple heterocyclic group to the C8-position of guanine can provide a powerful fluorescent probe for studying N7 binding and its impact on the structure, stability, and electronic properties of nucleic acids.

11.4 CONCLUSIONS

Fluorescence phenomena enable powerful and readily accessible technologies for probing biomolecule folding and activities.[10] Fluorescent nucleobase analogs provide a means for directly probing the unique photophysical and electronic properties of nucleic acid structures. Assays that utilize nucleobase analogs have historically been limited by relatively low sensitivity that results from fluorescence quenching of the probe by neighboring residues.[31–35] New probes such as **2PyG** and **StG** that remain highly emissive in the context of base-stacking interactions have provided new means for differentiating nucleic acid structures according to the differences in energy transfer efficiencies from endogenous residues. Notably, this can be accomplished by introducing a very small modification (e.g., a single styrene or pyridine group) into the oligonucleotide at a strategic location, such as 8 position of guanosine. Future challenges to this approach include the delivery and/or construction of such modifications in cellular environments to probe for DNA conformational changes *in vivo*.

11.5 PROSPECTS AND OUTLOOK

The emissive purine analogs reviewed here have been successfully applied in a variety of sensing applications *in vitro*. The ultimate value of these compounds as internal fluorescent probes, however, is dependent on their successful implementation in cellular environments. The incorporation of nucleoside analogs into synthetic oligonucleotides is routinely performed using phosphoramidite chemistry. Alternatively, nucleotide triphosphate building blocks can be assembled *in*

TABLE 11.8   Summary of Properties and Applications of Purine Analogs

Number	Name	Applications
P.2	2AP	High sensitivity to base-stacking interactions that have enabled assays for enzymatic activity,[35,46,51] DNA and RNA conformational changes,[59,60] DNA–protein binding interactions, G-quadruplex folding,[33,34] polymerase interactions,[52–54] hybridization,[49,50,65,66] base-pairing interactions,[48,49,61–64] RNA–aminoglycoside binding,[55–58] energy and charge transfer in DNA, [41,67–72] and molecular beacon construction[76]
P.4	**Etheno-A**	Substitute for ATP in enzymatic studies,[73,74] detection of DNA–protein binding interactions[75]
P.5	**Py-G**	Detection of DNA hybridization and single nucleotide polymorphisms; photoinducable donor for charge transfer studies[77,78]
P.6	^{PY}A	Base-discriminating fluorescence properties: C vs T detection[79]
P.7	A^{PY}	RNA hybridization assay[80,81]
P.8	A^{P}	Molecular beacon,[88] detection of B → Z DNA transitions,[84] single nucleotide polymorphisms,[82,83,87] G-quadruplex folding[85,86]
P.9	G^{P}	Molecular beacon, energy acceptor from 1-ethynylpyrene-modified cytosine,[90] detection of B → Z DNA transitions[91]
P.10	^{PyI}G	DNA hybridization assay[92]
P.11	dA^{pymcm}	DNA hybridization assay[93]
P.12	^{8Py}G	Distinction between single-strand, duplex, and G-quadruplex structures[94]
P.13	^{PGN}G	Detection of single nucleotide polymorphisms[95]
P.14	^{PGN}A	Detection of single nucleotide polymorphisms[95]
P.15	^{dan}G	Probing groove polarity[96,97]
P.16	^{VPy}G	Photoswitch (420 nm: $E \to Z$, $E{:}Z = 8{:}92$; 365 nm: $Z \to E$, $E{:}Z = 82{:}18$)[98]
P.17	^{8FV}G	Photoswitch (410 nm: $E \to Z$, $E{:}Z = 37{:}63$; 290 nm: $Z \to E$, $E{:}Z = 87{:}13$); photoregulation of duplex hybridization[99]
P.18	^{8FV}G	Photoswitch (420 nm: $E \to Z$, $E{:}Z = 23{:}77$; 310 nm: $Z \to E$, $E{:}Z = 77{:}23$); photoregulation of duplex hybridization,[99] and G-quadruplex folding[100]
P.19	dG^{Ru}TP	Detection of single nucleotide polymorphisms[101]
P.20	**3-MI**	Detection of DNA hybridization,[103] alkyl transferase activity,[104] DNA–protein binding interactions,[105] DNA–RNA polymerase interactions,[103] nucleoside cellular uptake,[103] RNA–aminoglycoside interactions,[106] and single molecule studies[107]
P.21	**6-MI**	Detection of DNA–protein,[75] DNA–RNA polymerase,[103] and RNA–aminoglycoside binding interactions[106]
P.22	**6-MAP**	Detection of DNA melting dynamics,[108] two-photon excitation[109]

TABLE 11.8 *(Continued)*

Number	Name	Applications
P.25	8-AzadG	pH-dependent fluorescence,[110] study of purine ionization state in RNA structures,[110] detection of mismatches,[111] and single nucleotide polymorphisms[112]
P.26	dxA	Reporting of base-pairing and -stacking interactions,[113] ring-expanded genetic code[114,115]
P.27	dxG	Reporting of base-pairing and -stacking interactions,[113] ring-expanded genetic code[114,115]
P.28	^{MD}I	Base-discriminating fluorescent probe[116]
P.29	^{MD}A	Base-discriminating fluorescent probe[116]
P.30	NDA	Base-discriminating fluorescent probe, FRET pair with fluorescein[117]
P.40	s	RNA structure and dynamics,[118–121] RNA aptamer labeling[119]
P.41	v	RNA structure and dynamics,[118,119] RNA aptamer labeling[119]
P.42	A-3CPh	Highly fluorescent in the context of DNA[122]
P.43	A-4CPh	Highly fluorescent in the context of DNA[122]
P.44	8vdA	Base-discriminating fluorescent probe,[123] DNA hybridization,[124] G-quadruplex topology[124]
P.49	A^T	Environmentally sensitive fluorescence[125]
P.57–P.59	—	pH-sensitive fluorescence[126]
P.60 and P.61	—	Sensitivity to base pairing[127]
P.62	2PyG	Solvatochromic, energy acceptor from dG, quantification of energy transfer in G-quadruplex DNA,[40] localization of metal ions to N7[128]
P.63	—	Base-discriminating fluorescent probe[129]
P.64 and P.65	—	Solvatochromic, quenched by polar solvents[130]
P.66	StG	Strong fluorescence in the context of DNA, energy acceptor from dG, detection of energy transfer in DNA,[131] photoswitch (370 nm: $E \rightarrow Z$, $E{:}Z = 6{:}94$; 254 nm: $Z \rightarrow E$, $E{:}Z = 80{:}20$)[132]
P.67	4PVG	Energy acceptor from dG, quantification of energy transfer in DNA[131]
P.68 and P.69	—	Solvatochromic[133]

vitro by DNA polymerases.[217,218] In both cases, however, the purified products would need to be transfected into living cells – a notoriously inefficient process. An alternative approach to this utilizes cellular kinases and polymerases for the metabolic incorporation of modified nucleosides into cellular DNA or RNA by whole biological systems.[219–221] The high fidelity of cellular enzymes involved in DNA/RNA synthesis greatly limits the size and structural diversity of modifications that can be introduced this way.[222] Purely synthetic methods, by comparison, can tolerate a wide variety of structures and functional groups. To combine the key advantages of both approaches, fluorogenic intracellular bioorthogonal reactions may provide a future means to synthesize fluorescent nucleobase analogs *in vivo*. In this approach, the cell's biosynthetic machinery will be used to incorporate a small, nonnative functional group into cellular DNA or RNA. Subsequently, a second component will be added to the cells such that a highly chemoselective reaction takes place to generate modified DNA or RNA molecules containing fluorescent nucleobase analogs. Since the impact metabolic labeling on the structure and biological stability of nucleic acids is expected to be inversely proportional to the size of the modification, the utilization of such labeling strategies may appear limited to very small substituents having relatively limited spectral properties (excitation in the far blue or UV). However, with the rapid developments in the field of two-photon imaging, the development of new probes having large two-photon cross sections may allow for the utilization of minimally disruptive probes for imaging studies *in vivo*. While no fluorogenic reactions according to this strategy have been demonstrated in cells, the purine analog 7-deaza-7-ethynyl-2'-deoxyadenosine (**EdA**) was found to be metabolically incorporated into living cells and whole animals, where it was used to visualize sites of new DNA synthesis following a copper-catalyzed "click" reaction with fluorescent azides.[223]

APPENDIX

See Table 11.8.

REFERENCES

1. Choi, J.; Majima, T. *Chem. Soc. Rev.* **2011**, *40*, 5893.
2. Patel, D. J.; Phan, A. T.; Kuryavyi, V. *Nucleic Acids Res.* **2007**, *35*, 7429.
3. Cahoon, L. A.; Seifert, H. S. *Science* **2009**, *325*, 764.
4. Smith, J. S.; Chen, Q.; Yatsunyk, L. A.; Nicoludis, J. M.; Garcia, M. S.; Kranaster, R.; Balasubramanian, S.; Monchaud, D.; Teulade-Fichou, M. P.; Abramowitz, L.; Schultz, D. C.; Johnson, F. B. *Nat. Struct. Mol. Biol.* **2011**, *18*, 478.
5. Biffi, G; Tannahill, D; McCafferty, J; Balasubramanian, S. *Nat. Chem.* **2013**, *5*, 182.
6. Elgar, G.; Vavouri, T. *Trends Genet.* **2008**, *24*, 344.
7. Eddy, S. R. *Nat. Rev. Genet.* **2001**, *2*, 919.
8. Montange, R. K.; Batey, R. T. *Annu. Rev. Biophys.* **2008**, *37*, 117.
9. Mattick, J. S.; Makunin, I. V. *Hum. Mol. Genet.* **2006**, *15*, R17.
10. Sinkeldam, R. W.; Greco, N. J.; Tor, Y. *Chem. Rev.* **2010**, *110*, 2579.
11. Phan, A. T.; Luu, K. N.; Patel, D. J. *Nucleic Acids Res.* **2006**, *34*, 5715.
12. Ambrus, A.; Chen, D.; Dai, J. X.; Jones, R. A.; Yang, D. Z. *Biochemistry* **2005**, *44*, 2048.
13. Phan, A. T.; Kuryavyi, V.; Burge, S.; Neidle, S.; Patel, D. J. *J. Am. Chem. Soc.* **2007**, *129*, 4386.

14. Dai, J. X.; Ambrus, A.; Hurley, L. H.; Yang, D. Z. *J. Am. Chem. Soc.* **2009**, *131*, 6102.

15. Jin, R. Z.; Gaffney, B. L.; Wang, C.; Jones, R. A.; Breslauer, K. J. *Proc. Natl. Acad. Sci. U. S. A.* **1992**, *89*, 8832.

16. Mergny, J. L.; Phan, A. T.; Lacroix, L. *FEBS Lett.* **1998**, *435*, 74.

17. Phan, A. T.; Mergny, J. L. *Nucleic Acids Res.* **2002**, *30*, 4618.

18. Parkinson, G. N.; Lee, M. P. H.; Neidle, S. *Nature* **2002**, *417*, 876.

19. He, F.; Tang, Y. L.; Wang, S.; Li, Y. L.; Zhu, D. B. *J. Am. Chem. Soc.* **2005**, *127*, 12343.

20. Mergny, J. L.; Maurizot, J. C. *ChemBioChem* **2001**, *2*, 124.

21. Nagatoishi, S.; Nojima, T.; Galezowska, E.; Juskowiak, B.; Takenaka, S. *ChemBioChem* **2006**, *7*, 1730.

22. Choi, J.; Kim, S.; Tachikawa, T.; Fujitsuka, M.; Majima, T. *J. Am. Chem. Soc.* **2011**, *133*, 16146.

23. Schaffitzel, C.; Berger, I.; Postberg, J.; Hanes, J.; Lipps, H. J.; Pluckthun, A. *Proc. Natl. Acad. Sci. U. S. A.* **2001**, *98*, 8572.

24. Callis, P. R. *Annu. Rev. Phys. Chem.* **1983**, *34*, 329.

25. Borresen, H. C. *Acta Chem. Scand.* **1963**, *17*, 921.

26. Wilhelmsson, L. M. *Q. Rev. Biophys.* **2010**, *43*, 159.

27. Srivatsan, S. G.; Weizman, H.; Tor, Y. *Org. Biomol. Chem.* **2008**, *6*, 1334.

28. Rist, M. J.; Marino, J. P. *Curr. Org. Chem.* **2002**, *6*, 775.

29. Greco, N. J.; Tor, Y. *Tetrahedron* **2007**, *63*, 3515.

30. Butler, R. S.; Cohn, P.; Tenzel, P.; Abboud, K. A.; Castellano, R. K. *J. Am. Chem. Soc.* **2009**, *131*, 623.

31. Hawkins, M. E.; Pfleiderer, W.; Mazumder, A.; Pommier, Y. G.; Balis, F. M. *Nucleic Acids Res.* **1995**, *23*, 2872.

32. Hawkins, M. E. *Fluoresc. Spectrosc.* **2008**, *450*, 201.

33. Kimura, T.; Kawai, K.; Fujitsuka, M.; Majima, T. *Tetrahedron* **2007**, *63*, 3585.

34. Gray, R. D.; Petraccone, L.; Trent, J. O.; Chaires, J. B. *Biochemistry* **2010**, *49*, 179.

35. Kirk, S. R.; Luedtke, N. W.; Tor, Y. *Bioorg. Med. Chem.* **2001**, *9*, 2295.

36. Wagenknecht, H. A. *Ann. N. Y. Acad. Sci.* **2008**, *1130*, 122.

37. Menacher, F.; Rubner, M.; Berndl, S.; Wagenknecht, H. A. *J. Org. Chem.* **2008**, *73*, 4263.

38. Gray, R. D.; Petraccone, L.; Buscaglia, R.; Chaires, J. B. G-Quadruplex DNA: Methods and Protocols, Methods in Molecular Biology, Springer, Vol. 608, **2010**, pp 121.

39. Shin, D.; Sinkeldam, R. W.; Tor, Y. *J. Am. Chem. Soc.* **2011**, *133*, 14912.

40. Dumas, A.; Luedtke, N. W. *J. Am. Chem. Soc.* **2010**, *132*, 18004.

41. Xu, D. G.; Nordlund, T. M. *Biophys. J.* **2000**, *78*, 1042.

42. Sen, D.; Gilbert, W. *Nature* **1988**, *334*, 364.

43. Sundquist, W. I.; Klug, A. *Nature* **1989**, *342*, 825.

44. Haran, T. E.; Mohanty, U. *Q. Rev. Biophys.* **2009**, *42*, 41.

45. Ward, D. C.; Reich, E.; Stryer, L. *J. Biol. Chem.* **1969**, *244*, 1228.

46. Nordlund, T. M.; Andersson, S.; Nilsson, L.; Rigler, R.; Graslund, A.; Mclaughlin, L. W. *Biochemistry* **1989**, *28*, 9095.

47. Jean, J. M.; Hall, K. B. *Proc. Natl. Acad. Sci. U. S. A.* **2001**, *98*, 37.

48. Fagan, P. A.; Fabrega, C.; Eritja, R.; Goodman, M. F.; Wemmer, D. E. *Biochemistry* **1996**, *35*, 4026.

49. Sowers, L. C.; Fazakerley, G. V.; Eritja, R.; Kaplan, B. E.; Goodman, M. F. *Proc. Natl. Acad. Sci. U. S. A.* **1986**, *83*, 5434.

50. Xu, D. G.; Evans, K. O.; Nordlund, T. M. *Biochemistry* **1994**, *33*, 9592.

51. Raney, K. D.; Sowers, L. C.; Millar, D. P.; Benkovic, S. J. *Proc. Natl. Acad. Sci. U. S. A.* **1994**, *91*, 6644.

52. Fedoriw, A. M.; Liu, H. Y.; Anderson, V. E.; deHaseth, P. L. *Biochemistry* **1998**, *37*, 11971.

53. Strainic, M. G.; Sullivan, J. J.; Velevis, A.; deHaseth, P. L. *Biochemistry* **1998**, *37*, 18074.

54. Sullivan, J. J.; Bjornson, K. P.; Sowers, L. C.; deHaseth, P. L. *Biochemistry* **1997**, *36*, 8005.

55. Tor, Y.; Tam, V. K.; Kwong, D. *J. Am. Chem. Soc.* **2007**, *129*, 3257.

56. Pilch, D. S.; Barbieri, C. M.; Kaul, M. *Tetrahedron* **2007**, *63*, 3567.

57. Shandrick, S.; Zhao, Q.; Han, Q.; Ayida, B. K.; Takahashi, M.; Winters, G. C.; Simonsen, K. B.; Vourloumis, D.; Hermann, T. *Angew. Chem. Int. Ed.* **2004**, *43*, 3177.

58. Kaul, M.; Barbieri, C. M.; Pilch, D. S. *J. Am. Chem. Soc.* **2004**, *126*, 3447.

59. Marino, J. P.; Zhao, C. *Tetrahedron* **2007**, *63*, 3575.

60. Johnson, N. P.; Baase, W. A.; von Hippel, P. H. *Proc. Natl. Acad. Sci. U. S. A.* **2004**, *101*, 3426.

61. Sowers, L. C.; Boulard, Y.; Fazakerley, G. V. *Biochemistry* **2000**, *39*, 7613.

62. Guest, C. R.; Hochstrasser, R. A.; Sowers, L. C.; Millar, D. P. *Biochemistry* **1991**, *30*, 3271.

63. Sagher, D.; Strauss, B. *Nucleic Acids Res.* **1985**, *13*, 4285.

64. Stivers, J. T. *Nucleic Acids Res.* **1998**, *26*, 3837.

65. Nakano, S.; Uotani, Y.; Uenishi, K.; Fujii, M.; Sugimoto, N. *Nucleic Acids Res.* **2005**, *33*, 7111.

66. Rachofsky, E. L.; Seibert, E.; Stivers, J. T.; Osman, R.; Ross, J. B. A. *Biochemistry* **2001**, *40*, 957.

67. Kelley, S. O.; Barton, J. K. *Science* **1999**, *283*, 375.

68. Nordlund, T. M. *Photochem. Photobiol.* **2007**, *83*, 625.

69. O'Neill, M. A.; Dohno, C.; Barton, J. K. *J. Am. Chem. Soc.* **2004**, *126*, 1316.

70. Kawai, M.; Lee, M. J.; Evans, K. O.; Nordlund, T. M. *J. Fluoresc.* **2001**, *11*, 23.

71. Davis, S. P.; Matsumura, M.; Williams, A.; Nordlund, T. M. *J. Fluoresc.* **2003**, *13*, 249.

72. Nordlund, T. M.; Xu, D. G.; Evans, K. O. *Biochemistry* **1993**, *32*, 12090.

73. Secrist, J. A.; Barrio, J. R.; Leonard, N. J. *Science* **1972**, *175*, 646.

74. Secrist, J. A.; Weber, G.; Leonard, N. J.; Barrio, J. R. *Biochemistry* **1972**, *11*, 3499.

75. Singleton, S. F.; Roca, A. I.; Lee, A. M.; Xiao, J. *Tetrahedron* **2007**, *63*, 3553.

76. Marti, A. A.; Jockusch, S.; Li, Z. M.; Ju, J. Y.; Turro, N. J. *Nucleic Acids Res.* **2006**, *34*, e50.

77. Valis, L.; Mayer-Enthart, E.; Wagenknecht, H. A. *Bioorg. Med. Chem. Lett.* **2006**, *16*, 3184.

78. Wagenknecht, H. A.; Wanninger-Weiss, C.; Valis, L. *Bioorg. Med. Chem.* **2008**, *16*, 100.

79. Saito, Y.; Miyauchi, Y.; Okamoto, A.; Saito, I. *Chem. Commun.* **2004**, 1704.

80. Engels, J. W.; Forster, U.; Lommel, K.; Sauter, D.; Grunewald, C.; Wachtveitl, J. *ChemBioChem* **2010**, *11*, 664.

81. Engels, J. W.; Grunwald, C.; Kwon, T.; Piton, N.; Forster, U.; Wachtveitl, J. *Bioorg. Med. Chem.* **2008**, *16*, 19.

82. Seo, Y. J.; Ryu, J. H.; Kim, B. H. *Org. Lett.* **2005**, *7*, 4931.

83. Seo, Y. J.; Lee, I. J.; Kim, B. H. *Mol. Biosyst.* **2009**, *5*, 235.

84. Seo, Y. J.; Kim, B. H. *Chem. Commun.* **2006**, 150.

85. Seo, Y. J.; Lee, I. J.; Yi, J. W.; Kim, B. H. *Chem. Commun.* **2007**, 2817.

86. Seo, Y. J.; Lee, I. J.; Kim, B. H. *Bioorg. Med. Chem. Lett.* **2008**, *18*, 3910.

87. Hwang, G. T.; Seo, Y. J.; Kim, B. H. *Tetrahedron Lett.* **2005**, *46*, 1475.

88. Seo, Y. J.; Hwang, G. T.; Kim, B. H. *Tetrahedron Lett.* **2006**, *47*, 4037.

89. Kim, B. H.; Seo, Y. J.; Rhee, H.; Joo, T. *J. Am. Chem. Soc.* **2007**, *129*, 5244.

90. Wagner, C.; Rist, M.; Mayer-Enthart, E.; Wagenknecht, H. A. *Org. Biomol. Chem.* **2005**, *3*, 2062.

91. Okamoto, A.; Ochi, Y.; Saito, I. *Chem. Commun.* **2005**, 1128.

92. Matsumoto, K.; Shinohara, Y.; Bag, S. S.; Takeuchi, Y.; Morii, T.; Saito, Y.; Saito, I. *Bioorg. Med. Chem. Lett.* **2009**, *19*, 6392.

93. Seio, K.; Mizuta, M.; Tasaki, K.; Tamaki, K.; Ohkubo, A.; Sekine, M. *Bioorg. Med. Chem.* **2008**, *16*, 8287.

94. Okamoto, A.; Kanatani, K.; Ochi, Y.; Saito, Y.; Saito, I. *Tetrahedron Lett.* **2004**, *45*, 6059.

95. Okamoto, A.; Tainaka, K.; Tanaka, K.; Ikeda, S.; Nishiza, K.; Unzai, T.; Fujiwara, Y.; Saito, I. *J. Am. Chem. Soc.* **2007**, *129*, 4776.

96. Kimura, T.; Kawai, K.; Majima, T. *Org. Lett.* **2005**, *7*, 5829.

97. Kimura, T.; Kawai, K.; Majima, T. *Chem. Commun.* **2006**, 1542.

98. Saito, Y.; Matsumoto, K.; Takeuchi, Y.; Bag, S. S.; Kodate, S.; Morii, T.; Saito, I. *Tetrahedron Lett.* **2009**, *50*, 1403.

99. Ogasawara, S.; Maeda, M. *Angew. Chem. Int. Ed.* **2008**, *47*, 8839.

100. Ogasawara, S.; Maeda, M. *Angew. Chem. Int. Ed.* **2009**, *48*, 6671.

101. Vrabel, M.; Horakova, P.; Pivonkova, H.; Kalachova, L.; Cernocka, H.; Cahova, H.; Pohl, R.; Sebest, P.; Havran, L.; Fojta, M.; Hocek, M. *Chem. - Eur. J.* **2009**, *15*, 1144.

102. Vrabel, M.; Pohl, R.; Votruba, I.; Sajadi, M.; Kovalenko, S. A.; Ernsting, N. P.; Hocek, M. *Org. Biomol. Chem.* **2008**, *6*, 2852.

103. Hawkins, M. E. *Cell Biochem. Biophys.* **2001**, *34*, 257.

104. Moser, A. M.; Patel, M.; Yoo, H.; Balis, F. M.; Hawkins, M. E. *Anal. Biochem.* **2000**, *281*, 216.

105. Wojtuszewski, K.; Hawkins, M. E.; Cole, J. L.; Mukerji, I. *Biochemistry* **2001**, *40*, 4892.

106. Hermann, T.; Parsons, J. *Tetrahedron* **2007**, *63*, 3548.

107. Sanabia, J. E.; Goldner, L. S.; Lacaze, P. A.; Hawkins, M. E. *J. Phys. Chem. B* **2004**, *108*, 15293.

108. Augustyn, K. E.; Wojtuszewski, K.; Hawkins, M. E.; Knutson, J. R.; Mukerji, I. *Biochemistry* **2006**, *45*, 5039.

109. Stanley, R. J.; Hou, Z. J.; Yang, A. P.; Hawkins, M. E. *J. Phys. Chem. B* **2005**, *109*, 3690.

110. Da Costa, C. P.; Fedor, M. J.; Scott, L. G. *J. Am. Chem. Soc.* **2007**, *129*, 3426.

111. Seela, F.; Jiang, D. W.; Xu, K. Y. *Org. Biomol. Chem.* **2009**, *7*, 3463.

112. Wierzchowski, J.; WielgusKutrowska, B.; Shugar, D. *Biochim. Biophys. Acta* **1996**, *1290*, 9.

113. Krueger, A. T.; Kool, E. T. *J. Am. Chem. Soc.* **2008**, *130*, 3989.

114. Krueger, A. T.; Lu, H. G.; Lee, A. H. F.; Kool, E. T. *Acc. Chem. Res.* **2007**, *40*, 141.

115. Delaney, J. C.; Gao, J. M.; Liu, H. B.; Shrivastav, N.; Essigmann, J. M.; Kool, E. T. *Angew. Chem. Int. Ed.* **2009**, *48*, 4524.

116. Okamoto, A.; Tanaka, K.; Fukuta, T.; Saito, I. *J. Am. Chem. Soc.* **2003**, *125*, 9296.

117. Okamoto, A.; Tanaka, K.; Fukuta, T.; Saito, I. *ChemBioChem* **2004**, *5*, 958.

118. Mitsui, T.; Kimoto, M.; Kawai, R.; Yokoyama, S.; Hirao, I. *Tetrahedron* **2007**, *63*, 3528.

119. Kawai, R.; Kimoto, M.; Ikeda, S.; Mitsui, T.; Endo, M.; Yokoyama, S.; Hirao, L. *J. Am. Chem. Soc.* **2005**, *127*, 17286.

120. Hirao, I.; Hikida, Y.; Kimoto, M.; Yokoyama, S. *Nat. Protoc.* **2010**, *5*, 1312.

121. Hirao, I.; Kimoto, M.; Mitsui, T.; Harada, Y.; Sato, A.; Yokoyama, S. *Nucleic Acids Res.* **2007**, *35*, 5360.

122. Zhao, Y.; Knee, J. L.; Baranger, A. M. *Bioorg. Chem.* **2008**, *36*, 271.

123. Kenfack, C. A.; Piemont, E.; Ben Gaied, N.; Burger, A.; Mely, Y. *J. Phys. Chem. B* **2008**, *112*, 9736.

124. Nadler, A.; Strohmeier, J.; Diederichsen, U. *Angew. Chem. Int. Ed.* **2011**, *50*, 5392.

125. Dierckx, A.; Diner, P.; El-Sagheer, A. H.; Kumar, J. D.; Brown, T.; Grotli, M.; Wilhelmsson, L. M. *Nucleic Acids Res.* **2011**, *39*, 4513.

126. Sun, K. M.; McLaughlin, C. K.; Lantero, D. R.; Manderville, R. A. *J. Am. Chem. Soc.* **2007**, *129*, 1894.

127. Schlitt, K. M.; Millen, A. L.; Wetmore, S. D.; Manderville, R. A. *Org. Biomol. Chem.* **2011**, *9*, 1565.

128. Dumas, A.; Luedtke, N. W. *Chem. - Eur. J.* **2012**, *18*, 245.

129. Shinohara, Y.; Matsumoto, K.; Kugenuma, K.; Morii, T.; Saito, Y.; Saito, I. *Bioorg. Med. Chem. Lett.* **2010**, *20*, 2817.

130. Saito, Y.; Matsumoto, K.; Takahashi, N.; Suzuki, A.; Morii, T.; Saito, I. *Bioorg. Med. Chem. Lett.* **2011**, *21*, 1275.

131. Dumas, A.; Luedtke, N. W. *Nucleic Acids Res.* **2011**, *39*, 6825.

132. Ogasawara, S.; Saito, I.; Maeda, M. *Tetrahedron Lett.* **2008**, *49*, 2479.

133. Saito, Y.; Koda, M.; Shinohara, Y.; Saito, I. *Tetrahedron Lett.* **2011**, *52*, 491.

134. Seidel, C. A. M.; Schulz, A.; Sauer, M. H. M. *J. Phys. Chem.* **1996**, *100*, 5541.

135. Driscoll, S. L.; Hawkins, M. E.; Balis, F. M.; Pfleiderer, W.; Laws, W. R. *Biophys. J.* **1997**, *73*, 3277.

136. Hawkins, M. E.; Pfleiderer, W.; Jungmann, O.; Balis, F. M. *Anal. Biochem.* **2001**, *298*, 231.

137. Hawkins, M. E.; Pfleiderer, W.; Balis, F. M.; Porter, D.; Knutson, J. R. *Anal. Biochem.* **1997**, *244*, 86.

138. Seela, F.; Chen, Y. M. *Nucleic Acids Res.* **1995**, *23*, 2499.

139. Wierzchowski, J.; Ogiela, M.; Iwanska, B.; Shugar, D. *Anal. Chim. Acta* **2002**, *472*, 63.

140. Seela, F.; Lampe, S. *Helv. Chim. Acta* **1994**, *77*, 1003.

141. Liu, H. B.; Gao, J. M.; Kool, E. T. *J. Org. Chem.* **2005**, *70*, 639.

142. Liu, H. B.; Gao, J. M.; Kool, E. T. *J. Am. Chem. Soc.* **2005**, *127*, 1396.

143. Hernandez, A. R.; Kool, E. T. *Org. Lett.* **2011**, *13*, 676.

144. Liu, H. B.; Gao, J. M.; Maynard, L.; Saito, Y. D.; Kool, E. T. *J. Am. Chem. Soc.* **2004**, *126*, 1102.

145. Gaied, N. B.; Glasser, N.; Ramalanjaona, N.; Beltz, H.; Wolff, P.; Marquet, R.; Burger, A.; Mely, Y. *Nucleic Acids Res.* **2005**, *33*, 1031.

146. Dyrager, C.; Borjesson, K.; Diner, P.; Elf, A.; Albinsson, B.; Wilhelmsson, L. M.; Grotli, M. *Eur. J. Org. Chem.* **2009**, 1515.

147. Dumas, A.; Luedtke, N. W. *ChemBioChem* **2011**, *12*, 2044.

148. Firth, A. G.; Fairlamb, I. J. S.; Darley, K.; Baumann, C. G. *Tetrahedron Lett.* **2006**, *47*, 3529.

149. Birnbaum, G. I.; Lassota, P.; Shugar, D. *Biochemistry* **1984**, *23*, 5048.

150. Dias, E.; Battiste, J. L.; Williamson, J. R. *J. Am. Chem. Soc.* **1994**, *116*, 4479.

151. Xu, Y.; Sugiyama, H. *Nucleic Acids Res.* **2006**, *34*, 949.

152. Uesugi, S.; Ikehara, M. *J. Am. Chem. Soc.* **1977**, *99*, 3250.

153. Stolarski, R.; Hagberg, C. E.; Shugar, D. *Eur. J. Biochem.* **1984**, *138*, 187.

154. He, G. X.; Krawczyk, S. H.; Swaminathan, S.; Shea, R. G.; Dougherty, J. P.; Terhorst, T.; Law, V. S.; Griffin, L. C.; Coutre, S.; Bischofberger, N. *J. Med. Chem.* **1998**, *41*, 2234.

155. Reichardt, C. *Chem. Rev.* **1994**, *94*, 2319.

156. Sinkeldam, R. W.; Tor, Y. *Org. Biomol. Chem.* **2007**, *5*, 2523.

157. Lakowicz, J. R. Principles of Fluorescence Spectroscopy; Second ed. Kluwer Academic/Plenum Publishers: New York, **1999**.

158. Wypych, G. Handbook of Solvents. ChemTec Publishing, **2001**.

159. Huppert, J. L.; Balasubramanian, S. *Nucleic Acids Res.* **2007**, *35*, 406.

160. Siddiqui-Jain, A.; Grand, C. L.; Bearss, D. J.; Hurley, L. H. *Proc. Natl. Acad. Sci. U. S. A.* **2002**, *99*, 11593.

161. Lane, A. N.; Chaires, J. B.; Gray, R. D.; Trent, J. O. *Nucleic Acids Res.* **2008**, *36*, 5482.

162. Davis, J. T. *Angew. Chem. Int. Ed.* **2004**, *43*, 668.

163. Neidle, S.; Parkinson, G. N. *Biochimie* **2008**, *90*, 1184.

164. Maizels, N. *Nat. Struct. Mol. Biol.* **2006**, *13*, 1055.

165. Oganesian, L.; Bryan, T. M. *Bioessays* **2007**, *29*, 155.

166. Hershman, S. G.; Chen, Q.; Lee, J. Y.; Kozak, M. L.; Yue, P.; Wang, L. S.; Johnson, F. B. *Nucleic Acids Res.* **2008**, *36*, 144.

167. Luedtke, N. W. *Chimia* **2009**, *63*, 134.

168. Sessler, J. L.; Sathiosatham, M.; Doerr, K.; Lynch, V.; Abboud, K. A. *Angew. Chem. Int. Ed.* **2000**, *39*, 1300.

169. Calzolari, A.; Di Felice, R.; Molinari, E.; Garbesi, A. *Appl. Phys. Lett.* **2002**, *80*, 3331.

170. Davis, J. T.; Spada, G. P. *Chem. Soc. Rev.* **2007**, *36*, 296.

171. Karimata, H.; Miyoshi, D.; Fujimoto, T.; Koumoto, K.; Wang, Z. M.; Sugimoto, N. *Nucleic Acids Symp. Ser.* **2007**, 251.

172. Borovok, N.; Iram, N.; Zikich, D.; Ghabboun, J.; Livshits, G. I.; Porath, D.; Kotlyar, A. B. *Nucleic Acids Res.* **2008**, *36*, 5050.

173. Betancourt, J. E.; Rivera, J. M. *J. Am. Chem. Soc.* **2009**, *131*, 16666.

174. Rivera-Sanchez Mdel, C.; Andujar-de-Sanctis, I.; Garcia-Arriaga, M.; Gubala, V.; Hobley, G.; Rivera, J. M. *J. Am. Chem. Soc.* **2009**, *131*, 10403.

175. Delaney, S.; Barton, J. K. *Biochemistry* **2003**, *42*, 14159.

176. Chinnapen, D. J. F.; Sen, D. *Proc. Natl. Acad. Sci. U. S. A.* **2004**, *101*, 65.

177. Huang, Y. C.; Cheng, A. K.; Yu, H. Z.; Sen, D. *Biochemistry* **2009**, *48*, 6794.

178. Alberti, P.; Bourdoncle, A.; Sacca, B.; Lacroix, L.; Mergny, J. L. *Org. Biomol. Chem.* **2006**, *4*, 3383.

179. Alberti, P.; Mergny, J. L. *Proc. Natl. Acad. Sci. U. S. A.* **2003**, *100*, 1569.

180. Lubitz, I.; Borovok, N.; Kotlyar, A. *Biochemistry* **2007**, *46*, 12925.

181. Eddy, J.; Maizels, N. *Nucleic Acids Res.* **2006**, *34*, 3887.

182. Hurley, L. H. *Nat. Rev. Cancer* **2002**, *2*, 188.

183. Mergny, J. L.; Riou, J. F.; Mailliet, P.; Teulade-Fichou, M. P.; Gilson, E. *Nucleic Acids Res.* **2002**, *30*, 839.

184. Neidle, S.; Parkinson, G. *Nat. Rev. Drug Discovery* **2002**, *1*, 383.

185. Pennarun, G.; Granotier, C.; Gauthier, L. R.; Gomez, D.; Boussin, F. D. *Oncogene* **2005**, *24*, 2917.

186. Sandin, P.; Wilhelmsson, L. M.; Lincoln, P.; Powers, V. E. C.; Brown, T.; Albinsson, B. *Nucleic Acids Res.* **2005**, *33*, 5019.

187. Zhao, Y.; Baranger, A. M. *J. Am. Chem. Soc.* **2003**, *125*, 2480.

188. Rai, P.; Cole, T. D.; Thompson, E.; Millar, D. P.; Linn, S. *Nucleic Acids Res.* **2003**, *31*, 2323.

189. O'Neill, M. A.; Barton, J. K. *Proc. Natl. Acad. Sci. U. S. A.* **2002**, *99*, 16543.

190. Markovitsi, D.; Gustavsson, T.; Sharonov, A. *Photochem. Photobiol.* **2004**, *79*, 526.

191. Gepshtein, R.; Huppert, D.; Lubitz, I.; Amdursky, N.; Kotlyar, A. B. *J. Phys. Chem. C* **2008**, *112*, 12249.

192. Miannay, F. A.; Banyasz, A.; Gustavsson, T.; Markovitsi, D. *J. Phys. Chem. C* **2009**, *113*, 11760.

193. Mendez, M. A.; Szalai, V. A. *Biopolymers* **2009**, *91*, 841.

194. Miyoshi, D.; Nakao, A.; Sugimoto, N. *Biochemistry* **2002**, *41*, 15017.

195. Kan, Z. Y.; Yao, Y. A.; Wang, P.; Li, X. H.; Hao, Y. H.; Tan, Z. *Angew. Chem. Int. Ed.* **2006**, *45*, 1629.

196. Nagatoishi, S.; Tanaka, Y.; Tsumoto, K. *Biochem. Biophys. Res. Commun.* **2007**, *354*, 837.

197. Zhou, J.; Wei, C. Y.; Jia, G. Q.; Wang, X. L.; Tang, Q.; Feng, Z. C.; Li, C. *Biophys. Chem.* **2008**, *136*, 124.

198. Doudna, J. A.; Cech, T. R. *Nature* **2002**, *418*, 222.

199. Muller, J. *Metallomics* **2010**, *2*, 318.

200. Schnabl, J.; Sigel, R. K. O. *Curr. Opin. Chem. Biol.* **2010**, *14*, 269.

201. Hud, N. Nucleic Acid-Metal Ion Interactions; RSC Biomolecular Sciences, Springer, **2008**.

202. Fournier, P.; Fiammengo, R.; Jaschke, A. *Angew. Chem. Int. Ed.* **2009**, *48*, 4426.

203. Feringa, B. L.; Boersma, A. J.; Megens, R. P.; Roelfes, G. *Chem. Soc. Rev.* **2010**, *39*, 2083.

204. Eastman, A. *Biochemistry* **1985**, *24*, 5027.

205. Waalkes, M. P.; Poirier, L. A. *Toxicol. Appl. Pharmacol.* **1984**, *75*, 539.

206. Kasprzak, K. S.; Waalkes, M. P.; Poirier, L. A. *Toxicol. Appl. Pharmacol.* **1986**, *82*, 336.

207. Sagripanti, J. L.; Goering, P. L.; Lamanna, A. *Toxicol. Appl. Pharmacol.* **1991**, *110*, 477.

208. Zhang, L. Z.; Cheng, P. *J. Inorg. Biochem.* **2004**, *98*, 569.

209. Gelagutashvili, E. *J. Therm. Anal. Calorim.* **2006**, *85*, 491.

210. Redon, S.; Bombard, S.; Elizondo-Riojas, M. A.; Chottard, J. C. *Biochemistry* **2001**, *40*, 8463.

211. Murphy, C. J.; Arkin, M. R.; Jenkins, Y.; Ghatlia, N. D.; Bossmann, S. H.; Turro, N. J.; Barton, J. K. *Science* **1993**, *262*, 1025.

212. Meade, T. J.; Kayyem, J. F. *Angew. Chem. Int. Ed. Engl.* **1995**, *34*, 352.

213. Dandliker, P. J.; Holmlin, R. E.; Barton, J. K. *Science* **1997**, *275*, 1465.

214. Shao, F. W.; Barton, J. K. *J. Am. Chem. Soc.* **2007**, *129*, 14733.

215. Weizman, H.; Tor, Y. *J. Am. Chem. Soc.* **2002**, *124*, 1568.

216. Hurley, D. J.; Tor, Y. *J. Am. Chem. Soc.* **2002**, *124*, 13231.

217. Srivatsan, S. G.; Tor, Y. *J. Am. Chem. Soc.* **2007**, *129*, 2044.

218. Holzberger, B; Strohmeier, J; Siegmund, V; Diederichsen, U.; Marx, A. *Bioorg. Med. Chem. Lett.* **2012**, *22*, 3136.

219. Salic, A.; Mitchison, T. J. *Proc. Natl. Acad. Sci. U. S. A.* **2008**, *105*, 2415.

220. Furman, P. A.; Fyfe, J. A.; Stclair, M. H.; Weinhold, K.; Rideout, J. L.; Freeman, G. A.; Lehrman, S. N.; Bolognesi, D. P.; Broder, S.; Mitsuya, H.; Barry, D. W. *Proc. Natl. Acad. Sci. U. S. A.* **1986**, *83*, 8333.

221. Poijarvi-Virta, P.; Lonnberg, H. *Curr. Med. Chem.* **2006**, *13*, 3441.

222. Sandin, P.; Stengel, G.; Ljungdahl, T.; Borjesson, K.; Macao, B.; Wilhelmsson, L. M. *Nucleic Acids Res.* **2009**, *37*, 3924.

223. Neef, A. B.; Samain, F.; Luedtke, N. W. *ChemBioChem* **2012**, *13*, 1750.

12

DESIGN AND PHOTOPHYSICS OF ENVIRONMENTALLY SENSITIVE ISOMORPHIC FLUORESCENT NUCLEOSIDES

RENATUS W. SINKELDAM AND YITZHAK TOR

Department of Chemistry and Biochemistry, University of California, San Diego CA 92093, USA

12.1 INTRODUCTION

Researchers have long recognized the utility of fluorescence spectroscopy for studying biomolecules (e.g., proteins, membranes, and nucleic acids). The sensitivity, relative ease of use, wide range of possible experiments (steady state, time-resolved, anisotropy, FRET, etc.), and continuous instrumental development (increased sensitivity, multiphoton, automation and high-throughput, etc.) make fluorescence spectroscopy one of the most valuable tools for the study of biomolecules and their often complex interactions with their microenvironment.[1–3] In contrast to proteins, nucleic acids lack inherently emissive building blocks. The canonical native nucleosides are characterized by very low fluorescence quantum yields and short excited state lifetimes (Fig. 12.1a).[4–8] To overcome this hurdle and to exploit opportunities offered by fluorescence spectroscopy, chemists have developed a plethora of synthetic emissive nucleoside analogs.[1,9,10]

When fluorescent nucleoside analogs are to be used as probes (molecules that can report on their local environment), as opposed to labels or tags (molecules able to reveal their presence, and thereby the presence of the molecules they are bound to), their shape and dimensions should closely mimic the size and shape of native

Fluorescent Analogs of Biomolecular Building Blocks: Design and Applications, First Edition.
Edited by Marcus Wilhelmsson and Yitzhak Tor.
© 2016 John Wiley & Sons, Inc. Published 2016 by John Wiley & Sons, Inc.

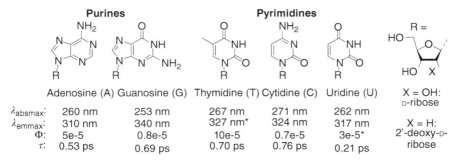

Figure 12.1 The canonical nonemissive biomolecular building blocks of nucleic acids and their selected photophysical parameters. *Values for the corresponding monophosphate. N.B. various sources reported slightly different values.

nucleosides to minimize probe-induced perturbations of base pairing and deformation of higher ordered structures. Such minimally perturbing fluorescent probes are classified as isomorphic fluorescent nucleosides.[1,11] In addition to this demanding structural requirement, their photophysical characteristics should ideally include (1) an exclusively excitable absorption band >300 nm, (2) sufficient quantum yield, and (3) a photophysically expressed sensitivity to changes in the microenvironment.

Arguably one of the most studied and used isomorphic fluorescent nucleosides is 2-aminopurine (**1**) (Fig. 12.2). The seminal 1969 paper by Stryer marks the inception of a lively research field, where numerous emissive nucleosides have been designed, synthesized, and implemented.[10a] Diverse isomorphic motifs include, in addition to 2-aminopurine (**1**)[10a], 8-vinyl-2′-deoxyadenosine (**2**),[12,13] 8-furan-2′-deoxyguanosine (**3**),[14] 8-aza-guanosine (**4**),[15–17] and thienoadenosine (**5**). Among the pyrimidines, selected examples include 5-(thiophene-2-yl)-uridine (**6**),[11] 5-(furan-2-yl)-cytidine (**7**),[18] 5-thiophene-2-yl)-6-azauridine (**8**),[19] quinazolines

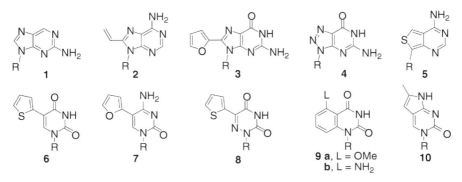

Figure 12.2 Selected examples of isomorphic fluorescent nucleoside analogs. R is either D-ribose, or 2′-deoxy-D-ribose.

(**9ab**),[20,21] as well as pyrrolo-C and its analogs (**10**),[22–24] (Fig. 12.2). Modifications typically involve extension of the native π-system in the 5-postion of the pyrimidines (**6**, **7**, and **8**) or 8-position of the purines (**2** and **3**). Alternatively, an additional (heterocyclic) aromatic ring is fused to the 5 and 6 (**9**) or 4 and 5 (**10**) positions of the pyrimidines. Yet another strategy is selective C-to-N substitution (**4** and **8**). Such modifications have proved to be structurally rather benign while endowing the derivatized nucleosides with desirable fluorescent properties.

In contrast to the development of nonperturbing probes for proteins and membrane components, the design of isomorphic fluorescent analogs is particularly challenging. Modified amino acids can frequently replace native residues without affecting protein folding and function. Similarly, decorating known fluorophores with lipophilic moieties will promote indiscriminate integration into lipophilic membranes when in aqueous media. Nucleic acid stability and function, however, are exceedingly sensitive to modification of the building blocks. Hence, developers of isomorphic fluorescent nucleosides have to navigate a tight structural landscape. To illustrate these constraints, two sets of native base pairs, essential for proper helix formation, are shown alongside two fully modified base pairs (Fig. 12.3(a) and (b)). The models demonstrate the limited opportunity for modification without seriously disrupting the hydrogen bonding interaction of a complimentary base pair as well as the projection of functional groups into the major and minor groove, which could easily impair intermolecular ligand recognition and binding. Furthermore, altering the electronic nature of the nucleobases, while enhancing their photophysical characteristics, could have severe detrimental effects on the chemical and photochemical stability of the resulting nucleoside and nucleotides, adding to the challenges associated with the development of isomorphic fluorescent nucleoside analogs.

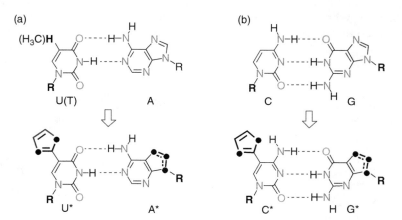

Figure 12.3 Watson–Crick base pairs of (a) native U(T)–A, and modified U*–A* and (b) native C–G, and modified C*–G*. The perfect duplex pairs show the limited opportunities for the introduction of fluorescent properties by structural modification. The solid circles indicate positions suitable for C to hetero atom, or N to C substitution.

When the isomorphic structural requirements are reconciled with an exclusively excitable absorption > 300 nm, and sufficient quantum yield, it is the sensitivity to environmental changes that determines the probes usefulness for biophysical applications. This key property, in the context of isomorphic fluorescent nucleoside analogs, is the focus of this chapter.

12.2 DESIGNING ENVIRONMENTALLY SENSITIVE EMISSIVE NUCLEOSIDES

12.2.1 Structural and Electronic Elements that Impart Environmental Sensitivity

The relationship between the molecular structure of a chromophore and its photophysical properties has been of interest since the advent of organic chemistry and spectroscopy. Despite decades of accumulated knowledge, this relation is not always apparent and predicting the magnitude of parameters associated with a molecule's excited state (e.g., excited state lifetime, quantum yield) is challenging. Fortunately, it is reasonably well understood which structural and electronic elements endow chromophores/fluorophores with specific susceptibility to environmental effects. Hence, the introduction of such motifs is widely utilized for the development of fluorescent nucleoside analogs. Indeed, designers of new sensitive probes can find ample inspiration in the great abundance of scientific literature on this topic.[1–3]

Environmental sensitivity is rooted in basic factors including polarity, viscosity (molecular crowding), acidity, and ionic strength. Carefully selected structural and electronic alterations of chromophores are known to induce a specific photophysical response (Table 12.1).

12.2.2 Sensitivity to Polarity

Environmental polarity impacts virtually all biomolecules.[1] In nucleic acids, changes in local polarity are associated with folding/unfolding, ligand binding, and lesion formation since such events can shield a specific region from or expose it to the surrounding medium. Hence, nucleic acid probes, which are sensitive to environmental polarity, have proved to be useful tools.[1] To endow a probe with sensitivity to polarity, a π-conjugated system is typically decorated with appropriately positioned push–pull functional groups (Table 12.1). The enhanced charge transfer and polarizability renders the fluorophore sensitive to the polarity of its environment. This is reflected by shifts in its absorption maximum, emission maximum, or both, and depends on differences in charge distribution between the ground and excited states as is illustrated with HOMO and LUMO calculations for prodan (Fig. 12.4), a well-known responsive fluorescent probe.

A common expression of polarity is the dielectric constant or relative permittivity. It represents a molecule's ability to attenuate an electric field generated between macroscopically distant electrodes relative to vacuum. Polar solvents such as water

TABLE 12.1 Structural Elements that Introduce Environmental Sensitivity

Sensitivity to:	Structural/Electronic Element
Polarity	Push–pull chromophore
Viscosity	Molecular rotor (Nonviscous / Viscous)
pH	(de)protonatable site part of π-system

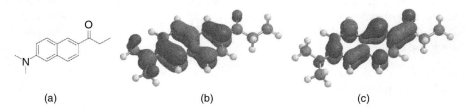

(a)   (b)   (c)

Figure 12.4 (a) Structure of prodan and its (b) HOMO and (c) LUMO surfaces. Structure has been modeled using the semiempirical AM1 method in Spartan'08.

($\varepsilon = 80.2$) and low MW alcohols (e.g., MeOH, $\varepsilon = 33.5$) are described by high ε values, whereas apolar alkanes (e.g., hexane, $\varepsilon = 1.9$) are described by low values relative to vacuum ($\varepsilon = 1$ by definition). The dielectric constant faithfully represents the ability of solvent molecules to reorganize, making it a useful bulk polarity parameter for pure solvents that can be viewed as a dielectric "continuum". Once a solute is introduced, however, dielectric constants fail to properly represent the first or second solvation spheres, or the environment of a (bio)molecular cavity. This becomes clear in the suboptimal correlation of the photophysical response of a fluorophore to environmental (i.e., solvent) polarity expressed as a dielectric constant.[25] In contrast, a microenvironmental polarity parameter derived from solvatochromic dyes, such as the $E_T(30)$ scale developed by Dimroth and Reichardt,[26–28] has proved to be very

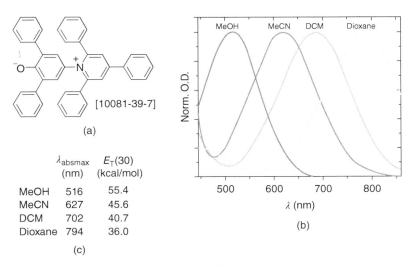

	λ_{absmax} (nm)	$E_T(30)$ (kcal/mol)
MeOH	516	55.4
MeCN	627	45.6
DCM	702	40.7
Dioxane	794	36.0

(c)

Figure 12.5 (a) Structure of Reichardt's dye, (b) its long-wavelength absorption spectra in solvents of various polarity, and (c) the conversion of absorption maxima to $E_T(30)$ values.

successful in describing a fluorophore's photophysical response (e.g., Stokes shift) toward solvent polarity changes.[25] In addition, while dielectric constants can be easily found for pure solvents, reliable values do not necessarily exist for diverse solvent mixtures, typically used to experimentally control sample polarity. The $E_T(30)$ value of any solvent or solvent mixture (as long as they are not highly acidic), however, can be determined easily by dissolving a minute amount of commercially available Reichardt's dye and by recording its absorption spectrum (Fig. 12.5a and b). Subsequently, the long-wavelength absorption maximum can be converted into the corresponding $E_T(30)$ value (Fig. 12.5c).

12.2.3 Sensitivity to Viscosity

Sensitivity to viscosity or molecular crowding is achieved by strategically incorporating a molecular rotor element, a single bond linking two π-systems equipped with donor and acceptor moieties (Table 12.1). Upon excitation in a nonviscous environment, molecular rotors can adopt a twisted excited state followed by rotational relaxation to return to the ground state (Fig. 12.6a).[3,29] This dominant radiationless decay mechanism has a profound limiting effect on the fluorescence quantum yield of such fluorophores. Exposing a molecular rotor to an environment that limits free rotation leads to rigidification of the chromophore[30,31] The molecular crowding of a medium, for example, a highly viscous solvent, hampers rotation thereby limiting population of a twisted excited state. By attenuating the contribution of the radiationless rotational relaxation pathway, the fluorescence quantum yield is typically enhanced. A well-studied example of such a viscosity probe is dicyanovinyl julolidine (**DCVJ**), shown in Figure 12.6b.[29,32-34] This molecular rotor molecule has clearly identifiable push–pull elements and a single bond linking two π-systems.

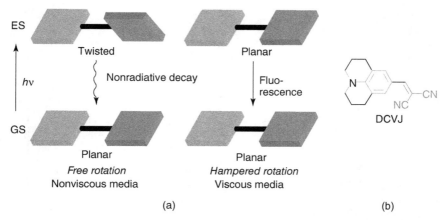

Figure 12.6 (a) Schematic representation explaining the viscosity sensitive fluorescence quantum yield of probes comprised of a molecular rotor element, (b) structure of known viscosity probe DCVJ. Push-pull elements and the single bond link are colored in blue, red, and grey, respectively. (*See color plate section for the color representation of this figure.*)

12.2.4 Sensitivity to pH

To impose sensitivity to pH changes, a fluorophore can be furnished with a (de)protonatable site conjugated to a π-system (Table 12.1). Protonation or deprotonation has a marked electronic effect but can also provoke a structural change if it concerns formation or breaking of an intramolecular hydrogen bond. In certain cases, such protonation/deprotonation events can be coupled to additional excited state processes, such as proton transfer, which could have significant effect on the photophysical parameters.

We also note that it is not uncommon to observe sensitivity to more than one environmental factor within a given fluorophore, as the distinct motifs outlined above, depending on their specific molecular structure, could be susceptible to multiple environmental perturbations, resulting in a cumulative response.[35,36]

12.3 TWO ISOMORPHIC ENVIRONMENTALLY SENSITIVE DESIGNS

Examples of isomorphic fluorescent nucleosides that incorporate the abovementioned structural elements are 5-(fur-2-yl)-2′-deoxy uridine (**11**), and 5-(pyr-2-yl)-uridine (**12**) (Fig. 12.7), which can be viewed as emissive mimics or surrogates of T. In structure **11**, high polarizability is recognizable in the conjugation of the electron-rich furan moiety to the electron-deficient pyrimidine core, elements prone to introduce sensitivity to polarity. The single bond linking the two π-systems resembles a molecular rotor element and should provide the probe with sensitivity to viscosity. The last two elements can also be identified in nucleoside **12**, and, thus, sensitivity toward polarity and viscosity is expected. In addition, the basic pyridine nitrogen in **12** also

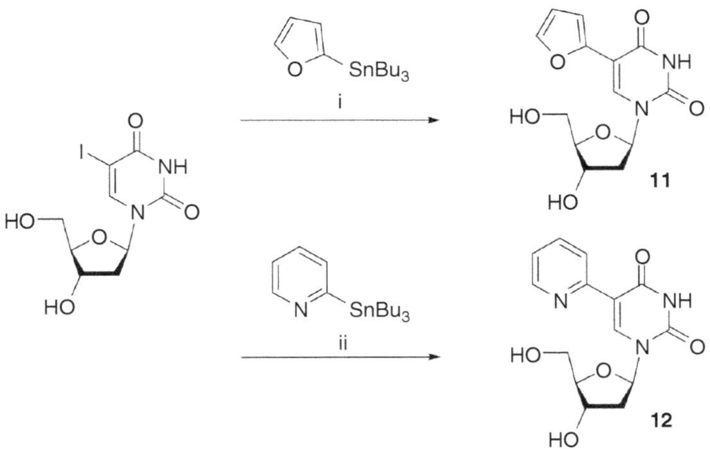

Figure 12.7   Synthesis of isomorphic fluorescent nucleosides **11** and **12**. Reagents and conditions: (i) PdCl$_2$(PPh$_3$)$_2$ (5 mol%), dioxane, 90 °C, 2 h, 94%; (ii) PdCl$_2$(PPh$_3$)$_2$ (5 mol%), dioxane, reflux, 2 h, 74%.

renders this probe sensitive to the acidity of its environment. Both **11** and **12** can be synthesized in a single high-yielding palladium-mediated Stille coupling from commercially available starting materials (Fig. 12.7).

Comparison of the crystal structures of the furan-modified **11**[14] and native T[37] shows retention of the anticonfiguration and almost identical sugar puckering (Fig. 12.8a). An overlay of the crystal structures, linking the atoms that make up the skeleton (i.e., the ribose unit and pyrimidine core), illustrates their structural similarity, which is corroborated by an RMS value of 0.0797 (Fig. 12.8b). To highlight their isomorphic character, crystal structure of the furan-modified **11** is docked onto the crystal structure of a 10-mer duplex DNA and compared to a native duplex (Fig. 12.8c). The models, with the methyl group of T and the furan moiety of **11** highlighted in light gray, reveal that the spacious major groove easily accommodates the furan modification as it is only slightly more space demanding than the methyl group of T. The nonperturbing nature of **11** was also established by comparative thermal denaturation studies.[11] Pyrimidines modified in the 5 position with a single five- or six-membered heterocycle are therefore minimally perturbing nucleoside analogs with, as is shown below, useful fluorescence properties virtually absent in the canonical native counterparts.

12.4   PROBING ENVIRONMENTAL SENSITIVITY

12.4.1   Probing Sensitivity to Polarity

The influence of polarity on the photophysical parameters of **11** is studied in binary mixtures of an apolar and polar solvent. The large difference in $E_T(30)$

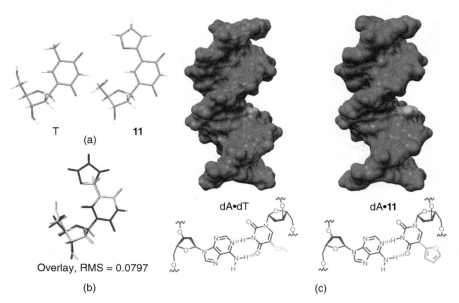

Figure 12.8 (a) Crystal structures for T (CCDC#: THYDIN04) and **11** (CCDC#: HEZWEP); (b) overlay of THYDIN04 (orange) and HEZWEP (blue); (c) models of DNA duplexes (blue) showing the location and surface area (orange) of the furan moiety of **11** (right) in comparison to the methyl group of T (left). Structural overlay of the crystal structures was performed in Mercury. (*See color plate section for the color representation of this figure.*)

values[28] and their miscibility in all ratios make dioxane ($E_T(30) = 36.0$ kcal/mol) and water ($E_T(30) = 63.1$ kcal/mol) ideal to control sample polarity.[25] Absorption and fluorescence spectra of samples of **11** in dioxane, water, and mixtures thereof show that the absorption parameters ($\lambda_{max} = 316$ nm, $\varepsilon = 9000$–11000 M^{-1} cm^{-1}) are practically unresponsive to changes in polarity (Fig. 12.9a). In contrast, the excited state appears heavily dependent on polarity as is reflected by changes in the emission profile. Upon going from dioxane to water, **11** expresses its sensitivity by a changing fluorescent intensity and, more importantly, a significant shift in emission maximum from 410 to 445 nm. The latter, in conjunction with an unchanging absorption maximum, makes the Stokes shift ($\nu_{abs} - \nu_{em}$) exquisitely sensitive to polarity. Extending the correlation to less polar media by taking the spectra in mixtures of methylcyclohexane ($E_T(30) = 32.2$ kcal/mol) and isopropanol ($E_T(30) = 48.4$ kcal/mol) required acetylation of the 3′ and 5′ hydroxyls in **11**, and the use of a lipophilic solvatochromic chromophore, a *tert*-butylated version of Reichardt's dye (Fig. 12.9a).[27] Stokes shift values were calculated for all solvent mixtures and plotted as a function of sample polarity revealing a linear correlation (Fig. 12.9b).[38] Importantly, the relationship, as expressed by the linearization shown, constitutes the dioxane–water and methylcyclohexane–isopropanol samples. This indicates that the microenvironmental $E_T(30)$ polarity values allow one to correlate

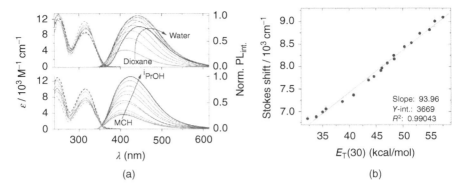

(a)   (b)

Figure 12.9 Sensitivity of **11** toward environmental polarity illustrated with (a) absorption (dashed lines) and emission (solid lines) spectra in samples of dioxane (red), water (blue), and mixtures thereof (black) water; and (b) a plot of the Stokes shift versus sample polarity (black circles) with a linear fit (orange line). (*See color plate section for the color representation of this figure.*)

Stokes shift changes in **11** regardless of the specific solvents used to control sample polarity.

The relationship between Stokes shift and environmental polarity serves as a polarity reference scale (Fig. 12.9b).[38] After incorporation of **11** into single-stranded DNA and various DNA duplexes, determination of its absorption and emission maxima now allows for estimation of the polarity of its local environment within folded oligonucleotides. To this end, **11** was incorporated in the central position of a 13-mer. Subsequently, the absorption and emission profile of the single strand and various duplexes were measured, the Stokes shifts calculated, and the microenvironmental polarity estimated by interpolation of the polarity reference scale (Fig. 12.9b).[38] The polarity of the major groove of B-DNA was estimated to have an $E_T(30)$ value of 46.2 kcal/mol, a value similar to the polarity of nitromethane $(E_T(30) = 46.3\,\text{kcal/mol})$ and pentanol $(E_T(30) = 46.5\,\text{kcal/mol})$. Importantly, this suggests the wall of the major groove to be much less polar than the bulk water $(E_T(30) = 63.1\,\text{kcal/mol})$ surrounding the duplex.[38]

This intriguing observation would benefit from placing it in historical perspective. After the influential paper by Breslauer and coworkers on estimating the polarity of the minor groove in B-DNA by using Hoechst 33258, an established minor groove binder,[39] others followed using native nucleobases tethered to known polarity probes (e.g., dansyl and DAN).[40–46] Their reported values for the polarity of the major groove, expressed as a dielectric constant, ranged from 55 to 70 (Table 12.2).[41,42,45] Comparison to the values obtained with our probe **11** was only possible after conversion of the published dielectric constant values to $v\%$ water in dioxane "scale" (Table 12.2), unconventional but in this case mutually applicable.[38]

Nucleoside **11** clearly senses a much lower polarity of the major groove compared to the other probes. This was attributed to the significant difference in probe size (Table 12.2). Being small and part of the WC pair, **11** is placed close to the major

TABLE 12.2   Comparison of Reported DNA Major Groove Polarity Values

Probe	Reported Value	v% Water in Dioxane[a]
	$E_T(30) = 46.2$[b]	10
	$\varepsilon = 55$[c]	75
	$\varepsilon = 61$[d]	83
	$\varepsilon = 70$[e]	92

[a] See Tor and coworkers[38] for details.
[b] Expressed in kilocalorie per mole, taken from Tor and coworkers[38].
[c] Taken from Ganesh and coworkers[41].
[d] Taken from Majima and coworkers[42].
[e] Taken from Saito and coworkers[45].

groove wall, whereas the larger dansyl and DAN-modified probes protrude far into the groove and likely sample the surrounding bulk water.[47] The probe-dependent differences in groove polarity values are in agreement with the calculated major groove polarity gradient,[47] which shows drastic changes over a short distance from very apolar to almost the polarity of bulk water as moving away from the wall of the major groove to the perimeter of the DNA helix.[47] Put in context, this analysis and the diverse values reported for groove polarity depending on the probe used underline the importance of probe design, selection, and placement.

12.4.2   Probing Sensitivity to Viscosity

To investigate the sensitivity of **11** to viscosity, absorption and emission spectra were recorded in binary mixtures of the nonviscous methanol and viscous glycerol.

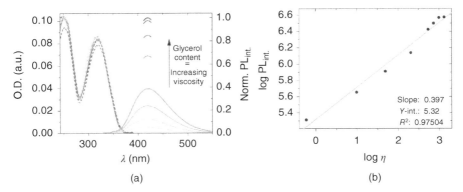

Figure 12.10 (a) Absorption (dashed lines) and emission (solid lines) spectra of **11** in methanol (light gray), glycerol (dark gray), and mixtures thereof (gray gradient) at 20 °C; (b) plot of log fluorescence intensity as a function of log sample viscosity according to Förster and Hoffmann.

These solvents were chosen for three key reasons: (1) methanol and glycerol differ more than 3 orders of magnitude in their viscosity values ($\eta_{20^\circ} = 0.583$ and 1317 cP, respectively), (2) they differ relatively little in polarity ($E_T(30) = 55.4$ and 57.0 kcal/mol, respectively), and (3) they are miscible in all ratios, thereby facilitating convenient control of the sample viscosity. The temperature dependency of viscosity requires all experiments to be performed under carefully controlled conditions. Besides minor changes in optical density, the variation of sample viscosity has virtually no influence on the absorption maxima of **11** (Fig. 12.10a). In contrast to the slightly responsive ground state, the strongly intensified fluorescence reveals a profoundly viscosity-sensitive excited state. The effect can be clearly visualized by plotting the fluorescence intensity as a function of sample viscosity. Replotting the data points on a double log plot gives a close to linear correlation, in accordance with the typical behavior observed for molecular rotors as was revealed by Förster and Hoffmann (Fig. 12.10b).[48]

To evaluate the sensitivity of **11** toward molecular crowding in an abasic site detection assay, **11** was incorporated in a central position of a 13-mer (Fig. 12.11). Placement of **11** opposite of dA, as in a perfect duplex, positions the furan appendage in the major groove of B-DNA resulting in a fluorescence signal of moderate intensity. When **11** was located opposite of an abasic site, a putative anti–syn base flip, as was corroborated by thermal denaturation studies,[11] positions the modified nucleobase in the intrahelical environment. Sandwiched in the confined space between two Watson–Crick base pairs, the furan's ability to freely rotate is restricted (Fig. 12.11). Similar to the observed fluorescence increase in samples of high viscosity (Fig. 12.10a), the molecular crowding imposed by the sandwiching base pairs results in a significant enhancement of the fluorescence intensity (Fig. 12.11).

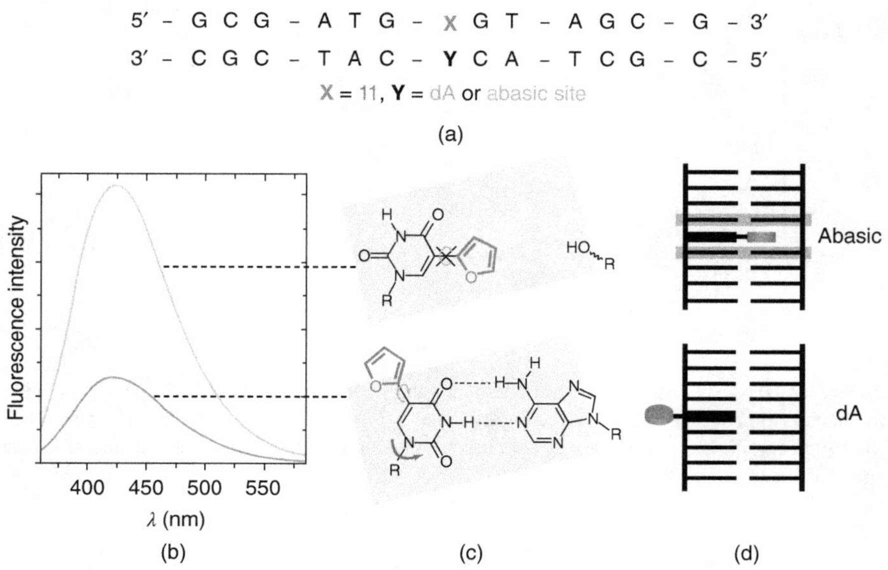
5′ - G C G - A T G - X G T - A G C - G - 3′
3′ - C G C - T A C - Y C A - T C G - C - 5′
X = 11, Y = dA or abasic site
(a)
Fluorescence intensity
400 450 500 550
λ (nm)
(b)
(c)
(d)
Abasic
dA

Figure 12.11 (a) B-DNA duplexes with **11** incorporated in a central position; (b) fluorescence spectra of **11** opposite of dA (green line) and across an abasic site (orange line); (c) top view of B-DNA duplex containing **11** opposite of dA (bottom) and opposite of an abasic site (top), (d) schematic side view of DNA helix with the furan modification (red disc) located in the major groove if across of dA, or in the free rotation restricting space between two Watson-Crick base pairs when opposite of an basic site. (*See color plate section for the color representation of this figure.*)

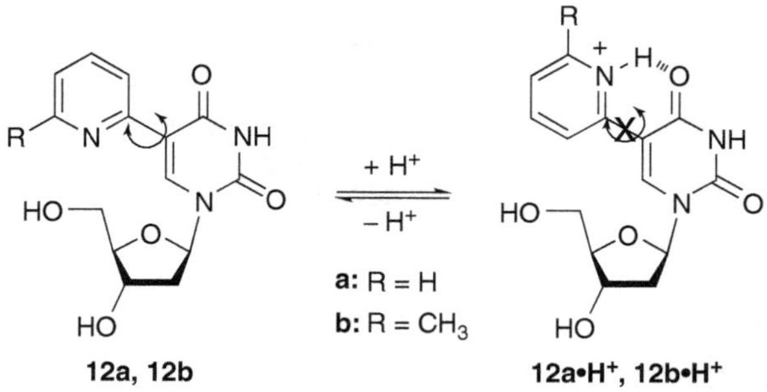
+ H+
– H+
a: R = H
b: R = CH3
12a, 12b
12a•H+, 12b•H+

Figure 12.12 Putative formation of an intramolecular hydrogen bond upon protonation of **12a** and **12b** in acidic media that would result in hampered free rotation of the pyridine moiety.

12.4.3 Probing Sensitivity to pH

Protonation of the basic pyridine nitrogen in **12a** is expected to result in a photophysical response. We postulated that in this specific case protonation would also favor formation of an intramolecular hydrogen bond (Fig. 12.12). This could lead to planarization of the π-system and, by deterred rotation around the single bond, to rigidification of the chromophore. The former is expected to cause a redshift in the absorption maximum and the latter to induce intensification of the fluorescence intensity. Both anticipated photophysical consequences of protonation make **12a** ideally suited for straightforward spectroscopic analyses.

To investigate the influence of protonation on the spectral properties of **12a**, buffered aqueous solution with pH values ranging from 8.4 to 2.18 were analyzed by absorption and emission spectroscopy (Fig. 12.13a). The ground-state absorption shows a redshift of the absorption maximum with a consecutive increase in the molar absorptivity going from basic to highly acidic conditions. All traces intersect at 295 nm, an isosbestic point, indicative of a two-species equilibrium. Fluorescence spectra were recorded after excitation at the isosbestic point. Very little influence of the sample pH on the emission maximum was observed, but, as predicted, the fluorescence intensity starkly increased with decreasing sample pH. A plot of the normalized fluorescence intensity versus sample pH was fit to a sigmoidal relationship (Fig. 12.13b), and a pK_a value of 4.42 was calculated for **12a**. The lower pK_a value compared to unsubstituted pyridine (pK_a = 5.14) is attributed to the conjugation of the pyridine ring to the electron-deficient pyrimidine core. Notably, the pyridine pK_a can be tuned by appropriate substitution.[36,49,50] To illustrate that this tunability is applicable to pH-sensitive nucleosides such as **12a**, a derivative with a methyl substituent ortho to the pyridine nitrogen **12b** was synthesized. Photophysical analysis as described for **12a**, yielded a pK_a value of 4.98, in agreement with the slightly electron-donating properties of the methyl group.[36]

Even though our hypothesis was corroborated by the photophysical analysis, we sought to structurally support the formation of an intramolecular hydrogen bond upon protonation of **12a** by crystallizing both neutral **12a** and its protonated form **12a•H$^+$**

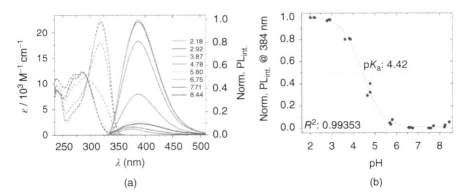

Figure 12.13   (a) Absorption (dashed lines) and emission spectra (solid lines) of **12** in aqueous phosphate buffers ranging in pH value from 8.44 (blue) to 2.18 (red) and intermediate values (black). (b) A plot of the normalized fluorescence intensity as function of sample pH (black circles) and a sigmoidal fit (orange line) to calculate the pK_a (4.42). (*See color plate section for the color representation of this figure.*)

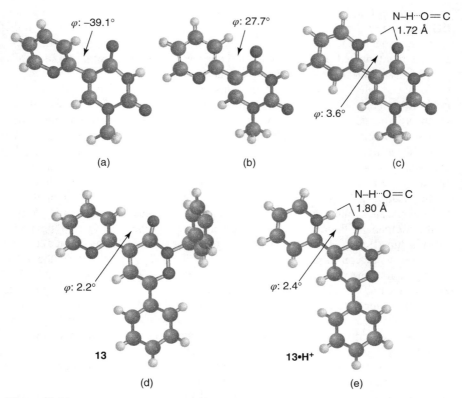

Figure 12.14 (a) crystal structure of **12a**, (b) modeled structure of **12a**, and (c) modeled structure of protonated **12a** (**12a•H$^+$**); (FN: Both **12a**, and **12a•H$^+$**, were modeled using molecular mechanics with the MMFFaq force field in Spartan) crystal structures of (d) neutral **13** (CCDC#: CIZNUV) and (e) protonated **13•H$^+$** (CCDC#: CIZPAD). Both **12a**, and **12a•H$^+$**, were modeled using molecular mechanics with the MMFFaq force field in Spartan. NB. The indistinguishable moiety in **13** is a benzyl group. (*See color plate section for the color representation of this figure.*)

(Fig. 12.12). The crystal structure of **12a** reveals a nonplanar orientation of the aromatic rings, with the pyridine nitrogen pointing away from the pyrimidine's carbonyl oxygen (Fig. 12.14). Unfortunately, attempts to obtain crystals of sufficient quality of protonated **12a** (**12a•H$^+$**) were unsuccessful. In lieu of a crystal structure, modeling was considered to provide additional structural insight. To asses if such an approach would provide a reliable structure of **12a•H$^+$**, **12a** was modeled first to compare the result to its crystal structure (Fig. 12.14). Similar to the crystal structure of **12a**, modeled **12a** has the pyridine nitrogen pointing away from the pyrimidine carbonyl and a similar deviation from a planar π-system. It is interesting to note, however, that the tetrahedral angle between the pyridine and the pyrimidine rings is opposite in sign for the crystal structure and the modeled **12a**. This possibly indicates the influence of parameters related to intermolecular packing in the crystal structure that are not taken into account in single-molecule modeled structures. The similarity between the crystal and the modeled structure of **12a** provided confidence to model **12a•H$^+$** as well. As hypothesized, the modeled **12a•H$^+$** has a planar π-system and a 180° pyridine ring flip brings the proton of the protonated pyridine to within hydrogen bonding proximity, 1.72 Å, with the pyrimidine carbonyl oxygen. Hence, both photophysical observations and modeling corroborate the putative structural implications induced upon protonation of **12a** (Fig. 12.12). To further strengthen our hypothesis, the published crystal structures of a related motif (**13**) in both the neutral and protonated states have been analyzed.[51] As with our nucleoside **12a**, **13** in its neutral form shows the pyridine nitrogen pointing away from the carbonyl (Fig. 12.14d). Protonation of the pyridine nitrogen to give **13•H$^+$** causes rotation of the pyridine ring. This positions the acidic proton within 1.80 Å from the carbonyl oxygen, allowing for an intramolecular hydrogen bond.

As was anticipated, based on its structural elements, **12a** also displayed sensitivity to viscosity and polarity,[36] making it an example of a fluorescent probe that shows responsiveness to multiple environmental parameters.

12.5   RECENT ADVANCEMENTS IN ISOMORPHIC FLUORESCENT NUCLEOSIDE ANALOGS

The nucleosides discussed in the previous section represent two prototypical emissive and responsive pyrimidine analogs. Related substituted pyrimidines with diverse aromatic rings conjugated at the 5 position have been extensively explored in our laboratory.[11,14] The resulting nucleosides emit in the visible range (390–443 nm), and have very large Stokes shifts (8400–9700 cm^{-1}), while their quantum efficiency is relatively low (Φ = 0.01–0.035). As discussed above, these nucleosides act like molecular rotors, where rotation around the biaryl bond provides an effective channel for nonradiative torsional relaxation (Fig. 12.6), thus leading to low emission quantum yield in nonviscous media. The quantum yield dramatically increases (>10-fold) in solutions of higher viscosity (Fig. 12.10).[35] This behavior, while leading to relatively low brightness in nonviscous media, does confer high responsiveness upon these nucleosides, which display sensitivity to both microenvironmental polarity and crowding effects.[35,36] As discussed above, the furan analog **11** has been used to signal the presence of abasic sites in DNA[11] and to estimate the polarity of major grooves in several distinct duplexes.[38] The corresponding ribonucleoside was used in antibiotic discovery assays that complement established 2AP-based constructs.[52] Additionally, The dC analog **7** (Fig. 12.2) was shown to provide a nondestructive method for *in vitro* detection of G, 8-oxoG (a mutagenic marker for cellular oxidative stress) and T, the downstream transverse mutation product of G oxidation.[18]

The minimally perturbing 5-modified pyrimidines are extremely responsive but suffer from relatively low emission quantum yields.[1] We hypothesized that altering the electronics by replacing the pyrimidine with the corresponding 1,2,4-triazine core would enhance the charge transfer character, ideally yielding a bathochromic shift as well as a hyperchromic effect and higher brightness. Compared to the uridine analog

11, the thiophen-conjugated 6-aza-U (**8**) (Fig. 12.2) displayed a redshifted emission (λ_{em} = 458 vs 434 nm for **11**) and a considerably higher (ca. 10-fold) quantum yield (20% in water for **8** vs 2% for **11**).[19] The opposite solvent effect on emission intensity, with elevated quantum efficiency in apolar solvents (80% in dioxane for **8**), suggests that N6 alters the excited state manifold.

Although singly labeled responsive oligonucleotides are indeed useful, robust biophysical assays typically benefit from the implementation of FRET pairs.[53–55] To photophysically match nonperturbing pyrimidine analogs with common fluorophores and generate useful FRET pairs using isomorphic nucleoside analogs, we designed a series of polarized quinazolines as expanded pyrimidines (examples **9a** and **9b**, Fig. 12.2). Modifications at the conjugated 5 and 7 positions with electron-donating groups (e.g., NH_2 (**9b**), OH, and OMe (**9a**)) give highly tunable and responsive fluorescent uridine analogs (Fig. 12.2). Their wide emission wavelength window, spanning 350–500 nm allowed us to identify ideal FRET partners and generate useful assays. For example, 5-methoxyquinazoline-2,4-($1H$,$3H$)-dione (**9a**) is an ideal donor for derivatives of 7-diethylaminocoumarin-3-carboxylic acid (Fig. 12.15a), which was used to generate robust discovery assays for antibiotics targeting the bacterial ribosome.[20] The 5-amino derivative (**9b**) was found to be a FRET acceptor for Trp emission (Fig. 12.15b), which allowed us to monitor RNA–peptide/protein interactions by relying on native, inherently fluorescent Trp residues with no need to label the peptide/protein.[21]

While numerous emissive pyrimidine analogs have been reported, highly emissive and responsive isomorphic purines are rare, with the classical 2-aminopurine, introduced in 1969, still being the most prevalently used. We have recently completed a fluorescent ribonucleoside alphabet, comprised of highly emissive purine (thA, thG) and pyrimidine (thU, thC) analogs, all derived from thieno[3,4-d]pyrimidine as the

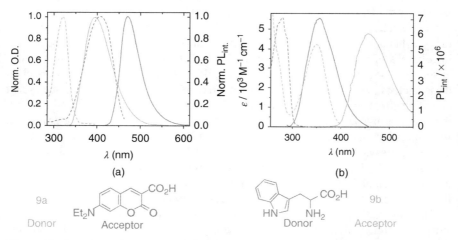

Figure 12.15   Absorption (dashed lines) and fluorescence (solid lines) of FRET donor and acceptor pairs for (a) **9a** and coumarin and (b) tryptophan and **9b**. (*See color plate section for the color representation of this figure.*)

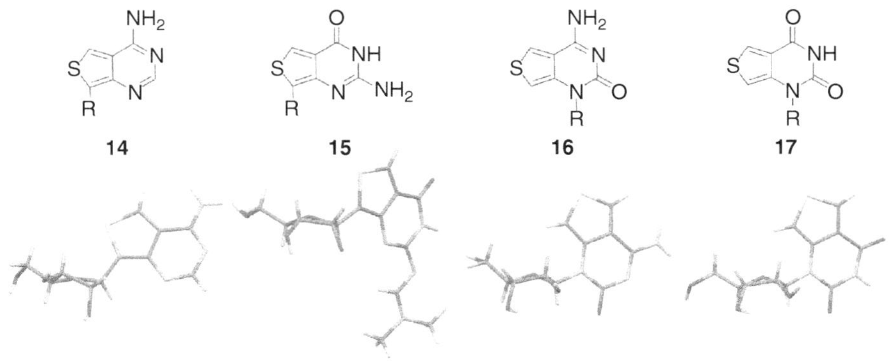

Figure 12.16   Crystal structures for emissive RNA alphabet [th]A (**14**), [th]G (**15**), [th]C (**16**), [th]U (**17**). Note: the crystal structure of **15** represents the DMF-protected nucleoside.

heterocyclic nucleus (Fig. 12.16).[36] The structural, biophysical, and spectroscopic characteristics of this emissive RNA nucleoside set illustrate highly desirable traits, including native Watson–Crick faces, unparalleled structural isomorphicity with respect to native nucleosides, minimal perturbation upon incorporation into duplexes, as well as intense visible emission. We have shown that the highly emissive [th]U, when incorporated into RNA oligonucleotides complementary to the rRNA α-sarcin/ricin loop, signals the depurination of A4324 in this highly conserved sequence by ribosome inactivating proteins (RIP, e.g., ricin) upon hybridization, via enhanced emission intensity.[56]

This alphabet of emissive nucleosides illustrates our underlying design principles: we aspire to identify emissive nucleobase-like heterocyclic motifs and synthetically elaborate them to furnish multiple, minimally perturbing emissive nucleosides. In the case of this emissive RNA alphabet, thieno[3,4-*d*]pyrimidine can be viewed as a precursor to 5,6-modified emissive pyrimidines and, at the same time, as a purine mimic with thiophene substituting the imidazole moiety. Glycosylation with D-ribose at N1 yields expanded and emissive pyrimidine nucleoside analogs, whereas *C*-glycosylation at the thiophene's C2 position provides fluorescent purine C-nucleoside analogs following elaboration of the H-bonding faces (Fig. 12.16).[57] The high isomorphicity of these emissive nucleoside analogs and their advantageous photophysical characteristics make them ideal candidates for implementation in diverse assays addressing fundamental questions in RNA biology (Fig. 12.16).[57]

12.6   SUMMARY

The first reported isomorphic fluorescent nucleoside analog, 2-aminopurine, was disclosed more than four decades ago.[10a] Ever since, a plethora of isomorphic fluorescent nucleosides have been developed and used in numerous applications,[1] illustrating that minimal structural perturbation can be reconciled with desirable photophysical properties. The significance and increasing popularity of isomorphic fluorescent nucleosides is further corroborated by a growing commercial availability.[58–60] Besides sufficient quantum yield and a redshifted absorption maximum, some isomorphic fluorescent nucleosides, as demonstrated herein with fluorescent T-mimics **11** and **12**, show a remarkable and versatile photophysically expressed environmental sensitivity. Using straightforward absorption and emission spectroscopy techniques, their sensitivity to polarity, pH, and viscosity can be qualified and quantified. Understanding the molecular factors that impact the nucleoside's responsiveness facilitates interpretation of spectroscopic observations after incorporation into oligomeric nucleic acids. This was illustrated by the ability of **11** to estimate the polarity in the major groove. Furthermore, the viscosity sensitivity of **11** supports intercalation as the mechanism for **11**'s ability to detect abasic sites when incorporated in an oligonucleotide opposite such DNA lesions.

12.7   PROSPECTS AND OUTLOOK

The study of biomolecules, including nucleic acids, and their biological functions has been shifting over the last decade. Straightforward spectroscopic analyses of nucleic acids under controllable conditions in a cuvette in aqueous buffer devoid of the complexities of a "real" biological medium have provided an invaluable basis for *in vitro* assays. Yet, challenging experiments in live cells and even living organisms using sophisticated instrumentation (e.g., fluorescence microscopy, fluorescence-activated cell sorting (FACS), single-molecule spectroscopy) have pushed the benchmark for fluorescent nucleoside analogs. Although more challenging, here too the isomorphic design principle will prove invaluable to ensure the fluorescence readout truly reflects native environments and biologically relevant events. The burden upon the developers of emissive fluorescent nucleoside analogs is to further shift the absorption and emission lines into the red. Perhaps equally important is the need to enhance the probe's brightness and photochemical stability. Advances in this field will not be accomplished by synthetic chemists alone; a tight collaboration between chemists, spectroscopists, biophysicists, and biologists will shape its future!

ACKNOWLEDGMENTS

We thank the National Institutes of Health for their generous support (GM 069773).

REFERENCES

1. Sinkeldam, R. W.; Greco, N. J.; Tor, Y., *Chem. Rev.* **2010**, *110*, 2579.
2. Lakowicz, J. R., Principles of Fluorescence Spectroscopy. 3rd ed.; Springer: New York, **2006**.

3. Valeur, B., Molecular Fluorescence, Principles and Applications. Wiley-VCH: Weinheim, **2002**.

4. Sprecher, C. A.; Johnson, W. C., *Biopolymers* **1977**, *16*, 2243.

5. Callis, P. R., *Annu. Rev. Phys. Chem.* **1983**, *34*, 329.

6. Peon, J.; Zewail, A. H., *Chem. Phys. Lett.* **2001**, *348*, 255.

7. Onidas, D.; Markovitsi, D.; Marguet, S.; Sharonov, A.; Gustavsson, T., *J. Phys. Chem. B* **2002**, *106*, 11367.

8. Cohen, B.; Crespo-Hernandez, C. E.; Kohler, B., *Faraday Discuss.* **2004**, *127*, 137.

9. Selected Reviews: (a) Hawkins, M. E.; Brand, L.; Johnson, M. L., *Methods Enzymol.* **2008**, *450*, 201; (b) Dodd, D. W.; Hudson, R. H. E., *Mini-Rev. Org. Chem.* **2009**, *6*, 378; (c) Tor, Y., *Pure Appl. Chem.* **2009**, *81*, 263; (d) Wilhelmsson, L. M., *Q. Rev. Biophys.* **2010**, *43*, 159; (e) Kimoto, M.; Cox, R. S. I.; Hirao, I., *Expert Rev. Mol. Diagn.* **2011**, *11*, 321.

10. Early individual contributions: (a) Ward, D. C.; Reich, E.; Stryer, L., *J. Biol. Chem.* **1969**, *244*, 1228; (b) Secrist, J. A.; Barrio, J. R.; Leonard, N. J., *Science* **1972**, *175*, 646.

11. Greco, N. J.; Tor, Y., *J. Am. Chem. Soc.* **2005**, *127*, 10784.

12. Gaied, N. B.; Glasser, N.; Ramalanjaona, N.; Beltz, H.; Wolff, P.; Marquet, R.; Burger, A.; Mely, Y., *Nucleic Acids Res.* **2005**, *33*, 1031.

13. Kenfack, C. A.; Burger, A.; Mely, Y., *J. Phys. Chem. B* **2006**, *110*, 26327.

14. Greco, N. J.; Tor, Y., *Tetrahedron* **2007**, *63*, 3515.

15. Roblin, R. O.; Lampen, J. O.; English, J. P.; Cole, Q. P.; Vaughan, J. R., *J. Am. Chem. Soc.* **1945**, *67*, 290.

16. Wierzchowski, J.; Wielgus-Kutrowska, B.; Shugar, D., *Biochim. Biophys. Acta, Gen. Subj.* **1996**, *1290*, 9.

17. Da Costa, C. P.; Fedor, M. J.; Scott, L. G., *J. Am. Chem. Soc.* **2007**, *129*, 3426.

18. Greco, N. J.; Sinkeldam, R. W.; Tor, Y., *Org. Lett.* **2009**, *11*, 1115.

19. Sinkeldam, R. W.; Hopkins, P. A.; Tor, Y., *ChemPhysChem* **2012**, *13*, 3350.

20. Xie, Y.; Dix, A. V.; Tor, Y., *J. Am. Chem. Soc.* **2009**, *131*, 17605.

21. Xie, Y.; Maxson, T.; Tor, Y., *J. Am. Chem. Soc.* **2010**, *132*, 11896.

22. Robins, M. J.; Barr, P. J., *Tetrahedron Lett.* **1981**, *22*, 421.

23. Robins, M. J.; Barr, P. J., *J. Org. Chem.* **1983**, *48*, 1854.

24. Liu, C. H.; Martin, C. T., *J. Mol. Biol.* **2001**, *308*, 465.

25. Sinkeldam, R. W.; Tor, Y., *Org. Biomol. Chem.* **2007**, *5*, 2523.

26. Dimroth, K.; Reichardt, C.; Siepmann, T.; Bohlmann, F., *Liebigs Ann. Chem.* **1963**, *661*, 1.

27. Reichardt, C.; Harbusch-Goernert, E., *Liebigs Ann. Chem.* **1983**, 721.

28. Reichardt, C., *Chem. Rev.* **1994**, *94*, 2319.

29. Haidekker, M. A.; Theodorakis, E. A., *Org. Biomol. Chem.* **2007**, *5*, 1669.

30. Stark, J.; Lipp, P., *Z. Phys. Chem.* **1913**, *86*, 36.

31. Oster, G.; Nishijima, Y., *J. Am. Chem. Soc.* **1956**, *78*, 1581.

32. Loutfy, R. O.; Law, K. Y., *J. Phys. Chem.* **1980**, *84*, 2803.

33. Loutfy, R. O.; Arnold, B. A., *J. Phys. Chem.* **1982**, *86*, 4205.

34. Laporte, S. L.; Harianawala, A.; Bogner, R. H., *Pharm. Res.* **1995**, *12*, 380.

35. Sinkeldam, R. W.; Wheat, A. J.; Boyaci, H.; Tor, Y., *ChemPhysChem* **2011**, *12*, 567.

36. Sinkeldam, R. W.; Marcus, P.; Uchenik, D.; Tor, Y., *ChemPhysChem* **2011**, *12*, 2260.

37. Hubschle, C. B.; Dittrich, B.; Grabowsky, S.; Messerschmidt, M.; Luger, P., *Acta Crystallogr., Sect. B: Struct. Sci., Cryst. Eng. Mater.* **2008**, *64*, 363.

38. Sinkeldam, R. W.; Greco, N. J.; Tor, Y., *ChemBioChem* **2008**, *9*, 706.

39. Jin, R.; Breslauer, K. J., *Proc. Natl. Acad. Sci. U. S. A.* **1988**, *85*, 8939.

40. Barawkar, D. A.; Ganesh, K. N., *Biochem. Biophys. Res. Commun.* **1994**, *203*, 53.

41. Barawkar, D. A.; Ganesh, K. N., *Nucleic Acids Res.* **1995**, *23*, 159.

42. Kimura, T.; Kawai, K.; Majima, T., *Org. Lett.* **2005**, *7*, 5829.

43. Kimura, T.; Kawai, K.; Majima, T., *Chem. Commun.* **2006**, 1542.

44. Jadhav, V. R.; Barawkar, D. A.; Ganesh, K. N., *J. Phys. Chem. B* **1999**, *103*, 7383.

45. Okamoto, A.; Tainaka, K.; Saito, I., *Bioconjugate Chem.* **2005**, *16*, 1105.

46. Tainaka, K.; Tanaka, K.; Ikeda, S.; Nishiza, K.; Unzai, T.; Fujiwara, Y.; Saito, I.; Okamoto, A., *J. Am. Chem. Soc.* **2007**, *129*, 4776.

47. Lamm, G.; Pack, G. R., *J. Phys. Chem. B* **1997**, *101*, 959.

48. Förster, T.; Hoffmann, G., *Z. Phys. Chem.* **1971**, *75*, 63.

49. Grandberg, I. I.; Faizova, G. K.; Kost, A. N., *Khim. Geterotsikl. Soedin.* **1966**, *2*, 561.

50. Borowiak-Resterna, A.; Szymanowski, J.; Voelkel, A., *J. Rad. Nucl. Chem.* **1996**, *208*, 75.

51. Clapham, K. M.; Batsanov, A. S.; Greenwood, R. D. R.; Bryce, M. R.; Smith, A. E.; Tarbit, B., *J. Org. Chem.* **2008**, *73*, 2176.

52. Srivatsan, S. G.; Tor, Y., *J. Am. Chem. Soc.* **2007**, *129*, 2044.

53. Börjesson, K.; Preus, S.; El-Sagheer, A. H.; Brown, T.; Albinsson, B.; Wilhelmsson, L. M., *J. Am. Chem. Soc.* **2009**, *131*, 4288.

54. Shi, Y.; Dierckx, A.; Wanrooij, P. H.; Wanrooij, S.; Larsson, N.-G. r.; Wilhelmsson, L. M.; Falkenberg, M.; Gustafsson, C. M., *Proc. Natl. Acad. Sci. U. S. A.* **2012**, *109*, 16510.

55. Stengel, G.; Gill, J. P.; Sandin, P.; Wilhelmsson, L. M.; Albinsson, B.; Nordén, B.; Millar, D., *Biochemistry* **2007**, *46*, 12289.

56. Srivatsan, S. G.; Greco, N. J.; Tor, Y., *Angew. Chem. Int. Ed.* **2008**, *47*, 6661.

57. Shin, D.; Sinkeldam, R. W.; Tor, Y., *J. Am. Chem. Soc.* **2011**, *133*, 14912.

58. Glen Research. www.glenresearch.com.

59. Jena Bioscience. www.jenabioscience.com.

60. Berry & Associates. www.berryassoc.com.

13

SITE-SPECIFIC FLUORESCENT LABELING OF NUCLEIC ACIDS BY GENETIC ALPHABET EXPANSION USING UNNATURAL BASE PAIR SYSTEMS

MICHIKO KIMOTO

Institute of Bioengineering and Nanotechnology (IBN), Singapore, Singapore

RIE YAMASHIGE

RIKEN Center for Life Science Technologies (CLST), Yokohama, Kanagawa, Japan

ICHIRO HIRAO

Institute of Bioengineering and Nanotechnology (IBN), Singapore, Singapore

13.1   INTRODUCTION

Site-specific fluorescent labeling of DNA and RNA molecules is a fundamental approach for the detection of specific base sequences and the analysis of the structural dynamics of functional nucleic acids. Site-specific labeling is mainly accomplished by chemical and biological processes, such as direct chemical synthesis of DNA and RNA, postsynthesis modifications of replicated or transcribed nucleic acids, and enzymatic incorporation, using primers containing fluorescent components for PCR or fluorescent cap analogs for transcription. Depending on the purpose, these methods are useful, but they are nevertheless cumbersome and restricted to

Fluorescent Analogs of Biomolecular Building Blocks: Design and Applications, First Edition.
Edited by Marcus Wilhelmsson and Yitzhak Tor.
© 2016 John Wiley & Sons, Inc. Published 2016 by John Wiley & Sons, Inc.

the lengths or the labeling positions of nucleic acid molecules. Thus, a simplified site-specific labeling method for large nucleic acid molecules is needed for a wide variety of applications, in response to recent advances in nucleic acid chemistry and biology.

The expansion of the genetic alphabet of DNA by an artificial extra base pair (unnatural base pair) would be a powerful tool for the site-specific incorporation of functional components, such as fluorophores and quenchers, into nucleic acids by polymerase reactions (Fig. 13.1). For the last 25 years, several research groups have developed several types of unnatural base pairs that function as a third base pair, along with the natural A–T and G–C pairs, in replication and transcription.[1–16] Some of these unnatural bases are modified with fluorophore residues, while others have intrinsically unique fluorescent properties. These fluorescently active components, as their triphosphate substrates, can be site-specifically incorporated into DNA and/or RNA, opposite their complementary unnatural bases in templates, by polymerases. A couple of unnatural base pairs were applied to real-time PCR and structural analysis of functional RNA molecules by site-specific fluorescent labeling. Here, we introduce unnatural base pair systems that can be practically used as a third base pair for *in vitro* replication and transcription, as a tool for the site-specific fluorescent labeling of functional DNA and RNA molecules.

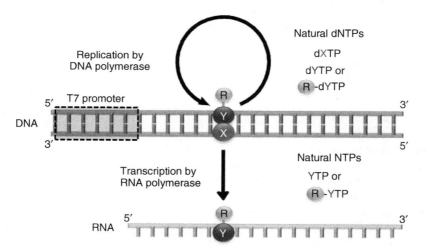

Figure 13.1 Expansion of the genetic alphabet of DNA by an artificial extra base pair (unnatural base pair). An unnatural base pair (X–Y), which functions together with the natural A–T and G–C pairs, enables the site-specific incorporation of extra functional components into nucleic acids at desired positions in replication and/or transcription. A functional group, such as a fluorophore and a quencher, can be attached to the unnatural base (Y) moiety via a linker.

13.2 DEVELOPMENT OF UNNATURAL BASE PAIR SYSTEMS AND THEIR APPLICATIONS

The initial idea of unnatural base pairs was proposed in Alexander Rich's review article in 1962, in which he described the possibility of a third base pair between 6-amino-2-ketopurine (**isoG**) and 2-amino-4-ketopyrimidine (**isoC**).[17] In 1989, Benner's group chemically synthesized nucleoside derivatives of **isoG** and **isoC** and tested their pairing abilities in replication and transcription.[18] The geometry of the hydrogen bond donor–acceptor pattern of the **isoG–isoC** pair is different from those of the A–T and G–C pairs (Fig. 13.2a and b), and thus the **isoG–isoC** pair was expected to function as a third base pair. Benner's group also designed and synthesized several candidate pairs with different hydrogen bonding patterns.[1] These unnatural base substrates (nucleoside 5′-triphosphates) were incorporated into DNA

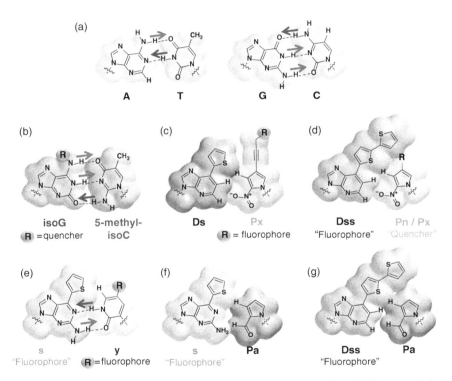

Figure 13.2 Chemical structures of natural and unnatural base pairs. (a) The natural A–T and G–C pairs. (b) The unnatural **isoG–5-methyl-isoC** pair. (c) The unnatural **Ds–Px** pair. (d) The unnatural **Dss–Pn** (R = H)/**Px** (R = propynyl) pair. (e) The unnatural **s–y** pair. (f) The unnatural **s–Pa** pair. (g) The unnatural **Dss–Pa** pair. The solid and dashed arrows indicate the hydrogen bond donor–acceptor patterns in the pairing bases. Functional groups of interest, such as fluorophores and quenchers, can be attached to the R positions (gray circles) of each unnatural base. The **s** and **Dss** bases are fluorescent, and the 2-nitropyrrole moiety (**Pn**) acts as a quencher.

and RNA, opposite their pairing partners in DNA templates, by some DNA and RNA polymerases. In addition, a modified **isoG** substrate, *N*6-(6-aminohexyl)isoguanosine 5′-triphosphate, was also incorporated into RNA, opposite 5-methyl-**isoC** in templates, by T7 RNA polymerase.[2] The aminohexyl residue of the incorporated **isoG** in RNA molecules is useful for posttranscriptional modification. The **isoC** nucleosides are rather chemically unstable in solution, and the methylation of the 5-position improves the chemical stability.[2,19,20] Furthermore, the **isoG**–5-methyl-**isoC** pair (Fig. 13.2b) was subjected to multiplex real-time PCR, to simultaneously detect several target sequences causing infectious diseases (see Section 13.2.1.1).[21,22] However, various shortcomings of the **isoG**–**isoC** pair, such as **isoG** tautomerism causing mispairing with T, have restricted further applications.[23,24]

Since Benner's pioneering studies toward the realization of genetic alphabet expansion systems, many unnatural base pairs have been designed and tested through proof-of-concept experiments, on the basis of several ideas and concepts, such as shape complementarity between the pairing bases, steric and electrostatic repulsion to prevent mispairing, and the hydrophobicity of unnatural bases, in addition to the hydrogen bonding geometry.[16,25–32] As a consequence, several groups have developed unnatural base pairs that can function as a third base pair with high efficiency and selectivity in PCR.[3,7,8,10,13] Among them, we created a hydrophobic unnatural base pair between **Ds** (7-(2-thienyl)-imidazo[4,5-*b*]pyridine) and **Px** (2-nitro-4-propynylpyrrole) (Fig. 13.2c) by fine-tuning the shape complementarity of the pairing surface.[8,10] To prevent the mispairing of the unnatural bases with the natural bases, we removed the atoms and residues involved in hydrogen-bonding interactions from the third base pair. In addition, the molecular shape of the **Ds** base is larger than those of the natural bases, due to the addition of the thienyl group. In contrast, the five-membered ring of the **Px** base is smaller than the six-membered rings of C and T. Therefore, these unnatural bases are spatially mismatched with the natural bases, but the shape of the **Ds** base fits closely with that of the **Px** base, like the pieces of a jigsaw puzzle. Furthermore, the nitro group of **Px** efficiently prevents the mispairing with A, by the electrostatic repulsion between the oxygen of the nitro group and the 1-nitrogen of A.[33] The **Ds**–**Px** pair exhibits extremely high selectivity in PCR. DNA fragments containing the **Ds**–**Px** pair are amplified 10^7–10^8-fold by 30–40 cycles of PCR using exonuclease-proficient DNA polymerases, and more than 97% of the **Ds**–**Px** pair was retained in the amplified DNA.[8,10]

The **Px** base can be modified with a wide variety of functional groups, such as fluorophores, quenchers, biotin, and amino-alkyl groups, and these modified **Px** substrates are also site-specifically incorporated into DNA, opposite **Ds** in templates, by PCR amplification (see Section 13.2.1.2).[8,10,34] In addition, a modified **Ds** base, 7-(2,2′-bithien-5-yl)-imidazo[4,5-*b*]pyridine (**Dss**), which was generated by the addition of an extra thiophene residue to the **Ds** base, exhibits strong fluorescence emission centered at 456 nm, upon excitation at 385 nm, in oligonucleotides.[35] During our studies, we found that the 2-nitropyrrole (**Pn**) moiety of **Px** has high quenching ability, and thus the **Dss**–**Pn** and **Dss**–**Px** pairs (Fig. 13.2d) are the unnatural base pairs between the fluorophore and quencher base analogs that function as a third base pair in replication.[36] These uniquely fluorescent unnatural base pairs were utilized as molecular beacons and in real-time PCR for the detection of target nucleic acid sequences (see Sections 13.2.1.3 and 13.2.1.4).[34,36]

Besides the **Ds–Px** and **Dss–Px** pairs, we have developed other types of unnatural base pairs that can be used for the site-specific incorporation of fluorescent components into RNA by transcription.[11] One of them is an unnatural base pair between 2-amino-6-(2-thienyl)purine (**s**) and pyridine-2-one (**y**) (Fig. 13.2e), which was designed by combining the concepts of hydrogen bonding geometry and steric hindrance, to prevent mispairing with the natural bases.[15,37] The extended shape of the thienyl group of **s** clashes with the 4-keto group of T, as well as the 4-amino group of C, efficiently preventing these mispairings. In contrast, the hydrogen of the **y** base can accommodate the Watson–Crick type geometry of the **s–y** pair. As a result, the substrate of **y** is site-specifically incorporated into RNA, opposite **s** in templates, with more than 95% selectivity by transcription using T7 RNA polymerase.[15] The **y** base can be modified with a wide variety of functional groups via a propynyl linker,[38–40] and fluorophore-linked **y** substrates are efficiently incorporated into RNA at desired positions, opposite **s** in templates, by T7 transcription (see Section 13.2.2.1).[39]

Another useful system for transcription is the unnatural base pair between **s** and pyrrole-2-carboaldehyde (**Pa**) (Fig. 13.2f).[12] We found that the five-membered ring of **Pa** also excludes the mispairing with the natural purine bases, similar to **Px**, but the pairing selectivity of **Px** with **Ds** in replication is higher than that of **Pa** with **Ds**. However, **Pa** is a favorable template base rather than **Px**, for the incorporation of not only **Ds** but also **s**.[13] These substrates of **Ds** and **s** are site-specifically and efficiently incorporated into RNA, opposite **Pa** in templates, by T7 RNA polymerase.[9,12,13] The incorporation efficiency and selectivity of the **s** substrate opposite **Pa** are superior to those opposite **y** in T7 transcription involving the **s–y** pair. Note that the **s** base is fluorescent, and its fluorescence intensity in RNA molecules changes depending on the microenvironment, such as a stacking interaction with neighboring bases. Thus, the site-specific **s** labeling is a useful tool for local structural analysis, based on the base stacking interactions in functional RNA molecules (see Section 13.2.2.2).[9,12] In addition, **Pa** is a suitable pairing partner for the site-specific incorporation of **Dss** into RNA (Fig. 13.2g),[35] compared to **Px**.

13.2.1   Site-Specific Fluorescent Labeling of DNA by Unnatural Base Pair Replication Systems

13.2.1.1   The isoG–5-Methyl-isoC Pair for Real-Time PCR: Prudent and coworkers developed a real-time PCR method utilizing the **isoG**–5-methyl-**isoC** pair (Fig. 13.2b) for the quantitative monitoring of amplified DNA fragments with a target sequence (Fig. 13.3a).[21,22] The system involves a triphosphate substrate of a modified **isoG** base with a quencher (dabcyl) (Fig. 13.3b) and a primer containing a fluorophore at the 5′-end and 5-methyl-**isoC** in close proximity to the 5′-end (Primer 1 in Fig. 13.3a). Thus, the initial reaction mixture produces fluorescence, and as the PCR amplification progresses, incorporation of the quencher-linked **isoG** opposite 5-methyl-**isoC** reduces the fluorescence intensity, depending on the amount of amplified double-stranded DNA fragments. Therefore, the progress of the PCR

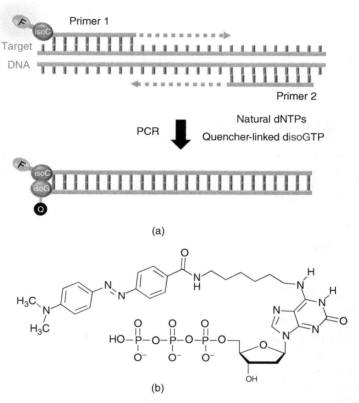

Figure 13.3 The Plexor system, using the **isoG**–5-methyl-**isoC** pair in real-time PCR. (a) The fluorophore (F) is attached proximally to 5-methyl-**isoC** in primer 1. The quencher (Q) is covalently attached to d**isoGTP**, and the modified **isoG** substrate is incorporated opposite 5-methyl-**isoC** during PCR. The **isoG**-5-methyl-**isoC** pairing brings the fluorophore and quencher close together in the PCR products, reducing the fluorescence intensity. (b) Chemical structure of dabcyl-linked d**isoGTP**.

amplification can be monitored by the reduction of the fluorescence intensity, in a quantitative real-time PCR apparatus. Furthermore, by using multiple primers with different fluorophores corresponding to each target sequence, a multiplex PCR detection system can be developed. This system is available as Plexor™ (Promega Corporation), and in combination with reverse transcription, several respiratory viral pathogens have been detected simultaneously.[21,41–43]

13.2.1.2   The Ds–Px Pair for Site-Specific Fluorescent Labeling of DNA through PCR: The **Ds–Px** pair is useful for the site-specific labeling of specific DNA fragments, by using modified **Px** base substrates linked with fluorophores (Fig. 13.4). Fluorophore-linked d**PxTP** substrates are site-specifically incorporated into

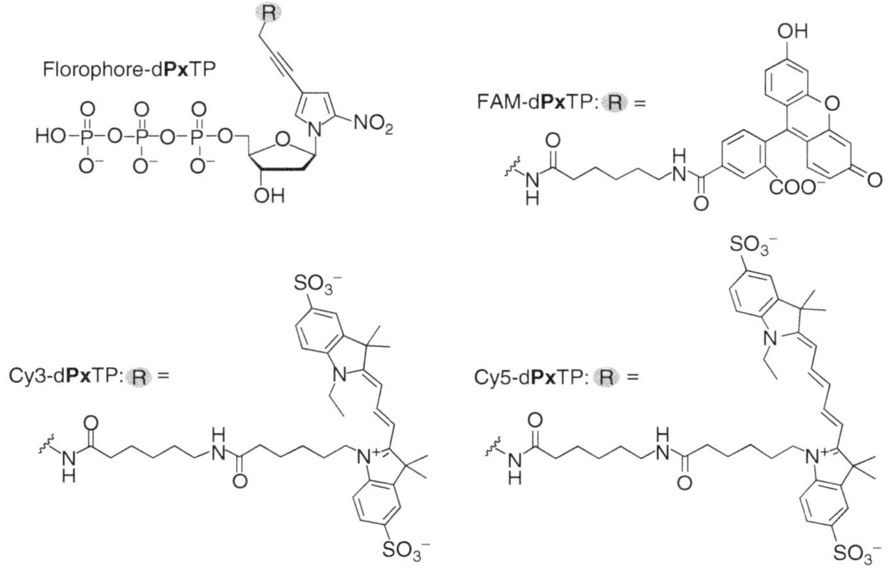

Figure 13.4 Chemical structures of fluorophore-linked d**Px**TPs.

DNA with high efficiency and selectivity, opposite **Ds** in templates, by PCR using exonuclease-proficient DNA polymerases, such as the Deep Vent and AccuPrime *Pfx* DNA polymerases.[8,10,34] **Ds**-containing DNA fragments are prepared by chemical synthesis using a phosphoramidite reagent of **Ds** (Glen Research Corporation).[12,13]

In a mixture with a large amount of foreign DNA fragments consisting of the natural bases, **Ds**-containing DNA fragments can specifically be detected by PCR involving the **Ds**–fluorophore-**Px** pair. This system could be applied to DNA authentication and steganography technologies for a security system. By performing PCR in the presence of d**Ds**TP and the fluorophore-linked d**Px**TP together with the natural dNTPs, the complementary strand of a target **Ds**-containing DNA fragment is site-specifically labeled with the fluorophore-linked **Px** base, and after removing the unreacted fluorophore-**Px** substrate, the existence of the **Ds**-containing DNA fragment can be identified in the mixture of amplified DNA fragments by its fluorescence. Furthermore, the amplified **Ds**-containing DNA fragments can be isolated from the amplified mixtures. For example, after 15-cycle PCR in the presence of d**Ds**TP and fluorescein-linked d**Px**TP (FAM-d**Px**TP in Fig. 13.4), a zeptomole (10^{-21})-scale **Ds**-containing DNA fragment was isolated from amplified mixtures with excess amounts (picomole (10^{-12}) scale) of foreign DNA fragments with random sequences by using the magnetic bead-bound antifluorescein antibody, and the **Ds**-containing sequence was identified. Without the isolation process, the target sequence was not identified because it was hidden in the random sequencing peaks derived from the foreign species.[10]

13.2.1.3   The Fluorophore–Quencher Base Pair, Dss–Px (Pn):   The fluorophore and quencher base pairs between **Dss** and **Pn** or **Dss** and **Px** were used in applications as molecular beacons and in real-time PCR. Around 80% of the **Dss** fluorescence in a DNA 12-mer was quenched by hybridization with its complementary 12-mer strand containing **Pn**. Although the double-stranded DNA 12-mer containing the **Dss–Pn** pair ($T_m = 45.4\,°C$ in a buffer containing 100 mM NaCl, 10 mM sodium phosphate (pH 7.0), and 0.1 mM EDTA) was less thermally stable relative to that with the A–T pair ($T_m = 48.6\,°C$), the **Dss–Pn** pair can be applied as a molecular beacon. The **Dss–Pn** pair was inserted into the stem region of a molecular beacon (Fig. 13.5a). In the absence of a target DNA molecule containing the complementary sequence to the loop region of the beacon, the fluorescence of **Dss** was quenched by pairing with **Pn** and was not detectable by the unaided eye (Fig. 13.5b). By adding its target DNA molecule to hybridize with the beacon, the fluorescence of **Dss** increased due to the separation of the stem region. This fluorescence change was detectable with the unaided eye by 375-nm irradiation (1 µM molecular beacon, 1 µM target DNA molecule at room temperature, Fig. 13.5b). This molecular beacon can discriminate single-nucleotide polymorphisms among target DNA sequences.[36]

The **Dss–Px** pair was employed for the quantitative detection of a target DNA sequence, using a real-time PCR apparatus. For the system, we designed a primer comprising a target hybridizing sequence and a tag sequence containing **Dss** (Fig. 13.6a). Through PCR amplification in the presence of dPxTP, the **Dss** fluorescence in the solution is reduced by the incorporation of the quencher **Px**

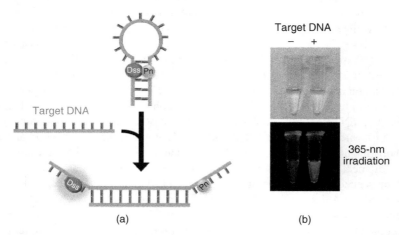

(a)   (b)

Figure 13.5   A molecular beacon containing the **Dss–Pn** pair. (a) Scheme for a target DNA detection system, using a molecular beacon containing the **Dss–Pn** pair in the stem region. Hybridization between the target DNA and the beacon loop region opens the stem region, which in turn allows **Dss** fluorescence. (b) Visible **Dss** fluorescence detection of the molecular beacon (1 µM) in the presence (+) and absence (−) of a target DNA (1 µM), in a solution containing 10 mM sodium phosphate (pH 7.0), 100 mM NaCl, and 0.1 mM EDTA, upon irradiation at 365 nm (lower picture) at room temperature.

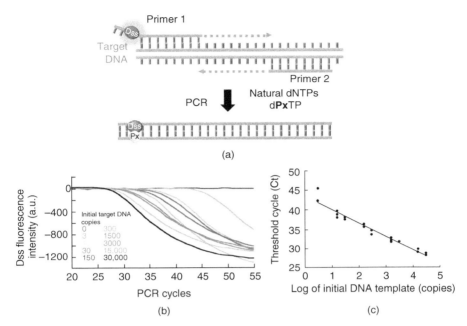

Figure 13.6 A real-time quantitative PCR amplification system involving the **Dss–Px** pair. (a) Scheme of the qPCR method using the **Dss–Px** pair, with a **Dss**-containing tag primer (Primer 1) and d**Px**TP as a quencher. In the amplification process, **Px** is incorporated opposite **Dss** in the primer, reducing the **Dss** fluorescence intensity. (b) Amplification plots of 55-cycle PCR, using 0–30,000 copies of the input DNA. (c) Standard curve for the real-time qPCR system involving the **Dss–Px** pair. The values of the threshold cycles (Ct) obtained from panel b were plotted against the log of the input DNA copy number. (*See color plate section for the color representation of this figure.*)

opposite **Dss**. Initial target DNA fragments ranging from 3 to 30,000 copies can be detected and quantitated by real-time PCR with a 350 nm excitation filter and a 440 nm emission filter, using TITANIUM Taq DNA polymerase (Takara Clontech) (Fig. 13.6b and c).

13.2.1.4 A Quantitative Real-Time PCR System Using the Ds–fluorophore-Px Pair:

Besides the real-time PCR system using the **Dss–Px** pair, we developed another simple real-time PCR system, using the unnatural base pair between **Ds** and fluorophore-linked **Px**.[34] The 2-nitropyrrole moiety of **Px** also partially quenches (about 40% at 20 °C) the fluorescence of a fluorophore group that is linked to the **Px** base because of the internal stacking between the fluorophore and the **Px** base in solution.[34] However, we found that the fluorescence of the fluorophore-linked **Px** was recovered by its incorporation into DNA strands. This might occur by the protrusion of the fluorophore group from the strand. This phenomenon can be used for target DNA detection by real-time PCR.

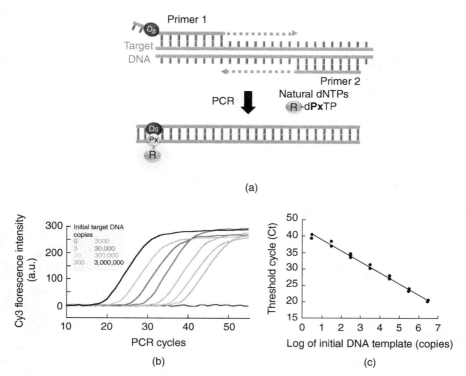

(a)

(b)   (c)

Figure 13.7 A simple, real-time quantitative PCR amplification system involving the **Ds–Px** pair in a primer region. (a) Scheme of the qPCR method using the **Ds**–fluorophore-**Px** pair, with a **Ds**-containing primer and fluorophore-linked dPxTP (**R**-dPxTP). In the amplification process, the fluorophore-**Px** is incorporated opposite **Ds** in the primer, which increases the fluorescence intensity of the fluorophore attached to the **Px** base. (b) Amplification plots of 55-cycle PCR with Cy3-dPxTP, using 0–3,000,000 copies of the input DNA. (c) Standard curve for the real-time qPCR system involving the **Ds**–Cy3-**Px** pair. The values of the threshold cycles (Ct) obtained from panel b were plotted against the log of the input DNA copy number. (*See color plate section for the color representation of this figure.*)

For real-time PCR, we used fluorophore-linked **Px** substrates, such as Cy3-, Cy5-, and FAM-linked dPxTPs (Fig. 13.4), and a **Ds**-containing primer, similar to the experiments with the **Dss–Px** pair (Fig. 13.7a). In this case, the fluorescence intensity of the fluorophore-linked **Px** increased by its incorporation into DNA opposite **Ds**. As with the real-time PCR using the **Dss–Px** pair, as few as three copies of the target DNA were detectable by the method (Fig. 13.7b), and a standard curve obtained from the amplification plot exhibited high linearity (Fig. 13.7c). Furthermore, little or no fluorescence change was observed using a primer comprising only natural bases (without **Ds**), and thus the fluorophore-linked **Px** was site-specifically incorporated opposite **Ds**. The primer regions of the amplified products are specifically labeled with the incorporated fluorophore-linked **Px**, and thus can be used for subsequent analyses or experiments. The spectral properties of the 2′-deoxyribonucleoside triphosphates of FAM-**Px**, Cy3-**Px** and Cy5-**Px** are summarized in Table 13.1.

TABLE 13.1   Spectral Properties of Unnatural NTPs

Triphosphates[a]	Absorption Maximum (nm)	Extinction coefficient $(M^{-1}\,cm^{-1})$	Emission Maximum (nm)
FAM-d**Px**TP	493	64,400	525
Cy3-d**Px**TP	550	180,000	563
Cy5-d**Px**TP	648	280,000	659
FAM-**y**TP	493	70,000	521
FAM-hx-**y**TP	493	62,000	522
s**TP**	347	14,000	435
DssTP	368	31,800	459

[a]In phosphate buffer (pH 7.0) at 25 °C.

    Ds-containing target DNA sequences can also be detected by real-time PCR, using the unnatural base pair between **Ds** and fluorophore-linked **Px**. The **Ds**-containing region of the target DNAs is amplified by PCR from both sides of the natural-base sequence regions, using Cy3-d**Px**TP and d**Ds**TP as unnatural substrates and AccuPrime *Pfx* DNA polymerase (Invitrogen) (Fig. 13.8).[34] This method is useful for DNA authentication and steganography technologies, as mentioned in Section 13.2.1.2. Since no fluorescence change was observed when using a DNA template comprising only the natural bases (without **Ds**) (Fig. 13.8c), the real-time PCR system can specifically detect only the **Ds**-containing DNA fragments in DNA solutions.

13.2.2   Site-Specific Fluorescent Labeling of RNA by Unnatural Base Pair Transcription Systems

13.2.2.1   The s–y Pair for Site-Specific Fluorescent RNA Labeling with Modified y: T7 transcription involving the **s**–**y** pair enables the site-specific incorporation of the modified-**y** base into RNA, opposite **s** in DNA templates, by T7 RNA polymerase (Fig. 13.9a). DNA templates containing **s** at desired positions can be prepared by several methods, involving chemical synthesis using a phosphoramidite derivative of **s** (Glen Research) with enzymatic ligation, primer extension, and PCR amplification. Templates (less than 80-mer including the T7 promoter) are directly synthesized using an automated DNA synthesizer. The **s**-containing template strand is partially hybridized with the nontemplate strand of the T7 promoter region for transcription. Templates (more than 100-mer) are prepared by enzymatic ligation[13,15] or primer extension, using chemically synthesized DNA fragments.[13,44] For the incorporation of **s** close to the 3′-terminus of the template, PCR amplification using primers containing **s** is useful. In the PCR amplification, the complementary unnatural base substrate, such as d**y**TP, is not required, because the misincorporation, opposite **s**, in the nontemplate strand does not affect transcription.[14,39,40] This method is useful for the simple 3′-terminal labeling of functional RNA molecules, such as aptamers and ribozymes, without any reduction of their RNA activity.[39] Fluorophore-linked **y** substrates, in which the fluorophore groups are linked at the 5-position of **y** via a

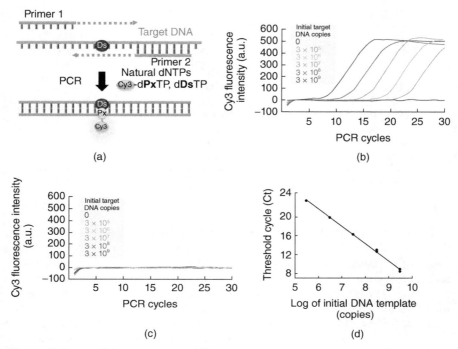

Figure 13.8 A real-time quantitative PCR amplification system involving the **Ds–Px** pair in a complementarily amplified region. (a) Scheme of the qPCR method using the **Ds–Cy3-Px** pair, with a **Ds**-containing target DNA, Cy3-linked d**Px**TP, and d**Ds**TP. In the amplification process, Cy3-**Px** is incorporated opposite **Ds** in the DNA template, which increases the fluorescence intensity of Cy3 attached to the **Px** base moiety. (b) Amplification plots of 30-cycle PCR, using 0 to 3×10^9 copies of the **Ds**-containing input DNA. (c) Amplification plots of 30-cycle PCR, using a DNA template comprising only natural bases. (d) Standard curve for the real-time qPCR system involving the **Ds–Cy3-Px** pair. The values of the threshold cycles (Ct) obtained from panel b were plotted against the log of the input DNA copy number. (*See color plate section for the color representation of this figure.*)

propargylamine linker (Fig. 13.9b), are chemically synthesized from the nucleoside derivative of 5-iodo-**y**.[39]

As the incorporation site of a fluorophore-linked **y** base, a uridine site in the functional RNA molecules is suitable, because the **y** base is a uracil analog lacking the 4-keto group. The incorporation selectivity of a fluorophore-linked **y** base into a specific position of RNA molecules can be adjusted by changing the ratio of the natural base and fluorophore-linked **y** substrate concentrations: an increase in the modified **y** substrate concentration enhances the incorporation selectivity of the modified **y** substrate opposite **s** but also increases the misincorporation of the modified **y** substrate opposite the natural bases. We generally use 0.25–2 mM fluorophore-linked **y**TP and 1–2 mM natural base NTPs, depending on the modification of the **y** base. To determine the optimal substrate concentrations, we usually test T7 transcription in the presence of fluorophore-linked **y**TP and the natural

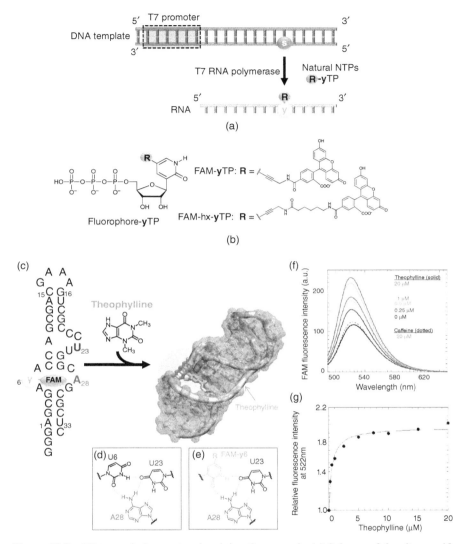

Figure 13.9 T7 transcription system involving the **s–y** pair. (a) Scheme of the site-specific incorporation of a modified **y** substrate (R-**y**TP) into RNA, opposite **s** in DNA templates, by T7 RNA polymerase. (b) Chemical structures of the fluorophore-linked **y**TPs. (c) Structure of an antitheophylline RNA aptamer. The FAM-**y** base was site-specifically incorporated into the aptamer, in place of U6. The FAM-**y** base was superimposed at the U6 position in the aptamer-theophylline complex (PDB 1O15). The FAM-**y**, U23, A28, and theophylline are shown in orange, blue, red, and green, respectively. (d) The original U6–U23–A28 base triplet. (e) The substituted FAM-**y**6–U23–A28 base triplet. (f) Fluorescence spectra of the antitheophylline aptamer labeled with FAM-**y** at position 6, in the presence or absence of theophylline (0–20 μM) or caffeine (20 μM), excited at 434 nm. (g) Fluorescence intensities at 522 nm were plotted against the theophylline concentrations, as relative ratios to that in the absence of theophylline. (*See color plate section for the color representation of this figure.*)

base NTPs at different concentrations, using an s-containing DNA template and its control DNA template containing only the natural bases, in which s is replaced with A, and analyze the transcripts by gel-electrophoresis. By comparing the fluorescence intensities between the transcript from the s-containing template and that from the control template, the incorporation rate of fluorophore-linked y opposite s and the misincorporation rate opposite the natural bases can be estimated. The substrate ratio should be determined depending on research purposes. For the internal incorporation of fluorophore-y into RNA (less than ~50-mer), the desired transcripts are isolated and purified by gel electrophoresis, by exploiting the lower mobility of the transcripts containing the fluorophore-y, relative to those of other transcripts, including those in which a natural base is misincorporated opposite s.

Since the fluorescence intensity differs, depending on the stacking features of the fluorophore moiety with other bases in a specific RNA structure, this site-specific labeling can be used to monitor the structural changes of functional RNA molecules, such as aptamers. In Section 13.3.1, we describe the site-specific FAM-linked y probing of an antitheophylline RNA aptamer, toward the development of a direct detection system for theophylline (Fig. 13.9c–g). The spectral properties of the ribonucleoside triphosphates of FAM-y and FAM-hx-y are summarized in Table 13.1.

13.2.2.2 The s–Pa and Dss–Pa Pairs for Site-Specific RNA Labeling with Fluorescent s and Dss Bases by Transcription:

T7 transcription involving the s–Pa and Dss–Pa pairs enables the site-specific incorporation of fluorescent s and Dss bases into RNA, opposite Pa in DNA templates (Fig. 13.10). The s base is an analog related to the well-known fluorescent 2-aminopurine (AP) base, and the ribonucleoside of s exhibits fluorescence excitation and emission centers at 352 and 434 nm, respectively,[12] which are shifted to longer wavelengths relative to those of the 2-aminopurine ribonucleoside (excitation: 303 nm, emission: 370 nm).[45] The Dss base also exhibits strong fluorescence and is slightly redshifted, in comparison with the fluorescence of s. The thienyl group at position 6 endows AP and Ds with a large Stokes shift.[35,46] The spectral properties of the ribonucleoside triphosphates of s and Dss are summarized in Table 13.1.

The fluorescence intensity of s sensitively varies depending on the strength of the stacking interactions with neighboring bases in a functional RNA molecule, and, thus, the intensity reflects the local structural features and conformational changes. In Section 13.3.2, we provide examples of the local structural analyses of a stable GNRA RNA hairpin and an L-shaped tRNA structure, by the site-specific fluorescent s labeling (Figs. 13.11 and 13.12).

13.3   IMPLEMENTATION

13.3.1   Fluorescence Sensor System Using an RNA Aptamer by Fluorophore-Linked y Labeling

Nucleic acid aptamers are generated by evolutionary engineering methods using nucleic acid libraries, and they specifically bind to a wide variety of target molecules,

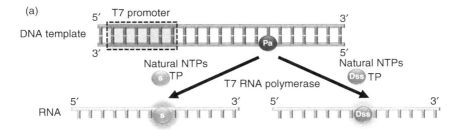

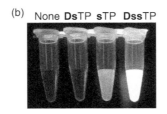

Figure 13.10 T7 transcription involving the **s–Pa** and **Dss–Pa** pairs. (a) Scheme for the site-specific incorporation of fluorescent unnatural bases, **s** and **Dss**, into RNA, opposite **Pa** in DNA templates, by T7 RNA polymerase. (b) Fluorescence of **s**TP and **Dss**TP, in comparison with that of **Ds**TP, upon irradiation at 365 nm.

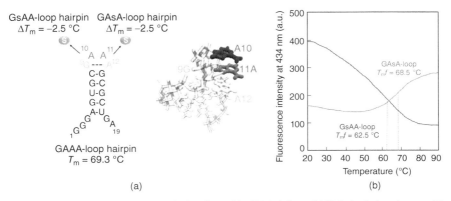

(a)   (b)

Figure 13.11 Local structural analysis of a stable GAAA-loop RNA hairpin by site-specific fluorescent **s** labeling. (a) Structure of the RNA hairpin with a GAAA loop (PDB 1ZIF). The second A (A10) and the third A (A11) in the loop are colored in dark gray and light gray, respectively. The difference between the T_m value for the original 19-mer RNA hairpin and that for the **s**-substituted RNA hairpin at A10 or A11 (ΔT_m), determined from their melting curves monitored by UV absorbance, are indicated. (b) Melting curves of the fluorescence intensity at 434 nm of RNA hairpins with **s** at position 10 or 11, excited at 352 nm. T_m values ($T_m f$), which were determined from the fluorescent profiles, are also indicated.

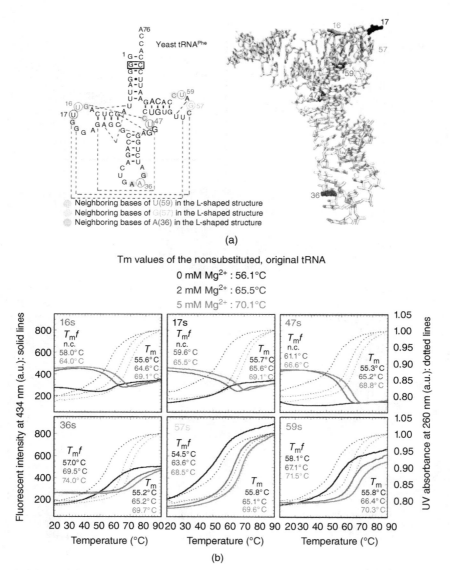

Figure 13.12 Site-specific fluorescent probing to analyze the dynamics of the local structural features of yeast tRNA^Phe. (a) Structure of yeast tRNA^Phe. The positions replaced with **s** are indicated by open circles in the secondary structure (left) and by the corresponding colors in the 3-D structure (right, PDB 1EVV). Left: The dotted lines indicate base–base interactions in the 3-D structure. The G–C pair that was changed from the original C–G pair is indicated in an enclosed box. Neighboring bases of positions U59 (red), G57 (orange), and A36 (purple) are indicated as filled circles. Right: yellow spheres represent Mg^{2+}. (b) Melting curves of the fluorescence intensity and UV absorption for the tRNA molecules containing **s** at various, specific positions, in the presence of different Mg^{2+} concentrations. Melting temperatures, determined from fluorescence profiles ($T_m f$) and UV absorbance profiles (T_m), are indicated within each graph. n.c.: Not calculated. (*See color plate section for the color representation of this figure.*)

such as metal ions, low-molecular-weight materials, peptides, proteins, viruses, and cells.[47–52] Upon binding their target molecules, the aptamers form a specific tertiary structure in the complex, and, thus, the site-specific fluorophore labeling of the aptamers could provide a system to monitor the structural differences before and after the binding by their fluorescence changes.[53–56] To demonstrate this system, we introduced the FAM-linked **y** base (FAM-**y**) into a specific position of an antitheophylline RNA aptamer (Fig. 13.9c).[39] When bound to theophylline, the RNA aptamer forms a specific tertiary structure with a unique base triplet, U6–U23–A28 (Fig. 13.9d), located proximally to the theophylline binding site.[57–62] In the base triplet, the 4-keto group of U6 is not directly involved in the triplet formation, and, thus, U6 can be replaced with FAM-**y**. The crystallographic structure of the aptamer–theophylline complex predicted that the FAM moiety of the unnatural base in the complex might protrude from the structure (Fig. 13.9e), resulting in an increase in the fluorescence intensity upon theophylline binding.

We prepared the transcript (41-mer) of the anti-RNA aptamer, in which FAM-**y** was introduced at the position corresponding to U6, by using T7 RNA polymerase with FAM-**y**TP (1 mM), natural base NTPs (1 mM), and the **s**-containing DNA template. The 41-mer RNA was purified by gel electrophoresis. The binding affinity of the FAM-**y**-containing RNA aptamer ($K_d = 206$ nM) to theophylline was as high as that of the unmodified aptamer ($K_d = 331$ nM), and thus there was no significant difference in the complex structures between the modified and unmodified aptamers. As expected, the fluorescence intensity of the FAM residue in the aptamer, upon 434 nm excitation, increased in response to higher concentrations of theophylline (0.25–20 μM) (Fig. 13.9f and g). The fluorescence alteration was specific for theophylline binding, and no significant intensity alteration of the aptamer was observed by the addition of 20 μM caffeine, a theophylline analog with an additional methyl group. Several tens of picomoles (~50 pmol) of theophylline could be detected by this system.

13.3.2 Local Structure Analyses of Functional RNA Molecules by s Labeling

The fluorescence intensity of **s** is quenched when the **s** base stacks with neighboring bases in RNA molecules, and, thus, the site-specific **s** labeling of RNA can be used to analyze the local structural features of functional RNA molecules, as well as their intra- or intermolecular interactions. Here, we describe the site-specific **s** incorporation into a GAAA-loop RNA hairpin and a tRNA molecule by T7 transcription mediated by the **s**–**Pa** pair, and show their characteristic profiles, depending on the labeling site, temperature, and Mg^{2+} concentration (Figs. 13.11 and 13.12).

The first example is the site-specific **s** labeling of the GAAA-loop region in an RNA hairpin (19-mer RNA). The GAAA-loop hairpin, one of the GNRA-loop hairpins (N = any natural bases, R = purines), exhibits high thermal stability relative to those of other tetra-loop hairpins. NMR studies indicated that the tetra-loop contains a sheared G–A pair between the first G and fourth A, the third base (R) stacks tightly with the fourth A, and the second base (N) is relatively flexible as compared to the third base.[63,64] According to the structural information, we prepared a 19-mer RNA, in which the second or third base of the GAAA loop was replaced with **s**, by T7 transcription using the **Pa**-containing DNA template with 2 mM **s**TP and 2 mM natural base NTPs. After 3 h transcription at 37 °C, the full-length transcripts were purified by denaturing gel-electrophoresis. In the fluorescence measurements at different temperatures, the fluorescence intensity melting curves of each RNA transcript were normalized by that of the nucleoside monomer of **s** or **s**TP, because the quenching of the **s** fluorescence by collision events with the solvent increases at higher temperatures, and such effects should be integrated in the data analysis.

The substitution of **s** in place of the second or third base did not affect the thermal stability of the hairpins and thus did not induce any large structural changes of the hairpins with the GsAA- and GAsA-loops (Fig. 13.11a). Although the T_m values of the s-containing loop hairpins are slightly reduced by 2.5°C, these hairpins are more stable than GCAA-loop hairpin, another thermostable GNRA-loop hairpin.[63] The fluorescence intensity alteration of each RNA hairpin with the GsAA- or GAsA-loop was measured at temperatures ranging from 20 to 90 °C, and revealed the characteristic fluorescence profiles (Fig. 13.11b), which reflected the structural features of the original GAAA-loop hairpin. At physiological temperatures, the fluorescence intensity of the GsAA-loop hairpin was much larger than that of the GAsA-loop hairpin and gradually diminished with increasing temperature (Type 1). This indicated that the second base is exposed to the solvent from the loop at physiological temperature, and it begins to interact with the neighboring bases when the structure denatures at higher temperatures. In contrast, the low fluorescence intensity of the GAsA-loop hairpin below the melting temperature quickly increased at around 60–76 °C (Type 2). Thus, the third base stacks with the fourth A base in the hairpin structure. The melting temperature ($T_m f$ = 68.5 °C) obtained from the fluorescence profile of the GAsA-loop hairpin was as high as the UV melting temperature (T_m = 69.3 °C) of the original GAAA-loop hairpin, while the $T_m f$ value (62.5 °C) of the GsAA-loop hairpin was much lower than the T_m value for the GAAA-loop hairpin. Since the UV melting temperature reflects the thermal stability of the entire RNA structure, these results suggested that the stacking between the third and fourth bases in the loop is one of the main stabilizing factors of the GAAA-loop hairpin structure. In contrast, the second A is flexible in the hairpin structure and has an insignificant effect on the thermal stability.

The second example is the site-specific **s** probing of a yeast phenylalanine tRNA, which forms the canonical L-shaped structure by the interaction between the D- and TΨC-loops in the presence of magnesium ion (76-mer) (Fig. 13.12a). The **Pa**-containing DNA templates were prepared by hybridization between chemically synthesized template (94-mer) and nontemplate (94-mer) fragments. We synthesized six tRNA molecules containing the **s** base at specific positions: U16, U17, A36, U47, G57, or U59. We chose these positions because they are not involved in any base pairing in the tertiary structure.[65] The changes in the fluorescence of each tRNA molecule, in the presence or absence of magnesium ion, were monitored while increasing the temperature from 20 to 90 °C (Fig. 13.12b). These tRNA molecules had similar melting temperatures (dotted lines in Fig. 13.12b), and, thus, the substitution of **s** at these positions did not affect the tRNA L-shaped structure in the presence of magnesium ion.

In the presence of magnesium ion, the fluorescence profiles (solid lines in Fig. 13.12b) of the tRNA molecules were categorized into two types (Type 1: 16s, 17s, and 47s modifications; Type 2: 36s, 57s, and 59s modifications), as shown for those of the GAAA-loop RNA hairpin. The local conformations around each base, as presumed by the characteristic fluorescent profiles, correlated closely with those from the X-ray crystallographic analyses, which revealed that U16, U17, and U47 are not stacked with any other bases, and A36, G57, and U59 are stacked with the neighboring bases in the tRNA L-shaped structure. In the absence of magnesium ion, the fluorescence profiles of 16s, 17s, and 47s showed the type 2 patterns, reaffirming the significance of magnesium ion for the formation of the L-shaped tRNA. A comparison of each $T_m f$ value with its T_m value allows an assessment of the stability of the local structural motif and the denaturing process of the L-shaped structure. The $T_m f$ values of 16s and 17s are significantly lower than their T_m values, and thus the stacking interaction at the $5'$-side of the D-loop is first weakened at around $59\,°C$ in the presence of 2 mM magnesium ion. The extra loop containing U47 then becomes unstable at around $61\,°C$, and this is followed by the denaturation of the interaction between the D- and TΨC-loops at $64–67\,°C$. The anticodon stem and loop structure is the most stable part, and the $T_m f$ value at the 36s position is as high as $70\,°C$. In this manner, the site-specific incorporation of fluorescent base analogs into RNA is useful for monitoring the local structural alterations of functional RNA molecules.

13.4 CONCLUSIONS

Recent advances in unnatural base pair studies have enabled the site-specific incorporation of extra components into DNA and RNA molecules by replication and transcription. In the early 2000s, unnatural base pair systems were applied to real-time multiplex PCR methods, such as the Plexor system using the **isoG–isoC** pair. At present, several types of unnatural base pairs, such as **Ds–Px**[8,10], **Z–P**[7], and **5SICS–MMO2**[3,4], can function as a third base pair with higher efficiency and selectivity, relative to those of the **isoG–isoC** pair, in PCR. Thus, highly sensitive real-time PCR systems with low backgrounds might be developed in the future, by the site-specific labeling of DNA mediated by the improved unnatural base pair systems.

Currently, the site-specific labeling of large RNA molecules mainly relies on a laborious and time-consuming method involving chemical synthesis and enzymatic ligation.[66] Alternative methods employing unnatural base pair systems have been developed as described here. Since the chemical synthesis of large DNA molecules is much easier than that of RNA, transcription mediated by unnatural base pairs using DNA templates, in which the complementary unnatural bases are embedded at desired positions, is an attractive method for the site-specific fluorescent labeling of large RNA molecules. In addition, some unnatural base pairs can function as a third base pair in PCR and thus facilitate DNA template preparation and amplification.

For large RNA molecules, the site-specific labeling with the unnatural base pair systems offers two types of uses:[39] (1) passive labeling of RNA molecules at a specific position that has an insignificant effect on the RNA functions and (2) active labeling for monitoring the fluorescent intensity alterations, as a fluorescent probe to understand the structures and dynamics of functional RNA molecules for detection as an analyte. For the passive labeling, a wide variety of fluorescent molecules with different excitation and emission wavelengths are attached to unnatural bases via an alkyne linker, and these fluorophore-linked unnatural base substrates can be incorporated into RNA, opposite their pairing partners in DNA templates, by RNA polymerases. The selectivity and efficiency of the unnatural base incorporation depend on the sequence context around the unnatural base, incorporation positions, and modification species.[67] Meanwhile, active labeling methods could confer new functionalities to RNA molecules, such as sensor systems of RNA aptamers. Unnatural bases that function as fluorescent analogs sensitively alter their fluorescence intensities, depending on the conformational changes around the incorporation site in functional RNA molecules. This is useful for local structural analyses of large, intricately folded RNAs or their complexes with other molecules. However, these applications are just the starting point of unnatural base pair technologies, and the application range is still limited. We hope that many researchers will help extending possible areas of applications of the new systems.

13.5   PROSPECTS AND OUTLOOK

We have described single labeling methods of nucleic acid molecules by genetic alphabet expansion. By using the complementary set or mixing of several different unnatural base pairs, the site-specific labeling of multiple sites with different fluorophores or fluorophore–quencher sets of a single RNA molecule could be possible and thus provide a FRET system for further analyses of the conformation and dynamics of large nucleic acid molecules.[46,68] Another challenge of unnatural base pair technology is to create a living cell with the expanded genetic system, in which the cellular DNA contains unnatural base pairs at specific positions. Using the new living cells, specific gene expression including the transcripts could be traced by fluorescent labeling through the unnatural bases. The living cells could proliferate with the replicated cellular DNAs by the addition of the unnatural base nucleosides into the culture medium. To phosphorylate the incorporated unnatural base nucleosides in the cell, a specific kinase might be required for the system.[69] The replication of DNA containing unnatural base pairs for the cell proliferation would rely on the intake of the unnatural base nucleosides, and thus the containment of the artificial cells would be achieved, since they require the unnatural base nucleoside nutrients. Although several breakthroughs are still needed to create such an artificial system, a living cell system with an increased genetic alphabet of DNA would provide a powerful synthetic biology tool to understand the complicated cellular systems of gene expression, modification, and regulation.

ACKNOWLEDGMENTS

This work was supported by Grants-in-Aid for Scientific Research (KAKENHI 19201046 to I. Hirao) and by the Targeted Proteins Research Program and the RIKEN Structural Genomics/Proteomics Initiative, the National Project on Protein Structural and Functional Analyses, from the Ministry of Education, Culture, Sports, Science, and Technology of Japan.

REFERENCES

1. Piccirilli, J. A.; Krauch, T.; Moroney, S. E.; Benner, S. A. *Nature* **1990**, *343*, 33.

2. Tor, Y.; Dervan, P. B. *J. Am. Chem. Soc.* **1993**, *115*, 4461.

3. Seo, Y. J.; Malyshev, D. A.; Lavergne, T.; Ordoukhanian, P.; Romesberg, F. E. *J. Am. Chem. Soc.* **2011**, *133*, 19878.

4. Malyshev, D. A.; Seo, Y. J.; Ordoukhanian, P.; Romesberg, F. E. *J. Am. Chem. Soc.* **2009**, *131*, 14620.

5. Seo, Y. J.; Matsuda, S.; Romesberg, F. E. *J. Am. Chem. Soc.* **2009**, *131*, 5046.

6. Yang, Z.; Sismour, A. M.; Sheng, P.; Puskar, N. L.; Benner, S. A. *Nucleic Acids Res.* **2007**, *35*, 4238.

7. Yang, Z.; Chen, F.; Alvarado, J. B.; Benner, S. A. *J. Am. Chem. Soc.* **2011**, *133*, 15105.

8. Yamashige, R.; Kimoto, M.; Takezawa, Y.; Sato, A.; Mitsui, T.; Yokoyama, S.; Hirao, I. *Nucleic Acids Res.* **2012**, *40*, 2793.

9. Hikida, Y.; Kimoto, M.; Yokoyama, S.; Hirao, I. *Nat. Protoc.* **2010**, *5*, 1312.

10. Kimoto, M.; Kawai, R.; Mitsui, T.; Yokoyama, S.; Hirao, I. *Nucleic Acids Res.* **2009**, *37*, e14.

11. Hirao, I. *Biotechniques* **2006**, *40*, 711.

12. Kimoto, M.; Mitsui, T.; Harada, Y.; Sato, A.; Yokoyama, S.; Hirao, I. *Nucleic Acids Res.* **2007**, *35*, 5360.

13. Hirao, I.; Kimoto, M.; Mitsui, T.; Fujiwara, T.; Kawai, R.; Sato, A.; Harada, Y.; Yokoyama, S. *Nat. Methods* **2006**, *3*, 729.

14. Kimoto, M.; Endo, M.; Mitsui, T.; Okuni, T.; Hirao, I.; Yokoyama, S. *Chem. Biol.* **2004**, *11*, 47.

15. Hirao, I.; Ohtsuki, T.; Fujiwara, T.; Mitsui, T.; Yokogawa, T.; Okuni, T.; Nakayama, H.; Takio, K.; Yabuki, T.; Kigawa, T.; Kodama, K.; Nishikawa, K.; Yokoyama, S. *Nat. Biotechnol.* **2002**, *20*, 177.

16. Kimoto, M.; Cox, R. S., 3rd; Hirao, I. *Expert Rev. Mol. Diagn.* **2011**, *11*, 321.

17. Rich, A. In Horizons Biochem.; Kasha, M.; Pullman, B., Eds.; Academic Press: New York, 1962; pp 103.

18. Switzer, C.; Moroney, S. E.; Benner, S. A. *J. Am. Chem. Soc.* **1989**, *111*, 8322.

19. Benner, S. A. *Acc. Chem. Res.* **2004**, *37*, 784.

20. Roberts, C.; Bandaru, R.; Switzer, C. *J. Am. Chem. Soc.* **1997**, *119*, 4640.

21. Prudent, J. R. *Expert Rev. Mol. Diagn.* **2006**, *6*, 245.

22. Sherrill, C. B.; Marshall, D. J.; Moser, M. J.; Larsen, C. A.; Daude-Snow, L.; Jurczyk, S.; Shapiro, G.; Prudent, J. R. *J. Am. Chem. Soc.* **2004**, *126*, 4550.

23. Sepiol, J.; Kazimierczuk, Z.; Shugar, D. *Z. Naturforsch.* **1976**, *31*, 361.

24. Switzer, C. Y.; Moroney, S. E.; Benner, S. A. *Biochemistry* **1993**, *32*, 10489.

25. Henry, A. A.; Romesberg, F. E. *Curr. Opin. Chem. Biol.* **2003**, *7*, 727.

26. Hunziker, J.; Mathis, G. *Chimia* **2005**, *59*, 780.

27. Bergstrom, D. E. Current Protocols in Nucleic Acid Chemistry. Unnatural Nucleosides with Unusual Base Pairing Properties. 2009 37:1.4:1.4.1–1.4.32.

28. Krueger, A. T.; Kool, E. T. *Chem. Biol.* **2009**, *16*, 242.

29. Hirao, I.; Kimoto, M.; Yamashige, R. *Acc. Chem. Res.* **2012**, *45*, 2055.

30. Hirao, I. *Curr. Opin. Chem. Biol.* **2006**, *10*, 622.

31. Matray, T. J.; Kool, E. T. *Nature* **1999**, *399*, 704.

32. Loakes, D.; Holliger, P. *Chem. Commun.* **2009**, 4619.

33. Hirao, I.; Mitsui, T.; Kimoto, M.; Yokoyama, S. *J. Am. Chem. Soc.* **2007**, *129*, 15549.

34. Yamashige, R.; Kimoto, M.; Mitsui, T.; Yokoyama, S.; Hirao, I. *Org. Biomol. Chem.* **2011**, *9*, 7504.

35. Kimoto, M.; Mitsui, T.; Yokoyama, S.; Hirao, I. *J. Am. Chem. Soc.* **2010**, *132*, 4988.

36. Kimoto, M.; Mitsui, T.; Yamashige, R.; Sato, A.; Yokoyama, S.; Hirao, I. *J. Am. Chem. Soc.* **2010**, *132*, 15418.

37. Fujiwara, T.; Kimoto, M.; Sugiyama, H.; Hirao, I.; Yokoyama, S. *Bioorg. Med. Chem. Lett.* **2001**, *11*, 2221.

38. Endo, M.; Mitsui, T.; Okuni, T.; Kimoto, M.; Hirao, I.; Yokoyama, S. *Bioorg. Med. Chem. Lett.* **2004**, *14*, 2593.

39. Kawai, R.; Kimoto, M.; Ikeda, S.; Mitsui, T.; Endo, M.; Yokoyama, S.; Hirao, I. *J. Am. Chem. Soc.* **2005**, *127*, 17286.

40. Moriyama, K.; Kimoto, M.; Mitsui, T.; Yokoyama, S.; Hirao, I. *Nucleic Acids Res.* **2005**, *33*, e129.

41. Moser, M. J.; Ruckstuhl, M.; Larsen, C. A.; Swearingen, A. J.; Kozlowski, M.; Bassit, L.; Sharma, P. L.; Schinazi, R. F.; Prudent, J. R. *Antimicrob. Agents Chemother.* **2005**, *49*, 3334.

42. Svarovskaia, E. S.; Moser, M. J.; Bae, A. S.; Prudent, J. R.; Miller, M. D.; Borroto-Esoda, K. *J. Clin. Microbiol.* **2006**, *44*, 4237.

43. Marshall, D. J.; Reisdorf, E.; Harms, G.; Beaty, E.; Moser, M. J.; Lee, W. M.; Gern, J. E.; Nolte, F. S.; Shult, P.; Prudent, J. R. *J. Clin. Microbiol.* **2007**, *45*, 3875.

44. Fukunaga, R.; Harada, Y.; Hirao, I.; Yokoyama, S. *Biochem. Biophys. Res. Commun.* **2008**, *372*, 480.

45. Ward, D. C.; Reich, E.; Stryer, L. *J. Biol. Chem.* **1969**, *244*, 1228.

46. Mitsui, T.; Kimoto, M.; Kawai, R.; Yokoyama, S.; Hirao, I. *Tetrahedron* **2007**, *63*, 3528.

47. Stoltenburg, R.; Reinemann, C.; Strehlitz, B. *Biomol. Eng.* **2007**, *24*, 381.

48. Shamah, S. M.; Healy, J. M.; Cload, S. T. *Acc. Chem. Res.* **2008**, *41*, 130.

49. Keefe, A. D.; Pai, S.; Ellington, A. *Nat. Rev. Drug Discov.* **2010**, *9*, 537.

50. Hermann, T.; Patel, D. J. *Science* **2000**, *287*, 820.

51. Ellington, A. D.; Szostak, J. W. *Nature* **1990**, *346*, 818.

52. Tuerk, C.; Gold, L. *Science* **1990**, *249*, 505.

53. Cho, E. J.; Lee, J. W.; Ellington, A. D. *Annu. Rev. Anal. Chem.* **2009**, *2*, 241.

54. Famulok, M.; Mayer, G. *Acc. Chem. Res.* **2011**, *44*, 1349.

55. Jhaveri, S.; Rajendran, M.; Ellington, A. D. *Nat. Biotechnol.* **2000**, *18*, 1293.

56. Nutiu, R.; Li, Y. *Methods* **2005**, *37*, 16.

57. Jenison, R. D.; Gill, S. C.; Pardi, A.; Polisky, B. *Science* **1994**, *263*, 1425.

58. Zimmermann, G. R.; Jenison, R. D.; Wick, C. L.; Simorre, J. P.; Pardi, A. *Nat. Struct. Biol.* **1997**, *4*, 644.

59. Zimmermann, G. R.; Shields, T. P.; Jenison, R. D.; Wick, C. L.; Pardi, A. *Biochemistry* **1998**, *37*, 9186.

60. Sibille, N.; Pardi, A.; Simorre, J. P.; Blackledge, M. *J. Am. Chem. Soc.* **2001**, *123*, 12135.

61. Zimmermann, G. R.; Wick, C. L.; Shields, T. P.; Jenison, R. D.; Pardi, A. *RNA* **2000**, *6*, 659.

62. Clore, G. M.; Kuszewski, J. *J. Am. Chem. Soc.* **2003**, *125*, 1518.

63. Heus, H. A.; Pardi, A. *Science* **1991**, *253*, 191.

64. Jucker, F. M.; Heus, H. A.; Yip, P. F.; Moors, E. H.; Pardi, A. *J. Mol. Biol.* **1996**, *264*, 968.

65. Jovine, L.; Djordjevic, S.; Rhodes, D. *J. Mol. Biol.* **2000**, *301*, 401.

66. Solomatin, S.; Herschlag, D. *Methods Enzymol.* **2009**, *469*, 47.

67. Morohashi, N.; Kimoto, M.; Sato, A.; Kawai, R.; Hirao, I. *Molecules* **2012**, *17*, 2855.

68. Kimoto, M.; Hikida, Y.; Hirao, I. *Isr. J. Chem.* **2013**, *53*, 450.

69. Wu, Y.; Fa, M.; Tae, E. L.; Schultz, P. G.; Romesberg, F. E. *J. Am. Chem. Soc.* **2002**, *124*, 14626.

14

FLUORESCENT C-NUCLEOSIDES AND THEIR OLIGOMERIC ASSEMBLIES

PETE CRISALLI AND ERIC T. KOOL

Department of Chemistry, Stanford University, Stanford CA 94305, USA

14.1 INTRODUCTION

In some ways, DNA is remarkably well suited to molecular sensing. Of course, a strand of DNA can bind with high affinity and selectivity to complementary nucleic acids, but the molecular recognition abilities of DNA go far beyond this. The individual DNA bases offer varied functional groups and aromatic structures that can bind both small and large molecular species. In addition, DNA can fold into complex shapes (such as in aptamers) that can recognize widely varied molecular epitopes with high selectivity.

However, natural DNA also has limitations in sensing. Although the backbone of DNA is inherently stable with a half-life on the order of billions of years,[1] it is rapidly degraded in biological fluids. In addition to the backbone, the natural heterocyclic nucleobases arrayed along the backbone are limited in their properties and stability. For example, DNA is prone to depurination in even mildly acidic conditions as a result of the *N*-glycosidic linkages. In addition, although they are aromatic, the nucleobases are not inherently fluorescent chromophores.[2–4] As a result, sensors built from DNA have classically required the attachment of one or more fluorophores to augment the natural structure.

Fluorescent Analogs of Biomolecular Building Blocks: Design and Applications, First Edition.
Edited by Marcus Wilhelmsson and Yitzhak Tor.
© 2016 John Wiley & Sons, Inc. Published 2016 by John Wiley & Sons, Inc.

DNA offers more than just its favorable recognition abilities to recommend its use in sensor designs. More than most polymers, it is highly water soluble and is very readily synthesized in almost any desired length or sequence. In addition, methods for conjugation of small and large molecules to DNA are well developed. Thus, if one could address some of the limitations of natural DNA by making modified components, DNA might be highly useful as a sensor of many species.

We and others have been approaching this problem by development of fluorescent C-nucleosides that can be used alongside, or even in place of, natural nucleobases. The use of the stable C–C glycosidic bond confers chemical and biological stability both to the monomer and in some cases to the DNA around it. Moreover, nucleoside-like structure enables the intimate placement of a fluorophore directly within the DNA, resulting in signals that can be more sensitive to chemical and enzymatic changes that occur within DNA. For example, certain fluorescent dyes can be employed assay the changes in the DNA environment upon even the most subtle alterations in DNA hybridization state. Thus, the C-glycosidic fluorophore adds useful new properties to the DNA as sensor. Furthermore, the charged backbone offers solubility to structures that might not be water soluble otherwise. This marriage of benefits allows DNA to be engineered into a sensor not just of nucleic acids or nucleic acid-processing enzymes but of broader classes of chemical and enzymatic species as well.

An important feature of DNA is its repeating structure that encourages intramolecular interactions of multiple aromatic nucleobases. The inherently π-stacked arrangements of natural DNA bases results in sensitive electronic interactions between them, and when fluorophores replace these bases, these interactions can result in dramatically different absorbance and emission properties, creating new fluorescence behavior not observed in the monomers alone. Such oligomeric dyes are being developed as tools for reporting and sensing in biology and environmental sciences. The repeating DNA-like structure makes thousands of sequences of fluorescent glycosides readily available, each with its unique electronic and optical properties. Sets of multispectral dyes have been developed, sharing a common excitation wavelength but emitting across the visible spectrum.[5] Oligomeric fluorophores often have delocalized excited states, conferring other unusual photophysical properties, such as highly efficient quenching and resonance energy transfer involving their delocalized states.[5–7] Their sensitive electronic states can be useful well beyond nucleic acids sensing, and many classes of analytes can be detected, including metal ions, vapor-phase organic molecules and toxic gases, and a variety of enzyme activities.

This chapter focuses on the development of fluorescent C-glycosides as monomers and oligomers, and their application in detecting an assortment of analytes. For more detailed reviews of DNA multichromophore systems and the attachment of fluorescent molecules to a DNA scaffold, the reader is directed to recently published monographs.[8–10]

14.2   DESIGN, SYNTHESIS, CHARACTERIZATION, AND PROPERTIES OF FLUORESCENT C-GLYCOSIDE MONOMERS

14.2.1   Design of Fluorescent C-Glycoside Monomers

The design of fluorescent C-glycosides is inevitably related to their desired uses. A large variety of fluorescent C-glycosides have been prepared and can be employed in a wide range of applications. As shown in Figure 14.1, fluorescent C-glycosides may be analogs or mimics of natural bases, such as the size expanded xDNA,[11] yDNA,[12] and xRNA[13] molecules described by Kool and coworkers or the thiophene-fused compounds prepared by Tor and coworkers.[14,15] Alternatively, completely nonnatural C-glycosides can be prepared, incorporating heterocyclic or hydrocarbon fluorophores as has been done by multiple groups (Fig. 14.2). These fluorophores are generally selected for their individual properties, such as environmental sensitivity, their ability to form excimers or exciplexes, or even their ability to stabilize triplex DNA structure, as is the case with the TRIPsides developed by Gold and coworkers.[16–19] Inclusion of a single fluorophore into a DNA sequence can be easily accomplished

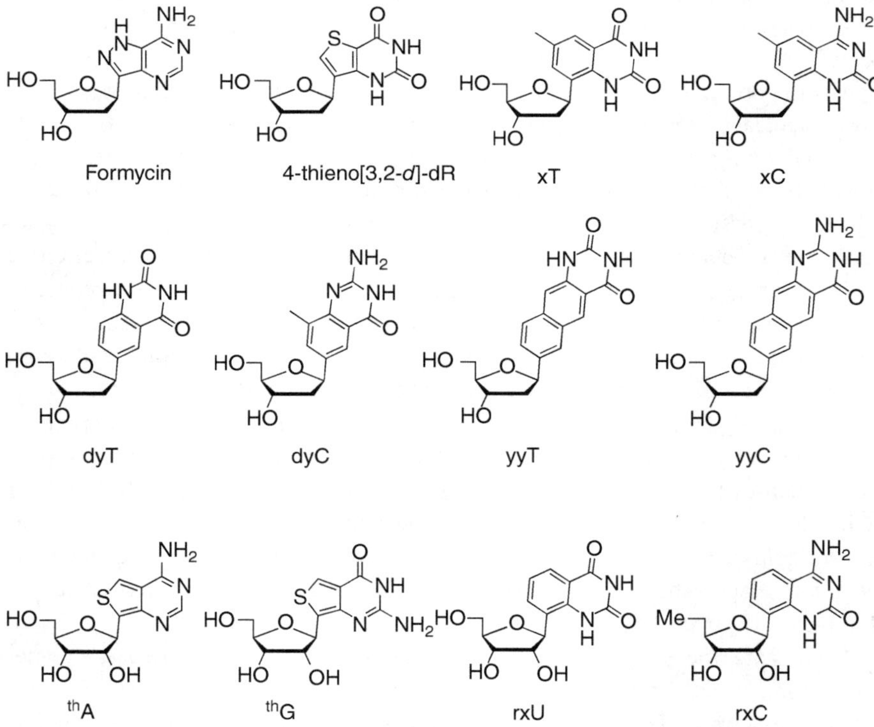

Figure 14.1   Structures of fluorescent C-glycosides that are isomorphic with regular RNA and DNA bases or maintain canonical Watson–Crick hydrogen bonding.

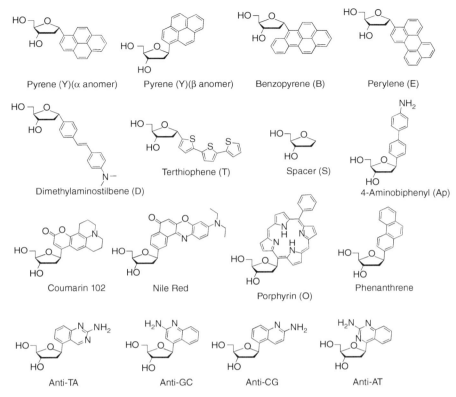

Figure 14.2   Fluorescent C-glycosides from inherently fluorescent compounds. Spacer (S) is included as a result of its common use for solubility or pairing opposite a large residue.

via synthesis of a phosphoramidite applicable to automated DNA synthesis or, in a few instances, a triphosphate derivative for enzymatic incorporation.

Although Figure 14.1 shows a variety of C-glycosidic fluorophores that have a close resemblance to biomolecules, this chapter focuses on properties and applications of C-glycosidic fluorophores from Figure 14.2. These fluorophores provide a more unique class of C-glycosides that, despite having very little structural similarity to nucleobases, are still extremely efficient at reporting on not only biomolecules such as DNA itself or enzymatic reactions and are being built into new sensors for a wide variety of biological and nonbiological analytes (Sections 14.4 and 14.5).

14.2.2   Synthesis of Fluorescent C-Glycoside Monomers

Two techniques are commonly used for the synthesis of fluorescent C-glycosides, and different anomers can be prepared depending on the procedure used. Upon successful preparation of the free nucleoside, standard techniques can be employed to prepare a phosphoramidite for standard automated DNA synthesis or monophosphates and triphosphates if enzymatic probes are preferred. For a thorough review of the synthesis of C-glycosides, the reader is directed to Reference 20.

An early study of conditions for C–C bond coupling showed that reaction of an organocadmium or organozinc species with Hoffer's chlorosugar was optimal for the formation of the α anomer of the desired fluorescent C-glycoside as the major product in moderate yield (Scheme 14.1).[21–24] ^{1}H NMR and NOE experiments confirmed formation of this anomer and a small amount of the β anomer (the approximate ratio of anomers varied from 3:1 to 8:1 α:β depending upon coupling partner). In many instances, the nonnatural α-anomer has been applied for the end applications. For the β anomer, one must either accept a small yield from this approach or take an additional equilibration step for conversion of some of the α product to the β anomer if such is desired. Treatment of the α anomer with benzenesulfonic acid in refluxing xylenes affords a new mixture of α and β anomers, but with the natural β anomer as the major product. Subsequent deprotection of the toluoyl-protected alcohols affords the free diol monomer, which is converted easily into a phosphoramidite for automated DNA synthesis via 4,4'-dimethoxytrityl (DMT) protection of the 5' alcohol and activation of the 3' alcohol as a phosphoramidite. Alternatively, should a mono- or triphosphate be desired, activation of the 5' alcohol with $POCl_3$ and trimethyl phosphate provides the 5'-monophosphate and further treatment with a pyrophosphate source can provide the desired triphosphate.[25]

In recent years, an alternative methodology has been developed that is both stereoselective and can afford higher yields of a desired β anomer than the above chemistry. Use of palladium catalyzed cross-coupling to activated alkenes has been an attractive synthetic reaction for many years, and an active sugar for cross-coupling to yield DNA-like products as pure β anomers has been developed and widely utilized (Scheme 14.2). In this reaction procedure, the Heck sugar precursor (1) reacts with

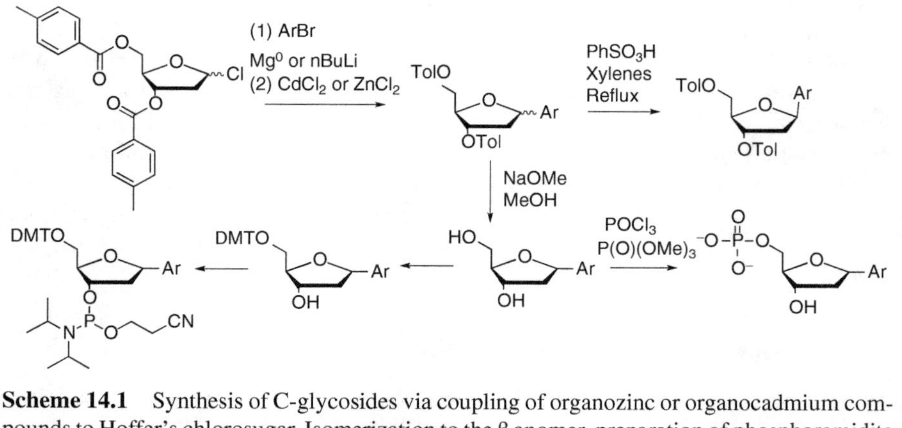

Scheme 14.1 Synthesis of C-glycosides via coupling of organozinc or organocadmium compounds to Hoffer's chlorosugar. Isomerization to the β anomer, preparation of phosphoramidite for automated DNA synthesis, and phosphorylation procedures are also provided.

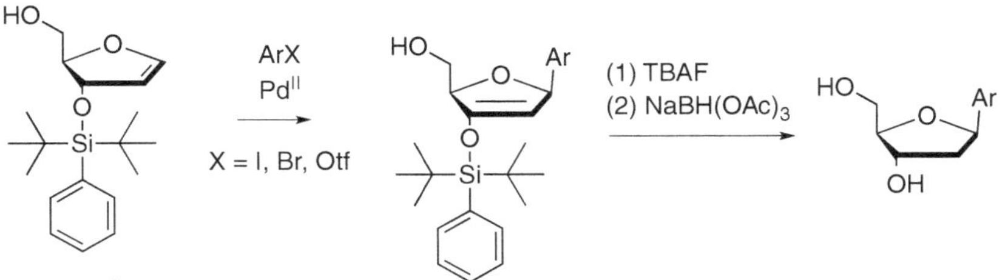

Scheme 14.2 Synthesis of C-glycosides via Heck coupling of an activated aryl compound to a sugar enol ether. Subsequent conversion to a phosphoramidite or monophosphate can be performed as in Scheme 14.1.

an aryl bromide, aryl iodide or, in limited cases, an aryl triflate in the presence of a palladium catalyst to afford the β anomer as the only product.[20,26,27] The downside of this approach is that, despite creating the biologically relevant β anomer, multiple steps are required to attain the Heck precursor. Moreover, additional steps are needed to prepare the C-glycoside diol; deprotection of the silyl group followed by stereoselective ketone reduction is required prior to DMT protection and phosphoramidite activation or phosphate synthesis discussed above.

14.2.3 Characterization and Properties of Fluorescent C-glycoside Monomers

Properties of new monomer nucleosides are routinely documented to provide a baseline for changes that are often observed when placed in DNA or RNA. Table 14.1

TABLE 14.1 Photophysical Properties of Isomorphic Fluorescent C-Glycoside Monomers (in Methanol Unless Otherwise Noted)

Fluorophore	Absorbance Maxima (nm)	Extinction Coefficient ($M^{-1}\,cm^{-1}$)	Emission Maxima (nm)	Quantum Yield
Formycin (water)[28]	280	800	338	0.058
4-thieno[3,2-d]-dR[14]	294	N.D.	351	0.037
dxT[11]	320	3400	377	0.30
dxC[11]	330	4100	388	0.52
dyT[12]	315	3040	379	0.54
dyC[12]	312	2780	394	0.40
dyyT (water)[12]	362	1510	446	0.67
dyyC (water)[12]	371	1690	433	0.35
[th]A (water)[15]	341	7440	420	0.21
[th]G (water)[15]	321	4150	453	0.46
rxU[13]	311	3900	369	0.27
rxC[13]	322	4500	379	0.48

TABLE 14.2   Photophysical Properties of Fluorescent C-Glycoside Monomers (in Methanol Unless Otherwise Noted)

Fluorophore	Absorbance Maxima (nm)	Extinction Coefficient $(M^{-1} cm^{-1})$	Emission Maxima (nm)	Quantum Yield
Pyrene nucleoside[29]	345	39,000	375, 395	0.12
Dimethylamino -stilbene nucleoside[29]	301	21,100	356	0.055
Porphyrin nucleoside[30]	400	32,400	629	0.110
Perylene nucleoside[29]	440	39,200	433, 472	0.88
Benzopyrene nucleoside[29]	394	28,200	408	0.98
Coumarin 102 nucleoside[10]	400	N.D.	515	N.D.
Nile Red nucleoside[27]	557	3380	632	0.09
Phenanthrene nucleoside[31]	252	73,200	N.D.	N.D.
4-Aminobiphenyl nucleoside[32]	278	14,000	370	N.D.
Anti-TA[17]	350	2580	425	N.D.
Anti-GC[16]	330	N.D.	N.D.	N.D.
Anti-CG[18]	330	N.D.	N.D.	N.D.
Anti-AT[19]	350	N.D.	430	N.D.

displays the optical properties of fluorescent C-glycosides that are structurally related to regular DNA and RNA, whereas Table 14.2 provides the properties of fluorescent C-glycosides unrelated to regular biomolecules. It is important to note that the photophysical properties of free monomers can vary strongly when incorporated into DNA, which can be a useful property as discussed in the sections below.

The interactions of fluorophores with natural nucleobases are pertinent to their use in sensors that combine natural structure with fluorescent reporters. Table 14.3 shows how a single neighboring nucleobase can affect fluorescence. A thorough study of the photophysical properties of pyrene, perylene, and benzopyrene when adjacent to each of the four natural bases was published.[33] Results showed that the nature of the nucleobase immediately adjacent to the fluorescent hydrocarbon can have strong effects on the properties of the fluorophore. If the fluorophore is placed immediately adjacent to thymine, potent fluorescence quenching can occur, with the quantum yield dropping to 0.003 for TY, 0.07 for TB, and 0.37 for TE.[33] Generally speaking, adenine has a negligible effect on the emission and quantum yield of pyrene (Y), perylene (E), and benzopyrene (B), with the fluorophore always

TABLE 14.3 Interactions of Fluorescent C-Glycosides[a] with Natural Nucleobases in Synthetic Dinucleotides

Sequence (5′ → 3′)	Emission Maxima (nm)	Quantum Yield
AY	378, 398	0.37
GY	378, 398	0.024
CY	378, 398	0.0006
TY	378, 398	0.0003
AE	453	0.94
GE	451	0.98
CE	450	0.48
TE	450	0.37
AB	415	0.73
GB	414	0.22
CB	413	0.34
TB	413	0.07

[a]Y is Pyrene, E is Perylene, and B is Benzopyrene.
Wilson et al.[33]. Reproduced with permission from John Wiley and Sons.

maintaining a high quantum yield (0.37, 0.94, and 0.73 for AY, AE, and AB, respectively) and emitting at the same maxima as the free monomer.[33] The other two nucleobases show pyrene-selective quenching among these three fluorophores (see Table 14.3).

14.3 IMPLEMENTATION OF FLUORESCENT C-GLYCOSIDE MONOMERS

14.3.1 Environmentally Sensitive Fluorophores

While many of the fluorescent C-glycosides in this chapter do not have strong dipoles, some, such as the coumarin-based dyes, do have strong dipoles, which render them sensitive to the polarity of their environment. Environmentally sensitive fluorophores have been employed in a wide variety of biological experiments to understand the changes in polarity of cellular microenvironments.[8] Ethidium bromide, thiazole orange, and SYBR dyes provide classical examples of molecules that change fluorescence properties upon undergoing close interactions with DNA, but only by associating with a duplex, usually by intercalation. Upon heating or alterations in hybridization state (e.g., DNA melting or in the presence of a mismatch), these dyes are insufficient to monitor small changes in the DNA structure and may simply dissociate. Incorporation of a fluorophore as a stable DNA "base" will ensure continual close interactions with the DNA structure as structure changes, allowing for better understanding of such alterations.

14.3.1.1 Coumarin 102 Nucleoside: One of the earliest uses of a fluorescent C-nucleoside in studying DNA structure was in the preparation of a coumarin 102 nucleoside for incorporation in DNA oligomers via automated synthesis (Fig. 14.2). Coumarin 102 is a well-known environmentally sensitive fluorophore, changing quantum yield, absorbance maximum, fluorescence lifetime, and fluorescence emission based upon its environment.[26] As a result of the size of the coumarin moiety, it is necessary to pair the fluorophore opposite an abasic or spacer (S) site to maintain duplex stability.[26,34–36] Although much of DNA macrostructure is rather well studied, DNA microstructure on a picosecond timescale is not well known and difficult to study. However, by the preparation of the coumarin 102 nucleoside, picosecond-scale fluorescence changes of the fluorophore could be observed based upon the altering microenvironment of the fluorophore.[34–36] As DNA hybridization changed, even subtly, the differences in fluorescence properties could easily be observed, allowing for a greater understanding of rapid alterations of DNA structure in the solution phase.

Indeed, time-resolved Stokes shifts experiments were clearly able to distinguish behavior of the coumarin moiety when the DNA environment altered (Fig. 14.3).[34] Upon incorporation into a stable duplex, the coumarin 102 fluorophore undergoes a redshift in absorbance and a blueshift in fluorescence emission compared to the free monomer, indicating shielding of the fluorophore from water as solvent as it is shielded by the nucleobases in the duplex.[34] If, however, the DNA is melted and the coumarin is more exposed to water in a single strand, the absorbance will shift back toward that of the free monomer as emission shifts red, increasing the Stokes shift relative to hybridized DNA. These and other experiments allowed for a more complete understanding of the changes in solvation of DNA bases and the changes in solvent exposure during melting over a six-decade timescale, which have inherent value in understanding DNA reactivity, organization, and duplex assembly.[35]

Further experiments also showed the ability of coumarin 102 incorporation as a stable DNA "base" to sense changes in DNA conformation upon protein binding[37]

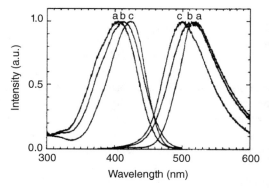

Figure 14.3 Changes of absorption and emission properties of coumarin 102 nucleoside. (a) Free nucleoside in solution, (b) in a melted, single-stranded oligonucleotide sequence at 90 °C, and (c) in a hybridized, double-stranded oligonucleotide sequence. Brauns *et al.*[34]. Reprinted with permission from American Chemical Society.

and DNA lesions[36] using both steady-state and time-resolved fluorescence techniques. By observing the changes in the fluorescence spectrum via these methods, alterations of DNA microstructures could be observed in nanosecond and even picosecond timescales, giving considerably greater resolutions than other existing techniques.

14.3.1.2 Nile Red Nucleoside: Another valuable environmentally sensitive nucleoside has been realized by attachment of the Nile Red fluorophore to deoxyribose.[27] Similar to free Nile Red, the prepared nucleoside displayed strong solvatochromism and solvatofluorochromism, allowing it to serve as a reporter for the microenvironment surrounding the fluorophore in a DNA context.[27] Similar to coumarin 102 above, the Nile Red nucleoside is preferentially paired opposite an abasic site to allow for the size of the fluorophore to properly fit into a DNA duplex without a strongly destabilizing perturbation, although pairing opposite of C or T is considerably less destabilizing and can even be tolerated compared to pairing opposite of A or G.[27]

Interestingly, only small fluorescence changes were observed when the Nile Red nucleoside was incorporated into a DNA strand and paired opposite A, G, C, or T, or the thermodynamically preferred abasic site, all showing absorbance maxima centered around 608 nm and emission at approximately 650 nm (Fig. 14.4a). These results indicate that the microenvironment of all possible pseudo base pairs with Nile Red (even pairing opposite an abasic site) is rather similar from a photophysical point of view, despite the differences observed in thermodynamics via melting temperatures, which varied across a 10 °C range. This might simply be an artifact of the

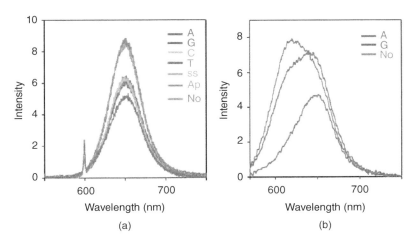

Figure 14.4 Environmental sensitivity of Nile Red nucleoside in DNA (a) with different bases opposite of the fluorophore and (b) of A, G, and no opposing base pairs in the presence of the addition of β-cyclodextrin. Okamoto *et al.*[27]. Reprinted with permission from American Chemical Society. (*See color plate section for the color representation of this figure.*)

sequences used in the experiments and the location of the fluorophore in the middle of the DNA sequence rather than at an end where it may undergo greater changes in microenvironment depending on flexibility and solvation. Since absorbance and emission properties were unchanged in the reported experiments, the results indicate a similar microenvironment of the DNA duplexes (at the same temperature), even with largely different duplex stabilities.

However, the use of the Nile Red nucleoside does display the expected alterations in absorbance and emission properties when β-cyclodextrin is added.[27] Nile Red is known to form an inclusion complex with β-cyclodextrin that provides a hydrophobic environment that induces significant changes in the absorbance and emission maxima as well as the quantum yield of the fluorophore. Intriguingly, when β-cyclodextrin was added to duplexes containing the Nile Red nucleoside strong differences in photophysical behavior were observed. When paired opposite an abasic site, C or T (which are the thermodynamically more stable pairing partners), little changes in fluorescence were observed upon addition of β-cyclodextrin, indicating that the Nile Red moiety remains stacked among the bases in the DNA duplex to "pair" with a partner rather than flipping out of the duplex to interact with the cyclodextrin.[27] If, however, the Nile Red nucleoside was paired opposite A or G, a blueshift in emission could be observed upon addition of β-cyclodextrin, indicating a bulged structure and preference of the Nile Red moiety to interact with the cyclodextrin instead of the DNA duplex (Fig. 14.4b).[27] Consequently, easily resolved preferences of the Nile Red for the DNA duplex could be observed based upon the size limitations of the duplex structure.

14.3.1.3 General Conclusions: Environmentally Sensitive Fluorophores in DNA: Environmentally sensitive fluorophores stably incorporated into DNA via a C-glycosidic bond provide considerably more valuable probes of DNA structure than standard use of ethidium bromide or other common DNA dyes that change photophysical properties upon simple intercalation. The coumarin 102 fluorophore gives easily observed changes in both steady-state and time-resolved methods that can allow changes in DNA microstructure on even the picosecond timescale or show changes upon interaction with a protein. The Nile Red nucleoside, on the other hand, gives information about internal duplex environment as well as stability. By differences in photophysical properties on the addition of β-cyclodextrin, information about duplex stability around the large fluorophore can easily be observed.

14.3.2 Pyrene Nucleoside in DNA Applications

The pyrene nucleoside monomer has been utilized in a wider variety of nucleic acid applications. This monomer has unique capabilities in DNA applications as it has been shown to be highly stabilizing in duplexes when paired opposite an abasic site, allowing it to be an excellent probe for DNA damage.[38,39] Further, the optical properties of pyrene in a DNA context provide useful advantages. Pyrene, as noted above, can be well quenched by thymine, which can allow activation of fluorescence in the presence of certain glycosylases. Moreover, the ability of pyrene to form a bright excimer when two monomers are placed in close proximity can provide for excellent spectral resolution when the Stokes shift of the fluorophore increases from 30 nm to approximately 140 nm.

14.3.2.1 Pyrene Deoxynucleoside 5'-Triphosphate (dPTP) as a Reporter of TdT Activity:

A simple, common enzyme with important use in biology is terminal deoxynucleotidyl transferase (TdT).[40] TdT is a template-independent polymerase that catalyzes the addition of random nucleotides to the 3' end of a DNA strand. Biologically, the enzyme serves to increase diversity in the immune system by creating randomized N regions during V(D)J recombination.[40] The enzyme has gained large analytical use for its ability to polymerize random extension of the 3' hydroxyl of a DNA sequence, serving as an effective means of adding radio- or fluorescently labeled nucleotides.

Since TdT is template-independent polymerase and has a large, flexible active site, it can be of great use in the addition of nonnatural nucleosides with interesting fluorescent properties. Particularly, the preparation of a pyrene nucleoside triphosphate (dPTP) and utilization of the properties of TdT created a new, highly efficient means of detecting the activity of TdT while also serving as a biological means of oligomerizing a nonnatural nucleotide.[40] As pyrene is known to undergo a change in emission from 375 and 395 nm as a monomer to 480 nm for excimer emission when two or more pyrene moieties are placed in close proximity to each other, incorporation of oligomers of dPTP to the end of a nucleic acid strand can easily be visualized by excimer formation (Fig. 14.5a).[40] Experiments showed that TdT is able to successfully incorporate multiple dPTP residues and the products display characteristic

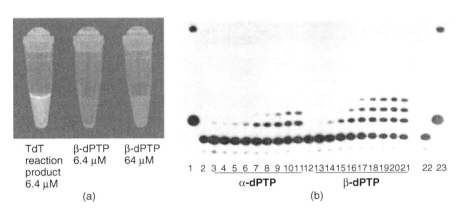

TdT reaction product 6.4 μM	β-dPTP 6.4 μM	β-dPTP 64 μM

(a)

1 2 3 4 5 6 7 8 9 101112131415161718192021   22 23

α-dPTP   β-dPTP

(b)

Figure 14.5 Enzymatic oligomerization of both anomers of dPTP by the template -independent TdT polymerase. (a) Visual observation of excimer formed by TdT incorporation of β-dPTP compared to unreacted monomer. (b) Time course studies of enzymatic incorporation. Lanes 1 and 23: 10 bp markers, lanes 2 and 12: template, lanes 3–11: incorporation of α-dPTP with increasing incubation time (from 10 s to 50 min), lanes 12–21: incorporation of β-dPTP with increasing incubation time (from 10 s to 50 min). Cho et al.[40]. Reproduced with permission from John Wiley and Sons.

excimer emission with a broadened peak centered around 480 nm. Gel experiments indicated that as many as four pyrene residues could be incorporated into the 3′ end of the growing DNA strand by TdT before the enzyme would apparently stall (Fig. 14.5b).[40] Moreover, the enzyme could incorporate both the α and β anomers of dPTP, also showing the large flexibility of the active site of the TdT enzyme itself.[40]

This multiincorporation of dPTP offers an appealing alternative for performing a TUNEL assay for apoptosis, which requires the use of TdT in conjunction with nick-end labeling. While normal TUNEL assays would use fluorescein, a powerful fluorophore that only has an approximately 20-nm Stokes shift and is self-quenched by the incorporation of multiple dyes, use of dPTP incorporation provides a Stokes shift of approximately 130 nm and increases emission intensity as more monomers are incorporated into the growing oligomer. These large spectral changes show the great value of using fluorescent C-glycosides as reporters for enzymatic activity compared to traditional fluorophores.

Other unnatural C-glycosides were also recently found to be substrates for TdT. Both dxTTP and dxCTP have also been employed in enzymatic incorporation by TdT and show different fluorescence behavior than pyrene upon incorporation.[41] While pyrene showed excimer formation and a significant redshift in emission, dxC oligomerization by TdT simply results in a fluorescence enhancement at the regular emission of dxC.[41] Both dxC and dxT were incorporated almost as efficiently as the natural dTTP and dCTP substrates, and more efficiently than dPTP, with as many as 30 monomers being incorporated as compared to the 3 or 4 pyrene monomers.[41]

14.3.2.2 dPTP Monomer as a Probe for Polymerase Structure: The utilization of dPTP is not limited to the formation of the fluorescent excimer by template-independent polymerases. Although fluorescence experiments were not employed greatly in this experiment, it is highly valuable to note that the large, nonhydrogen bonding pyrene moiety could be utilized by multiple polymerases even more efficiently than regular nucleobases by certain enzymes in the presence of damaged sequences.

Early research employing the *exo*-Klenow fragment (Kf) of *Escherichia coli* DNA polymerase I showed that when the template strand contained either an abasic or spacer site, dPTP was incorporated with considerably greater efficiency than any of the naturally occurring bases.[42] As expected, the large size of dPTP resulted in very poor incorporation opposite of any other nucleobase, and only selective incorporation opposite the damaged site, as the size of the pyrene moiety is approximately equal to the size of a regular A–T pair. In principle, the dPTP incorporation can allow for a simple means of observing DNA abasic site damage by the emission of pyrene. Further studies with dPTP helped to elucidate the repair mechanisms utilized by multiple DNA polymerases (T7, Yeast Pol η, L868F pol α) based upon the ability to incorporate the large pyrene moiety opposite a transient abasic site.[43–45]

14.3.2.3 Pyrene Monomer/Excimer as a Sensor for Detection of Single Nucleotide Polymorphisms: Incorporation of monomeric pyrene nucleoside into single strands of DNA has also allowed for a simple means of detecting relevant sequences of natural DNA. As noted above, the pyrene monomer has emission maxima at 375 and 395 nm but, upon the close interaction of two pyrene monomers, excimer formation can occur, resulting in a redshift of emission to approximately 490 nm. Taking advantage of this change in emission, probes can be designed that will target nucleic acid sequences with probes containing pyrene monomers (P) that, upon proper hybridization, will form the pyrene excimer to serve as a simple reporter of a desired nucleic acid sequence. As a simple proof of principle, two probes, d(GGC GCC GTC GP) and d(PGT GGG CAA GA), that hybridize to a target sequence d(CGC ACT CTT GCC CAC ACC GAC GGC GCC CAC), placing the two pyrene residues in close proximity to allow for excimer formation.[46] In the unhybridized state, the pyrenes would be free in solution and only display monomer emission at 375 and 395 nm, which was observed. Upon addition of the target sequence, hybridization of probes induced the desired formation of the excimer, which could readily be monitored by emission at 490 nm (Fig. 14.6a).[46] After optimization of probe design for hybridization geometry, a 40-fold increase in emission at 490 nm could be obtained by excimer formation, whereas a single mismatch introduced only a negligible increase in excimer emission (Fig. 14.6b).[46]

14.3.2.4 Pyrene in Sensors of Oxidative DNA Damage: Since pyrene and similar hydrocarbon fluorophore monomers have displayed excellent and variable quenching properties depending on the neighboring nucleobase (Table 14.3), the removal of a quenching nucleobase could provide a means of activating a fluorescence signal. Pyrene has been shown to be well quenched by thymine, but this information lacks application as thymine is one of the four natural bases and very common in DNA, thus does not represent DNA damage.[33] However, the electronically and structurally similar uracil is only present in DNA as a result of deamination of cytosine and

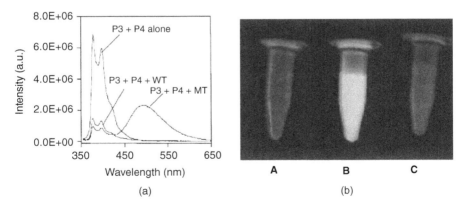

Figure 14.6 (a) Pyrene excimer formation in the presence of a target sequence (MT) versus a singly mismatched WT sequence. (b) Visual imaging of excimer formation. Probes in the absence of target (left), probes in the presence of target (center), and probes in the presence of singly mismatched WT. Paris *et al.*[46]. Reprinted with permission from Oxford University Press.

can result in mutations via misreading of a U–G base pair.[47,48] Biologically, multiple enzymes exist to remove uracil from a DNA strand via hydrolysis of the N-glycosidic linkage and allow for proper replication. Fortunately, uracil was found to be an efficient quencher of pyrene in a similar manner to thymine, which is unsurprising when considering the close electronic properties of the two nucleobases.[47,48]

Designing probes that take advantage of the quenching of pyrene by uracil allows for a simple, highly efficient means of detecting DNA damage and the expression of base repair enzymes. Uracil DNA glycosylase (UDG) is a common bacterial enzyme that catalyzes the removal of uracil from damaged DNA, leaving an abasic site for other repair enzymes to correct with the proper base.[47,48] If a pyrene is placed in close proximity to the uracil, a quenched probe is obtained, which will yield a fluorescence increase upon treatment with UDG. Stivers and coworkers initially reported on the ability of a uracil–pyrene "base pair" to serve as a fluorescent reporter for UDG that could also be used to understand the mechanism of UDG in excising uracil from double-stranded DNA.[48] Importantly, UDG requires base flipping of the quenching uracil to be excised from double-stranded DNA, but mutation of an important residue, Leu191, greatly lowers enzymatic activity by decreasing base flipping by the enzyme.[48] However, incorporation of pyrene opposite uracil allows for a spatial reorganization induced by the size of the pyrene, effectively forcing the uracil to flip out and also stabilizes the resulting abasic site after cleavage of the uracil. Results showed that the uracil–pyrene "base pair" managed to completely rescue activity of L191G and L191A mutants, with base excision of the quenching uracil allowing for pyrene fluorescence.[48] Understanding of this mechanism also allowed for utilization of a quenched pyrene–cytosine pair to be utilized for rational engineering of a mutant DNA glycosylase for the unnatural repair (glycosylase of cytosine) with recovery of pyrene fluorescence upon cytosine cleavage.[49]

While Stivers showed the ability for pyrene to be utilized in double-stranded assays with UDG, the later development of a single-stranded application for UDG allowed for cellular imaging of oxidative damage and subsequent expression of UDG. Exposure of a simple sequence containing two quenching uracils (5′-AAUYUAA-3′) to UDG *in vitro* results in a strong activation of fluorescence upon excitation at 340 nm, with a 90-fold increase in emission signal after only 60 min of exposure to enzyme (Fig. 14.7).[47] The control probe, replacing uracil with thymine (5′-AATYTAA-3′), shows no such fluorescence activation, confirming the specificity of the enzyme for uracil and allowing for direct sensing of UDG activity.[47] Although other methods exist for detection of UDG activity, they rely upon indirect detection and only have poor signal activation with high background, neither of which are problems with this design. Moreover, this single-stranded probe system has been successfully applied to image UDG activity *in vivo*.

14.3.2.5 General Conclusions about Monomeric Pyrene: Pyrene has enjoyed wide application as a fluorescent monomer. Despite the size of the aromatic moiety and complete lack of hydrogen bonding abilities, the monomer can be readily and stably incorporated by a variety of enzymes to give either excimeric emission via oligomerization (by TdT) or as a simple fluorophore that can detect abasic site

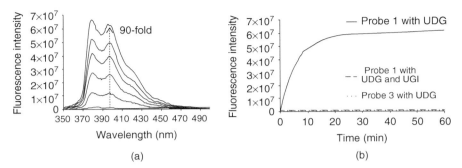

Figure 14.7 Fluorescence activation of a quenched pyrene containing probe by UDG. (a) Fluorescence spectra time course and (b) increase in emission at 395 min for uracil probe 1 (solid line), thymine-based probe 3 (dotted line), and probe 1 in the presence of a UDG inhibitor (dashed line). Ono *et al.*[47]. Reproduced with permission from John Wiley and Sons.

damage in DNA by halting activity of multiple polymerases at the damaged residue. Further, the recent development of direct enzyme sensors for oxidatively damaged bases (e.g., UDG activity) helps to show the promising future uses of this unique fluorophore.

14.4   OLIGOMERS OF FLUORESCENT C-GLYCOSIDES: DESIGN, SYNTHESIS, AND PROPERTIES

14.4.1   Design of Fluorescent C-Glycoside Oligomers

Sometimes, it may be more beneficial to use fluorescent C-glycosides in their oligomeric forms, as use of the DNA scaffold forces fluorophores into direct electronic interactions that results in new optical properties, as already discussed with formation of oligomeric pyrene nucleosides by TdT. However, with the increased complexity of the oligomeric sequences, greater consideration of the design is required. In recent work, oligodeoxyfluorosides (ODFs) have been prepared to intentionally utilize assembled structures that place the aromatic fluorophores in direct contact with each other (Fig. 14.8). This design relies upon close aromatic and electronic interactions as the faces of the fluorophores are separated by as little as 3.4 angstroms, the van der Waals contact distance that separates bases in natural DNA. Either α or β anomers of the fluorescent C-glycoside monomers can be used in synthesis of libraries of oligomeric fluorophores.[5,50] The use of randomized libraries can be helpful in finding properties resulting from dye–dye interactions that would be difficult to predict *de novo*.

The design of larger libraries of ODF sequences is considerably more complicated than the simple design of a single fluorescent C-glycoside and can result in noticeably different properties of the resulting oligomers. However, this complexity gives rise to a key advantage of the design of ODFs compared to regular fluorophores as the resulting interactions can generate greatly different optical behavior than free

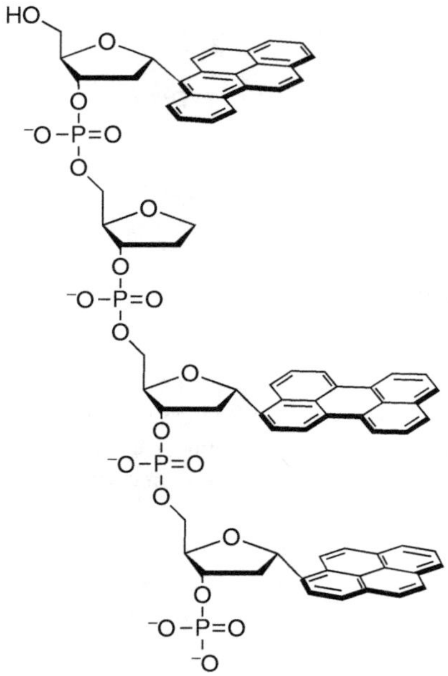

Figure 14.8   Structure of a representative ODF tetramer. The sequence shown is 5′-BSEY-3′.

monomers. Consequently, entirely new sensing techniques can be designed utilizing fluorescent monomers that, by themselves, lack the sensing capabilities that are found in the created oligomers.

Accordingly, this combinatorial design affords multiple advantages over existing techniques for designing fluorescent probes as sensors for a variety of conditions. Utilization of common split-and-pool methods with oligomers built on polystyrene beads (Fig. 14.9) allows for the ability to perform high-throughput assays for a wide variety of desired analytes, and the same library can easily be used to sense a diverse set of targets, adding a further multiplex advantage to the design. The resulting libraries can easily be imaged by fluorescence microscopy or exposed to an analyte of interest while still attached to the polystyrene bead to observe any fluorescence changes.

14.4.2   Synthesis of Fluorescent C-Glycoside Oligomers

Attachment of a fluorescent molecule to the deoxyribose moiety is a sufficiently nontrivial problem with modern synthetic techniques, but in the case of developing combinatorial libraries, a greater degree of complexity is necessary to create the desired compounds. After synthesis of monomers of the fluorescent glycosides, DNA synthesis of oligomers combining split-and-pool methodology is necessary to develop the desired library in a manner that allows sequence determination.[10] By performing synthesis of the library on polystyrene beads, rhodium-catalyzed carbene insertion of a tagged compound can be used in conjunction with addition of each monomer of

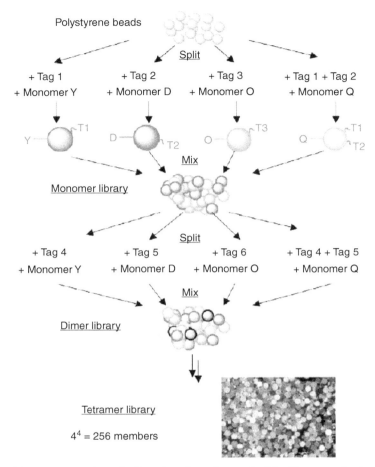

Polystyrene beads

Split

+ Tag 1   + Tag 2   + Tag 3   + Tag 1 + Tag 2
+ Monomer Y   + Monomer D   + Monomer O   + Monomer Q

Y   T1   D   T2   O   T3   Q   T1
                                                                          T2

Mix

Monomer library

Split

+ Tag 4   + Tag 5   + Tag 6   + Tag 4 + Tag 5
+ Monomer Y   + Monomer D   + Monomer O   + Monomer Q

Mix

Dimer library

Tetramer library

4^4 = 256 members

Figure 14.9   Representation of split and pool tagging method for library synthesis of ODFs and fluorescence microscopy of a 256-member library under single wavelength excitation. Wilson *et al.*[10]. Reproduced with permission from Royal Society of Chemistry.

DNA synthesis. After selecting a single polystyrene bead that changes fluorescence properties upon analyte exposure, subsequent deprotection of tags by cerium ammonium nitrate oxidation and sequencing by gas chromatography allows for simple and convenient sequencing of any member of the developed library, even with libraries of greater than 14,000 members (Fig. 14.9). After selection of a desired sequence, resynthesis can be easily performed by standard automated DNA synthesis.

14.4.3   Characterization and Properties of Fluorescent C-Glycoside Oligomers

In the case of ODFs, further characterization is necessary depending upon the state in which the molecules will serve as a sensor. If the new oligomer is to serve as a sensor in the solution state, absorbance maxima, emission maxima, and quantum yield are determined, as well as the fluorescence changes upon the desired molecular interaction.[10,29] If the sensor is instead to be utilized via attachment to solid support, imaging via fluorescence microscopy is needed followed by image processing to quantify changes in photophysical properties upon analyte exposure.

14.4.3.1 Short Designed Sequences:

An interesting complexity of oligomeric fluorophores is that even small variations in sequence can result in considerably different fluorescence properties (Table 14.4).[50] This complexity is part of the design of the ODFs as it allows for a greater variety of photophysical properties and sensing mechanisms. A thorough exploration of ODFs containing only different combinations of pyrene (Y) and perylene (E) as fluorophores showed that the order of the sequence and number of each fluorophore created significant changes in spectral properties of the resulting oligomers, as can be observed in Table 14.4. Pyrene and perylene monomers display absorbance maxima at 342 and 440 nm, respectively, with emission maxima at 375 nm/394 nm and 443 nm/472 nm/503 nm, respectively.[50] However, once complex electronic interactions are incorporated into the system by adding multiple fluorophore moieties, interesting absorbance and fluorescence results are obtained as excimers and exciplexes and other forms of energy transfer occur, resulting in greatly altered emission spectra. Addition of three consecutive pyrene residues, for example, results in an emission spectrum lacking any peaks characteristic of monomer emission and displaying emission of excimer only at 492 nm (Table 14.4, entry 4)[50] Further interesting results are obtained when heterooligomers are created

TABLE 14.4 Photophysical Properties of Simple ODF Sequences in Water

Sequence $(5' \rightarrow 3')$	Absorbance Maxima (nm)[a]	Extinction Coefficient	Emission Maxima (nm)	Quantum Yield
Y (MeOH)	**342**, 326, 312	47,000	375, 394	0.12
YSSS	**344**, 327	47,000	377, 396	0.35
YYSSS	343, **329**	44,000	377, 396, 484	0.22
YYYSSS	343, **329**	67,000	492	0.12
E (MeOH)	**440**, 415, 393	39,000	443, 472, 503	0.88
ESSS	**440**, 415	39,000	450, 477	0.95
EESSS	443, **420**	43,000	450, 478, 563	0.06
EEESSS	447, **422**	62,000	450, 478, 557	0.03
EYSSS	**448**, 421	39,000	490	0.39 (ex = 330 nm)
	348, 327	42,000	490	0.66 (ex = 420 nm)
ESYSS	**448**, 421	39,000	490	0.35 (ex = 330 nm)
	348, 327	42,000	490	0.61 (ex = 420 nm)
ESSSY	**448**, 421	39,000	465, 490	0.40 (ex = 330 nm)
	348, 327	42,000	465, 490	0.67 (ex = 420 nm)

[a]Bold values indicate highest maxima.[49] "S" is a nonfluorescent tetrahydrofuran (abasic) spacer. Wilson et al.[50]. Reprinted with permission from Elsevier.

by the mixture of different fluorescent C-glycosides into sequences. 5'-EYSSS-3' displays the expected mixed absorbance spectrum containing the absorbance maxima of both the Y and E monomers at 348 nm (from pyrene) and 448 nm (from perylene), but the emission spectrum is characteristic of exciplex formation between the two fluorophores.[50] (Note that "S" is a nonchromophoric spacer added to enhance solubility and discourage aggregation.) Changing the order of the sequence, however, affords some notable differences. Addition of a single spacer between pyrene and perylene to give 5'-ESYSS-3', a simple anagram of 5'-EYSSS-3' displays a very similar absorbance and emission spectrum, whereas addition of a second spacer to give 5'-ESSSY-3' displays decreased exciplex emission with increased fluorophore separation.[50] Figure 14.10 shows the ability of ODFs to be utilized for broad emission properties upon UV excitation, as a small set of just 11 oligomers (composed of only pyrene, perylene, and spacer monomers) shows distinct emission colors ranging from blue to dark yellow-orange.[50]

As more fluorescent monomers are incorporated into ODF sequences, greater complexity and changes in fluorescence properties are observed between different sequence arrangements (Table 14.5). 5'-EDSY-3' serves as a violet-blue fluorophore, with emission maxima at 375, 395, and 459 nm with a quantum yield of 0.44.[51] The anagramic 5'-ESYD-3' serves as a blue-cyan fluorophore with emission maxima at 460 and 488 nm and a quantum yield of 0.19.[51] 5'-DSEY-3' offers a third anagram with increased complexity, showing emission maxima at 375, 463, and 490 nm and a quantum yield of 0.30.[51] Table 14.5 also shows the differences in emission properties between anagrams containing D, E, B, and Y fluorophores, showing the closer similarity of these anagrams in terms of absorbance, emission, and quantum yield compared to the anagrams discussed above.

Interestingly, delocalized excimers and exciplexes of two fluorophores interact differently with the natural bases than the monomers do (as discussed in Section 14.2).[33] While TY is highly quenched (Table 14.6), TYY has a comparatively higher quantum yield of 0.02 with emission at 383 and 482 nm (compared to emission at 375 and 395 nm for TY).[33] Exciplexes containing mixtures of different fluorescent C-glycosides are poorly quenched, with TYE, for example, maintaining a quantum yield of 0.63, a much higher quantum yield than any sequence containing pyrene

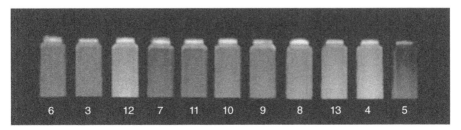

Figure 14.10 Real-color emission of 11 different ODF sequences comprised of only 2 fluorophore monomers upon UV excitation. Wilson *et al.*[50]. Reprinted with permission from Elsevier. (*See color plate section for the color representation of this figure.*)

TABLE 14.5   Photophysical Properties of Anagramic Sequences of ODFs

Sequence	Absorbance Maxima (nm)	Emission Maxima (nm)	Quantum Yield
5'-EDSY-3'	264, 342, 451	375, 395, 459	0.443
5'-ESYD-3'	257, 346, 349, 451	460, 488	0.191
5'-DSEY-3'	257, 343, 349, 451	375, 463, 490	0.303
5'-DEBY-3'	257, 347, 399, 448	375, 395, 464, 489	0.229
5'-EDBY-3'	258, 349, 403, 452	461, 490	0.188
5'-DYBE-3'	256, 349, 403, 442	462, 492	0.130

Teo et al.[51]. Reproduced with permission from John Wiley and Sons.

TABLE 14.6   Interactions of Fluorescent C-Glycosides with Natural
Nucleobases

Sequence (5' → 3')	Emission Maxima (nm)	Quantum Yield
TY	378, 398	0.0003
TE	450	0.37
TYY	383, 482	0.02
TYE	461, 484	0.63

Wilson et al.[33]. Reproduced with permission from John Wiley and Sons.

alone.[33] This can be, however, a preferable property if the ODF is to be used in a DNA sequence, as the large Stokes shifts provided by excimer and exciplex formation can greatly reduce background in any fluorescence-based assay. If needed, other quenchers can be matched to the delocalized excited states (see, for example, Section 14.5.1.2).

This complex photophysical behavior illustrates how it can be difficult to intentionally design a sequence of an ODF to have discrete absorbance or emission properties. Although some trends are obvious, such as excimer or exciplex formation or the ability of nucleobases to quench certain monomeric fluorescent C-glycosides, general prediction methods are still lacking. As a result, it can be expedient to prepare libraries of ODFs on a solid support for high-throughput assays and search for the desired properties among many possible sequences.

When the fluorescent oligomers are to be used on a solid support and imaged via fluorescence microscopy, as is becoming more common in ODF libraries, a mixture of characterization techniques must be employed. Naturally, one must confirm that the desired oligomer has been created on the solid surface that will be used in sensor applications (generally polystyrene beads).[10] Although split-and-pool methodology can assure that a monomer has been incorporated at a step during synthesis and allows for sequencing, it does not show yield or purity of the sequence on the polymer bead. For this, partial synthesis of the same sequence on controlled pore glass (CPG) utilizing standard automated DNA synthesis can be performed, with regular cleavage/deprotection conditions and purification by HPLC, affording information about yield and purity of the desired sequence and MALDI-TOF data to confirm mass of the desired compound. If desired, absorbance maxima, extinction coefficients, emission maxima, and quantum yields can further be determined of the free sequences in solution or in the presence of an analyte. If ODFs are to be employed in applications on a solid support (see Section 14.5.2), quantitative color imaging can be performed to analyze subtle changes in fluorescence intensity and wavelength.

14.4.3.2 ODF Sequences Discovered from Libraries: As a result of the above-mentioned difficulties in designing a discrete oligomeric fluorophore with tuned emission properties, library approaches have become more valuable in discovering sequences with desirable fluorescent properties. To evaluate this approach, Wilson *et al.* prepared a 4096-member library of tetramers starting from seven fluorescent monomers and an abasic tetrahydrofuran spacer (S).[5] The library was prepared on polystyrene beads and imaged by epifluorescence microscopy using a single excitation wavelength filter (340–380 nm) and monitoring emission beyond 400 nm (Fig. 14.11). From this library, 80 different beads were selected based upon their emission properties and sequenced. Of the 80 sequences, 23 were found to contain a subset of only 4 fluorescent monomers (Y, E, B, and K, an O-glycosidic fluorophore). This set displayed emission across the entire visible spectrum (Fig. 14.12) upon single wavelength excitation.[5] This multispectral property allows multiple analytes to be visualized simultaneously with a simple color camera, even in moving systems.

Resynthesis of the selected sequences by standard automated DNA synthesis allowed for more complete characterization of the fluorescent properties. In some cases, more spacer (S) monomers were added to increase the solubility of the discovered fluorophores, but no other changes to fluorophore sequence

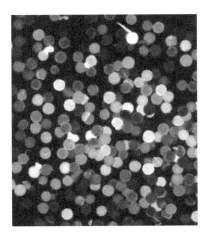

Figure 14.11 Emission of a library of ODFs containing seven fluorescent monomers attached to polystyrene beads upon single wavelength excitation. Teo *et al.*[5]. Reprinted with permission from American Chemical Society. (*See color plate section for the color representation of this figure.*)

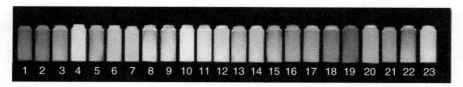

Figure 14.12   Emission of 23 different ODF oligomers upon excitation at 354 nm. Only four fluorophore monomers, arrayed in different sequences, comprise the oligomers. Teo et al.[5]. Reprinted with permission from American Chemical Society. (*See color plate section for the color representation of this figure.*)

were required after library screening.[5] Emission of the new fluorophore library ranged from the violet-blue (5′-SY-3′, emission at 376 and 396 nm) to the red (5′-SSBK-3′, 412, 633 nm) or even white emission (5′-SSSYKYK-3′, 380, 482, 629 nm).[5] The discovered fluorophores also maintained moderate-to-high quantum yields and brightness, all under single-wavelength UV excitation. Importantly, the solution-state fluorescence emission was approximately the same as the colors observed on the beads, further validating the library approach to allow for rapid discovery of fluorophores with valuable fluorescence properties.

Implementation of the oligomeric forms of fluorescent C-glycosides is entirely dependent on the system that will be studied. If a simple fluorescent reporter for an enzymatic reaction is to be used, a known ODF sequence with discrete fluorescence properties can be prepared and attached to the enzyme substrate and utilized in the solution state. If, however, discovery of a new sensor for an analyte of interest is desired, then combinatorial library synthesis and imaging via microscopy can provide a considerably more rapid means of realizing such. As before, the following discussion is meant to highlight some of the existing research, and for a more complete understanding, the reader is directed to the primary literature.

14.5   IMPLEMENTATION OF FLUORESCENT C-GLYCOSIDE OLIGOMERS

14.5.1   ODFs as Chemosensors in the Solution State

The most explored class of oligomeric fluorescent C-glycosides are the oligodeoxyfluorophores (ODFs), and multiple libraries have been created that allow for a wide variety of sensors for greatly disparate analytes to be created. To date, ODFs have served as reporters for enzymes (esterases, proteases), light, volatile organic compounds (VOCs), and toxic gases, among others. Library synthesis and simple fluorescence imaging have allowed for high-throughput means of developing small combinations of sensors that are capable of detecting and discriminating multiple analytes.

14.5.1.1   ODFs as Light Sensors (Chemodosimeters):   Among the earliest applications of ODFs to serve as sensors in a manner that the free monomers are not capable of, simple exposure of tetramers containing the aromatic dyes to light resulted in simple, obvious changes in absorbance, emission, and quantum yield properties of certain ODF sequences.[29] After creation of a library containing 14,641 members (a tetramer library starting from 11 different monomers, 11^4), exposure of the library to different wavelengths of light for varying amounts of time resulted in strong spectral changes for certain sequences (Fig. 14.13). For example, 5'-SBBB-3' displayed a broad emission band centered at 510 nm consistent with an excimer emission of benzopyrene.[29] After light exposure, however, emission shifted to 412 nm, consistent with the free monomer emission of benzopyrene.[29] Although the exact reason for the change in fluorescence properties is uncertain, it is possible that the light was able to selectively react with and, in some manner, quench (or photobleach) one of the fluorophores uniquely. Addition of an oxygen scavenger reduced the rate of change in emission, indicating a possible source of the photoreactivity of the sensor, hinting at photoinduced oxidation as the mechanism for fluorescence changes.[29] Nonetheless, this and multiple other sequences displayed a new means of sensing light exposure not available in the free monomers composing the tetramer library.

14.5.1.2 ODFs as Sensors for Enzymatic Reactions: Another valuable use of ODFs is as reporter molecules for enzymatic reactions that separate a fluorophore from a fluorescence quencher. Placing a fluorophore in close proximity to a fluorescence quencher is a commonly used technique for developing a fluorometric assay, as contact quenching can allow for potent quenching and subsequent fluorescence activation and has been used for a variety of fluorogenic assays. ODFs offer a unique class of fluorescent molecules, however, as multiple fluorescent monomers are joined together to create a fluorescent oligomer. Studies of the ability to quench these oligomeric fluorophores indicated that, interestingly, the compounds could be quenched with considerably high efficiency, likely as a result of their long fluorescence lifetimes.[51,52] Consequently, attaching a quencher to an ODF in a manner that could be released by an enzymatic reaction allows another method for fluorogenic enzymatic reactions. Further, the ability of ODFs to emit across the entire visible spectrum with only a single excitation wavelength allows for simple multiplex applications with multiple sensors in a sample to sense a single analyte.

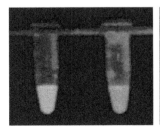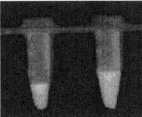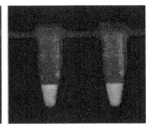

Figure 14.13 Changes in three different ODF sequences before (left) and after (right) light exposure. Gao *et al.*[29]. Reprinted with permission from American Chemical Society. (*See color plate section for the color representation of this figure.*)

ODFs as Lipase/Esterase Sensors: Esterases and lipases are common enzymes that are useful for organic synthesis, prodrug activation, and can even be targeted by certain pharmaceuticals.[53] Multiple designs have been created for esterase-activated fluorescent sensors, but these generally suffer from a variety of problems including poor aqueous solubility, complicated synthetic procedures, and low fluorescence turn-on ratios related to high background signals. Use of ODFs can help to alleviate many of these problems, as the phosphodiester backbone endows the new sensors with greatly increased water solubility, automated DNA synthesis can be employed and certain ODF sequences are known to be efficiently quenched.

Synthesis of multiple different scaffolds allowed for a simple means of detecting the activity of a variety of different classes of esterases and lipases based upon the type of ester bond that is cleaved.[53] 5'-YYYY-3' was used as a simple fluorescence reporter, particularly with the large Stokes shift resulting from excimer emission, as well as one that is strongly quenched by dabcyl. Reaction of an ODF linked to dabcyl via an aliphatic ester showed quick activation of fluorescence with up to a 110-fold turn-on in fluorescence emission at 480 nm with 1.0 μM probe concentration in the presence of porcine liver esterase (PLE) or human liver esterase (HLE).[53] Although the ODF-based probes show slower fluorescence activation than commercially available fluorescein diacetate (FDA), they portrayed significantly greater stability and lower background emission from spontaneous hydrolysis in buffer. After 24 h of exposure to buffer alone, FDA showed large degradation and subsequent fluorescence activation, whereas the ODF-based probe gave negligible hydrolysis and maintained low fluorescence background, only activating in the presence of certain esterases and lipases.[53] The new design afforded considerably potent quenching with excellent signal activation and low background, attaining better properties than existing esterase and lipase probes.

ODFs as Protease Sensors: The efficient ability of ODF sequences to be quenched by common fluorescence quenchers also allows for excellent probes to be designed for protease cleavage. A particularly valuable class of proteases is found in the caspase family, which plays pertinent roles in cellular processes of necrosis, apoptosis, and inflammation, depending on which caspase is active.[54] Since different caspase activities signal different cellular processes, designing multiple probes that can be easily prepared and spectrally resolved are greatly desired. ODFs offer this ability via their single wavelength excitation and resulting complex electronic interactions that give rise to large Stokes shifts. The use of automated DNA synthesis allows for rapid synthesis of any ODF sequence, easily allowing for conjugation of any desired ODF selected for the desired emission properties.

With this in mind, three different caspase sensors were designed to sense three different caspase proteases that show different activity. Probe 1 (5'-SSEE-Val-Asp-Val-Ala-Asp-Dabcyl) serves as a sensor for caspase 2 with emission at 550 nm. Probe 2 (5'-SSYYYY-Asp-Glu-Val-Asp-Dabcyl) serves as a sensor for caspase 7 with emission at 480 nm and probe 3 (5'-SSYKY-Leu-Glu-His-Asp-Dabcyl') serves as a sensor for caspase 9 with emission at 620 nm.[54] All sequences can easily be excited with a single filter (330–380 nm) and observed for cleavage only in the

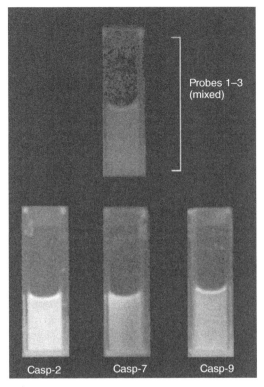

Figure 14.14   Use of ODFs as sensor cocktails for different caspases. Top: Probes 1–3 mixed in the same fluorescence cuvette. Bottom: same probe mixture in the presence of only one caspase shows selective probe activation by color of fluorescence under UV excitation. Dai *et al.*[54]. Reproduced with permission from John Wiley and Sons. (*See color plate section for the color representation of this figure.*)

presence of their target caspase. After 4 h of treatment of each sensor with the appropriate caspase, fluorescence activation can easily be observed, although probe 3 shows a considerably slower rate than probes 1 and 2.[54] More valuable is the ability to use these sensors as cocktails, where all sensors can be combined and only one will be activated in the presence of the appropriate caspase. A mixture of equimolar amounts of probes 1–3 in a fluorescence cuvette followed by the addition of only one caspase and incubation showed fluorescence activation for only the desired signal matching caspase activity (Fig. 14.14). Addition of caspase 2 to a sensor cocktail resulted in the expected increase in emission at 580 nm with negligible increase in emission at 480 or 620 nm, showing the specificity of each probe.[54] Similar results were obtained by the addition of caspase 7 or 9 to the sensor cocktail mixture, with activation of only the desired emission, emphasizing the value of the photophysical properties of ODFs under single wavelength excitation.

14.5.1.3   ODFs in Cellular Applications: The ability of ODFs to provide a wide range of fluorescence emission with single wavelength excitation provides a valuable tool for cellular imaging. In this context, ODFs or even individual monomers can be applied as sensors of a desired analyte, as noted in Section 14.3.2.4, or can be utilized simply to serve as a set of multicolor fluorescence reporters applied to common cellular assays. Although the design of the ODF scaffold being reliant upon the phosphate backbone of DNA endows the molecules with multiple negative charges, the hydrophobic character of the aromatic fluorophores allows for a certain degree of amphiphilicity that appears to allow ODFs to cross the cell membrane.

Initial experiments with two ODF sequences (5′-SSYB-3′ and 5′-SSSBEY-3′) showed that simple incubation for 3 h with HeLa cells allowed for permeation of the dye-based oligonucleotide across the plasma membrane.[5] Comparison with a commercially available membrane-labeling dye showed that the ODFs were not associated with the cell membrane and did not appear to enter the nucleus. More valuable, no change in dye emission was observed, even after extended incubation, indicating the biostability of the ODFs based upon the C-glycosidic bond.[5]

With the simple permeability of ODFs into HeLa cells, examining the possibility to utilize these dyes for multispectral applications in dynamic systems becomes possible. Four different oligomers (5′-SB-3′, 5′-SSEY-3′, 5′-SSBB-3′, and 5′-SSSYYK-3′) were selected based upon different emission properties (blue, cyan, yellow, and orange-red) under UV excitation (354 nm) and incubated with Zebrafish embryos (Fig. 14.15).[5] After 24 h, the embryos showed localization of dyes to the chorion, with partial penetration to the body of the developing fish. After 48 h of incubation, even hatched fish could be observed with dye localization allowing for visualization of internal structures of the head. Excitation allowed for imaging of all dye containing Zebrafish embryos in the same sample under microscopy to provide real-color images without the need for changing any filter sets or false-coloring image reconstruction. Most importantly, the dyes displayed excellent

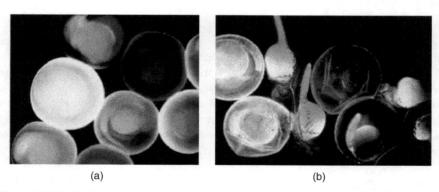

(a)   (b)

Figure 14.15   Real-color images of Zebrafish embryos after incubation with different ODF dyes for (a) 24 h and (b) 48 h under UV excitation. Teo *et al.*[5]. Reprinted with permission from American Chemical Society. (*See color plate section for the color representation of this figure.*).

photostability and biostability while not being toxic at concentrations as high as 5 μM.[5]

With the ability of ODF-based structures to easily permeate cells, multiple probes have been designed and applied to different cell lines to obtain activated signals for a variety of enzymes. Experiments with HeLa cells have confirmed that the previously discussed esterase, lipase, and caspase probes can permeate cells and provide a strong fluorescence signal indicating activity of the desired enzyme in cells.[53,54] Moreover, experiments in *E. coli* have displayed the efficiency of ODF-based probes for activity of base excision repair enzymes (such as UDG discussed previously) directly in cells by simple probe incubation.[47]

14.5.2 ODFs as Sensors in the Solid State

While many early and some current applications of ODFs as sensors have been applied to the solution phase, development of simple solid-phase sensors offers an intriguing, easier means of detecting a greater variety of analytes of interest while also allowing for new interactions of the sensors with analytes. If the ODF sequence can be attached to a solid support, which is simple given the utilization of automated DNA synthesis technologies, then exposure of the solid support containing the ODF to an analyte of interest and fluorescence imaging via microscopy can provide access to a rapid means of detecting multiple analytes. Further, combinatorial synthesis of ODF libraries can be applied to give high-throughput means of rapidly screening for the sensors that undergo the greatest changes upon exposure to analytes.

14.5.2.1 ODFs as Sensors of Volatile Organic Compounds:

An elegant example of application of this approach is the use of a library of ODFs prepared on polystyrene beads to be exposed to vapors of VOCs and observing any changes that may occur in fluorescence emission.[55] Preparation of a library containing 2401 members (tetramers composing of 7 monomer units, 7^4) displayed large changes in fluorescence emission properties upon exposure to such VOCs as acrolein, mesitylene, propionic acid, and nitrobenzene for as little as 2 min (although a 30-min exposure time is preferable for sufficient changes in emission properties to be observed).[55] Figure 14.15 displays the mixed gray-scale images of eight different sequences against four analytes.

From the library, eight of the most responsive sequences were selected for further screening and cross screening to observe selectivity of the ODFs to sense the VOCs. An example is the exposure of 5′-YEHH-3′ (H is dihydrothymine) to a variety of VOCs. Without any analyte present, the ODF displays bright green emission characteristic of the pyrene–perylene exciplex.[55] Upon exposure to acrolein, emission enhances and shifts to turquoise, as presented by a bright blue difference image (Fig. 14.16).[55] On exposure to nitrobenzene, the same sequence shows both a color change and quenching as shown by a maroon difference map.[55]

Examining the changes in red, green, and blue (RGB) values from the fluorescence images provides even greater information about the changes that occur upon exposure to an analyte. For example, 5′-YEHH-3′ displays strong quenching of the red channel

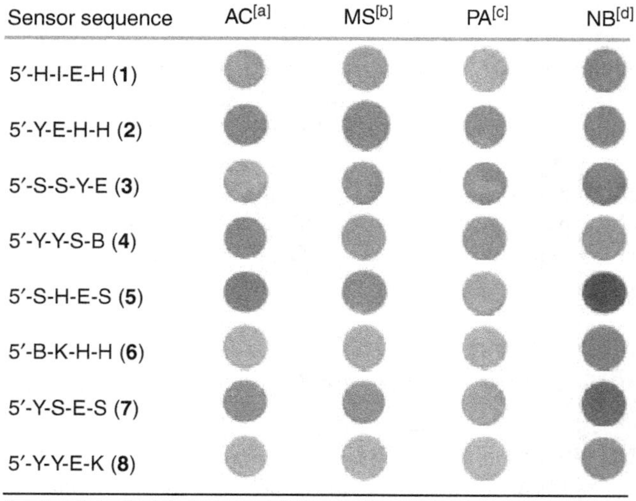

Sensor sequence	AC[a]	MS[b]	PA[c]	NB[d]
5'-H-I-E-H (1)				
5'-Y-E-H-H (2)				
5'-S-S-Y-E (3)				
5'-Y-Y-S-B (4)				
5'-S-H-E-S (5)				
5'-B-K-H-H (6)				
5'-Y-S-E-S (7)				
5'-Y-Y-E-K (8)				

[a] Acrolein. [b] Mesitylene. [c] Propionic acid. [d] Nitrobenzene.

Figure 14.16 Use of ODFs appended to polystyrene beads as sensors for volatile organic compounds. Blended difference images show qualitative color changes (gray indicates no change upon exposure). Samain *et al.*[55]. Reproduced with permission from John Wiley and Sons. (*See color plate section for the color representation of this figure.*)

with minimal change in the green and blue channels upon exposure to acrolein, providing a simple explanation for the observed color change. 5'-YYSB-3', in contrast, shows strong quenching of the green channel and moderate quenching of the blue channel upon exposure to the same analyte, providing a completely different color change observed as maroon in the difference map. 5'-SSYE-3' displays yet another distinct change upon exposure to acrolein with enhancement of fluorescence in the blue channel, whereas the red and green channels remain relatively unchanged. Representative changes in RGB values for eight different ODF sequences upon exposure to acrolein are provided in Figure 14.17.[55]

The ability of these scaffolds to undergo strongly different changes based upon sequence indicates a greater use for ODFs as sensor arrays for multiple analytes of interest. A more complete examination of the 2401-member library in the presence of 10 VOCs containing a wide variety of functional groups (acrolein, mesitylene, nitrobenzene, propionic acid, ethylisocyanate, 2,6-lutidine, *N,N*-dimethylaniline, methyliodide, dimethylmethyl phosphonate, and acrylonitrile) revealed that a set of just 4 sensors could distinguish among the 10 VOCs with high statistical significance.[56] A single sequence, 5'-SHES-3', displayed discernible differences by the naked eye to 6 of the 10 analytes used in the study even without the need for image processing.[56] These responses were easily visualized from a tetramer library containing only three inherently fluorescent monomers (Y, E, B) as well as four commercially available nonfluorescent monomers, also indicating the speed with which analytical diversity can be created by this approach.

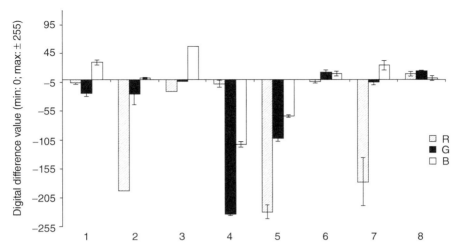

Figure 14.17 Representative changes in RGB values of eight different ODF sequences (on polystyrene solid support) upon exposure to acrolein: (1) 5′-HIEH-3′ (2) 5′-YEHH-3′ (3) 5′-SSYE-3′ (4) 5′-YYSB-3′ (5) 5′-SHES-3′ (6) 5′-BKHH-3′ (7) 5′-YSES-3′ (8) 5′-YYEK-3′. Samain *et al.*[55]. Reproduced with permission from John Wiley and Sons.

The sensing mechanism for the ODFs to these different analytes is as yet unclear although multiple possible binding mechanisms exist. ODFs are comprised of planar, aromatic fluorophores that undergo multiple π interactions with each other to provide the observed fluorescence signals. It is likely that the analytes containing aromatic moieties or extended regions of π conjugation (e.g., acrolein, mesitylene, nitrobenzene, 2,6-lutidine, *N,N*-dimethylaniline, and acrylonitrile) interact in a π-stacking mechanism, altering the electronic interactions between fluorophores in the tetramers. Other analytes contain hydrogen bonding residues (e.g., acrolein, propionic acid, ethylisocyanate, 2,6-lutidine, *N,N*-dimethylaniline, dimethylmethyl phosphonate, and acrylonitrile) that may interact with the dihydrothymine monomer contained in the tetramers.[56]

The ability of ODFs to serve as sensors of different VOCs allowed for the development of a further application for the ability to sense bacterial species that can be characterized by their biological synthesis of such VOCs.[57] It is known that multiple pathogenic bacteria biosynthesize certain combinations of VOCs, usually fatty acid esters, alcohols, amines, aromatic hydrocarbons, and sulfur compounds that can serve as a fingerprint of the bacterial species. Traditionally, gas chromatography and mass spectrometry methods are used to determine the volatile analytes present, but the need for careful sample preparation and expensive instrumentation serves as an impediment to broad application of these methods.[57] Fluorescence sensors, such as ODF scaffolds, provide a simpler means of bacterial species identification via VOC emission through simple optical imaging, particularly with the use of solid support as above, as no sample preparation will be necessary when the sensor need only be exposed to the headspace gas emissions of bacterial species.

Screening of the same 2401-member ODF library in the air above three medically relevant bacterial species (*E. coli, Mycobacterium tuberculosis,* and *Pseudomonas putida*) identified 14 ODF sequences that could be used to differentiate species based upon VOC fingerprints after only 30-min exposure to bacterial colonies, when simply placed in the same covered Petri dish.[57] As a representative example, the sequence 5'-IYSH-3', a sequence containing only one fluorophore, pyrene (I is 5-nitroindole), gave an increase in all RGB values upon exposure to *M. tuberculosis*, while decreasing all RGB values on exposure to *P. putida*.[57] A second sensor, 5'-EYHE-3', showed only small changes in RGB values on exposure to *M. tuberculosis* and *P. putida*, while showing an increase in RGB values, with a particularly large increase in the red channel upon exposure to *E. coli*.[57] Use of these two sensors in conjunction with statistical analysis (PCA) could correctly classify among the three bacterial species with 100% accuracy (Fig. 14.18).[57] These results are particularly intriguing as headspace gas of VOCs from bacterial species is estimated in the 30–50,000 ppb range, showing the high sensitivity of the methodology.

14.5.2.2   ODFs as Sensors of Toxic Gases: As ODFs displayed a powerful ability to sense VOCs, the transition to sensing toxic gases is a natural next step, as the design already showed the capacity to sense in the gas phase. For this purpose, a small library of only 256 members comprising of four fluorescent monomers (Y, E, D, and a zinc porphyrin) was prepared on polystyrene beads and exposed to 1000 ppm concentrations of eight different gases (SO_2, H_2S, MeSH, NH_3, $NHMe_2$, HCl, Cl_2, and BF_3) for 15 min and changes in RGB values were monitored as usual.[58] Fifteen candidate sequences were found that showed changes in emission, fluorescence enhancement, or quenching upon exposure to the gases and resynthesized for further characterization.[58]

Responses of the sequences to different toxic gases varied depending upon the gas, showing a slightly different behavior than the larger, more varied changes

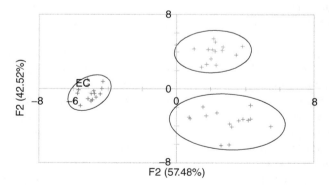

Figure 14.18   The use of ODF sensors to identify three bacterial species in Petri dishes by sensing their volatile bacterial metabolites. Shown is well-separated statistical clustering of fluorescence changes of four polystyrene-supported ODF sensors. Koo *et al.*[57]. Reproduced with permission from Royal Society of Chemistry.

upon exposure to VOCs. For detection of NH_3 and $NHMe_2$, 12 of the 15 sequences showed approximately the same response to both gases, although the remaining three sequences did show strongly different results.[58] Similarly, exposure to MeSH and H_2S resulted in similar changes for most sequences, indicating a greater difficulty in differentiating among the structurally similar gases. Nonetheless, some interesting sequences were shown to provide strongly different responses to gas exposure, despite sequence similarity. 5'-EDDY-3' and 5'-DEDY-3', two anagramic sequences that provided strong changes during library screening, displayed surprisingly different behavior upon exposure to different toxic gases.[58] Even with the close structural similarity of the two sensors, completely different response patterns were observed upon exposure to BF_3, H_2S, MeSH, and $NHMe_2$, indicating that two simple sensors are capable of differentiating among four different analytes. Further statistical analysis using PCA indicated that a set of three sensors could easily identify six different gases with high confidence.[58] Without any optimization, the ODF sensor system was capable of detecting seven of the eight toxic gases assayed at levels below immediately dangerous to life or health (IDLH) concentrations,[58] indicating the efficiency of the approach to sensor design, starting from a small library of 256 members.

14.5.3 Alternative Designs of Oligomeric Fluorescent Glycosides

Other architectures for arranging multiple fluorophores on a DNA backbone are also possible and may confer some benefits. In this vein, Inouye and coworkers have prepared alkynyl-linked alpha anomeric monomers of pyrene, perylene, and anthracene (Fig. 14.19).[59] Preparation of the anomerically pure compounds allows for analysis of the spectral properties to be examined in a different context than commonly found in the above ODF libraries, while alkynyl linkages provide a redshift in both the absorbance and emission properties.

The resulting monomers provide excellent fluorophores, with the absorbance and emission properties provided in Table 14.7.[59] Formation of short oligomers show similar behavior to regular ODFs, with the expected redshift in emission resulting from

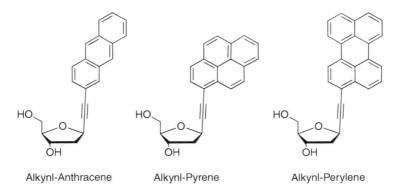

Alkynl-Anthracene   Alkynl-Pyrene   Alkynl-Perylene

Figure 14.19 β-Alkynyl monomers of fluorescent C-glycosides.

TABLE 14.7 Photophysical Properties of α-Alkynyl Fluorescent C-glycosides in Water

Sequence (5' → 3')	Absorbance Maxima (nm)	Extinction Coefficient (M^{-1} cm^{-1})	Emission Maxima (nm)	Quantum Yield
Alkynyl pyrene (Py)	360	54,100	385	0.66
Alkynyl perylene (Pe)	454	27,000	468	0.66
Alkynyl anthracene (An)	387	4700	398	0.37
S-Py-Py-S	365	36,200	510	N.D.
S-Pe-Pe-S	462	32,300	595	N.D.
S-An-An-S	391	8000	525	N.D.
S-Py-Pe-S	464	24,800	522	N.D.
S-Py-An-S	365	19,500	500	N.D.
S-Pe-An-S	491	28,000	526	N.D.

Chiba et al.[59]. Reproduced with permission from John Wiley and Sons.

extended conjugation of the fluorophore to the sugar via the alkynyl linkage, with the pyrene excimer emitting at 510 nm compared to 490 nm in the regular ODF design.[59] However, utilization of the anomerically pure, alkynyl-linked monomers can, in certain instances, reduce the electronic interactions of monomeric fluorophores. While the pyrene and perylene derivatives showed strong excimer formation, the anthracene derivative displayed only small formation of an excimer.[59] However, oligomers containing anthracene with pyrene or perylene displayed strong exciplex emission, and the resulting heteroexciplexes can provide valuable emission properties.[59] Library synthesis and sensing properties of this design for ODFs has yet not been reported.

Leumann and coworkers have also shown that incorporation of a variety of biphenyl and phenanthrenyl C-glycosides into DNA structures can afford new fluorescent properties. The development of multiple monomers allowed for design of sequences that could either enhance fluorescence at the same wavelength, form excimers with shifted emission properties, or undergo fluorescence quenching in DNA duplexes.[29,30] Development of nitro-based biphenyl and phenanthrenyl C-nucleosides has also allowed for orthogonal fluorophore–quencher pairs that form stable duplexes when incorporated as oligomers into natural DNA strands and can serve as reporters for DNA hybridization studies.[29,30,60]

14.5.4 General Conclusions: Oligomers of Fluorescent C-glycosides

Preparing oligomers of fluorescent C-glycosides results in new derivatives with greatly varied photophysical properties available with single UV excitation. Even closely related sequences can show surprisingly different emission and quantum yield properties, allowing for the development of new sensors with properties not available in any commercial dye system. The resulting sequences have been successfully utilized as probes for enzymes, even being able to permeate cells without the need for transfection reagents. ODF libraries have also proved invaluable in allowing for high-throughput screening and development of new sensors that regular DNA scaffolds would not be able to readily or simply detect.

14.6 CONCLUSIONS

Fluorescent C-glycosides offer the ability to incorporate bright dyes in DNA sequences with extreme stability. The fluorophores can be easily employed to provide valuable information of DNA structure and dynamics or create entire libraries of nonnatural molecules that can sense a wide variety of structures in both solution and on solid support. The latter approach further offers the ability to use just a small number of fluorophores to create new fluorescence properties, with single wavelength UV excitation allowing emission across the entire range of the visible spectrum.

Moreover, the hydrophobic properties of many of the dyes used as C-glycosides appear to allow for increased cell permeability, even with the strong negative charge of a natural DNA backbone. As a result, simple enzyme sensors that are simply incubated with cells and provide rapid fluorescence activation in the presence of target enzymes therein. In just a few years of development, successful applications have been provided for lipases, esterases, caspases, and enzymes for DNA damage without using highly complex instrumentation. Even using VOC fingerprints, preliminary sensor arrays have displayed the ability to distinguish among multiple medically relevant bacterial species with high confidence.

14.7 PROSPECTS AND OUTLOOK

Current developments of fluorescent C-glycosides have already provided many new valuable tools, but further development is needed to make full use of the capabilities of the scaffold. Although library synthesis and screening has allowed for moderately rapid identification of small sets of sequences, development of sensor arrays is still of great interest. Designing new fluorophores, particularly with redshifted emission properties and creating new means of predicting fluorescence properties of ODFs and similar molecules will help to further enhance biological and analytical applications. Moreover, designing new fluorophores may help to further broaden the analytes that can be sensed by oligomeric scaffolds or provide new monomers for understanding DNA structures and interactions with other biomolecules.

Simple synthesis of sensor arrays may allow the easy application of the ability for multiple analytes to be detected by just a small subset of sensors in the current design. In the present form, ODFs have not yet been applied in great scale to sensing multiple different analytes in the same sample. Developing an approach to allow the sensors to readily distinguish among analytes in the same sample would provide great viability for many industrial and medical applications by simple observance of different patterns of signal responses.

ACKNOWLEDGMENTS

The authors gratefully acknowledge support from the U.S. National Institutes of Health (GM067201 and GM072705).

REFERENCES

1. Williams, N. H., Takasaki, B., Wall, M., Chin, J. *Acc. Chem. Res.*, **1999**, *32*, 485.
2. Peon, J.; Zewail, A. H. *Chem. Phys. Lett.* **2001**, *348*, 255.
3. Sprecher, C. A.; Johnson, W. C. *Biopolymers* **1977**, *16*, 2243.
4. Callis, P. R. *Annu. Rev. Phys. Chem.* **1983**, *34*, 329.
5. Teo, Y. N.; Wilson, J. N.; Kool, E. T. *J. Am. Chem. Soc.* **2009**, *131*, 3923.
6. Teo, Y. N.; Kool, E. T. *Bioconjugate Chem.* **2009**, *20*, 2371.
7. Teo, Y. N.; Kool, E. T. *Nucleic Acids Symp. Ser.* **2008**, *52*, 233.
8. Teo, Y. N.; Kool, E. T. *Chem. Rev.* **2012**.
9. Sinkeldam, R. W.; Greco, N. J.; Tor, Y. *Chem. Rev.* **2010**, *110*, 2579.
10. Wilson, J. N.; Kool, E. T. *Org. Biomol. Chem.* **2006**, *4*, 4265.
11. Krueger, A. T., Kool, E. T. *J. Am. Chem. Soc.*, **2008**, *130*, 3989.
12. Lee, A. H. F., Kool, E. T. *J. Am. Chem. Soc.*, **2006**, *128*, 9219.
13. Hernández, A. R., Kool, E. T. *Org. Lett.*, **2011**, *13*, 676.
14. Tor, Y., Del Valle, S., Jaramillo, D., Srivatsan, S. G., Rios, A., Weizman, H. *Tetrahedron*, **2007**, *63*, 3608.
15. Shin, D., Sinkeldam, R. W., Tor, Y. *J. Am. Chem. Soc.*, **2011**, *133*, 14912.
16. Li, J. S., Shikiya, R., Marky, L. A., Gold, B. *Biochemistry* **2004**, *43*, 1440.
17. Li, J. S., Fan, Y. H., Zhang, Y., Marky, L. A., Gold, B. *J. Am. Chem. Soc.*, **2003**, *125*, 2084.
18. Li, J. S., Chen, F. A., Shikiya, R., Marky, L. A., Gold, B. *J. Am. Chem. Soc.*, **2005**, *127*, 12657.
19. Li, J. S., Gold, B., *J. Org. Chem.*, **2005**, *70*, 8764.
20. Štambaský, J.; Hocek, M.; Kočovský, P. *Chem. Rev.* **2009**, *109*, 6729.
21. Chaudhuri, N. C.; Ren, R. X. F.; Kool, E. T. *Synlett*, **1997**, 341.
22. Chaudhuri, N. C.; Kool, E. T. *Tetrahedron Lett.* **1995**, *36*, 1795.
23. Ren, R. X. F.; Chaudhuri, N. C.; Paris, P. L.; Rumney IV, S.; Kool, E. T. *J. Am. Chem. Soc.* **1996**, *118*, 7671.
24. Strässler, C.; Davis, N. E.; Kool, E. T. *Helv. Chim. Acta* **1999**, *82*, 2160.
25. Ludwig, J. *Acta Biochim. Biophys. Acad. Sci. Hung.*, **1981**, *16*, 131.
26. Coleman, R.S.; Madaras, M. L. *J. Org. Chem.* **1998**, *63*, 5700.
27. Okamoto, A.; Tainaka, K.; Fujiwara, Y. *J. Org. Chem.* **2006**, *71*, 3592.
28. Wierzchowski, J., Shugar, D. *Photochem. Photobiol.*, **1982**, *35*, 445.
29. Gao, J.; Watanabe, S.; Kool, E. T. *J. Am. Chem. Soc.* **2004**, *126*, 12748.
30. Morales-Rojas, H.; Kool, E. T. *Org. Lett.* **2002**, *4*, 4377.
31. Grigorenko, N. R., Leumann, C. J. *Chem. - Eur. J.* **2009**, *15*, 639.
32. Zahn, A., Leumann, C. J. *Chem. - Eur. J.* **2008**, *14*, 1087.

33. Wilson, J. N.; Cho, Y.; Tan, S.; Cuppoletti, A.; Kool, E. T. *ChemBioChem* **2008**, *9*, 279.

34. Brauns, E. B.; Madaras, M. L.; Coleman, R. S.; Murphy, C. J.; Berg, M. J. *J. Am. Chem. Soc.* **1999**, *121*, 11644.

35. Andreatta, D.; Lusteres, J. P. L.; Kovalenk, S. A.; Emsting, N. P.; Murphy, C. J.; Coleman, R. S.; Berg, M. A. *J. Am. Chem. Soc.* **2005**, *127*, 7270.

36. Somoza, M. M.; Andreatta, D.; Murphy, C. J.; Coleman, R. S.; Berg, M. A. *Nucleic Acids Res.* **2004**, *32*, 2494.

37. Sen, S.; Paraggio, N. A.; Gearheart, L. A.; Connor, E. E.; Issa, A.; Coleman, R. S.; Wilson III, D. M.; Wyatt, M. D.; Berg, M. A. *Biophys. J.* **2005**, *89*, 4129.

38. Singh, I.; Hecker, W.; Prasad, A. K.; Parmar, V. S.; Seitz, O. *Chem. Commun.* **2002**, 500.

39. Singh, I.; Beuck, C.; Bhattacharya, A.; Hecker, W.; Parmar, V. S.; Weinhold, E.; Seitz, O. *Pure Appl. Chem.* **2004**, *76*, 1563.

40. Cho, Y.; Kool, E. T. *ChemBioChem* **2006**, *7*, 669.

41. Jarchow-Choy, S. K., Krueger, A. T., Liu, H., Gao, J., *Nucleic Acids Res.*, **2011**, *39*, 1586.

42. Matray, T. J., Kool, E. T., *Nature*, **1999**, *399*, 704.

43. Niimi, A., Limsirichaikul, S., Yoshida, S., Iwai, S., Masutani, C., Hanaoka, F., Kool, E. T., Nishiyama, Y., Suzuki, M. *Mol. Cell Biol.*, **2004**, *24*, 2734.

44. Sun, L., Zhang, K., Zhou, L., Hohler, P., Kool, E. T., Yuan, F., Wang, Z., Taylor, J. S., *Biochemistry* **2003**, *42*, 9431.

45. Sun, L., Wang, M., Kool, E. T., Taylor, J. S. *Biochemistry*, **2000**, *39*, 14603.

46. Paris, P. L.; Langenhan, J. M.; Kool, E. T. *Nucleic Acids Res.* **1998**, *26*, 3789.

47. Ono, T.; Wang, S.; Koo, C. K.; Engstrom, L.; David, S. S.; Kool, E. T. *Angew. Chem. Int. Ed.* **2012**, *51*, 1689.

48. Jiang, Y. L., Kwon, K., Stivers, J. T., *J. Biol. Chem.*, **2001**, *276*, 42347.

49. Kwon, K., Jiang, Y. L., Stivers, J. T., *Chem. Biol.*, **2003**, *10*, 351.

50. Wilson, J. N.; Gao, J.; Kool, E.T. *Tetrahedron* **2007**, *63*, 3427.

51. Teo, Y. N.; Wilson, J. N.; Kool, E. T. *Chem. - Eur. J.* **2009**, *15*, 11551.

52. Wilson, J. N.; Teo, Y. N.; Kool, E. T. *J. Am. Chem. Soc.* **2007**, *129*, 15426.

53. Dai, N.; Teo, Y. N.; Kool, E. T. *Chem. Commun.* **2010**, *46*, 1221.

54. Dai, N.; Guo, J.; Teo, Y. N.; Kool, E. T. *Angew. Chem. Int. Ed.* **2011**, *50*, 5105.

55. Samain, F.; Ghosh, S.; Teo, Y. N.; Kool, E. T. *Angew. Chem. Int. Ed.* **2010**, *49*, 7025.

56. Samain, F.; Dai, N.; Kool, E. T. *Chem. - Eur. J.* **2011**, *17*, 174.

57. Koo, C. K.; Wang, S.; Gaur, R. L.; Samain, F.; Banaei, N.; Kool, E. T. *Chem. Commun.* **2011**, *47*, 11435.

58. Koo, C. K.; Samain, F.; Dai, N.; Kool, E. T. *Chem. Sci.* **2011**, *2*, 1910.

59. Chiba, J.; Takeshima, S.; Mishima, K.; Maeda, H.; Nanai, Y.; Mizuno, K.; Inouye, M. *Chem. - Eur. J.* **2007**, *13*, 8124.

60. Stoop, M., Zahn, A., Leumann, C. J., *Tetrahedron* **2007**, *63*, 3440.

15

MEMBRANE FLUORESCENT PROBES: INSIGHTS AND PERSPECTIVES

AMITABHA CHATTOPADHYAY, SANDEEP SHRIVASTAVA, AND ARUNIMA CHAUDHURI

Centre for Cellular and Molecular Biology, Council of Scientific and Industrial Research, Uppal Road, Hyderabad, 500 007, India

ABBREVIATIONS

2-AS: 2-(9-anthroyloxy)stearic acid

12-AS: 12-(9-anthroyloxy)stearic acid

25-NBD-cholesterol: 25-[*N*-[(7-nitrobenz-2-oxa-1,3-diazol-4-yl)-methyl]amino]-27- norcholesterol

DOPC: dioleoyl-*sn*-glycero-3-phosphocholine

MβCD: methyl-β-cyclodextrin

NBD: 7-nitrobenz-2-oxa-1,3-diazol-4-yl

NBD-PE: *N*-(7-nitrobenz-2-oxa-1,3-diazol-4-yl)-1,2-dipalmitoyl-*sn*-glycero-3-phosphoethanolamine

REES: red edge excitation shift

SDS: sodium dodecyl sulfate

Fluorescent Analogs of Biomolecular Building Blocks: Design and Applications, First Edition.
Edited by Marcus Wilhelmsson and Yitzhak Tor.
© 2016 John Wiley & Sons, Inc. Published 2016 by John Wiley & Sons, Inc.

15.1   INTRODUCTION

Biological membranes are two-dimensional, anisotropic supramolecular assemblies consisting of lipids, proteins, and carbohydrates. Cellular membranes allow compartmentalization of individual cells and act as the interface necessary for cells to sense their environment and communicate with other cells. Most importantly, cellular membranes provide an appropriate environment for function of membrane proteins.[1] It has been estimated that ~50% of all biological processes occur at the cell membrane.[2]

Membrane lipids represent crucial components of cell membranes since they carry out a variety of cellular functions along with membrane proteins. Monitoring lipid molecules in the crowded membrane constitutes an experimental challenge. In this context, membrane lipid probes assume relevance.[3,4] Fluorescent lipid probes offer advantages in monitoring membrane organization and dynamics due to their high sensitivity, suitable time resolution, and multitude of measurable parameters. Many fluorescent lipid probes have the extrinsic fluorophore covalently attached to the parent lipid molecule. The popularity of these probes arises from the fact that the user has a choice of the fluorescent tag to be used, and, therefore, specific lipid probes with appropriate spectral characteristics can be designed depending on the type of application. Figure 15.1 shows the molecular structures of a few representative membrane fluorescent probes. These probes have sensitive fluorescent groups such as

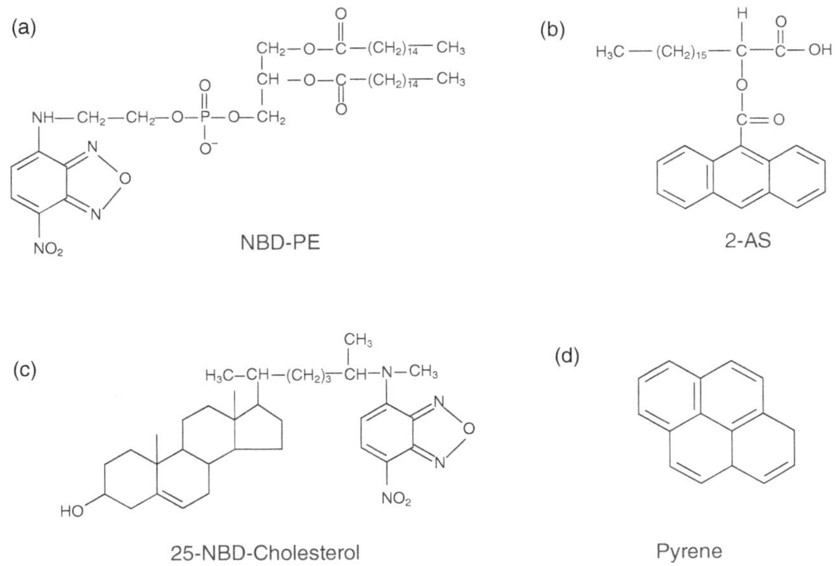

Figure 15.1 Chemical structures of representative membrane fluorescent probes. (a) NBD-PE (the fluorescent NBD group is covalently attached to the polar lipid head group in this molecule); (b) 2-AS (a representative member of anthroyloxy stearic acid probes); (c) 25-NBD-cholesterol (the NBD moiety is attached to the flexible acyl chain of cholesterol); (d) pyrene (a polycyclic aromatic hydrocarbon).

NBD (7-nitrobenz-2-oxa-1,3-diazol-4-yl) or anthroyloxy moiety attached to various positions of phospholipids, fatty acids, and cholesterol. One of the probes shown in the figure is the polycyclic aromatic hydrocarbon pyrene. It is to be noted that pyrene does not have a fatty acyl chain, characteristic of lipid probes (although it can be conjugated to lipid probes). Pyrene partitions into the membrane bilayer and its use in monitoring membrane environment and dynamics is based on its spectral characteristics (vibronic peaks; see Section 15.4).

In this review, we focus on the application of membrane fluorescent probes to obtain information on environment, organization, and dynamics in membranes (or membrane-mimetic media) with representative examples taken from previous work from our group. Readers interested in a more detailed information about some of these probes are referred to our earlier reviews for such information.[4,5]

15.2   NBD-LABELED LIPIDS: MONITORING SLOW SOLVENT RELAXATION IN MEMBRANES

An extensively used fluorophore in studies of model and cellular membranes is the NBD group.[4,5] The NBD group is very weakly fluorescent in water but displays intense fluorescence in a hydrophobic medium. NBD fluorescence is in the visible range and is characterized by sensitivity to its immediate environment. In addition, fluorescence lifetime of the NBD group is also sensitive to environmental polarity. For these reasons, NBD-labeled lipids are widely used in studies of model and biological membranes.[4,5] We describe below the application of NBD-labeled lipids to monitor slow solvent relaxation in the membrane utilizing red-edge excitation shift (REES).

An interesting consequence of membrane organization is the restriction imposed on the dynamics of the constituent structural components in the membrane. Importantly, this kind of confinement leads to coupling of the motion of solvent molecules with the slow-moving molecules in the membrane.[6] In such a case, REES represents a sensitive approach that can be used to monitor the environment and dynamics around the fluorophore in membranes or membrane-mimetic media such as micelles.[7–11] REES is operationally defined as the shift in the wavelength of maximum fluorescence emission toward higher wavelengths, caused by a shift in the excitation wavelength toward the red edge of the absorption band. The origin of REES lies in slow (relative to fluorescence lifetime) rates of solvent relaxation around an excited-state fluorophore. As a consequence, REES depends on the environment-induced motional restriction imposed on solvent molecules in the immediate vicinity of the fluorophore. A striking feature of REES is that it allows to assess the rotational mobility of the environment itself (represented by the relaxing solvent molecules) utilizing the fluorophore *merely* as a reporter group (it should be noted here that "solvent" in this context could include the host dipolar matrix such as the peptide backbone in proteins).[12]

The biological membrane, with its viscous interior and characteristic motional gradient along its vertical axis (z-axis), is an ideal molecular assembly for the application of REES.[8,10] The interfacial region in membranes display unique motional and dielectric characteristics, different from the bulk aqueous phase and the more isotropic hydrocarbon-like deeper regions of the membrane. Since the membrane interface offers slow rates of solvent relaxation, it is most likely to exhibit REES. However, it is important to choose a suitable membrane probe that displays appropriate properties (such as site of localization in the membrane and appreciable change in dipole moment upon excitation).[8,10] The NBD group in membrane-bound NBD-PE (see Fig. 15.2a) fulfills these criteria.[14] The fluorescent NBD group is covalently attached to the head group of phosphatidylethanolamine in NBD-PE. The orientation and location of the NBD group in membrane-bound NBD-PE has previously been well worked out.[15–21] The NBD group in NBD-PE is localized at the membrane interface and is suitable for monitoring REES. Interestingly, the NBD group displays a relatively large change in dipole moment upon excitation ($\sim$4 D),[22] a necessary condition for REES.[10] Fig. 15.2b and c shows REES of NBD-PE in dioleoyl-*sn*-glycero-3-phosphocholine (DOPC) membranes.[14] Since the localization of the fluorescent NBD group in membrane-bound NBD-PE is interfacial,[16–21] REES of NBD-PE implies that the interfacial region of the membrane offers considerable restriction to the reorientational motion of the solvent dipoles around the excited-state NBD group. This property of the membrane interface has huge implications in membrane protein conformation and function. This is due to the fact that tryptophan residues in membrane-spanning proteins and peptides are usually localized at the membrane interface[23] and, therefore, offer the possibility of REES as a novel tool to explore membrane protein conformation.[11,24,25]

15.3   *n*-AS MEMBRANE PROBES: DEPTH-DEPENDENT SOLVENT RELAXATION AS MEMBRANE DIPSTICK

Membranes display considerable anisotropy along the axis perpendicular to the membrane plane (see Fig. 15.3a).[8,10] The center of the membrane bilayer is nearly isotropic. However, the upper portion of the bilayer, located only a few angstroms away toward the membrane surface, is highly ordered. As a result of such an anisotropic transmembrane environment, the mobility of solvent (water) molecules is differentially retarded at varying membrane depths compared to their mobility in the bulk aqueous phase.[26] In such a scenario, REES can be effectively used to monitor the dynamics of differentially localized reporter fluorophores along the membrane *z*-axis. This has been validated by demonstrating that chemically identical fluorescent probes, differing solely in depths at which they are localized in the membrane, experience different local environments, as monitored by REES.[13,26] This was achieved by the use of anthroyloxy stearic acid (*n*-AS) derivatives in which the anthroyloxy group has earlier been shown to be either located at a shallow [2-(9-anthroyloxy)stearic acid (2-AS)] or a deeper location [12-(9-anthroyloxy)stearic acid (12-AS)] in the bilayer (see Fig. 15.3a). Anthroyloxy fatty acids have been shown to be located at a graded series of depths in the bilayer, depending on the position of attachment of the anthroyloxy group to the fatty acyl chain.[27] Depth analysis using the parallax method has earlier shown

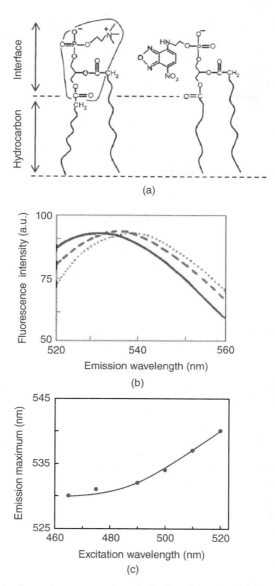

Figure 15.2 (a) A schematic representation of a leaflet of the phosphatidylcholine membrane bilayer showing the localization of the NBD group of NBD-PE. The NBD group of NBD-PE localizes at the membrane interfacial region. The horizontal line at the bottom indicates the center of the bilayer. Chattopadhyay and Mukherjee[13]. Reprinted with permission from American Chemical Society. (b) Intensity-normalized fluorescence emission spectra of NBD-PE in DOPC vesicles at increasing excitation wavelengths: 465 (——), 500 (— —), and 510 (-----) nm. Chattopadhyay and Mukherjee[14]. Reprinted with permission from American Chemical Society. (c) REES of NBD-PE in DOPC membranes. Chattopadhyay and Mukherjee[14]. Reprinted with permission from American Chemical Society. See text for other details.

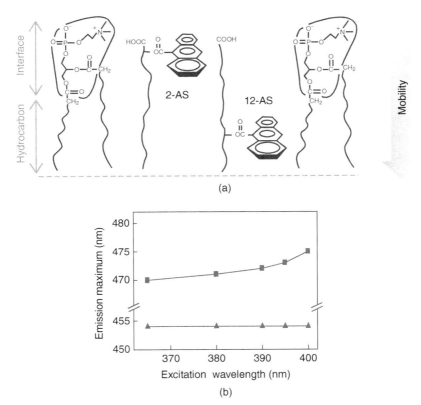

Figure 15.3   (a) A schematic representation of a leaflet of the membrane bilayer showing the localizations of the anthroyloxy groups of 2- and 12-AS in phosphatidylcholine bilayer. The anthroyloxy group of 2-AS localizes at the interfacial region while that of 12-AS resides at the nonpolar hydrocarbon region. A mobility gradient along the bilayer normal is set up (shown in the figure) due to differential dynamics at varying depths in the membrane. The dotted horizontal line at the bottom indicates the center of the bilayer. Chattopadhyay and Mukherjee[26]. Reprinted with permission from American Chemical Society. (b) Depth-dependent REES of 2-AS (■) and 12-AS (▲). Chattopadhyay and Mukherjee[26]. Reprinted with permission from American Chemical Society.

that the anthroyloxy probes in 2-AS (the shallow probe) and 12-AS (the deep probe) are localized at 15.8 and 6 Å from the center of the bilayer, respectively (see Fig. 15.3a).[20] REES experiments with these depth-dependent membrane probes show that the anthroyloxy group of 2- and 12-AS experience different local membrane microenvironments, as reflected by depth-dependent variation of REES. The shallow anthroyloxy group in 2-AS displays REES of 5 nm, whereas the deeper anthroyloxy group in 12-AS does not exhibit REES.[26] These results are attributed to differential rates of solvent reorientation in the immediate vicinity of the anthroyloxy group as the membrane penetration depth changes, that is, slower solvent relaxation at the membrane interface relative to deeper regions. These results show that REES offers a suitable approach to monitor depth-dependent membrane dynamics, that is, as a membrane dipstick. Very recently, we have extended the use of such anthroyloxy membrane probes to monitor depth-dependent heterogeneity in the membrane by analysis of fluorescence lifetime distribution width.[28]

15.4 PYRENE: A MULTIPARAMETER MEMBRANE PROBE

The polycyclic aromatic hydrocarbon pyrene has been widely used as a fluorescent probe in membranes and membrane-mimetic media such as micelles. The emission spectrum of pyrene is sensitive to environmental polarity[29] (see Fig. 15.4a). Pyrene is localized predominantly in the interfacial region in micelles and membranes.[31,32] Interestingly, this is the region of the membrane or micelle that is sensitive to polarity changes due to water penetration. Figure 15.4b and c shows the application of polarity-sensitive vibronic peaks of pyrene to monitor environmental changes in micelles and membranes under various conditions. Structural transition (shape change) can be induced in charged micelles by increasing ionic strength (salt concentration).[33–35] For example, spherical micelles of sodium dodecyl sulfate (SDS) that exist in water at concentrations higher than critical micelle concentration assume an elongated rod-like (prolate) shape in the presence of high salt concentrations. In such a case, utilizing changes in the ratio of polarity-sensitive vibronic peak intensities (I_1/I_3; see Fig. 15.4b), the apparent polarity in spherical- and rod-shaped micelles could be determined.[34] These results showed that the apparent polarity was less in rod-shaped micelles relative to the polarity experienced in spherical micelles. Figure 15.4c shows the change in the ratio of vibronic peak intensities (I_1/I_3) in pyrene emission spectra in neuronal hippocampal membranes with decreasing cholesterol content. Hippocampal membranes represent a convenient natural source for exploring the interaction of neuronal receptors, such as the serotonin$_{1A}$ receptor, with membrane lipids.[30,36,37] Methyl-β-cyclodextrin (MβCD) is a water-soluble compound and has previously been shown to selectively and efficiently extract cholesterol from hippocampal membranes by including it in a central nonpolar cavity.[36] Figure 15.4c shows that hippocampal membranes treated with increasing concentrations of MβCD (i.e., with increasing extents of cholesterol depletion) resulted in an increase in the vibronic peak intensity ratio. This implies an increase in apparent polarity experienced by pyrene in cholesterol-depleted hippocampal membranes, due to an increase in water penetration in the membrane upon cholesterol depletion. This is in agreement with our earlier results using fluorescence lifetime of the hydrophobic probe Nile Red.[38]

15.5 CONCLUSION AND FUTURE PERSPECTIVES

Tracking lipid molecules in a crowded cellular milieu poses considerable challenge. Fluorescent membrane probes offer a sensitive way to achieve this. Lipid probes have proved to be useful in membrane and cell biology due to their ability to monitor a wide variety of properties such as polarity, rotational dynamics, and diffusion in a depth-dependent fashion. Continuous improvement in instrumentation has allowed detection of fluorescent lipid probes with increasing spatiotemporal resolution.[39] Along with this, there have been new approaches to design membrane probes with specific properties. For example, the use of novel probes to detect membrane

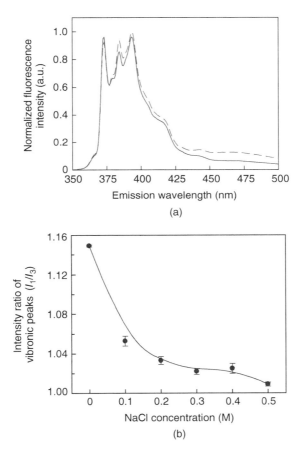

(a)

(b)

Figure 15.4 Change in environmental polarity monitored using pyrene fluorescence. (a) Intensity-normalized fluorescence emission spectra of pyrene in SDS micelles in the absence (——) and presence (— —) of NaCl. The polarity-sensitive vibronic peaks are clearly seen in the spectra. Chaudhuri *et al.* [34]. Reprinted with permission from Elsevier. (b) Change in fluorescence intensity ratio of the first (373 nm) and third (384 nm) vibronic peaks of pyrene (I_1/I_3) in SDS micelles as a function of increasing NaCl concentration. Chaudhuri *et al.* [34]. Reprinted with permission from Elsevier. (c) Change in fluorescence intensity ratio of the first (373 nm) and third (384 nm) vibronic peaks of pyrene (I_1/I_3) in hippocampal membranes with decreasing membrane cholesterol content (i.e., increasing MβCD concentration). Chaudhuri *et al.* [34]. Reprinted with permission from Elsevier. See text for more details.

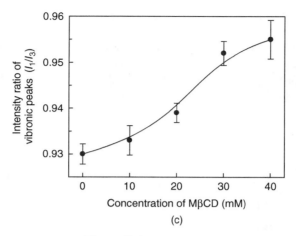

Figure 15.4 (*Continued*)

phases could be useful.[40] Future exciting applications include monitoring amyloid fibril formation utilizing NBD-labeled lipids and mapping cellular viscosity using fluorescent "molecular rotors".[41–43] Another important and exciting area is the determination of membrane dipole potential in which electrochromic membrane probes are used.[44–47] We envision that with novel and intelligently designed probes, and increasing instrument resolution, membrane probes will provide insightful information in the context of organization and dynamics of cellular membranes.

ACKNOWLEDGMENTS

Work in A.C.'s laboratory was supported by the Council of Scientific and Industrial Research, and Department of Science and Technology, Government of India. Some of the work described in this chapter was carried out by former members of A.C.'s research group whose contributions are gratefully acknowledged. A.C. thanks the Council of Scientific and Industrial Research for the award of a Senior Research Fellowship. A.C. is an Adjunct Professor at the Special Centre for Molecular Medicine of Jawaharlal Nehru University (New Delhi, India) and Indian Institute of Science Education and Research (Mohali, India), and Honorary Professor of the Jawaharlal Nehru Centre for Advanced Scientific Research (Bangalore, India). A.C. gratefully acknowledges J.C. Bose Fellowship (Department of Science and Technology, Government of India). We thank members of our laboratory for critically reading the manuscript.

REFERENCES

1. Jafurulla, M.; Chattopadhyay, A. *Curr. Med. Chem.* **2013**, *20*, 47.
2. Zimmerberg, J. *Curr. Biol.* **2006**, *16*, R272.

3. Chattopadhyay, A. (Ed.) *Chem. Phys. Lipids* **2002**, *116*, 1.

4. Haldar, S.; Chattopadhyay, A. In Fluorescent Methods to Study Biological Membranes; Mely, Y. and Duportail, G., Eds.; Springer, Heidelberg, **2013**; pp 37.

5. Chattopadhyay, A. *Chem. Phys. Lipids* **1990**, *53*, 1.

6. Bhattacharyya, K.; Bagchi, B. *J. Phys. Chem. A* **2000**, *104*, 10603.

7. Mukherjee, S.; Chattopadhyay, A. *J. Fluoresc.* **1995**, *5*, 237.

8. Chattopadhyay, A. *Chem. Phys. Lipids* **2003**, *122*, 3.

9. Demchenko, A. P. *Methods Enzymol.* **2008**, *450*, 59.

10. Haldar, S.; Chaudhuri, A.; Chattopadhyay, A. *J. Phys. Chem. B* **2011**, *115*, 5693.

11. Chattopadhyay, A.; Haldar, S. *Acc. Chem. Res.* **2014**, *47*, 12.

12. Haldar, S.; Chattopadhyay, A. *J. Phys. Chem. B* **2007**, *111*, 14436.

13. Chattopadhyay, A.; Mukherjee, S. *J. Phys. Chem. B* **1999**, *103*, 8180.

14. Chattopadhyay, A.; Mukherjee, S. *Biochemistry* **1993**, *32*, 3804.

15. Haldar, S.; Chattopadhyay, A. In Reviews in Fluorescence 2010; Geddes, C. D., Ed.; Springer, New York, **2012**; pp 155.

16. Chattopadhyay, A.; London, E. *Biochemistry* **1987**, *26*, 39.

17. Chattopadhyay, A.; London, E. *Biochim. Biophys. Acta* **1988**, *938*, 24.

18. Mitra, B.; Hammes, G. G. *Biochemistry* **1990**, *29*, 9879.

19. Wolf, D. E., Winiski, A. P.; Ting, A. E.; Bocian, K. M.; Pagano, R. E. *Biochemistry* **1992**, *31*, 2865.

20. Abrams, F. S.; London, E. *Biochemistry* **1993**, *32*, 10826.

21. Mukherjee, S.; Raghuraman, H.; Dasgupta, S.; Chattopadhyay, A. *Chem. Phys. Lipids* **2004**, *127*, 91.

22. Mukherjee, S.; Chattopadhyay, A.; Samanta, A.; Soujanya, T. *J. Phys. Chem.* **1994**, *98*, 2809.

23. Kelkar, D. A.; Chattopadhyay, A. *J. Biosci.* **2006**, *31*, 297.

24. Rawat, S. S.; Kelkar, D. A.; Chattopadhyay, A. *Biophys. J.* **2004**, *87*, 831.

25. Raghuraman, H.; Kelkar, D. A.; Chattopadhyay, A.. In Reviews in Fluorescence 2005; Geddes, C. D., Lakowicz, J. R. Eds.; Springer, New York, **2005**; pp. 199.

26. Chattopadhyay, A.; Mukherjee, S. *Langmuir* **1999**, *15*, 2142.

27. Abrams, F. S.; Chattopadhyay, A.; London, E. *Biochemistry* **1992**, *31*, 5322.

28. Haldar, S.; Kombrabail, M.; Krishnamoorthy, G.; Chattopadhyay, A. *J. Phys. Chem. Lett.* **2012**, *3*, 2676.

29. Dong, D. C.; Winnik, M. A. *Photochem. Photobiol.* **1982**, *35*, 17.

30. Saxena, R.; Shrivastava, S.; Chattopadhyay, A. *J. Phys. Chem. B* **2008**, *112*, 12134.

31. Shobha, J.; Srinivas, V.; Balasubramanian, D. *J. Phys. Chem.* **1989**, *93*, 17.

32. Hoff, B.; Strandberg, E.; Ulrich, A. S.; Tieleman, D. P.; Posten, C. *Biophys. J.* **2005**, *88*, 1818.

33. Rawat, S. S.; Chattopadhyay, A. *J. Fluoresc.* **1999**, *9*, 233.

34. Chaudhuri, A.; Haldar, S.; Chattopadhyay, A. *Biochem. Biophys. Res. Commun.* **2009**, *390*, 728.

35. Chaudhuri, A.; Haldar, S.; Chattopadhyay, A. *Chem. Phys. Lipids* **2012**, *165*, 497.

36. Pucadyil, T. J.; Chattopadhyay, A. *Biochim. Biophys. Acta* **2004**, *1663*, 188.

37. Singh, P.; Tarafdar, P. K.; Swamy, M. J.; Chattopadhyay, A. *J. Phys. Chem. B* **2012**, *116*, 2999.

38. Mukherjee, S.; Kombrabail, M.; Krishnamoorthy, G.; Chattopadhyay, A. *Biochim. Biophys. Acta* **2007**, *1768*, 2130.

39. Eggeling, C.; Ringemann, C.; Medda, R.; Schwarzmann, G.; Sandhoff, K.; Polyakova, S.; Belov, V. N.; Hein, B.; von Middendorff, C.; Schönle, A.; Hell, S. W. *Nature* **2009**, *457*, 1159.

40. Kucherak, O. A.; Oncul, S.; Darwich, Z.; Yushchenko, D. A.; Arntz, Y.; Didier, P.; Mély, Y.; Klymchenko A. S. *J. Am. Chem. Soc.* **2010**, *132*, 4907.

41. Ryan, T. M.; Griffin, M. D. W. *Biochemistry* **2011**, *50*, 9579.

42. Kuimova, M. K.; Yahioglu, G.; Levitt, J. A.; Suhling, K. *J. Am. Chem. Soc.* **2008**, *130*, 6672.

43. Hosny, N. A.; Mohamedi, G.; Rademeyer, P.; Owen, J.; Wu, Y.; Tang, M.-X.; Eckersley, R. J.; Stride, E.; Kuimova, M. K. *Proc. Natl. Acad. Sci. U. S. A.* **2013**, *110*, 9225.

44. Clarke, R. J. *Adv. Colloid Interface Sci.* **2001**, *89–90*, 263.

45. Wang, L. *Annu. Rev. Biochem.* **2012**, *81*, 615.

46. Haldar, S.; Kanaparthi, R. K.; Samanta, A.; Chattopadhyay, A. *Biophys. J.* **2012**, *102*, 1561.

47. Singh, P.; Haldar, S.; Chattopadhyay, A. *Biochim. Biophys. Acta* **2013**, *1828*, 917.

16

LIPOPHILIC FLUORESCENT PROBES: GUIDES TO THE COMPLEXITY OF LIPID MEMBRANES

MAREK CEBECAUER AND RADEK ŠACHL

Department of Biophysical Chemistry, J. Heyrovský Institute of Physical Chemistry of the Academy of Sciences of the Czech Republic, Dolejskova 3, Prague 8, Czech Republic

16.1 INTRODUCTION

Membranes surround cellular interiors in order to preserve their content and protect cells from sudden environmental changes. In addition, membranes function as the interface and two-dimensional surface for structural organization and signaling in cells.[1] Due to these essential functions, membranes were found attractive, even though uneasy, subject of biophysical, biochemical, and biological studies. For example, the understanding of membrane properties is essential for drug design – their efficient delivery to cells is a prerequisite for successful treatment of patients.

How can we investigate structural and dynamic properties of cellular membranes? While biochemical techniques provide indispensable information on membrane composition and specific interactions of the molecules in these platforms, fluorescence techniques were developed for the characterization of the dynamics of membranes and the membrane-associated processes. Recently, we have witnessed dramatic development in the field of fluorescence spectroscopy and imaging.[2–11] As the title of this book states, we will focus on the fluorescent probes that enable a deeper view into the world of membranes. There are at least two areas of membrane

Fluorescent Analogs of Biomolecular Building Blocks: Design and Applications, First Edition.
Edited by Marcus Wilhelmsson and Yitzhak Tor.
© 2016 John Wiley & Sons, Inc. Published 2016 by John Wiley & Sons, Inc.

research, where fluorescent probes can be helpful. Firstly, the focus can be on individual membrane-associated molecules (e.g., diffusion or clustering of particular lipids and proteins) and, secondly, on local membrane environment (e.g., lipid acyl chain ordering, viscosity, polarity). Some probes are used for a variety of experiments, whereas others were developed for a specific purpose. In this chapter, we offer an insight into the use of the most common membrane fluorescent probes. We try to provide not only the information on advantages but also limitations associated with the use of probes for specific experimental setup. But let's start with a brief description of (biological) membranes and their elementary components, lipids.

16.2   LIPIDS, LIPID BILAYERS, AND BIOMEMBRANES

Lipids, specifically glycerophospholipids, form the basis of biological membranes. These amphiphilic molecules can generate lamellar structures called lipid bilayers (Fig. 16.1) by orientating their hydrophilic head groups toward the aqueous environment and "hiding" hydrophobic acyl chains in the center. The other frequent lipid species of biological membranes are sphingolipids and sterols. Sphingolipids and glycerolipids share the common structure, but the glycerol base is replaced with the sphingoid base in sphingolipids (Fig. 16.1a). The structure of sterols differs from that of glycerolipids in means these are flat and relatively rigid molecules. Sterols can form up to 40% of total lipids in some membranes (the plasma membrane of eukaryotes; Fig. 16.1b). Sterols are essential for eukaryotes by preserving the liquid character of highly rigid plasma membrane.

Model lipid bilayers, such as vesicles and planar membranes, are useful tools for elementary biophysical studies of membrane components and properties. Small, large, and giant unilamellar vesicles (SUVs, LUVs, GUVs; Fig. 16.1d) represent free-standing membranes where unrestrained mobility and organization of lipids can be tested. On the contrary, supported planar bilayers, as the name states, are generated on the glass or mica support, therefore, providing a flat and stable model membrane for imaging-based experiments. There are arguments about the use of supported membranes for biophysical studies due to the impact of support on the mobility of lipids. Of note, many biological membranes represent semisupported membranes due to a dense cortical actin network underlying the plasma membrane. Due to their relative simplicity (composed of 1–5 lipid species) model membranes are essential for biophysical studies of basic processes such as lipid segregation, lipid ordering, or the mobility of molecules (diffusion). The use of model membranes allows a good control of lipid composition, size and curvature (vesicles), and access to a number of functionalized lipids (e.g., fluorescent lipid species) that cannot be easily incorporated into biological membranes *in vivo*. A majority of the experiments documenting the use of fluorescent membrane probes in the following section were performed using model membranes. On the other hand, one has to be careful when extrapolating model membrane-derived data to interpret observations achieved by imaging cellular membranes (E. Gratton, P. Šebo, personal communication). The plasma membrane

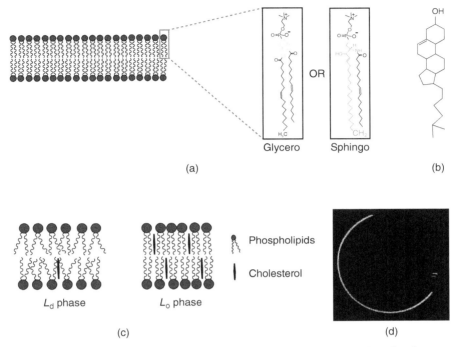

Glycero   Sphingo

(a)

(b)

L_d phase L_o phase

Phospholipids

Cholesterol

(c)

(d)

Figure 16.1 (a) Schematic illustration of a lipid bilayer with two examples of main membrane constituents: glycerophospholipids (glycero) and sphingolipids (sphingo). Glycerol base is indicated in red, sphingosine in green. (b) Planar structure of cholesterol. (c) Schematic illustration of L_d and L_o phase with lower and higher content of cholesterol, respectively. (d) Image of phase separation in GUV composed of DOPC:DPPC:cholesterol (1:1:1). BODIPY-based tri-alkylated molecular rotor (compound **3a** in Ref. 12) segregates out of L_o phase (dark part) into L_d phase (bright part). (*See color plate section for the color representation of this figure.*)

of eukaryotic cells is a nonequilibrium, asymmetric, dynamic structure supported by cortical actin and strengthened by membrane potential. No model membrane to date can mimic even a part of these properties.[13]

Biological membranes are composed of a large number of different lipid species, which differ in their head group, acyl chain length, and saturation and, at the cell surface, in glycosylation status. In addition, biological membranes contain proteins that are embedded in the bilayer (integral membrane proteins). Other proteins are attached to or peripherally interact with the membrane. This illustrates a complex character of biological membranes, which, at physiological temperature, are dynamic structures with high mobility (diffusion coefficients 0.01–2 μm²/s) and large turnover ($t_{1/2} \sim 30$ min) of components. These are general numbers and cannot be applied to all molecules, but underline the need for fast and robust techniques in membrane research. Fluorescent probes and relevant techniques offer a good tool to study rapid processes of such multicompound systems.

16.3   LIPID PHASES, PHASE SEPARATION, AND LIPID ORDERING

An intrinsic membrane property studied extensively in model membranes are lipid phases, their properties, and phase separation (Fig. 16.1c and d). Due to the variable structure and physical properties, lipids can exist in the liquid or solid state. In unilamellar membranes, 2 liquid phases – disordered (L_d) and ordered (L_o) – with the high lateral mobility of molecules have been demonstrated. The L_d phase is composed of lipids with unsaturated acyl chains missing the capacity for self-organization and supporting extensive freedom for the movement of lipid acyl chains. In the absence of sterols, glycerophospholipids and sphingolipids with long and highly saturated acyl chains form solid (S_o) phase. Sterols help to preserve the liquid character of membranes with high content of lipids with long, saturated acyl chains – the L_o phase. Acyl chain movement in the L_o phase is limited due to a tighter lipid packing compared to the L_d phase. Sterols also rigidify membranes with high content of unsaturated lipids. Of note, the capacity of sterols to influence the glycerophospholipid membranes varies with cholesterol and ergosterol of higher eukaryotes and yeast having the strongest impact.[14]

In model membranes containing unsaturated phospholipid, saturated phospholipid, and sterol, phase separation into domains was visualized using fluorescence confocal microscopy.[15] More recently, formation of nanoscopic domains in membranes with lipid compositions was believed to form homogenous L_d phase were demonstrated.[2,16,17] In living cells, lipid phases cannot be formed due to their nonequilibrium character. Therefore, membrane physical property – lipid ordering – is analyzed to better understand differences in cell membranes, particularly lipid segregation.[18,19] The advantage of lipid ordering (general polarization (GP) value; see Section "Solvatochromic Dyes") measurements is that this membrane parameter does not depend on equilibrium character of the subject and can be analyzed in membranes with unlimited compositional complexity. Of note, much smaller differences of lipid ordering were detected between "phases" in cell-derived vesicles compared to those in model membranes.[20]

In the following parts, we will characterize selected fluorescent probes and their use in membrane research. Their general use is described; cited references should guide the reader to the works, where membrane fluorescence probes were applied, but we put a lot of effort to describe the limitations of the existing probes, first, to indicate those with respect to the interpretation of data and, second, to stimulate interest in the synthesis of novel membrane fluorescent probes. To cite the reviewer of this book chapter: "Suitable probes for the membrane research are very scarce and development in this direction is highly needed."

16.4   FLUORESCENT PROBES FOR MEMBRANE STUDIES

Fluorescent membrane probes can be classified according to a variety of criteria. There are probes that are obtained by linking a chromophore moiety to a lipid molecule generating fluorescent lipid variants. Other probes that do not resemble any particular lipid species associate with membranes based on their hydrophobicity. This hydrophobicity is achieved by the addition of alkyl chains to the chromophore core, as in the case of commercially available dyes such as DiI, DiO, or DiD (Molecular Probes). Highly hydrophobic polycyclic aromatic hydrocarbons, such as naphthopyrene, perylene, or rubicene to name just a few, are fluorescent molecules integrating into the membranes. Fluorescently labeled lipids provide species-specific information (e.g., mobility or aggregation), whereas more general membrane properties (e.g., lipid ordering or membrane potential) are investigated using nonlipid-derived probes.

All of these probes were applied in a number of different studies, and we will describe here only their most common use based on the application.

16.4.1   Fluorescently Labeled Lipids

Lipids are prevalently nonfluorescent molecules. In order to investigate the physical behavior of specific lipid species (or lipid families) in membranes, a lipid moiety needs to be directly labeled with a fluorescent dye. Obviously, the addition of a relatively large chromophore to the lipid can influence its properties. It is important whether the chromophore is attached to the alkyl chains localized in the hydrophobic core of the lipid bilayer or to the head group exposed to the aqueous environment.

In the case of head group-labeled lipids, a dye is usually attached to the phosphoethanolamine (PE) group by means of NH–CO bond. To our knowledge, the availability of other head group labeled lipid families is still very limited (e.g., www.avantilipids.com). Indeed, labeled phosphatidylcholines (PC), the most common lipids in biological membranes of eukaryotes,[21] are still missing probably due to difficulties in attaching a label to the $-N^+(CH_3)_3$ group. Although modification of the head group is supposed to disturb the bilayer to a smaller extent, an effect is still detectable. For instance, it has been shown that replacement of the PC head group by PE labeled by boron-dipyrromethene (BODIPY) changes redistribution of diacyl/monoacylpalmitoyl lipids in various lipid bilayers.[22] It is important to mention here that some head group-labeled lipids lose capacity to interact with their cytosolic partners (e.g., labeled phosphoinositides with PH domains of proteins). Head group-labeled fluorescent lipids were successfully applied in the studies focused on the diffusion[23,24] or segregation of specific lipids.[16,25]

Similarly, acyl chain-labeled lipids were used to their lateral and transbilayer diffusion[26,27] and lipid segregation.[25,28] When using a chromophore attached to the acyl chain region attention should be paid, since a chromophore does not need to stay at the corresponding depth of the bilayer and loops back to the lipid/water interface. This occurs for instance in the case of NBD[29] or BODIPY[30] moieties. Interestingly, methylation of BODIPY at four positions of the ring does not prevent the BODIPY chromophore attached to differently long sn-2 acyl chains from looping back to the lipid/water interface.[31] As has been shown, a chromophore that stays at corresponding depths of the bilayer is the anthroyloxy (AS) moiety in n-(9-anthroyloxy)-stearic acid (9AS), where the chromophore is attached to the nth carbon of the stearic acid (Fig. 16.2).[32,33] This probe is environment sensitive

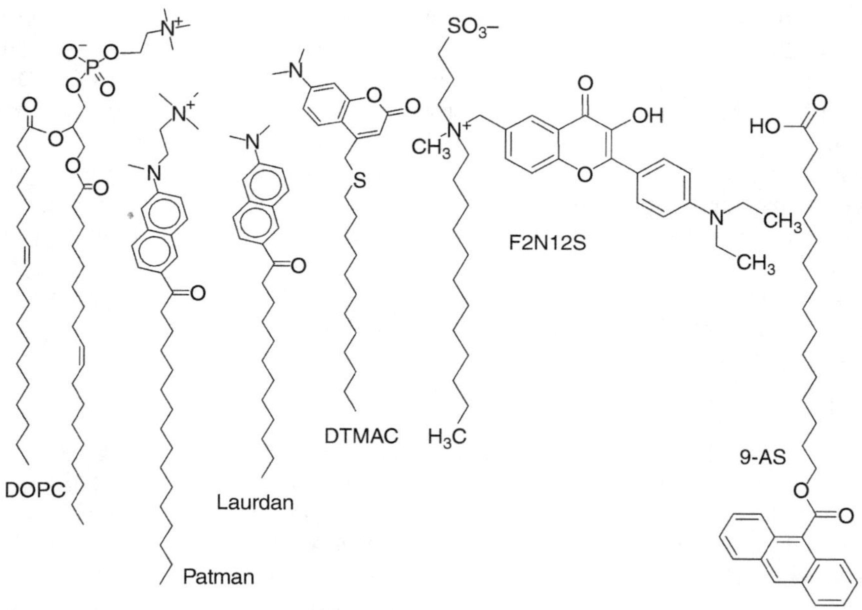

Figure 16.2 Position of membrane fluorescent dyes along the z-vertical axis with respect to a dioleoylphosphatidylcholine (DOPC) molecule. The following fluorescent probes in the DOPC bilayer are shown: 6-hexadecanoyl-2-(((2-(trimethylammonium)ethyl) methyl)amino)naphth alene chloride (Patman), 6-lauroyl-2-dimethylaminonaphthalene (Laurdan), 4-[(n-dode cylthio)methyl]-7-(N,N-dimethylamino)-coumarin (DTMAC), 4'-N,N-diethylamino-6-(N-dod ecyl-N-methyl-N-(3-sulfopropyl))ammoniomethyl-3-hydroxyflavone (F2N12S) and (9-(9-an throyloxy)stearic acid) (9-AS).

(see Section 16.4.2) and thus allows for probing physicochemical properties of the bilayer.

Since cholesterol influences physics of the lipid bilayer to a large extent,[34] interaction of cholesterol with the lipid bilayer has been intensely studied in the past. Insertion of cholesterol into the lipid bilayer gives rise to unfavorable interactions with water because of a small hydrophilic and a large hydrophobic part. These interactions are screened by tighter packing of surrounding lipid molecules in the presence of unsaturated phospholipids but the more liquid character of areas with saturated lipids.[34] Consequently, diffusion of lipid components in the bilayer is influenced. A number of fluorescent cholesterol derivatives are commercially available (e.g. Avanti, Lipids, Cayman Chemical, Life Technologies, Sigma-Aldrich), and a few more were synthesized for membrane studies by the groups of Schroeder[35] and Molotkovsky[36]. Numerous studies used these fluorescent sterols to investigate their membrane behavior.[25,37,38] Similar to the fluorescently labeled phospholipids, one should be careful regarding the interpretation of data. For instance, cholesterol linked to NBD by its hydroxyl group cannot form hydrogen bonds, whereas cholesterol labeled by NBD at the ring position 22 is reported not to partition to the L_o phase.

Both derivatives exhibit large deviation from cholesterol, as actually do cholesterol analogs labeled by BODIPY. A way how to circumvent this problem is to use sterol variants that intrinsically emit light. Dehydroergosterol contains, in addition to cholesterol, three double bonds and a methyl group, and its physical properties differ from cholesterol only a little.[39] This derivative has the absorption and the emission maximum in the UV region, which could complicate the measurements, especially the microscopy in living cells. Another fluorescent cholesterol variant, cholestatrienol, was found to partition into the L_o phase comparably to the unmodified cholesterol.[25,40] However, its use, at least for microscopy, is limited due to its poor fluorescence properties. In such cases, studying the effects of cholesterol on physical properties of lipid bilayers needs to be circumvented by using other probes that are localized in the same region of the lipid bilayer. But then one is analyzing local environment and not specific behavior of cholesterol, a topic described in the following section.

Sphingolipids are essential for numerous functions such as intracellular signaling[41,42] and are basic constituents of cellular membranes.[43] Diffusion of lipid molecules decreases with increased levels of sphingomyelin in a two-component lipid system, but interestingly, diffusion of lipid molecules in the L_o domains is not as sensitive to sphingomyelin concentration as in the case of cholesterol.[44] Although sphingolipids labeled by NBD and BODIPY in the acyl region are available, effect of these lipids on the membrane biophysics has rather been studied indirectly by using other fluorescently labeled membrane components. In cells, Tyteca and coworkers found nonhomogeneous and nonoverlapping distribution in microdomains of various acyl chain-labeled sphingo- and glycosphingolipids.[45,46] How this resembles distribution of specific unlabeled sphingolipids in cellular membranes needs to be tested.

G_{M1}, member of glycosphingolipid family, consists of sphingosine, a fatty acid and a bulky sugar head group, and is involved in various biological processes.[47,48] In model systems, G_{M1} segregates to L_o domains[49–51] and exhibits self-aggregation.[52] These molecules have been successfully labeled by BODIPY both in the polar and nonpolar region.[53] BODIPY chromophore attached to the bulky sugar head group has no observable effect on the aggregative behavior.[52] However, a possible influence of the chromophore on the redistribution of G_{M1} between the L_o and the L_d phase cannot be excluded (for the distribution coefficient of labeled G_{M1} between the L_o and the L_d phase, see the next section).

16.4.2 Environment-Sensitive Membrane Probes

16.4.2.1 Probes to Determine Lipid Phases and Their Separation: While previous section described probes used for the characterization of individual lipid molecules, the following sections will focus on local membrane environment and its properties. Probably the most studied property of membranes is the determination of the lipid phase and the formation of membrane microdomains due to the popularity of lipid rafts.[54] In such experiments, the partitioning of specific fluorescent probes

to liquid (L_d, L_o) and gel (S_o) phases is monitored. In addition, variation of lipid diffusion coefficient between phases can be used for the characterization of heterogeneity of membrane phases. Even though the S_o phase was intensely studied along L_d phase in early experiments,[55–59] we will focus on liquid phases due to a poor relevance of gel phase for biological membranes and, to some extent, S_o similarity to the L_o phase.

For studying lipid phases, polar fluorescent moiety should be exclusively attached to the head group region of lipids or other hydrophobic molecules. Most of the fluorescently labeled lipids were demonstrated to segregate out of the more rigid L_o phase into L_d phase in model membranes, where the labeled molecules better fit due to a higher flexibility of the disordered, unsaturated acyl chains. This phenomenon is rather independent of whether high T_m lipid was used as a basis for the synthesis.[25,60] Evidently, it is the chromophore part that influences the partitioning the most (at least in tertiary lipid composed model membranes). For the future development, the use of lipids with longer, saturated acyl chains (>C20:0) is expected to reduce the influence of chromophore for the probe's phase preference and improve their partitioning into L_o phase as shown using pyrene-labeled glycerophospholipids and sphingomyelins.[61] To date, the only exception found was nitrobenzoxadiazole-labeled dipalmitoylphosphatidylethanolamine (NBD-DPPE) that partitioned into the L_o phase in one model system[60,62–64] but was largely segregated out of this phase when different lipid composition was used.[25,60] This indicates a unique character (e.g., lipid ordering) of each individual phase formed using particular lipid composition and the fact that all probes need to be thoroughly tested for each individual system. Supporting this statement, the partitioning of probes is significantly increased for the "L_o-like phase" in cell-derived GPMVs with lipid and protein composition similar to that of plasma membrane showing reduced difference in lipid ordering between separated domains.[20]

No universal fluorescent marker of lipid phases has been established yet, even though the L_d phase was reproducibly labeled using Texas Red (TR)-DOPE lipid-based probe. Another generally accepted and commercially available markers of L_d phase are Liss-Rho-DOPE (*Lissamine* rhodamine B-labeled low T_m lipid) and some dialkylcarbocyanines (see below). Unfortunately, probes that would efficiently and reproducibly label the L_o (or L_o-like) phase are still missing (Table 16.1).[69] The highest affinity for L_o phase in model membranes so far exhibits labeled cholera toxin subunit B (ChTB). ChTB is a large pentameric proteinaceous structure that binds to G_{M1} gangliosides from outside of the lipid bilayer and, therefore, cannot be qualified as a small fluorescent probe of the membranes. ChTB being a pentamer causes cross-linking of G_{M1} and rather large changes within the membrane organization.[16] ChTB, therefore, cannot be applied as an "unbiased" sensor of membrane environment.

To find an efficient small molecule – L_o marker remains still as a challenge for the following years. Such molecules could represent head group labeled lipids that are abundant in the L_o domains, as for instance sphingomyelins, saturated phosphatidylcholines, or gangliosides, respectively, with long (>20 carbons) acyl chains. Unfortunately, even when the probe is attached in the head group region, it reduces its affinity for the L_o phase. Unlabeled G_{M1} preferentially resides in the L_o

TABLE 16.1 List of Fluorescent Probes that Have Increased Affinity to the Liquid Ordered (L_o) Phase

Fluorescent Probe	Lipid Composition of Tested Membrane (mol %; 1:1:1 If Not Stated)	L_o Phase Partition Coefficient with References
Perylene	DOPC/Sph/Chol/ DOPG/DPPE-biotin (29/39/25/5/2)	0.8 ± 0.2[65]
Perylene	DOPC/Sph/Chol	>1[66]
Cholera toxin	DOPC/Sph/Chol/DOPG/ DPPE-biotin (44/24/25/5/2)	6 ± 3[16]
Cholera toxin	DOPC/Sph/Chol	11 ± 5[67]
NBD-DPPE	POPC/Sph/Chol	4.3 ± 1.2[68]
DPH	DOPC/Sph/Chol	>1[25]
Terrylene	DOPC/Sph/Chol	>1[25]
Naphthopyrene	DOPC/Sph/Chol	>1[25]
Rubicene	DOPC/Sph/Chol	>1[25]

phase[70,71] but when labeled by FL-BODIPY partition coefficient of G_{M1} is slightly reduced.

The other group of fluorescent probes applied for the analysis of lipid phases are alkylated chromophores mimicking the structure of lipids. Outstanding examples are dialkylcarbocyanines such as DiI or DiD from Molecular Probes. Depending on the length of alkyl groups attached to the chromophore, these probes partition into or out of the L_o phase. DiI species with short, saturated alkyl chains (C12:0 and C16:0) partition out of L_o phase, whereas those with C18:0 or longer chains prefer more ordered environment of the L_o phase.[25,72] Here, again some level of variation in phase preference has been observed since all DiI species partitioned out of L_o phase in less stringent model system composed of DOPC:sphingomyelin:cholesterol at 50:27:23 ratio.[25] Similar inconsistent phase preference for $DiIC_{18}$ was observed by others when lipid composition or the use of two or more probes has been applied in biphasic model membranes.[44,62,73]

Probably the most promising markers of the L_o phase are polycyclic aromatic probes such as perylene, rubicene, terrylene, or naphthopyrene.[62,74–79] Equal or even highly L_o preferential distribution of these probes was demonstrated in a variety of lipid mixtures.[25,62,78,79] Nonpreferential distribution of smaller polycyclic aromatic probes (perylene and rubicene) can be used as an advantage for the characterization of general membrane shape and optical properties of individual membrane parts. Larger molecules, naphthopyrene and terrylene, were shown to be highly accumulated in L_o (and S_o) phase[25,80] and are to date the best small molecule markers of ordered membranes. Disadvantage of these dyes is complicated photophysics including excimer formation.

Preference of the abovementioned fluorescent probes for a particular lipid phase was prevalently determined qualitatively (visually) based on the intensity of fluorescence acquired in separated domains of biphasic model membranes. Few experimental data led to the quantitative values describing behavior of the probes in membranes. Partitioning of fluorescent probes to the L_o phase (or out of this phase) can be quantitatively characterized by (phase) partition coefficient defined as $K = $ [probe in L_o]/[probe in L_d]. For NBD-DPPE, $K = 4$ was found in POPC/Chol/Sph bilayer (Table 16.1).[68] For instance, labeled cholera toxin shown $K = 6-11$[16,67,68] depending on the bilayer composition, to date the highest measured values. At the other end of spectrum is DiD (C12:0) with $K = 0.004$ and high preference for the L_d phase. But only a handful of quantitative data is available, and systematic characterization of probes in different systems will require enormous effort. It is important to note here that there are serious known issues limiting the quantitative determination of probes' partitioning in membranes: (1) fluorescence quantum yield of a fluorophore in disordered phase can significantly differ from more ordered environment; (2) probes with aromatic residues may form ground or excited state dimers more effectively in the L_o phase, which would affect the emission of the fluorophore; and (3) more rigid orientation of probes in ordered phases influences data acquired using polarized excitation source in the fluorescence microscope with laser light source. Moreover, one needs to be careful when using "markers" of certain phase when studying different systems and in combination with additional components (e.g., second probe or proteins). The K value can vary for individual probes in systems composed of different lipids; for example, the use of dipalmitoylphosphatidylcholine (DPPC) or sphingomyelin as high T_m lipid in ternary, biphasic model lipid membranes. Incorporation of fluorescent probes (even at a low concentration such as 0.1 mol%) also modifies the physical properties of model membranes such as the miscibility transition temperature T_{mix}.[62] There is always need to characterize these probes in the model membranes planned for specific experiments.

16.4.2.2 Lipid Ordering-Sensitive Probes:

16.4.2.2 Lipid Ordering-Sensitive Probes: While lipid phases can be investigated in model membranes, no phases exist in nonequilibrium systems such as living cells. Physical properties of membranes are therefore characterized indirectly, for example, as a higher or lower lipid ordering.[18] In contrast to standard dyes used for direct labeling of lipids, probes sensitive to changes in lipid ordering must exhibit extensive asymmetry in the so-called "push–pull" chromophore responsive to the hydration of the local environment (solvatochromicity) or charge mobility due to changes of potential at the membrane (electrochromicity).[81–83] Both processes are, to some extent, defined by lipid ordering of the membrane.

Solvatochromic Dyes: Hydration-responsive fluorescent probes (also called solvatochromic dyes) exhibit a large change in the dipole moment after excitation, which leads to a relatively large Stokes shift. Surrounding water molecules interacting with a chromophore modulate the relaxation process from excited to more favorable equilibrium state. This is captured (i.e., the kinetics and the extent of relaxation) in the time-resolved fluorescence spectra and depends on the viscosity and polarity of the environment. In the case of the commonly used fluorescent probe Laurdan (Fig. 16.2), the large dipole moment change is mainly attributed to the charge transfer stirred by the dialkylamino group attached to one side of the naphthalene ring (Fig. 16.3), which works as the donor of electrons, and a carbonyl group attached to the other side of the ring, electron acceptor.[84] The presence of polar solvent molecules (e.g., water) causes a redshift of the spectrum of Laurdan (70 nm). Nowadays, a number of dyes working on similar principle and probing the entire bilayer region are available: while Laurdan and Patman are well localized at the lipid/water interface,[85] a set of probes consisting of a stearic acid and an anthroyloxy chromophore attached at different carbons of the stearyl chain can sense almost the entire bilayer interior (Fig. 16.2).[32] These were successfully employed to characterize the model membrane composed of a single lipid (POPC; Ref. 86).

Laurdan is currently the most extensively used membrane environment-sensitive probe in biological systems,[87–89] including whole small animals.[90] It was characterized in a high detail using model membranes and general polarization (GP) value was defined to provide a quantitative tool for cell membrane experiments.[87,88] GP is calculated from two spectral channels simultaneously measured in a sample:

$$GP = \frac{I_{500-593} - I_{600-680}}{I_{500-593} + I_{600-680}}$$

A novel derivative of Laurdan, C-Laurdan, with turn-on fluorescence for more ordered membranes was synthesized to improve the GP contrast and was

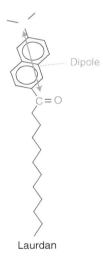

Laurdan

Figure 16.3 Illustration of the electric dipole moment in Laurdan solvatochromic chromophore. (*See color plate section for the color representation of this figure.*)

successfully applied to image ordered parts of model membranes and membranes in living cells.[91–93] Of note, C-Laurdan has a slower flip–flop kinetics compared to its original version limiting its transfer to the cytoplasmic leaflet of the plasma membrane in cells (Kai Simons, personal communication).

Electrochromic Dyes: Another group of membrane biosensors has been originally generated to provide the information on the function of ion channels in cells of neuronal system and membrane potential – potentiometric probes or electrochromic dyes.[94] Electrochromic dyes exhibit shift in the excitation and emission spectrum upon changes in the membrane polarity (electric field).[95] Emission of these dyes is also influenced by solvent relaxation processes similar to solvatochromic dyes.

Loew and coworkers, in probably the first systematic work on fluorescent chemical compounds,[83] uncovered styryl dyes sensitive to membrane potential. Of these early biosensors, di-4-ANEPPS was successfully applied for imaging localized depolarization events in growing neurites[96] and is in use to date. Further development led to novel probes such as di-4-ANEPPDHQ with a higher signal-to-noise ratio (SNR) and improved plasma membrane partitioning.[97] Due to its partial solvatochromic properties and redshifted excitation wavelength compared to Laurdan, it was applied for lipid ordering characterization in model[98,99] and cellular membranes.[100] Currently, styryl dyes with even more redshifted spectrum to avoid signal overlap with the autofluorescence of cells are synthesized.[101–103] The use of these dyes for experiments focused on membrane lipid ordering and its changes during, for example, cell activation still require systematic characterization in model membranes and in cells.

The group of Demchenko, Mély, and colleagues developed a family of 3-hydroxyflavone dyes with a large dipole moment.[95,104] These dyes allow simultaneous characterization of membrane polarization and hydration (lipid acyl chain packing).[105] Parallel capacity to sense lipid ordering and membrane polarization was proven advantageous in model system and living cells using F2N12S dye (Fig. 16.2), for example, enabling *in situ* analysis of apoptotic cells.[106]

Diphenylhexatriene (DPH): Diphenylhexatriene (DPH) is a frequently used membrane probe due to its low solubility and quenched emission in water. The probe is localized in the interior of the bilayer but can also be localized at the lipid–water interface when tetramethylammonium group is attached. Because of its preferential localization in the acyl chain region, it has been used to study lipid ordering or viscosity of the bilayer (see below) and the overall bilayer dynamics. It was assumed for a long time that DPH is due to its prolonged shape preferentially oriented with the long axis perpendicularly to the bilayer surface.[110,111] However, as shown by van der Heide,[112] a huge fraction of the dye may be oriented parallel to the bilayer and resides between the sheets in the middle of the bilayer, whereas a minor fraction is oriented as assumed previously. This phenomenon probably depends on actual lipid composition and may have consequences on the interpretation of the data acquired employing DPH as lipid ordering-sensitive probe.

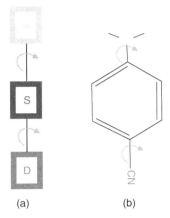

(a)   (b)

Figure 16.4 Illustration of molecular rotor function. A scheme of rotor (a) and example molecule – julolidine DCVJ (b). An acceptor (A) is connected to the donor (D) via flexible spacer (S). Higher viscosity of the environment reduces rotations around indicated bonds leading to increased quantum yields of fluorescence. (*See color plate section for the color representation of this figure.*)

16.4.2.3 Membrane Viscosity – Molecular Rotors: In contrast to solvato- and electrochromic dyes, molecular rotors sense local membrane viscosity due to the orientation-dependent dipole of the chromophore. In a push–pull chromophore of a molecular motor, the electron donor unit and acceptor can rotate relative to each other (Fig. 16.4) depending on local viscosity. The relative orientation of these two units defines emission quantum yield.[113]

Viscosity of the membranes differs from that of commonly known viscosity of bulk solutions. Viscosity is by definition a macroscopic parameter. Its use on the nanoscale (membrane) is an approximation that merely provides a best estimate. Herein, the term membrane viscosity refers to lipid packing and mobility in membranes, especially, how they affect rotation of fluorescent probes – molecular membrane rotors.[12]

Viscosity of cell compartments was historically investigated using fluorescence anisotropy[114,115] and fluorescence recovery after photobleaching.[116,117] In order to improve the precision of local membrane viscosity measurements, fluorescent molecular rotors with affinity for membranes were developed.[118] A natural dye, Nile Red, shows viscosity-dependent fluorescence,[119] but early studies indicated the need of more sensitive probes for in-cell studies. A family of julolidine rotors was developed by Haidekker *et al.*[120,121]. The fluorescent molecular rotor, farnesyl-(2-carboxy-2-cyanovinyl)-julolidine (FCVJ), was established as a highly sensitive viscosity sensor in model membranes.[122] In parallel, Kuimova *et al.* explored the viscosity-dependent fluorescence of porphyrin dimers[123] and meso-substituted BODIPY.[124] While both rotors provide good contrast in glycerol solutions with varying viscosity, more experimental evidence for the use in cells is required. Similarly, ROBOD, another BODIPY-based molecular rotor still awaits its full characterization in model and cell systems.[125] Our recent work has focused on the optimization of BODIPY-based rotor properties,[12] but dramatic changes in the orientation of probes and their localization within the lipid bilayer caused by the variation of lipid composition of membranes prevented to turn these probes into standards for biophysical and live-cell measurements.

16.4.2.4   Redox/Proton Transfer Sensing Probes: Oxidoreductive and proton transfer processes stay at the heart of living cells (energy and metabolism). Some oxidoreductive and proton transfer processes of cell signaling were described to be localized at the surface of cellular membranes. It was, therefore, important to monitor redox potential at the membrane. Imaging helps to define local changes in cellular compartments but requires a specific, in most cases fluorescent, probe(s).

Ever since the pH-sensitive character of fluorescein was implemented to membrane studies,[126] covalent attachment of pH-sensitive chromophores to lipid head groups was attempted for the visualization of pH changes at or close to the membrane. Fluorescein (Fl) undergoes protonation in bulk solution according to $Fl^{2-} + H^+ \leftrightarrow FlH^-$. While di-anionic form fluoresces in the solution, the protonated form does not emit any light. This property can be used to measure the proton transfer kinetics by FCS.[127,128] The fluorescence fluctuations of fluorescein, which are monitored by this technique, are not caused only by diffusion of the probe through the focal spot and conversion into the triplet state but also by the protonation/deprotonation of the di-anionic form. This approach has been applied to study protonation kinetics close to the lipid bilayer, where the kinetics depends strongly on the ionic strength and the properties of the bilayer.[127] Of note, care must be taken due to fast photobleaching rate of the fluorescein.[128,129] Since the experiments should be carried out at a pH close to the pK_a of the corresponding dye, we list here a few probes with different pK_a values. For the instance, pK_a for carboxynapthofluorescein equals 7.6,[130] fluorescein 6.4,[127] Oregon green 4.7,[127] erythrosine 4.1,[131] and for eosine Y 2.9.[131]

Dramatic development of green fluorescent protein (GFP) variants led to the investigations of their pH sensitivity.[132] The apparent success of this approach and, at the same time, poor sensitivity and photophysical properties of pH-sensitive small molecule probes, reduced the interest in novel pH-sensitive small molecule dyes.[133,134]

16.4.3   Specialized Techniques Using Fluorescent Probes to Investigate Membrane Properties

16.4.3.1   Förster Resonance Energy Transfer (FRET): Homo-FRET Probes and Suitable FRET Pairs: FRET is used to measure the distance between two fluorescent molecules in model systems and cellular environment. In addition, it can provide useful information of nanoscale clustering of membrane molecules undetectable using standard fluorescence microscopy techniques with resolution limited by Abbe's law.[135] On the other hand, FRET requires careful consideration of the chromophores employed and *in silico* modeling is recommended prior to the experiment to evaluate possibilities of the method for the particular question. Due to its potential, membrane fluorescent probes were successfully applied to measure intra- and intermolecular distances in membranes, lipid domain sizes or to provide qualitative information on aggregative behavior of membrane molecules.[16,31,52,136] Here, we describe a few important issues relevant to a successful FRET experiment (in membranes).

In order to have an efficient DA/DD pair, the donor emission spectrum should effectively overlap with the acceptor emission spectrum. While this is easy to accomplish for a DA pair, it is more difficult for a DD pair, where the spectra belonging to the same molecule must overlap. BODIPY dyes exhibit such a large overlap and are ideal for (homo)-FRET for several reasons: (1) the absorption and fluorescence spectra as well as the fluorescence lifetime are insensitive to pH, polarity, and viscosity; (2) the high values of the extinction coefficient ($\varepsilon_{BODIPY-FL} = 90,000 \, dm^3/mol \, cm$, $\varepsilon_{BODIPY \, 564/570} = 138,000 \, dm^3/mol \, cm$); (3) they are relatively stable even when exposed to the illumination with intense lasers; and (4) they form efficient DD pairs with R_0 (BODIPY-FL) ≈ 5.7 nm and R_0 (BODIPY 564/570) ≈ 6.8 nm.[137,138]

It is less well known that BODIPY can form dimers, which may influence the photophysics of the BODIPY monomer (M). There are two types of BODIPY dimers: D_I and D_{II}. The so-called D_I dimers were observed only when two BODIPY monomers had been attached to the same macromolecule.[139,140] D_I dimers do not exhibit fluorescence and their absorption spectrum is slightly blueshifted with the peak maximum at 477 nm. In D_I dimers, the BODIPY rings stack to each other with almost parallel $S_0 \leftrightarrow S_1$ transition dipoles moments. The structure of the so-called D_{II} dimers is different. The BODIPY rings reside in the same plane with the angle of $S_0 \leftrightarrow S_1$ transition dipoles moments of $55°$.[140] The D_{II} dimers form spontaneously when BODIPY molecules are solubilized at sufficiently high concentration in lipid bilayers. Under certain conditions, evidence for D_{II} dimer formation exists even at relatively low probe to lipid ratios (1:1000). The D_{II} absorption spectrum is redshifted with the peak maximum centered at 580 nm. In contrast to D_I dimers, D_{II} dimers fluoresce with a broad emission maximum at around 630 nm. One has to consider the following photophysical processes when exciting BODIPY in lipid vesicles:[140]

$$M + h\nu \rightarrow M^*; D_{II} + h\nu \rightarrow D_{II}^*; M^* + M \rightarrow M + M^*; M^* + D_{II} \rightarrow M + D_{II}^*;$$
$$D_{II}^* + D_{II} \rightarrow D_{II} + D_{II}^*.$$

Similarly to BODIPY dyes, probes from ATTO, Abberior and Alexa family exhibit substantial overlap of absorption with emission spectra and can be used in homo-FRET studies as well.

On the contrary to suitable DD pairs, there exist a vast number of optimal DA pairs. In recent years, FRET has become popular in detection of L_o domains in lipid bilayers and, to date, is the most powerful technique in the size determination of L_o nanodomains.[16,65,69,141] Demands that are put on an ideal DA pair have been mentioned above. Probes that are efficiently segregated into one of the phases are necessary for the estimation of L_o domain sizes by FRET. This can be exemplified by FRET combined with Monte Carlo simulations (Figs. 16.5 and 16.6). Donors and acceptors that are excluded from the domains with $K > 0.01$ cannot resolve domains within a broad range of domain radii 0–5 R_0. On the other hand, when both donors and acceptor reside in the domains, K higher than 10 is necessary for

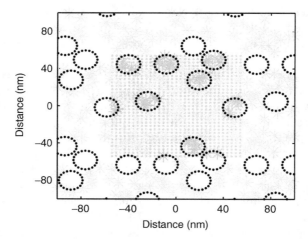

Figure 16.5 Illustration of a typical Monte Carlo experimental setup for the calculation of FRET efficiency for circular nanodomains (black dotted rings) in the bilayer. The partition coefficients for donors (green stars) and acceptors (red stars) distributed between the domains and the remaining bilayer were $K_D = 1000$ and $K_A = 0.001$. The replicated part of the bilayer is displayed by means of the blue dots. (*See color plate section for the color representation of this figure.*)

the visualization of nanodomains with $R \approx R_0$. The last possible case when donors and acceptors are segregated into the opposite phases appears as the most favorable, since already existing donor/acceptors pairs with realistic K can resolve domain sizes comparable to R_0 (for more details see Ref. 65). To our knowledge, the best DA pair for nanodomain studies so far seems to be cholera toxin subunit B labeled by Alexa Fluor 488 and DiD. This pair could reveal nanodomains formed by cholera toxin that had a 5 nm radius and occupied only a few percent of the entire bilayer area.[16,65]

A less well-known DA pair represents tryptophan transferring the excitation energy to BODIPY-FL. The latter probe has a weak transition at about 360 nm ($\varepsilon = 5000 \, \text{dm}^3/\text{mol cm}$) belonging to the $S_0 \leftrightarrow S_2$ transition, whose absorption spectrum effectively overlaps with the emission band of tryptophan. Due to the low $R_0 = 3.46 \, \text{nm}$, this pair is convenient especially for probing intramolecular distances and direct molecular interactions (membrane protein folding or protein/peptide membrane interaction, etc.).[142]

Typical distances that can be determined by FRET range between 1 and 10 nm. Distances beyond this range lead inevitably to small FRET efficiencies, and consequently to very similar time-resolved decays of donors for DD and DA systems, respectively. This makes the distance determination less precise. One way how to overcome this problem is to use luminescent lanthanides with long fluorescence lifetimes (milliseconds), large Stokes shifts, and R_0 up to 9 nm.[143] However, due to a low photostability and difficulties in attaching the lanthanides to molecules with biological relevance, their application is still limited.[144] An elegant solution to a low

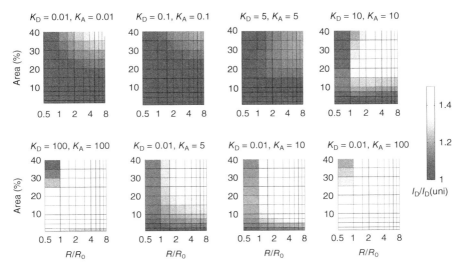

Figure 16.6 The resolution of FRET simulated by Monte Carlo simulations as a function of the domain area and the relative domain radius (i.e., R/R_0). The FRET resolution is given by the *ratio* I_D/I_D(uni) for donors (D:s) and acceptors (A:s) preferring different phases, or by the *ratio* I_D(uni)/I_D for D:s and A:s preferring the same phase. I_D denotes the steady-state intensity of donors for probe distributions determined by K_D and K_A, whereas I_D(uni) denotes the steady-state intensity for probes that are homogeneously distributed over the entire bilayer surface. The D-lifetime was 6 ns in the L_o and L_d phase and the probes were localized at the lipid–water interface. The colors correspond to the following conditions: (dark gray) the domains are beyond the resolution of TR-FRET and SS-FRET for the *ratio* < 1.05; (gray) the domains are close to the detection limit of TR-FRET and beyond the limit of SS-FRET when 1.05 < *ratio* < 1.1; (bright gray) for the range 1.1 < *ratio* < 1.2 the domains can be resolved by TR-FRET and are close to the detection limit of SS-FRET; (white) finally, for *ratio* > 1.2, the domains are detectable by TR-FRET and SS-FRET. Values exceeding 1.5 are displayed with the same color at the limiting value.

FRET sensitivity is to make use of the triplet state, which is common to the most popular fluorescent dyes (e.g., rhodamine, Alexa Fluor, ATTO dyes). The population of the triplet state depends on the excitation rate. Since the triplet state is long-lived, even small changes in the excitation rate of acceptors (=FRET rate) may lead to large changes in the occupancy of the acceptor triplet state. Consequently, measuring the triplet state kinetics of acceptors (by FCS) may extend the range of distances that are accessible by FRET using common fluorescent dyes.[145]

16.4.3.2 *Membrane Curvature Sensing Probes:* Morphology is together with lateral organization and physical properties of membranes at the center of interest of cell membrane biologists. Compared to the flat surface, curvature and tubulation provide different environment for biological processes and are indispensable for large-scale membrane dynamics such as membrane fission and fusion.[146] *In vitro*, lipid monolayer composed of lipids with conical or inverted-conical shape

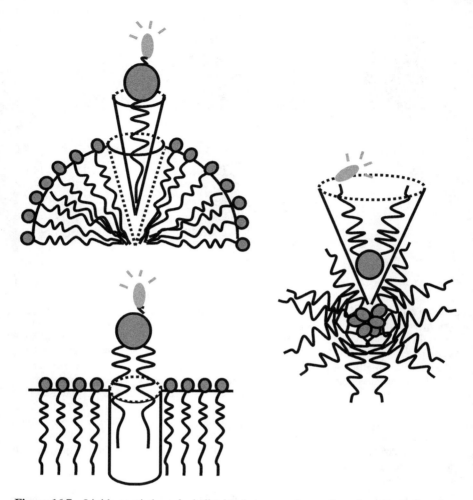

Figure 16.7 Lipids consisting of a bulky head group and a small acyl part (e.g., lysophos-phatidylcholines) self-assemble into structures with positive curvature, exemplified by the pic-ture of a micelle, whereas cylindrically shaped lipids (e.g., diacylphosphatidylcholines) form planar structures such as a lipid bilayer (bottom left). The lipid molecules containing a small head group and a bulky acyl tail form inverted structures with negative curvature (top right). Green structure indicates putative fluorescent label for the visualization of such membrane morphologies. (*See color plate section for the color representation of this figure.*)

(Fig. 16.7) spontaneously form curvature.[147,148] In lipid bilayers, unconvincing rela-tion between shape of lipids and their sorting into or out of curved membranes was found.[149,150]

Live cell imaging is preferred for studies of membrane bending and the involve-ment of curvature in biological processes (e.g., phagocytosis). This opens up demand for curvature-specific fluorescent probes since submicron size of curved membranes excludes direct visualization of membrane curvature by transmitted light microscopy. At the same time, more precise electron microscopy suffers from chemical and physical treatment of samples, which may dramatically influence membrane morphology. Unfortunately, and in agreement with theoretical predictions,[151,152] no shape-determined sorting of lipids to curved membranes was observed.[149,150,153] It seems difficult to design a small fluorescent probe with strong preference for curved membranes. The use of amphipathic helical peptides as membrane curvature sensors was therefore suggested.[154] Curvature-sensitive peptides cannot be classified as small fluorescent biosensors, which are at the center of this book. And these fascinating structures were recently described in the extensive review.[155] Therefore, just briefly, ALPS motifs and the other protein structure, BAR domains, adhere to curved membranes via relaxed lipid packing and electrostatic forces, respectively. The advantage of these structures is their selectivity for different bending, for example, <100 nm in diameter for ALPS motif- and >100 nm for BAR domain-based probes.[154,156,157] A serious limitation is the impact of lipid species on the efficiency of the probe binding to the curved membrane, for example, positively charged phospholipids are essential for membrane association of BAR domains.[158–161] Even though the role of ALPS motifs and BAR domains in biological processes is well documented,[159,162,163] their potential to function as more general membrane curvature probes have not been explored in detail yet.

Edge regions of planar model membranes and in pores represent another example of membrane curvature. Such curved membranes have been investigated using different lipid-based membrane probes with variable success.[22,164] Bicelles, that is, molecular assemblies resembling a disc, may serve as an illustration for the influence of the molecular shape on the way the molecules are self-assembled in a model system. While the cylindrical DPPC molecules constitute the planar regions of bicelles, the 1,2-distearoyl-*sn*-glycero-3-phosphoethanolamine-*N*-[amino(polyethylene glycol)-2000] (DSPE-PEG2000) lipid derivatives with a bulky head group and conical shape form preferentially the highly curved bicellar rim.[165,166] Similarly, conical monopalmitoyl phosphatidylethanolamine (mPE) molecules tend to pack into the outer leaflet of small unilamellar vesicles at the ratio of 3:1.[167] Development of fluorescent probes that would exhibit increased/decreased affinity to curved bilayer regions seems thus possible at least for these extremely curved structures. In Ref. 22, it has been shown that bicelles represent a suitable model for testing the probe's preferences regarding curvature by means of FRET. In this study, affinity of conically shaped mPE molecules labeled in the head group region by either BODIPY-FL-C5 (donor) or 564/570 BODIPY (acceptor) and cylindrically shaped diacylphosphatidylethanolamine (dPE) molecules labeled in the head group region by the same donor or acceptor has been investigated. Redistribution of mPE and dPE molecules on the bicellar surface has been shown to be comparable with the partition coefficient (the rim vs the plane) of less than 10. This study demonstrates just other difficulties in designing lipid-based fluorescent probes that would have substantial affinity (i.e., $K > 100$) to the highly curved bilayer regions.

Finally, fluorescence polarization microscopy[168] was employed to investigate the orientation of standard membrane probes (e.g., DiI-C18 or di-8-ANEPPS) in curved membranes.[169-171] Again, the sensitivity of this setup is inadequate for general membrane morphology studies. But all aforementioned methods of membrane curvature imaging are still in their early stage of development and may turn up to be successful in the future.

16.5 CONCLUSIONS

A panel of useful fluorescent probes for membrane studies has been generated over the years. Using such lipophilic fluorescent dyes, the behavior of specific lipids can be approximated from diffusion measurements in model and cell membranes. Their organization in higher order assemblies such as nanodomains can be tested. In addition, local membrane environment such as lipid ordering, lipid phase, pH, or curvature can be investigated selecting the appropriated probe. Using fluorescence, these features can be studied in model membranes but also compared to living cells. Currently, efforts are made to correlate membrane properties with data from assays focused on the function of cellular processes. To date, all studies underline the importance of heterogeneous membrane environment for biological processes, features that can almost exclusively be tested using fluorescence in living cells.

16.6 PROSPECTS AND OUTLOOK

Herein presented hydrophobic fluorescent probes provide excellent tools to study membrane properties and dynamics of model and cell membranes. Current versions already represent very advanced forms of the probes for most of the described applications, but their limitations are significant (see above). Development of improved or novel membrane fluorescent probes is still extremely needed especially since a number of probes has been tested and successfully applied in model membrane systems but are unsuitable for biological samples.

The other direction that awaits its fulfillment is systematic comparison of model and cellular membranes. It is known that lipid (and protein) composition of membranes differs between various cell types. It would be interesting to see how much lipid ordering or membrane curvature varies between cells such as neurons, muscle, or lymphoid cells. Can we find unifying principles? Or have cells evolved significantly different environment to accomplish their specific tasks in, for example, the human body? Have cancer cells similar, but at the same time unique, organization of their membranes as suggested by proteo- and lipidomics? These and many other exciting questions concerning membranes can be investigated using existing or future lipophilic fluorescent dyes and we hope this book chapter will help to accelerate membrane research.

ACKNOWLEDGMENTS

This work was financially supported by the Czech Science Foundation via grants: 15-06989S (MC) and 14-03141 (RŠ). MC would also like to acknowledge Purkyne Fellowship of the Czech Academy of Sciences.

REFERENCES

1. Cebecauer, M.; Spitaler, M.; Serge, A.; Magee, A. I. *J. Cell Sci.* **2010**, *123*, 309.

2. Eggeling, C.; Ringemann, C.; Medda, R.; Schwarzmann, G.; Sandhoff, K.; Polyakova, S.; Belov, V. N.; Hein, B.; von Middendorff, C.; Schonle, A.; Hell, S. W. *Nature* **2009**, *457*, 1159.

3. Digman, M. A.; Sengupta, P.; Wiseman, P. W.; Brown, C. M.; Horwitz, A. R.; Gratton, E. *Biophys. J.* **2005**, *88*, L33.

4. Kapusta, P.; Wahl, M.; Benda, A.; Hof, M.; Enderlein, J. *J. Fluoresc.* **2007**, *17*, 43.

5. Axelrod, D. *J. Cell Biol.* **1981**, *89*, 141.

6. Bastos, A. E.; Scolari, S.; Stockl, M.; Almeida, R. F. *Methods Enzymol.* **2012**, *504*, 57.

7. Frigault, M. M.; Lacoste, J.; Swift, J. L.; Brown, C. M. *J. Cell Sci.* **2009**, *122*, 753.

8. Gustafsson, M. G. *Proc. Natl. Acad. Sci. U. S. A.* **2005**, *102*, 13081.

9. Oddos, S.; Dunsby, C.; Purbhoo, M. A.; Chauveau, A.; Owen, D. M.; Neil, M. A.; Davis, D. M.; French, P. M. *Biophys. J.* **2008**, *95*, L66.

10. Owen, D. M.; Neil, M. A.; French, P. M.; Magee, A. I. *Semin. Cell Dev. Biol.* **2007**, *18*, 591.

11. Schutz, G. J.; Sonnleitner, M.; Hinterdorfer, P.; Schindler, H. *Mol. Membr. Biol.* **2000**, *17*, 17.

12. Olsinova, M.; Jurkiewicz, P.; Poznik, M.; Sachl, R.; Prausova, T.; Hof, M.; Kozmik, V.; Teply, F.; Svoboda, J.; Cebecauer, M. *Phys. Chem. Chem. Phys.* **2014**, *16*, 10688.

13. Sezgin, E.; Schwille, P. *Mol. Membr. Biol.* **2012**, *29*, 144.

14. LaRocca, T. J.; Pathak, P.; Chiantia, S.; Toledo, A.; Silvius, J. R.; Benach, J. L.; London, E. *PLoS Pathog.* **2013**, *9*, e1003353.

15. Veatch, S. L.; Keller, S. L. *Biophys. J.* **2003**, *85*, 3074.

16. Stefl, M.; Sachl, R.; Humpolickova, J.; Cebecauer, M.; Machan, R.; Kolarova, M.; Johansson, L. B. A.; Hof, M. *Biophys. J.* **2012**, *102*, 2104.

17. van Zanten, T. S.; Gomez, J.; Manzo, C.; Cambi, A.; Buceta, J.; Reigada, R.; Garcia-Parajo, M. F. *Proc. Natl. Acad. Sci. U. S. A.* **2010**, *107*, 15437.

18. Owen, D.; Gaus, K.; Magee, A.; Cebecauer, M. *Immunology* **2010**, *131*, 1.

19. Lingwood, D.; Ries, J.; Schwille, P.; Simons, K. *Proc. Natl. Acad. Sci. U. S. A.* **2008**, *105*, 10005.

20. Sezgin, E.; Levental, I.; Grzybek, M.; Schwarzmann, G.; Mueller, V.; Honigmann, A.; Belov, V. N.; Eggeling, C.; Coskun, U.; Simons, K.; Schwille, P. *Biochim. Biophys. Acta* **2012**, *1818*, 1777.

21. Sampaio, J. L.; Gerl, M. J.; Klose, C.; Ejsing, C. S.; Beug, H.; Simons, K.; Shevchenko, A. *Proc. Natl. Acad. Sci. U. S. A.* **2011**, *108*, 1903.

22. Šachl, R.; Mikhalyov, I.; Gretskaya, N.; Olzynska, A.; Hof, M.; Johansson, L. B.-Å. *Phys. Chem. Chem. Phys.* **2011**, *13*, 11694.

23. Schmidt, T.; Schutz, G. J.; Baumgartner, W.; Gruber, H. J.; Schindler, H. *Proc. Natl. Acad. Sci. U. S. A.* **1996**, *93*, 2926.

24. Bag, N.; Yap, D. H.; Wohland, T. *Biochim. Biophys. Acta* **2014**, *1838*, 802.

25. Baumgart, T.; Hunt, G.; Farkas, E. R.; Webb, W. W.; Feigenson, G. W. *Biochim. Biophys. Acta* **2007**, *1768*, 2182.

26. Vats, K.; Knutson, K.; Hinderliter, A.; Sheets, E. D. *ACS Chem. Biol.* **2010**, *5*, 393.
27. Homan, R.; Pownall, H. J. *Biochim. Biophys. Acta* **1988**, *938*, 155.
28. Feigenson, G. W.; Buboltz, J. T. *Biophys. J.* **2001**, *80*, 2775.
29. Raghuraman, H.; Shrivastava, S.; Chattopadhyay, A. *Biochim. Biophys. Acta* **2007**, *1768*, 1258.
30. Menger, F. M.; Keiper, J. S.; Caran, K. L. *J. Am. Chem. Soc.* **2002**, *124*, 11842.
31. Šachl, R.; Boldyrev, I.; Johansson, L. B.-Å. *Phys. Chem. Chem. Phys.* **2010**, *12*, 6027.
32. Abrams, F. S.; Chattopadhyay, A.; London, E. *Biochemistry* **1992**, *31*, 5322.
33. Hutterer, R.; Schneider, F. W.; Lanig, H.; Hof, M. *Biochim. Biophys. Acta* **1997**, *1323*, 195.
34. de Meyer, F.; Smit, B. *Proc. Natl. Acad. Sci. U. S. A.* **2009**, *106*, 3654.
35. Fischer, R. T.; Stephenson, F. A.; Shafiee, A.; Schroeder, F. *Chem. Phys. Lipids* **1984**, *36*, 1.
36. Grechishnikova, I. V.; Bergstrom, F.; Johansson, L. B. A.; Brown, R. E.; Molotkovsky, J. G. *Biochim. Biophys. Acta* **1999**, *1420*, 189.
37. Bjorkbom, A.; Rog, T.; Kaszuba, K.; Kurita, M.; Yamaguchi, S.; Lonnfors, M.; Nyholm, T. K.; Vattulainen, I.; Katsumura, S.; Slotte, J. P. *Biophys. J.* **2010**, *99*, 3300.
38. Robalo, J. R.; do Canto, A. M.; Carvalho, A. J.; Ramalho, J. P.; Loura, L. M. *J. Phys. Chem. B* **2013**, *117*, 5806.
39. Pourmousa, M.; Rog, T.; Mikkeli, R.; Vattulainen, L.; Solanko, L. M.; Wustner, D.; List, N. H.; Kongsted, J.; Karttunen, M. *J. Phys. Chem. B* **2014**.
40. Scheidt, H. A.; Muller, P.; Herrmann, A.; Huster, D. *J. Biol. Chem.* **2003**, *278*, 45563.
41. Varma, R.; Mayor, S. *Nature* **1998**, *394*, 798.
42. Hannun, Y. A.; Obeid, L. M. *Nat. Rev. Mol. Cell Biol.* **2008**, *9*, 139.
43. de Almeida, R. F.; Fedorov, A.; Prieto, M. *Biophys. J.* **2003**, *85*, 2406.
44. Kahya, N.; Scherfeld, D.; Bacia, K.; Poolman, B.; Schwille, P. *J. Biol. Chem.* **2003**, *278*, 28109.
45. D'Auria, L.; Van der Smissen, P.; Bruyneel, F.; Courtoy, P. J.; Tyteca, D. *PLoS One* **2011**, *6*, e17021.
46. Tyteca, D.; D'Auria, L.; Der Smissen, P. V.; Medts, T.; Carpentier, S.; Monbaliu, J. C.; de Diesbach, P.; Courtoy, P.J. *Biochim. Biophys. Acta* **2010**, *1798*, 909.
47. Curatolo, W. *Biochim. Biophys. Acta* **1987**, *906*, 111.
48. Hakomori, S. *Curr. Opin. Hematol.* **2003**, *10*, 16.
49. Dietrich, C.; Volovyk, Z. N.; Levi, M.; Thompson, N. L.; Jacobson, K. *Proc. Natl. Acad. Sci. U. S. A.* **2001**, *98*, 10642.
50. Masserini, M.; Palestini, P.; Venerando, B.; Fiorilli, A.; Acquotti, D.; Tettamanti, G. *Biochemistry* **1988**, *27*, 7973.
51. Yuan, C.; Furlong, J.; Burgos, P.; Johnston, L. J. *Biophys. J.* **2002**, *82*, 2526.
52. Marushchak, D.; Gretskaya, N.; Mikhalyov, I.; Johansson, L. B.-Å. *Mol. Membr. Biol.* **2007**, *24*, 102.
53. Mikhalyov, I.; Gretskaya, N.; Johansson, L. B.-Å. *Chem. Phys. Lipids* **2009**, *159*, 38.
54. Edidin, M. *Annu. Rev. Biophys. Biomol. Struct.* **2003**, *32*, 257.
55. Klausner, R. D.; Wolf, D. E. *Biochemistry* **1980**, *19*, 6199.

56. Huang, N. N.; Florine-Casteel, K.; Feigenson, G. W.; Spink, C. *Biochim. Biophys. Acta* **1988**, *939*, 124.

57. Spink, C. H.; Yeager, M. D.; Feigenson, G. W. *Biochim. Biophys. Acta* **1990**, *1023*, 25.

58. Beck, A.; Heissler, D.; Duportail, G. *Chem. Phys. Lipids* **1993**, *66*, 135.

59. Loura, L. M.; Fedorov, A.; Prieto, M. *Biochim. Biophys. Acta* **2000**, *1467*, 101.

60. Dietrich, C.; Bagatolli, L. A.; Volovyk, Z. N.; Thompson, N. L.; Levi, M.; Jacobson, K.; Gratton, E. *Biophys. J.* **2001**, *80*, 1417.

61. Koivusalo, M.; Alvesalo, J.; Virtanen, J. A.; Somerharju, P. *Biophys. J.* **2004**, *86*, 923.

62. Juhasz, J.; Davis, J. H.; Sharom, F. J. *Biochem. J.* **2010**, *430*, 415.

63. Ayuyan, A. G.; Cohen, F. S. *Biophys. J.* **2006**, *91*, 2172.

64. Crane, J. M.; Tamm, L. K. *Biophys. J.* **2004**, *86*, 2965.

65. Šachl, R.; Humpolíčková, J.; Štefl, M.; Johansson, L. B.-Å.; Hof, M. *Biophys. J.* **2011**, *101*, L60.

66. Hammond, A. T.; Heberle, F. A.; Baumgart, T.; Holowka, D.; Baird, B.; Feigenson, G. W. *Proc. Natl. Acad. Sci. U. S. A.* **2005**, *102*, 6320.

67. Chiantia, S.; Ries, J.; Kahya, N.; Schwille, P. *ChemPhysChem* **2006**, *7*, 2409.

68. de Almeida, R. F. M.; Loura, L. M. S.; Fedorov, A.; Prieto, M. *J. Mol. Biol.* **2005**, *346*, 1109.

69. de Almeida, R. F. M.; Loura, L. M. S.; Prieto, M. *Chem. Phys. Lipids* **2009**, *157*, 61.

70. Harder, T.; Scheiffele, P.; Verkade, P.; Simons, K. *J. Cell Biol.* **1998**, *141*, 929.

71. Brown, R. E. *J. Cell Sci.* **1998**, *111*, 1.

72. Scherfeld, D.; Kahya, N.; Schwille, P. *Biophys. J.* **2003**, *85*, 3758.

73. Kahya, N.; Scherfeld, D.; Schwille, P. *Chem. Phys. Lipids* **2005**, *135*, 169.

74. Lakowicz, J. R.; Knutson, J. R. *Biochemistry* **1980**, *19*, 905.

75. Chong, P. L.; van der Meer, B. W.; Thompson, T. E. *Biochim. Biophys. Acta* **1985**, *813*, 253.

76. Smet, M.; Shukla, R.; Fulop, L.; Dehaen, W. *Eur. J. Org. Chem.* **1998**, 2769.

77. Moerner, W. E.; Plakhotnik, T.; Irngartinger, T.; Croci, M.; Palm, V.; Wild, U. P. *J. Phys. Chem.* **1994**, *98*, 7382.

78. Levental, I.; Byfield, F. J.; Chowdhury, P.; Gai, F.; Baumgart, T.; Janmey, P. A. *Biochem. J.* **2009**, *424*, 163.

79. Morales-Penningston, N. F.; Wu, J.; Farkas, E. R.; Goh, S. L.; Konyakhina, T. M.; Zheng, J. Y.; Webb, W. W.; Feigenson, G. W. *Biochim. Biophys. Acta* **2010**, *1798*, 1324.

80. Farkas, E. R.; Webb, W. W. *Rev. Sci. Instrum.* **2010**, *81*, 093704.

81. Parasassi, T.; Ravagnan, G.; Rusch, R. M.; Gratton, E. *Photochem. Photobiol.* **1993**, *57*, 403.

82. Weber, G.; Farris, F.J. *Biochemistry* **1979**, *18*, 3075.

83. Fluhler, E.; Burnham, V. G.; Loew, L. M. *Biochemistry* **1985**, *24*, 5749.

84. Balter, A.; Nowak, W.; Pawelkiewicz, W.; Kowalczyk, A. *Chem. Phys. Lett.* **1988**, *143*, 565.

85. Sýkora, J.; Kapusta, P.; Fidler, V.; Hof, M. *Langmuir* **2002**, *18*, 571.

86. Jurkiewicz, P.; Olzynska, A.; Langner, M.; Hof, M. *Langmuir* **2006**, *22*, 8741.

87. Parasassi, T.; Gratton, E.; Yu, W. M.; Wilson, P.; Levi, M. *Biophys. J.* **1997**, *72*, 2413.

88. Gaus, K.; Gratton, E.; Kable, E. P.; Jones, A. S.; Gelissen, I.; Kritharides, L.; Jessup, W. *Proc. Natl. Acad. Sci. U. S. A.* **2003**, *100*, 15554.

89. Owen, D. M.; Rentero, C.; Magenau, A.; Abu-Siniyeh, A.; Gaus, K. *Nat. Protoc.* **2012**, *7*, 24.

90. Owen, D. M.; Magenau, A.; Majumdar, A.; Gaus, K. *Biophys. J.* **2010**, *99*, L7.

91. Klemm, R. W.; Ejsing, C. S.; Surma, M. A.; Kaiser, H. J.; Gerl, M. J.; Sampaio, J. L.; de Robillard, Q.; Ferguson, C.; Proszynski, T. J.; Shevchenko, A.; Simons, K. *J. Cell Biol.* **2009**, *185*, 601.

92. Kim, H. M.; Jeong, B. H.; Hyon, J. Y.; An, M. J.; Seo, M. S.; Hong, J. H.; Lee, K. J.; Kim, C. H.; Joo, T. H.; Hong, S. C.; Cho, B. R. *J. Am. Chem. Soc.* **2008**, *130*, 4246.

93. Kaiser, H. J.; Lingwood, D.; Levental, I.; Sampaio, J. L.; Kalvodova, L.; Rajendran, L.; Simons, K. *Proc. Natl. Acad. Sci. U. S. A.* **2009**, *106*, 16645.

94. Demchenko, A. P.; Yesylevskyy, S. O. *Chem. Phys. Lipids* **2009**, *160*, 63.

95. Demchenko, A. P.; Mely, Y.; Duportail, G.; Klymchenko, A. S. *Biophys. J.* **2009**, *96*, 3461.

96. Bedlack, R. S.; Wei, M. D.; Loew, L. M. *Neuron* **1992**, *9*, 393.

97. Obaid, A. L.; Loew, L. M.; Wuskell, J. P.; Salzberg, B. M. *J. Neurosci. Methods* **2004**, *134*, 179.

98. Jin, L.; Millard, A. C.; Wuskell, J. P.; Dong, X.; Wu, D.; Clark, H. A.; Loew, L. M. *Biophys. J.* **2006**, *90*, 2563.

99. Owen, D. M.; Lanigan, P. M.; Dunsby, C.; Munro, I.; Grant, D.; Neil, M. A.; French, P. M.; Magee, A. I. *Biophys. J.* **2006**, *90*, L80.

100. Owen, D. M.; Oddos, S.; Kumar, S.; Davis, D. M.; Neil, M. A.; French, P. M.; Dustin, M. L.; Magee, A. I.; Cebecauer, M. *Mol. Membr. Biol.* **2010**, *27*, 178.

101. Matiukas, A.; Mitrea, B. G.; Qin, M.; Pertsov, A. M.; Shvedko, A. G.; Warren, M. D.; Zaitsev, A. V.; Wuskell, J. P.; Wei, M. D.; Watras, J.; Loew, L. M. *Heart Rhythm* **2007**, *4*, 1441.

102. Wuskell, J. P.; Boudreau, D.; Wei, M. D.; Jin, L.; Engl, R.; Chebolu, R.; Bullen, A.; Hoffacker, K. D.; Kerimo, J.; Cohen, L. B.; Zochowski, M. R.; Loew, L. M. *J. Neurosci. Methods* **2006**, *151*, 200.

103. Yan, P.; Xie, A. F.; Wei, M. D.; Loew, L. M. *J. Org. Chem.* **2008**, *73*, 6587.

104. Klymchenko, A. S.; Demchenko, A. P. *Phys. Chem. Chem. Phys.* **2003**, *5*, 461.

105. Klymchenko, A. S.; Mely, Y.; Demchenko, A. P.; Duportail, G. *Biochim. Biophys. Acta* **2004**, *1665*, 6.

106. Oncul, S.; Klymchenko, A. S.; Kucherak, O. A.; Demchenko, A. P.; Martin, S.; Dontenwill, M.; Arntz, Y.; Didier, P.; Duportail, G.; Mely, Y. *Biochim. Biophys. Acta* **2010**, *1798*, 1436.

107. Lentz, B. R. *Chem. Phys. Lipids* **1993**, *64*, 99.

108. Veatch, W. R.; Stryer, L. *Biophys. J.* **1977**, *17*, A69.

109. Ho, C.; Slater, S. J.; Stubbs, C. D. *Biochemistry* **1995**, *34*, 6188.

110. Parasassi, T.; Conti, F.; Glaser, M.; Gratton, E. *J. Biol. Chem.* **1984**, *259*, 4011.

111. Barrow, D. A.; Lentz, B. R. *Biophys. J.* **1985**, *48*, 221.

112. van der Heide, U. A.; van Ginkel, G.; Levine, Y. K. *Chem. Phys. Lett.* **1996**, *253*, 118.

113. Grabowski, Z. R.; Rotkiewicz, K.; Rettig, W. *Chem. Rev.* **2003**, *103*, 3899.

114. Binenbaum, Z.; Klyman, E.; Fishov, I. *Biochimie* **1999**, *81*, 921.

115. Dix, J. A.; Verkman, A. S. *Biophys. J.* **1990**, *57*, 231.

116. Vaz, W. L. C.; Stumpel, J.; Hallmann, D.; Gambacorta, A.; Derosa, M. *Eur. Biophys. J.* **1987**, *15*, 111.

117. Swaminathan, R.; Bicknese, S.; Periasamy, N.; Verkman, A. S. *Biophys. J.* **1996**, *71*, 1140.

118. Haidekker, M. A.; Theodorakis, E. A. *J. Biol. Eng.* **2010**, *4*, 11.

119. Dutta, A. K.; Xu, C.; Reith, M. E. *J. Med. Chem.* **1996**, *39*, 749.

120. Haidekker, M. A.; Ling, T. T.; Anglo, M.; Stevens, H. Y.; Frangos, J. A.; Theodorakis, E. A. *Chem. Biol.* **2001**, *8*, 123.

121. Loutfy, R. O.; Teegarden, D. M. *Macromolecules* **1983**, *16*, 452.

122. Nipper, M. E.; Majd, S.; Mayer, M.; Lee, J. C. M.; Theodorakis, E. A.; Haidekker, M. A. *Biochim. Biophys. Acta* **2008**, *1778*, 1148.

123. Kuimova, M. K.; Balaz, M.; Anderson, H. L.; Ogilby, P. R. *J. Am. Chem. Soc.* **2009**, *131*, 7948.

124. Kuimova, M. K.; Yahioglu, G.; Levitt, J. A.; Suhling, K. *J. Am. Chem. Soc.* **2008**, *130*, 6672.

125. Alamiry, M. A. H.; Benniston, A. C.; Copley, G.; Elliott, K. J.; Harriman, A.; Stewart, B.; Zhi, Y. G. *Chem. Mater.* **2008**, *20*, 4024.

126. Thelen, M.; Petrone, G.; Oshea, P. S.; Azzi, A. *Biochim. Biophys. Acta* **1984**, *766*, 161.

127. Sandén, T.; Salomonsson, L.; Brzezinski, P.; Widengren, J. *Proc. Natl. Acad. Sci. U. S. A.*, *107*, 4129.

128. Ojemyr, L.; Sanden, T.; Widengren, J.; Brzezinski, P. *Biochemistry* **2009**, *48*, 2173.

129. Sanden, T.; Salomonsson, L.; Brzezinski, P.; Widengren, J. *Proc. Natl. Acad. Sci. U. S. A.* **2010**, *107*, 4129.

130. Haugland, R. P. Handbook of Fluorescent Probes and Research Chemicals; Molecular Probes Inc., **1992**.

131. Sabnis, R. W. Handbook of Biological Dyes and Stains; John Wiley &Sons., Inc.: New Jersey, **2010**.

132. Miesenbock, G.; De Angelis, D. A.; Rothman, J. E. *Nature* **1998**, *394*, 192.

133. Li, Y.; Tsien, R. W. *Nat. Neurosci.* **2012**, *15*, 1047.

134. Choi, W.-G.; Swanson, S. J.; Gilroy, S. *Plant J.* **2012**, *70*, 118.

135. Abbe, E. *Arch. F. Mikr. Anat.* **1873**, *9*, 413.

136. Loura, L. M.; Fedorov, A.; Prieto, M. *Biophys. J.* **2001**, *80*, 776.

137. Karolin, J.; Johansson, L. B.-Å.; Strandberg, L.; Ny, T. *J. Am. Chem. Soc.* **1994**, *116*, 7801.

138. Marushchak, D.; Kalinin, S.; Mikhalyov, I.; Gretskaya, N.; Johansson, L. B.-Å. *Spectrochim. Acta A Mol. Biomol. Spectrosc.* **2006**, *65*, 113.

139. Bergström, F.; Mikhalyov, I.; Hagglöf, P.; Wortmann, R.; Ny, T.; Johansson, L. B.-Å. *J. Am. Chem. Soc.* **2002**, *124*, 196.

140. Mikhalyov, I.; Gretskaya, N.; Bergström, F.; Johansson, L. B.-Å. *Phys. Chem. Chem. Phys.* **2002**, *4*, 5663.

141. Loura, L. M. S.; Fernandes, F.; Prieto, M. *Eur. Biophys. J.* **2010**, *39*, 589.

142. Olofsson, M.; Kalinin, S.; Zdunek, J.; Oliveberg, M.; Johansson, L. B.-Å. *Phys. Chem. Chem. Phys.* **2006**, *8*, 3130.

143. Selvin, P. R. *Annu. Rev. Biophys. Biomol. Struct.* **2002**, *31*, 275.

144. Rajapakse, H. E.; Gahlaut, N.; Mohandessi, S.; Yu, D.; Turner, J. R.; Miller, L. W. *Proc. Natl. Acad. Sci. U. S. A.*, *107*, 13582.

145. Hevekerl, H.; Spielmann, T.; Chmyrov, A.; Widengren, J. *J. Phys. Chem. B*, *115*, 13360.

146. McMahon, H. T.; Gallop, J. L. *Nature* **2005**, *438*, 590.

147. Helfrich, W. *Z. Naturforsch. C* **1973**, *28*, 693.

148. Gruner, S. M. *J. Phys. Chem.* **1989**, *93*, 7562.

149. Sorre, B.; Callan-Jones, A.; Manneville, J. B.; Nassoy, P.; Joanny, J. F.; Prost, J.; Goud, B.; Bassereau, P. *Proc. Natl. Acad. Sci. U. S. A.* **2009**, *106*, 5622.

150. Tian, A.; Baumgart, T. *Biophys. J.* **2009**, *96*, 2676.

151. Risselada, H. J.; Marrink, S. *J. Phys. Chem. Chem. Phys.* **2009**, *11*, 2056.

152. Cooke, I. R.; Deserno, M. *Biophys. J.* **2006**, *91*, 487.

153. Kamal, M. M.; Mills, D.; Grzybek, M.; Howard, J. *Proc. Natl. Acad. Sci. U. S. A.* **2009**, *106*, 22245.

154. Bigay, J.; Casella, J. F.; Drin, G.; Mesmin, B.; Antonny, B. *EMBO J.* **2005**, *24*, 2244.

155. Antonny, B. *Annu. Rev. Biochem.* **2011**, *80*, 101.

156. Drin, G.; Casella, J. F.; Gautier, R.; Boehmer, T.; Schwartz, T. U.; Antonny, B. *Nat. Struct. Mol. Biol.* **2007**, *14*, 138.

157. Frost, A.; Unger, V. M.; De Camilli, P. *Cell* **2009**, *137*, 191.

158. Gruner, S. M. *Proc. Natl. Acad. Sci. U. S. A.* **1985**, *82*, 3665.

159. Middleton, E. R.; Rhoades, E. *Biophys. J.* **2010**, *99*, 2279.

160. Zimmerberg, J.; Kozlov, M. M. *Nat. Rev. Mol. Cell Biol.* **2006**, *7*, 9.

161. Arkhipov, A.; Yin, Y.; Schulten, K. *Biophys. J.* **2009**, *97*, 2727.

162. Drin, G.; Morello, V.; Casella, J. F.; Gounon, P.; Antonny, B. *Science* **2008**, *320*, 670.

163. Attard, G. S.; Templer, R. H.; Smith, W. S.; Hunt, A. N.; Jackowski, S. *Proc. Natl. Acad. Sci. U. S. A.* **2000**, *97*, 9032.

164. Smith, A. M.; Vinchurkar, M.; Gronbech-Jensen, N.; Parikh, A. N. *J. Am. Chem. Soc.* **2010**, *132*, 9320.

165. Lundquist, A.; Wessman, P.; Rennie, A. R.; Edwards, K. *Biochim. Biophys. Acta* **2008**, *1778*, 2210.

166. Sandström, M. C.; Johansson, E.; Edwards, K. *Biophys. Chem.* **2008**, *132*, 97.

167. Bhamidipati, S. P.; Hamilton, J. A. *Biochemistry* **1995**, *34*, 5666.

168. Axelrod, D. *Biophys. J.* **1979**, *25*, A92.

169. Anantharam, A.; Axelrod, D.; Holz, R. W. *Cell. Mol. Neurobiol.* **2010**, *30*, 1343.

170. Anantharam, A.; Onoa, B.; Edwards, R. H.; Holz, R. W.; Axelrod, D. *J. Cell Biol.* **2010**, *188*, 415.

171. Greeson, J. N.; Raphael, R. M. *Biophys. J.* **2009**, *96*, 510.

17

FLUORESCENT NEUROTRANSMITTER ANALOGS

JAMES N. WILSON

Department of Chemistry, University of Miami, Coral Gables, FL, USA

17.1 INTRODUCTION

Neurotransmitters are chemical messengers found in both the central nervous system (CNS) and periphery. They are responsible for a broad range of functions linked to human behavior, locomotion, cognition, and development as well as neurological diseases, addiction, and mood disorders. While neurotransmission has been studied for well over a century, the development of new methods and imaging technologies propels research into the function, distribution, and regulation of neurotransmitters. The fluorescent analogs of neurotransmitters and other neuroactive compounds discussed in this chapter are important molecular tools in this effort; when combined with imaging and detection platforms, they can provide superb spatial and temporal resolution of biomolecular processes governing neurotransmitters.

17.1.1 Structure of Neurotransmitters

Neurotransmitters and neuromodulators exhibit great structural diversity; this chapter focuses on small-molecule neurotransmitters and their fluorescent analogs or replacements. Before discussing fluorescent analogs, a brief review of the basic structure of neurotransmitters as well as the biomacromolecules that regulate their function is warranted. Neurotransmitters may be divided into two subclasses based on their chemical structures as shown in Figure 17.1. The first group, whose members

Fluorescent Analogs of Biomolecular Building Blocks: Design and Applications, First Edition.
Edited by Marcus Wilhelmsson and Yitzhak Tor.

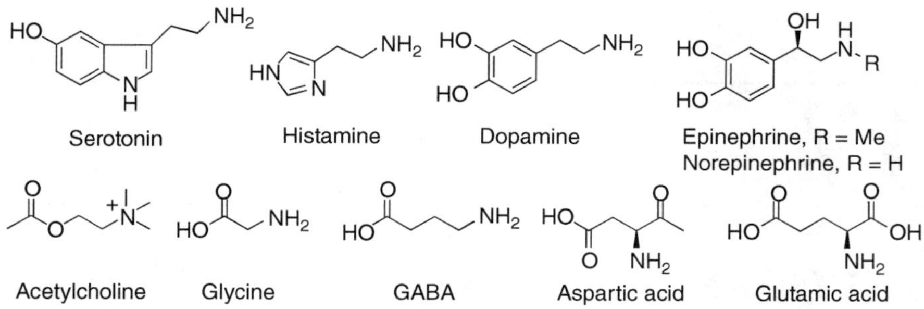

Figure 17.1   Chemical structures of selected small-molecule neurotransmitters.

possess aromatic cores, includes the tyrosine-derived catecholamines, dopamine, epinephrine (adrenaline), and norepinephrine (noradrenaline). Histamine, which is synthesized from histidine, and serotonin, which is produced from tryptophan, also belong to this group. A second group lacking aromatic units includes the amino acids, aspartic acid, glutamic acid, and glycine, and two additional neurotransmitters, acetylcholine and γ-aminobutyric acid (GABA), that are derived from serine[1] and glutamic acid, respectively. With the exception of acetylcholine, all of these neurotransmitters possess a primary or secondary amine, which is likely protonated under most conditions at physiological pH; in the case of acetylcholine, the nitrogen is quaternized via methylation. Thus, each of these structures presents a cationic element that contributes to recognition and binding to the appropriate biomolecular target. From a design perspective, the first group possesses structures more amenable to the generation of fluorescent neurotransmitter analogs as the aromatic core can be expanded and/or substituted. Indeed, this is the basis of histochemical techniques described here that are employed to visualize neurotransmitter distribution. For the second group lacking aromatic cores, the generation of fluorescent analogs may be more challenging as the addition of a fluorescent moiety represents a significant increase in steric bulk; however, several fluorescent analogs have been reported that effectively mimic these neurotransmitters.

17.1.2   Regulation of Neurotransmitters

In addition to the enzymes required for biosynthesis and degradation, neurotransmitters are regulated via several proteins that are responsible for their trafficking in and around cellular gap junctions known as synapses (Fig. 17.2). The generation of fluorescent neurotransmitter analogs requires some knowledge of the function of these biomacromolecules. Neurotransmitters are packaged into vesicles in the axons of presynaptic neurons by vesicular monoamine transporters (VMATs). During synaptic transmission, vesicles release a fraction of their neurotransmitters following fusion to the membrane and opening of a pore in a process described as "kiss-and-run."[2] The pool of released neurotransmitters binds to receptors on the postsynaptic cell and may induce either excitatory or inhibitory responses. There are

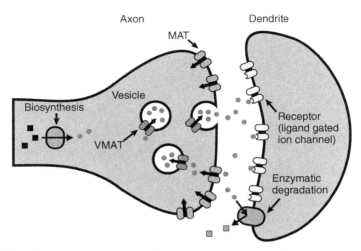

Figure 17.2 Cartoon representation of a synapse. Neurotransmitters in the presynaptic neuron are packaged into vesicles that may release a fraction of their contents into the synapse following stimulation. Neurotransmitters bind to receptors on the postsynaptic neurons; receptors may be coupled to ion channels (as depicted) or act through another signaling mechanism (e.g., a G-coupled protein receptor). Neurotransmitters may be recycled through reuptake via monoamine transporter (MAT) or enzymatically degraded (e.g., hydroxymethyltransferase, acetylcholinesterase).

numerous isoforms of neurotransmitter receptors, which include both ligand-gated ion channels and G-protein coupled receptors. Crystal structures and derived homology models of neurotransmitter receptors reveal binding pockets lined with aromatic residues that presumably stabilize the aromatic portion of neurotransmitters, as well as a conserved aspartate residue, that likely interacts with the cationic amine of the neurotransmitter.[3–5] Monoamine transporters (MATs) are responsible for the termination of neurotransmission by removing neurotransmitters from the synaptic cleft. Neurotransmitters may be recycled by transport via MATs into the presynaptic neuron and repackaged into vesicles. Examples of specific MATs include the dopamine transporter (DAT), norepinephrine transporter (NET), or the serotonin transporter (SERT). At present, crystal structures of MATs are not known, but homology models[6,7] derived from the bacterial leucine transporter[8,9] show conserved structural elements including several aromatic residues and an aspartate residue that form the binding pocket. These structural elements guide the development of fluorescent analogs of neurotransmitters and provide a means of assessing binding modes.

17.1.3 Native Fluorescence of Neurotransmitters

Several neurotransmitters, including the catecholamines and serotonin, exhibit native fluorescence; however, the relatively short excitation and emission wavelengths (see Table 17.1) of these neurotransmitters makes detection somewhat challenging

TABLE 17.1   Optical Properties of Neurotransmitters and Selected Analogs

Compound	$\lambda_{max, abs}$ (nm)	$\lambda_{max, em}$ (nm)	Notes
Neurotransmitters with endogenous fluorescence:			
Catecholamines[10]	280	320–350	Undergo photodecomposition[11]
Serotonin[10–18]	250, 315	350	$\Phi_{em} = 0.28$ (pH 7, MES buffer)
Cationic fluorescent analogs:			
5, 6[19]	480,[a] 505[b]	620,[b] 650[a]	Φ_{em} varies, up to 6.5 × brighter when bound
8[20,21]	485	620,[a] 595[b]	Φ_{em} varies, up to 20 × brighter when bound
9[22], J. N. Wilson *et al.*, Unpublished Results	391[a]	559,[a] 545[b]	Φ_{em} varies, up to 16 × brighter when bound
Alkylamine fluorescent analogs:			
Propranolol[23]	319	340,[a] 356[b]	–
7[24]	340	384	–
11[25]	391–406[c]	447–501[c]	For probe in vesicles, $\lambda_{max, em} \approx 475$ nm
12[26]	350	480	0.32 (pH 7.2, PBS)
13[27]	324, 365	460	–

[a]Values for unbound probe.
[b]Values for bound probe.
[c]Varies with solvent conditions.

against the ubiquitous background autofluorescence of cell or tissue preparations. The high concentration of neurotransmitters in specific cell types and subcellular structures aids in their visualization via UV or multiphoton excitation.[28] The absorption and emission profiles of serotonin and the catecholamines are dependent on the chemical microenvironment,[12] but in general, peak absorption of the lowest energy transition is in the range of 280–310 nm, with the emission maxima ranging from 320 to 340 nm. To achieve longer wavelength excitation and emission profiles, several strategies have been employed, including chemical modification of the native substrates *in situ*, that is, histochemical techniques, as well as the development of novel fluorescent analogs.

17.1.4   Fluorescent Histochemical Techniques

Histochemical techniques have been developed to generate fluorescent derivatives of catecholamines, serotonin, and histamine.[13–16] These methods utilize mixtures of formaldehyde, glutaraldehyde, or glycolic acid to cyclize the pendant ethylamine of the neurotransmitters generating polycyclic intermediates that are then slowly oxidized to produce conjugated products. A number of fluorescent compounds result from these reaction conditions in the complex chemical environment of whole tissue preparations; some representative examples, **1** and **2**, are depicted in

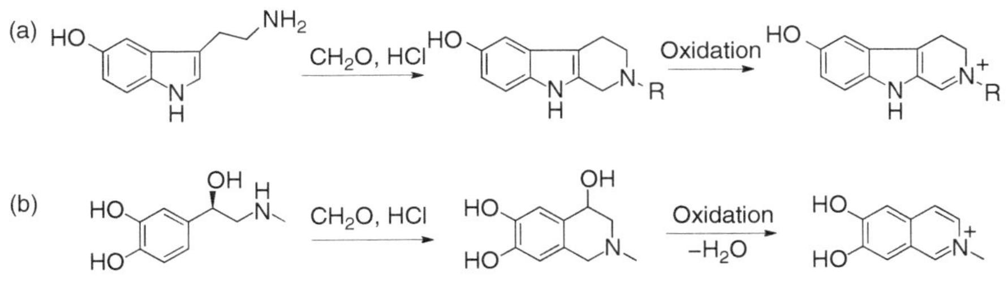

Figure 17.3 Proposed scheme for generation of a fluorescent serotonin derivative (a) and a fluorescent epinephrine derivative (b) via the formaldehyde immunohistochemical method.[13,17]

Figure 17.3.[13,17] The emission profiles of the resulting fluorophores are redshifted significantly relative to the parent compounds producing blue, green, and even yellow emission that is readily distinguished against the background autofluorescence.[16] The notion of extending conjugation through chemical modification is present in current generation probes with the aim of producing fluorophores with emission profiles well segregated from background autofluorescence that are compatible with live cell, live tissue, and possibly whole organism imaging studies.

17.2   DESIGN AND OPTICAL PROPERTIES OF FLUORESCENT NEUROTRANSMITTERS

17.2.1   Early Examples

When considering the design of fluorescent analogs of neurotransmitters, it is useful to divide them into two groups based on their chemical structure. Members of the first group possess a quaternized nitrogen and therefore bear a permanent positive charge. Members of the second group possess an amine functionality that can be protonated depending on pH or chemical microenvironment and therefore do not bear a permanent positive charge. Examples of both of these groups are depicted in Figure 17.4. Dimer **3**, containing a 3-aminopyridinium fluorophore, was reported by Maelicke *et al.* Despite the donor–acceptor substitution, **3** absorbs and emits at relatively short wavelengths (Table 17.1); however, the absorption maximum of

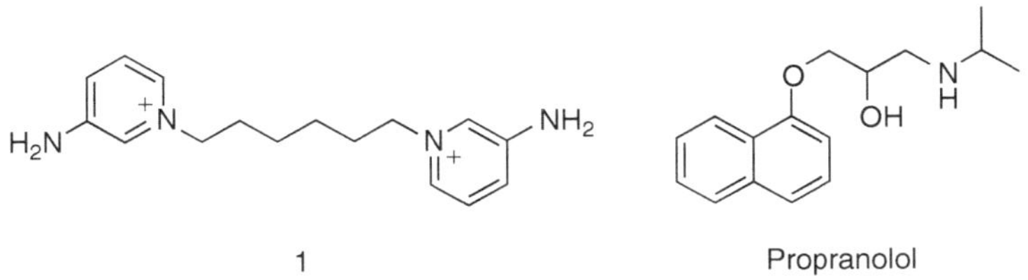

1   Propranolol

Figure 17.4 Examples of fluorescent neurotransmitter analogs including those that bear permanent cations (**1**) and those that possess alkylamines (propranolol) that may be protonated at physiological pH.

the 3-aminopyridinium probes is slightly redshifted relative to the catecholamines, enabling selective excitation of this probe in the presence of these neurotransmitters. Subsequent members of this class have expanded the donor-π-acceptor framework to push the absorption and emission maxima to lower energies. Propranolol (Fig. 17.4) is a representative member of the second class of fluorescent probes possessing amines that are protonated at physiological pH. Propranolol functions as a fluorescent analog of the epinephrine and norepinephrine and has been utilized to study the binding site topology and chemical microenvironment of β-adrenergic receptors.[29] The bicyclic alkoxynaphthalene core of propranolol results in an absorption maxima around 290 nm, with emission maxima between 340 and 360 nm, which largely overlaps with the absorption and emission bands of neurotransmitters.

17.2.2   Recent Examples

The design and synthesis of new generation fluorescent analogs of neurotransmitters has produced probes with improved optical properties such as longer wavelength absorption and emission. In addition, many of these probes exhibit optical switching, that is, binding or interaction with their biomolecular target results in an emission enhancement or wavelength shift. Propidium (**4**), hexidium (**5**), and decidium (**6**) (Fig. 17.5), reported by Johnson *et al.*, are an excellent example of probes with relatively low-energy excitation bands.[19] Furthermore, these probes exhibit attractive chromic shifts upon binding to their biomolecular target, acetylcholinesterase. A bathochromic shift, from 480 nm in aqueous buffer to 505 nm when bound, was observed for decidium in the absorption spectrum, whereas the emission underwent a hypsochromic shift from 645 nm in solution to 610 nm when bound. The optical

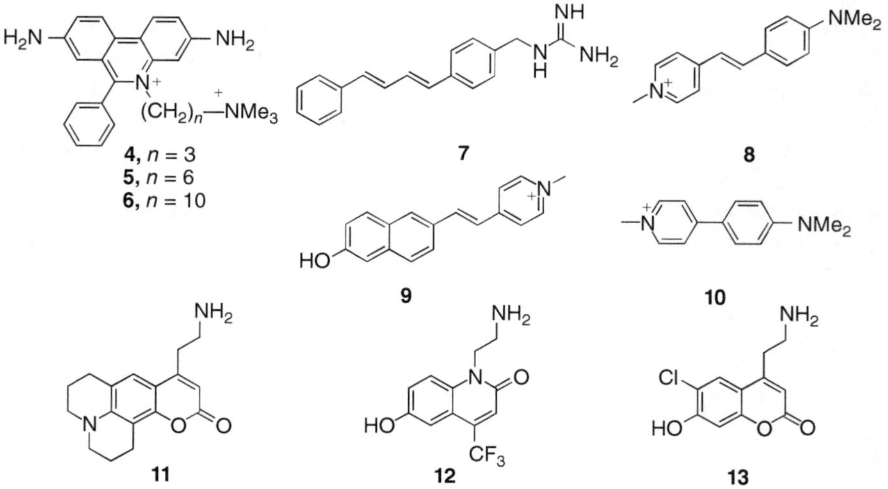

Figure 17.5   Recent examples of fluorescent neurotransmitter analogs and strategies to achieve long wavelength absorption/emission. Donor-π-acceptor motifs (**4–6, 8–13**) are a common strategy as is extension of the aromatic framework (**7–10**).

behavior of these probes is a result of the donor–acceptor-substituted phenylphenanthridium moiety that is appended to the trimethylammonium recognition unit via an alkyl tether.

Other examples of fluorescent neurotransmitter analogs have been generated in which the fluorophore is embedded in the recognition unit. Hadrich et al.[24] reported a series of benzyl guanidinium derivatives (e.g., **7**, Fig. 17.5) as NET ligands in which the fluorophore unit replaced the catechol moiety of norepinephrine. Though excitation of these analogs remained in the UV, emission was observed in the violet region of the visible spectrum. DeFelice and coworkers examined 4-(4-(dimethylamino)-styrl)-N-methylpyridinium (**8**) as a fluorescent analog of norepinephrine. The relatively compact construction of this analog allows it to function as a substrate of the NET. Despite its small size, **8** possesses relatively low-energy optical transitions, with absorption maxima between 480 and 500 nm and emission maxima ranging from 550 to 600+ nm depending on solvent or environment. The donor-π-acceptor architecture of **8** is responsible for the low-energy optical transitions and imparts optical switching between bulk solution and bound states. In aqueous solution, **8** is essentially nonemissive ($\Phi_{em} < 0.001$); however, binding to NET "turns-on" emission and bound **8** is readily visualized. A structurally related stilbazolium dye, **9**, which incorporates a hydroxynaphthalene unit in place of the dimethylaniline of **8**, was described by Brown et al.[22] The optical transitions are shifted to slightly higher energies, owing to the somewhat weaker electron-donating ability of the hydroxynaphthalene, while the emission enhancement upon binding to NET is conserved. The switchable emission is an attractive design feature of stilbazolium dyes and related donor-π-acceptor fluorophores. Following photoexcitation, these molecules may relax to twisted intramolecular charge transfer excited (TICT) state. Polar media stabilizes this state leading to emission quenching, while less polar (or more restrictive) environments limit access to this dark state. Thus, binding of **8** or **9** results in an emission enhancement allowing monitoring of binding events. Another relative of **8**, 4-(4-(dimethylamino)phenyl)-1-methylpyridinium (**10**), does not exhibit the characteristic "turn-on" emission associated with binding to its target, SERT, though accumulation of this analog in SERT expressing cells does lead to a detectable signal.[30] Thus, the donor-π-acceptor design strategy may not be universally applicable, but many probes successfully incorporate this motif and exhibit attractive optical properties and emission switching.

The donor–acceptor motif is also employed in the second group of fluorescent neurotransmitter analogs, that is, those possessing alkylamine arms. A recent example of this class was reported by Sames and coworkers, who generated a coumarin 102 derivative (**11**, Fig. 17.5) that targets VMATs.[25] The pairing of the electron-donating amine opposite the electron-withdrawing carbonyl produces a fluorophore with the lowest energy absorption band centered near 400 nm and an emission peak between 450 and 500 nm, depending on solvent. Fluorescent analogs of neurotransmitters have also been generated based on carbostyrils, the amido-homologs of coumarins. Micotto et al. reported a 1-ethylamino-7-hydroxycarbostyril (**12**, Fig. 17.5) that serves as a MAT substrate.[26] Here again, the donor-π-acceptor motif enables a relatively compact chromophore core to be excited at wavelengths spectrally

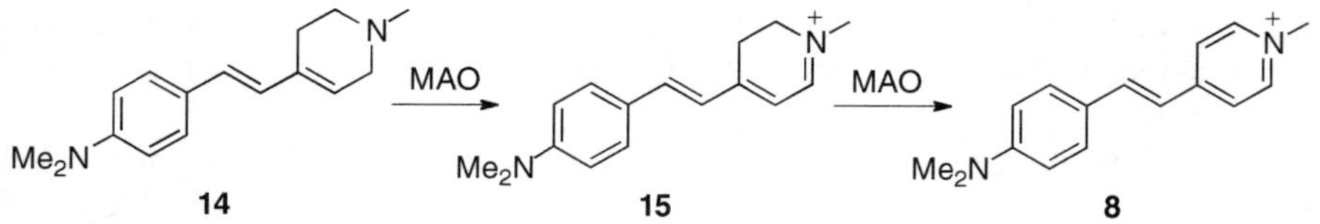

Figure 17.6 Example of a profluorescent substrate, **14**, that can be oxidized by monoamine oxidase (MAO) to yield fluorescent compounds **15** and **8**.

segregated from the native neurotransmitters. pH-responsive hydroxy-coumarins have also been reported by the Sames lab (**13**, Fig. 17.5).[27] The design of these probes allows for manipulation of their pK_a through substitution with electron-withdrawing groups (—Cl or —F) at the coumarin 6 position.

In addition to the turn-on probes described above, several examples of profluorescent neurotransmitter analogs have also been reported.[21,31–33] Figure 17.6 depicts one such example, **14**,[31] the tetrahydrostilbazole analog of **8**. In this design, the reduction of the pyridinium moiety results in the loss of the donor-π-acceptor system, which eliminates the low-energy optical transitions characteristic of **8**; **14** exhibits negligible absorption or emission above 500 nm. The action of monoamine oxidase (MAO) results in aromatization of the pyridinium ring, restoring the donor-π-acceptor conjugation.

At present, the design and synthesis of fluorescent neurotransmitter analogs has largely focused on MAT substrates. This may be the result of the inherent advantages of targeting MATs as the probes can be continually accumulated via MAT transport thereby giving a high signal against background fluorescence or nonspecific binding. A few fluorescent ligands targeting neurotransmitter receptors have been developed[19,34–38]; however, many of these are more likely to be considered fluorescently *labeled* probes as opposed to fluorescent *analogs* of neurotransmitters. One advantage to the labeling strategy is that the recognition unit may remain largely unperturbed thereby preserving the selectivity and typically high affinity (nM) of these ligands. Another advantage is the ability to choose existing fluorophores such as BODIPY, rhodamines, and fluoresceins, which have robust chemical and photochemical properties as well as standardized filter sets.

17.3   APPLICATIONS OF FLUORESCENT NEUROTRANSMITTERS

17.3.1   Probing Binding Pockets with Fluorescent Neurotransmitters

Fluorescent analogs have played a key role in exploring the structure and function of neurotransmitter transporters, receptors as well as the enzymes responsible for neurotransmitter metabolism. By employing probes that are sensitive to their chemical microenvironment or that interact with residues in the binding pocket, several groups have explored the topology of these important biomacromolecules. These studies have enabled mapping of substrate binding pockets shedding light on the supramolecular chemistry governing ligand–receptor interactions as well as the role of inhibitors or antagonists. One of the earliest fluorescent analogs, propranolol, was utilized by Cherksey and coworkers to investigate the topology of β-adrenergic receptors establishing the presence of tryptophan residues in the hydrophobic binding pocket.[23] A subsequent study established a linkage between β-adrenergic receptors and the cytoskeleton through fluorescence depolarization experiments.[29] The high sensitivity of the fluorescent phenanthridium derivatives, decidium, hexidium, and propidium enabled accurate characterization of the acetylcholinesterase active site binding pocket[19]; the presence of an unusually long gorge leading to the active site confirmed by a recent crystal structure.[39]

In addition to the examples highlighted here, which possess relatively flexible structures, several rigid fluorescent neurotransmitter analogs have been used to probe binding pockets of biomacromolecular targets of neurotransmitters. Hadrich and coworkers demonstrated that NET may accommodate ligands at least 16 Å in length in their investigation of fluorescent derivatives incorporating guanidine, cocaine, or nisoxetine.[24] The binding of 7 to NET suggested a relatively deep binding pocket, which has been supported by recent homology models[7] based on the bacterial leucine transporter.[8] The binding of another rigid fluorescent analog at NET was described by DeFelice and coworkers.[20,40] Taking advantage of the attractive "turn-on" emission of 8, they detailed the binding kinetics of this stilbazolium probe; the transport of 8 was investigated as well and is discussed below. NET is predicted to possess two binding sites, one deep binding site for the transporton and a second site that may bind inhibitors. Through colocalization experiments with GFP-fused NET, 8 was found to bind to NET with 1:1 stoichiometry and given the fact that 8 is transported via NET, it can be assumed that binding takes place at the transporton binding site. We have extended the work of Hadrich and DeFelice by exploring the interaction of several stilbazolium dyes with NET. The high sensitivity of these donor–acceptor-substituted probes makes them ideal for monitoring binding as well as for characterizing the microenvironment of the substrate pocket. Thus, the "turn-on" emission observed not only reports the binding of the ligands to NET but can also provide information about the conformation of the ligand and the overall polarity or solvent accessibility of the binding pocket. Probes based on both 8 and 9 were synthesized incorporating bulky "heads" and "tails" with varying dimensions; several dimers were generated as well in order to explore the effect of orientation and spacing of charge (see examples **16–19**, Fig. 17.7). It was determined that NET may bind stilbazolium probes as long as 19 Å, suggesting that a "head-first" orientation is the likely mode of insertion into the binding pocket. Dimers that are greater than 30 Å in length may also bind to NET, provided they also assume a "head-first" orientation. The polarity-dependent emission of the stilbazolium dyes allows a qualitative assessment of the overall binding environment. Based on Reichardt's $E_T(30)$ scale, the NET binding pocket appears to have a polarity somewhere between 34 and 37 kcal/mol (J. N. Wilson *et al.*, Unpublished Results).

17.3.2 Imaging Transport and Release of Fluorescent Neurotransmitters

Fluorescent analogs of neurotransmitters also serve as substrates for a number of MATs demonstrating the functional elasticity of these important biomacromolecules.

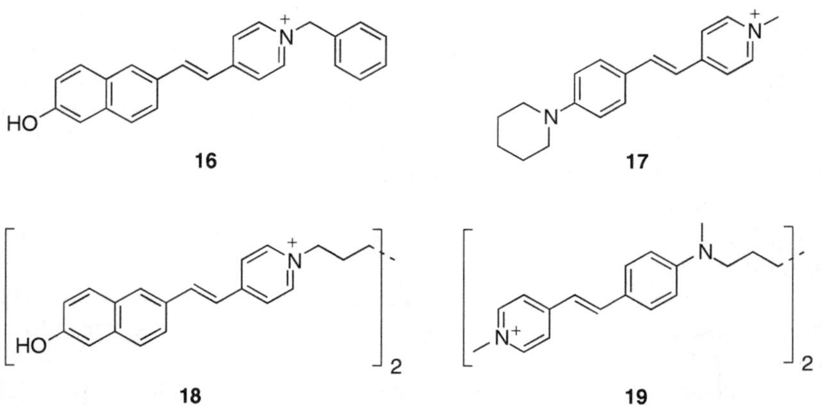

Figure 17.7 Expanded analogs of **8** and **9** have been utilized to probe the binding pocket of the norepinephrine transporter. Despite the extended dimensions, monomers **16** and **17** and dimer **18** exhibit binding at NET; **19** does not bind, indicating that a "charge first" orientation is disfavored.

An early example of a MAT substrate (or transporton[41]), **8**, was reported by DeFelice and coworkers. Probe **8** is a stilbazolium dye that exhibits high sensitivity to its chemical microenvironment. Binding to a target results in emission, whereas unbound probe remains essentially spectroscopically silent. The binding, transport, and intracellular accumulation of **8** can thus be monitored by fluorescence microscopy with excellent spatial and temporal resolution. Excitation of **8** can be achieved with fluorescein or GFP filter sets as well as the 476 or 488 nm lines of argon lasers. Emission of **8** can be captured between 500 and 625 nm depending on its chemical environment. Probe **8** was found to rapidly associate with NET stably expressed on HEK293 cells following introduction into cell media; binding to NET is readily visualized by emission localized in the plasma membrane. Binding of **8** could be reversed through the introduction of desipramine, a NET inhibitor; binding of **8** was not observed for cells that do not express NET. Probe **8** is transported via NET as evidenced by the intracellular accumulation of **8** in mitochondria of NET-expressing HEK293 cells. Pretreatment of cells with desipramine inhibits the uptake and cells that do not express NET (or the related catecholamine transporter, DAT) see severely attenuated uptake. While **8** serves as a substrate for NET, and to a lesser extent DAT, uptake via SERT is minimal. A phenylpyridinium analog, **10**, was found to be a superior SERT substrate allowing monitoring of SERT transport activity.[30] Interestingly, "turn-on" emission upon binding to the transporter was not observed for **10**, yet accumulation in mitochondria produced the same emission enhancement observed for related stilbazolium probes such as **8** and **9**.

More recently, Sames and coworkers have reported several fluorescent analogs of neurotransmitters based upon alkylamine-functionalized coumarins. Probes **11** and **13** (Fig. 17.5) are VMAT transportons that allow monitoring of vesicular uptake and release. In an elegant study, Probe **11** was demonstrated as an optical tracer for dopamine accumulation, trafficking, and release in cells and live tissue preparations.[37] Probe **11** accumulates into large dense core vesicles in murine chromaffin cells through the action of VMAT; following stimulation by K$^+$, the release of **11** could be monitored by fluorescence microscopy. In tissue preparations, **11** allows mapping, with high spatial and temporal resolution, of heterogeneities in dopamine uptake, trafficking, and release.

pH-responsive fluorescent neurotransmitter analogs, such as **13** (Fig. 17.5), have also been described.[27]. The structure of these analogs is more compact than that of **13** as the julolidine moiety of **13** is replaced by a phenol equivalent. This chemical modification influences the optical spectra of **13**. The hydroxyl functionality is responsive to changes in pH; at low pH it remains protonated, whereas at higher pH it may be deprotonated. The pK_a of the probe can be modulated by halide substitution at the 6 position of the coumarin. The absorption spectra of the protonated and deprotonated species differ substantially; the absorption maximum of the protonated species is 335 nm and that of the unprotonated species is 370 nm. Deprotonation not only produces a bathochromic shift but also results in hyperchromicity. Regardless of the protonation state, the absorption and emission maxima are at higher energies than those observed for **13**, reflecting the somewhat weaker electron-donating ability of the oxygen. The sensitivity of this hydroxycoumarin allows assessment of pH in subcellular compartments in which the probes accumulate based on ratiometric emission from 2-photon excitation; the pH of large dense core vesicles in PC-12 cells was determined to be 5.9 (Fig. 17.8).

17.3.3 Enzyme Substrates

Neurotransmitters are regulated by several enzymes that are responsible for their biosynthesis and catabolism. The action of acetylcholine is terminated by acetylcholinesterase, while the catecholamines and serotonin may be inactivated through the action of MAO. The distribution, activity, and topology of these important enzymes can be probed with fluorescent analogs of neurotransmitters. In addition

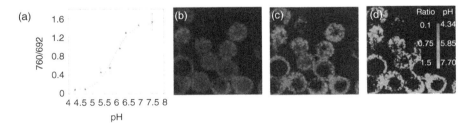

Figure 17.8 A pH-sensitive probe, **13**, exhibiting ratiometric excitation allows interrogation of intracellular pH: (a) calibration curve; (b) emission from 760 nm excitation; (c) emission from 692 excitation; (d) calibrated pH map. From Ref. [28]. Reprinted with permission from American Chemical Society. (*See color plate section for the color representation of this figure.*)

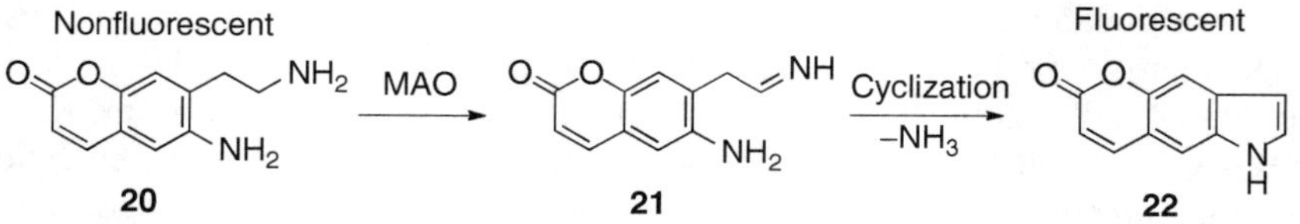

Figure 17.9 A probe of MAO activity: **20** is quenched via a TICT excited state. Oxidation via MAO results in cyclization to produce a fluorescent derivative, **22**.

to the fluorescent acetylcholinesterase ligands described above, examples of enzymatically active substrates have also been described. Meyers *et al.* synthesized a series of NBD-*n*-acylcholines that function as hydrolysable substrates of acetylcholinesterase.[32] Their studies provided additional evidence pointing to a deep hydrophobic pocket leading to the active site.[42] Unlike other fluorescent analogs that exhibit "turn-on" emission upon binding, these NBD-based analogs were quenched, possibly due to hydrogen bonding interactions with residues lining the binding pocket.

Examples of "turn-on" fluorescent neurotransmitter analogs have also been described. A reduced form of ASP+, 1-methyl-4-[4- dimethylaminophenylethenyl]-1, 2,3,6-tetrahydropyridine, **14**, was examined as a substrate of MAO by two groups.[21,31] **14** bears structural similarities to 1-methyl-4-phenyl-1,2,5,6-tetrahydropyridine (MPTP), a by-product of illicit drug synthesis that can produce Parkinsonism when oxidized by MAO.[43] **14** does not exhibit the long-wavelength absorption or emission profiles of **8** as the donor-π-acceptor motif is disrupted through the reduction of quaternized *N*-methylpyridinium. The action of MAO-A or MAO-B restores the donor-π-acceptor structure and serves to turn on the long-wavelength absorption and emission. Thus, enzyme activity may be assessed by a fluorometric assay in solution, cellular, or whole tissue assays. More recently, the group of Sames has reported profluorescent MAO substrates, such as **20** (Fig. 17.9), based upon alkylamine substituted coumarins.[33] The mechanism of emission enhancement differs from that of **14** (Fig. 17.6): **20** is likely quenched by a TICT mechanism, while the conjugation of **14** is truncated. Following oxidation of the amine, a condensation with the aniline amine produces a new fluorescent heterocycle, **22**. Compared to the previous examples, profluorescent MAO substrates more closely mimic the structures of neurotransmitters and may also serve as substrates for MATs or receptors.

17.4 CONCLUSIONS

Fluorescent neurotransmitter analogs serve as important tools that enable detailed investigations of the biomolecular machinery regulating neurotransmitter trafficking, signaling, and metabolism. Great strides have been made in the design and synthesis of these fluorescent probes with respect to optical properties, specificity, and function. It is remarkable that such a broad spectrum of synthetic structures can target the receptors, enzymes, and transporters that interact with the relatively compact architectures of native neurotransmitters. Three conclusions can be drawn from these examples presented in this chapter. First, the two key design elements necessary to generate a fluorescent neurotransmitter analog are a cationic recognition unit coupled to an aromatic cycle, which serves both as a recognition unit and fluorescent reporter. Second, neurotransmitter binding pockets may accommodate significantly expanded analogs of the endogenous substrates or ligands. Third, it is possible to tune the optical properties of fluorescent neurotransmitter analogs through chemical modification and extension of the aromatic cores. In doing so, chemists have successfully generated analogs that possess relatively long-wavelength absorption and emission profiles allowing for selective excitation and imaging of probes against the complex background of biological samples. In many cases, these analogs simultaneously incorporate switchable emission outputs that enable detection of specific binding events or neurotransmitter release.

Fluorescent neurotransmitter analogs complement existing strategies such as site-directed mutations and protein crystallography that allow us to evaluate the structure and function of the molecular machinery governing neurotransmission. In their role as molecular probes, these analogs allow mapping of binding site topology, polarity, and solvent accessibility. Most importantly, they allow for studies of biomolecular function in live cell and tissue preparations, which is key for unraveling the complex molecular processes involved in neurotransmitter signaling, synaptogenesis, and diseases of the nervous system.

17.5 PROSPECTS AND OUTLOOK

The efforts of many groups over the past few decades have demonstrated not only the feasibility, but also the utility of fluorescent neurotransmitter analogs as biochemical research tools. However, there remain several challenges and opportunities to improve the performance of these probes. At present, fluorescent neurotransmitter analogs capable of discriminating between major receptor types, that is, dopaminergic versus serotonergic, let alone receptor subtypes (e.g., $5HT_1$ vs $5HT_2$), have not been demonstrated. Small-molecule probes compatible with live cell or tissue preparations offer distinct advantages over immunohistochemical techniques, so the generation of receptor-selective fluorescent neurotransmitter analogs would be of great utility. For MATs, some progress has been made toward generating selective transporter substrates; however, it is not clear if several probes can be imaged simultaneously to comprehensively characterize a sample. Certainly, there are opportunities to fine-tune optical outputs to enable multitarget analyses. Further advances in the development of fluorescent neurotransmitter analogs will require contributions from chemistry, microscopy, pharmacology, and related disciplines making this a rich collaborative field.

ACKNOWLEDGMENTS

J. N. Wilson acknowledges the support of the James & Esther King Biomedical Research Program (1KF08).

REFERENCES

1. Rontein, D.; Nishida, I.; Tashiro, G.; Yoshioka, K.; Wu, W. I.; Voelker, D. R.; Basset, G.; Hanson, A. D. *J. Biol. Chem.* **2001**, *276*, 35523.

2. Fesce, R.; Grohovaz, F.; Valtorta, F.; Meldolesi, J. *Trends Cell Biol.* **1994**, *4*, 1.

3. Rasmussen, S. G.; Choi, H. J.; Rosenbaum, D. M.; Kobilka, T. S.; Thian, F. S.; Edwards, P. C.; Burghammer, M.; Ratnala, V. R.; Sanishvili, R.; Fischetti, R. F.; Schertler, G. F.; Weis, W. I.; Kobilka, B. K. *Nature* **2007**, *450*, 383.

4. Cherezov, V.; Rosenbaum, D. M.; Hanson, M. A.; Rasmussen, S. G.; Thian, F. S.; Kobilka, T. S.; Choi, H. J.; Kuhn, P.; Weis, W. I.; Kobilka, B. K.; Stevens, R. C. *Science* **2007**, *318*, 1258.

5. Nichols, D. E.; Nichols, C. D. *Chem. Rev.* **2008**, *108*, 1614.

6. Ravna, A. W.; Sylte, I.; Dahl, S. G. *J. Mol. Model.* **2009**, *15*, 1155.

7. Gabrielsen, M.; Sylte, I.; Dahl, S. G.; Ravna, A. W. *BMC Res. Notes* **2011**, *4*, 559.

8. Singh, S. K.; Piscitelli, C. L.; Yamashita, A.; Gouaux, E. *Science* **2008**, *322*, 1655.

9. Krishnamurthy, H.; Gouaux, E. *Nature* **2012**, *481*, 469.

10. Chang, H. T.; Yeung, E. S. *Anal. Chem.* **1995**, *67*, 1079.

11. de Mol, N. J.; Beyersbergen van Henegouwen, G. M. J.; Gerritsma, K. W. *Photochem. Photobiol.* **1979**, *29*, 7.

12. Chattopadhyay, A.; Rukmini, R.; Mukherjee, S. *Biophys. J.* **1996**, *71*, 1952.

13. Falck, B.; Hillarp, N. A.; Thieme, G.; Torp, A. *J. Histochem. Cytochem.* **1962**, *10*, 348.

14. Falck, B.; Torp, A. *Med. Exp. Int. J. Exp. Med.* **1962**, *6*, 169.

15. Ronnberg, A. L.; Hansson, C.; Drakenberg, T.; Hakanson, R. *Anal. Biochem.* **1984**, *139*, 329.

16. Furness, J. B.; Costa, M.; Wilson, A. *J. Histochemistry* **1977**, *52*, 159.

17. Bjorklund, A.; Falck, B.; Lindvall, O.; Svensson, L. A. *J. Histochem. Cytochem.* **1973**, *21*, 17.

18. Tan, W.; Parpura, V.; Haydon, P. G.; Yetang, E. S. *Anal. Chem.* **1995**, *67*, 5.

19. Berman, H. A.; Decker, M. M.; Nowak, M. W.; Leonard, K. J.; McCauley, M.; Baker, W. M.; Taylor, P. *Mol. Pharmacol.* **1987**, *31*, 610.

20. Schwartz, J. W.; Novarino, G.; Piston, D. W.; DeFelice, L. J. *J. Biol. Chem.* **2005**, *280*, 19177.

21. Song, X.; Ehrich, M.; Flaherty, D.; Wang, Y.-X.; Castagnoli, N. *Toxicol. Appl. Pharmacol.* **1996**, *137*, 163.

22. Brown, A. S.; Bernal, L. M.; Micotto, T. L.; Smith, E. L.; Wilson, J. N. *Org. Biomol. Chem.* **2011**, *9*, 2142.

23. Cherksey, B. D.; Murphy, R. B.; Zadunaisky, J. A. *Biochemistry* **1981**, *20*, 4278.

24. Dirk Hadrich, F. B.; Steckhan, E.; Bonisch, H. *J. Med. Chem.* **1999**, *42*, 8.

25. Gubernator, N. G.; Zhang, H.; Staal, R. G. W.; Mosharov, E. V.; Pereira, D. B.; Yue, M.; Balsanek, V.; Vadola, P. A.; Mukherjee, B.; Edwards, R. H.; Sulzer, D.; Sames, D. *Science* **2009**, *324*, 1441.

26. Micotto, T. L.; Brown, A. S.; Wilson, J. N. *Chem. Commun.* **2009**, 7548.

27. Lee, M.; Gubernator, N. G.; Sulzer, D.; Sames, D. *J. Am. Chem. Soc.* **2010**, *132*, 8828.

28. Maiti, S.; Shear, J. B.; Williams, R. M.; Zipfel, W. R.; Webb, W. W. *Science* **1997**, *275*, 530.

29. Cherksey, B. D.; Zadunaisky, J. A.; Murphy, R. B. *Proc. Natl. Acad. Sci. U. S. A.* **1980**, *77*, 6401.

30. Solis, E., Jr.; Zdravkovic, I.; Tomlinson, I. D.; Noskov, S. Y.; Rosenthal, S. J.; De Felice, L. J. *J. Biol. Chem.* **2012**, *287*, 8852.

31. Sablin, S. O.; Tkachenko, S. E.; Bachurin, S. O. *Dokl. Akad. Nauk SSSR* **1991**, *318*, 228.

32. Meyers, H. W.; Jurss, R.; Brenner, H. R.; Fels, G.; Prinz, H.; Watzke, H.; Maelicke, A. *Eur. J. Biochem.* **1983**, *137*, 399.

33. Chen, G.; Yee, D. J.; Gubernator, N. G.; Sames, D. *J. Am. Chem. Soc.* **2005**, *127*, 4544.

34. Johnson, D. A.; Brown, R. D.; Herz, J. M.; Berman, H. A.; Andreasen, G. L.; Taylor, P. *J. Biol. Chem.* **1987**, *262*, 14022.

35. Allen, T. A.; Narayanan, N. S.; Kholodar-Smith, D. B.; Zhao, Y.; Laubach, M.; Brown, T. H. *J. Neurosci. Methods* **2008**, *171*, 30.

36. Monsma, F. J., Jr.; Barton, A. C.; Kang, H. C.; Brassard, D. L.; Haugland, R. P.; Sibley, D. R. *J. Neurochem.* **1989**, *52*, 1641.

37. Ariano, M. A.; Monsma, F. J., Jr.; Barton, A. C.; Kang, H. C.; Haugland, R. P.; Sibley, D. R. *Proc. Natl. Acad. Sci. U. S. A.* **1989**, *86*, 8570.

38. Berque-Bestel, I.; Soulier, J.-L.; Giner, M.; Rivail, L.; Langlois, M.; Sicsic, S. *J. Med. Chem.* **2003**, *46*, 2606.

39. Dvir, H.; Silman, I.; Harel, M.; Rosenberry, T. L.; Sussman, J. L. *Chem. Biol. Interact.* **2010**, *187*, 10.

40. Schwartz, J. W.; Blakely, R. D.; DeFelice, L. J. *J. Biol. Chem.* **2003**, *278*, 9768.

41. Phanstiel, O.; Archer, J. *J. RSC Drug Discovery Ser.* **2012**, *17*, 162.

42. Jurss, R.; Maelicke, A. *J. Biol. Chem.* **1981**, *256*, 2887.

43. Langston, J. W.; Ballard, P.; Tetrud, J. W.; Irwin, I. *Science* **1983**, *219*, 979.

INDEX

Fluorescent Analogs of Biomolecular Building Blocks: Design and Applications, First Edition.
Edited by Marcus Wilhelmsson and Yitzhak Tor.
© 2016 John Wiley & Sons, Inc. Published 2016 by John Wiley & Sons, Inc.

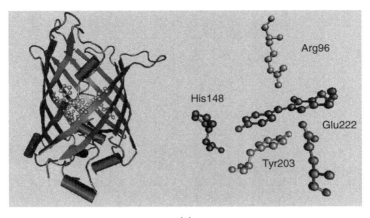

(a)

Neutral chromophore

Anionic chromophore

(b)

Figure 4.1

Fluorescent Analogs of Biomolecular Building Blocks: Design and Applications, First Edition.
Edited by Marcus Wilhelmsson and Yitzhak Tor.
© 2016 John Wiley & Sons, Inc. Published 2016 by John Wiley & Sons, Inc.

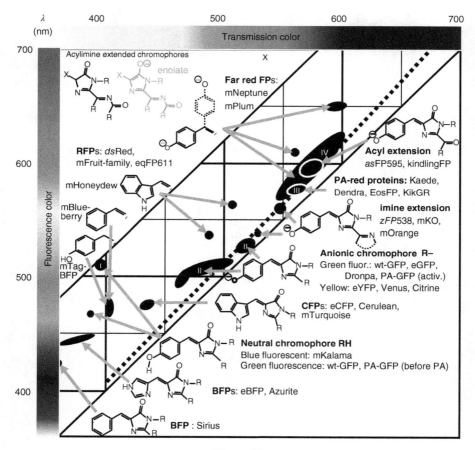

Figure 4.2

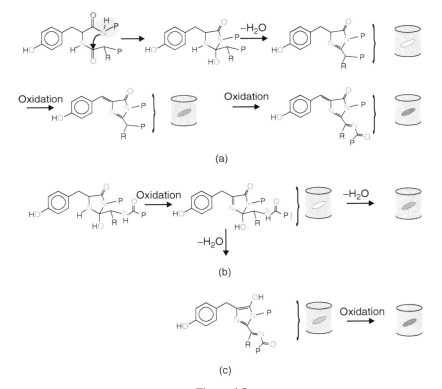

(a)

(b)

(c)

Figure 4.5

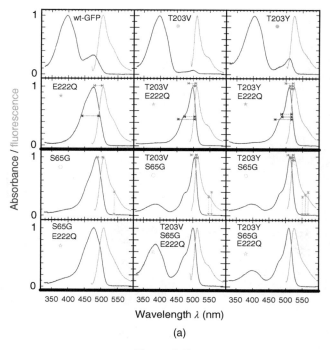

(a)

Figure 4.6A

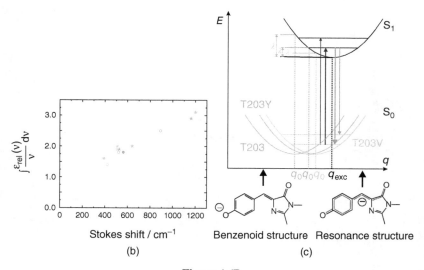

(b)   (c)

Figure 4.6B

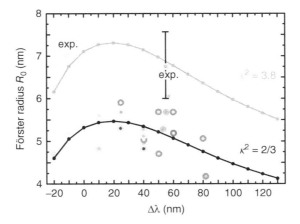

Figure 4.8

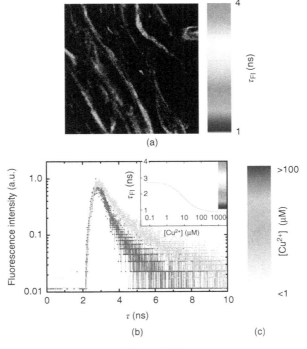

(a)

(b)   (c)

Figure 4.10

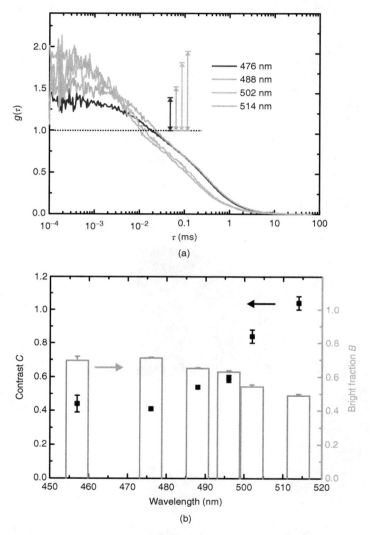

Figure 4.12

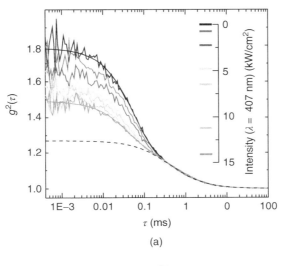

(a)

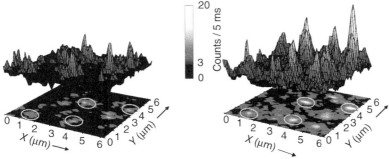

Excitation at $\lambda = 476$ nm   Double resonance excitation at $\lambda = 407$ and 476 nm

(b)

Figure 4.13

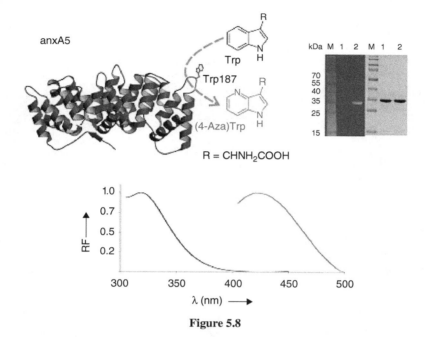

anxA5

Trp

Trp187

(4-Aza)Trp

R = CHNH₂COOH

$R = CHNH_2COOH$

Figure 5.8

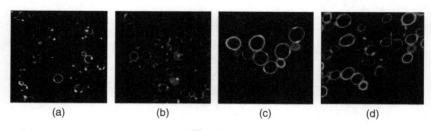

(a)   (b)   (c)   (d)

Figure 6.3

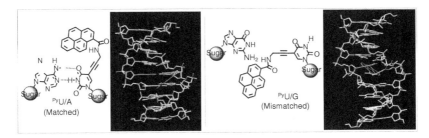

Figure 7.8

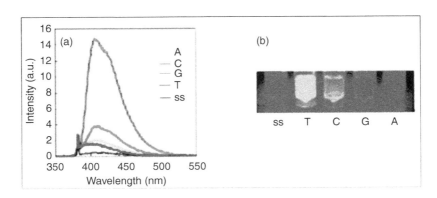

Figure 7.9

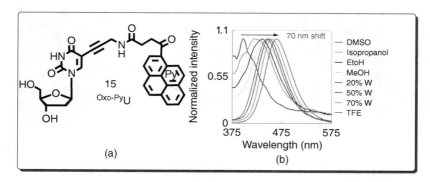

Figure 7.12

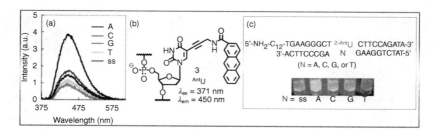

Figure 7.15

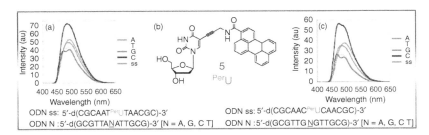

ODN ss: 5′-d(CGCAAT^{Per}UTAACGC)-3′
ODN N : 5′-d(GCGTTANATTGCG)-3′ [N = A, G, C T]

ODN ss: 5′-d(CGCAAC^{Per}UCAACGC)-3′
ODN N : 5′-d(GCGTTGNGTTGCG)-3′ [N = A, G, C T]

Figure 7.16

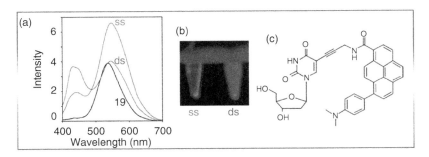

Figure 7.19

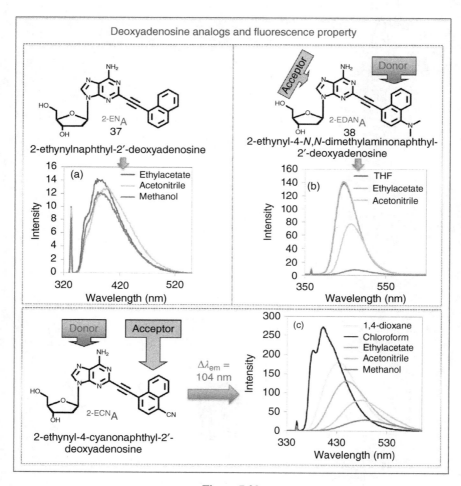

Figure 7.23

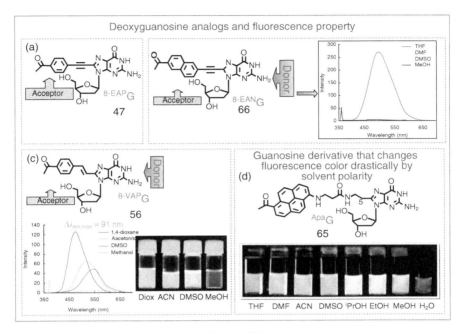

Figure 7.24

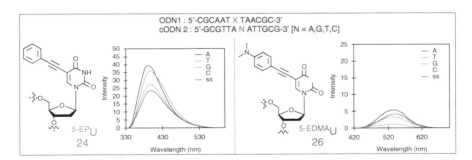

Figure 7.25

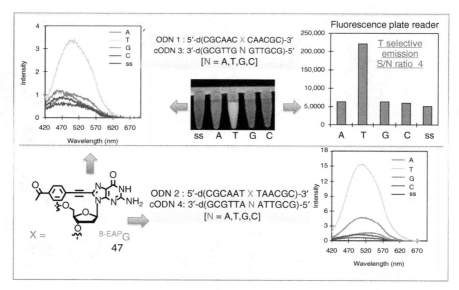

Figure 7.26

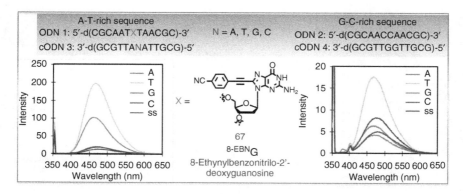

Figure 7.27

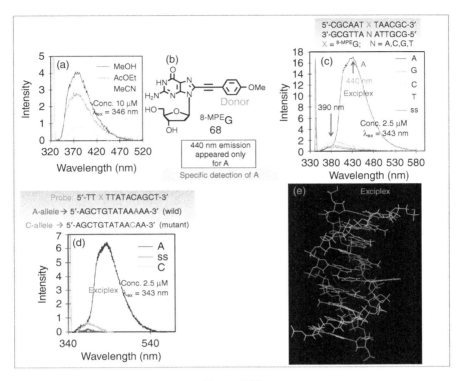

Figure 7.29

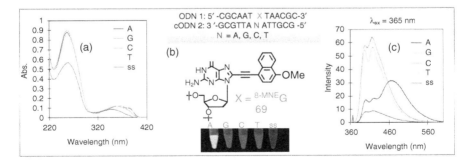

Figure 7.30

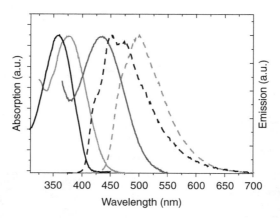

Figure 10.5

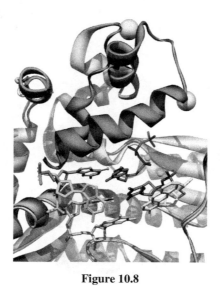

Figure 10.8

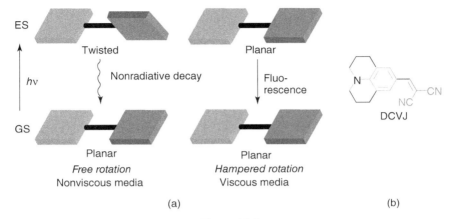

Figure 12.6

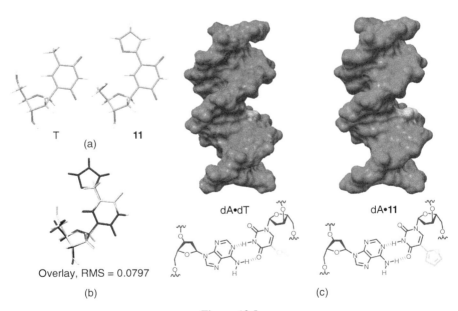

Figure 12.8

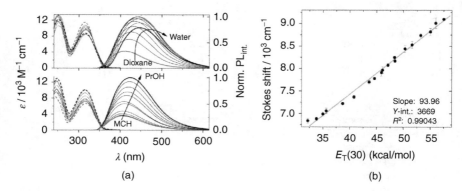

Figure 12.9

5′ – G C G – A T G – X G T – A G C – G – 3′
3′ – C G C – T A C – Y C A – T C G – C – 5′

X = 11, Y = dA or abasic site

(a)

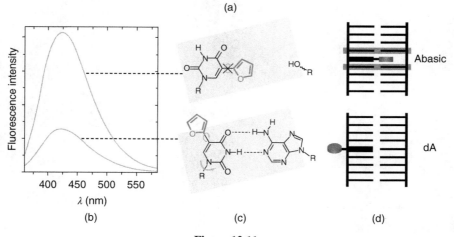

Figure 12.11

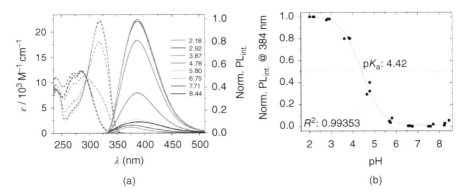

(a)

(b)

Figure 12.13

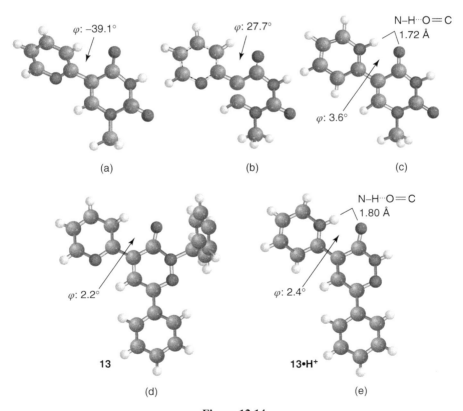

φ: −39.1°

φ: 27.7°

N–H···O=C
1.72 Å

φ: 3.6°

(a)

(b)

(c)

φ: 2.2°

13

N–H···O=C
1.80 Å

φ: 2.4°

13•H⁺

(d)

(e)

Figure 12.14

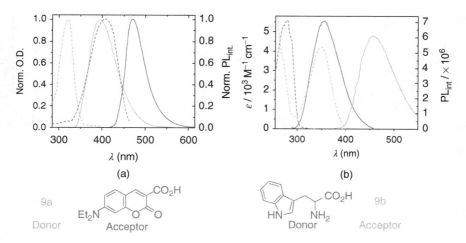

9a

Donor

Et₂N — Acceptor

Acceptor

Donor

9b

Acceptor

Figure 12.15

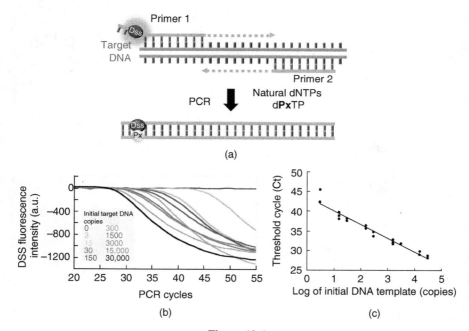

Figure 13.6

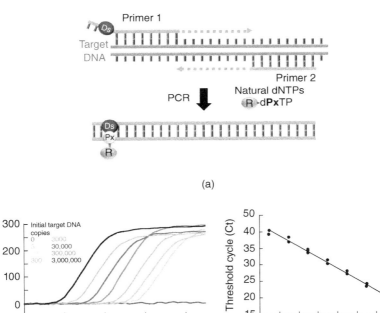

(a)

(b)

(c)

Figure 13.7

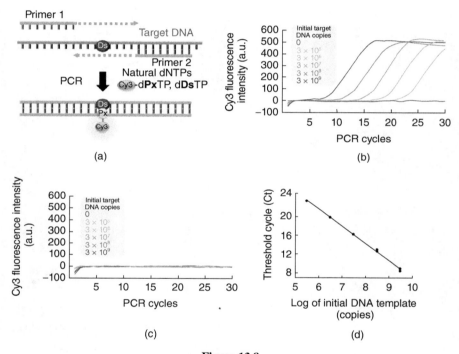

Figure 13.8

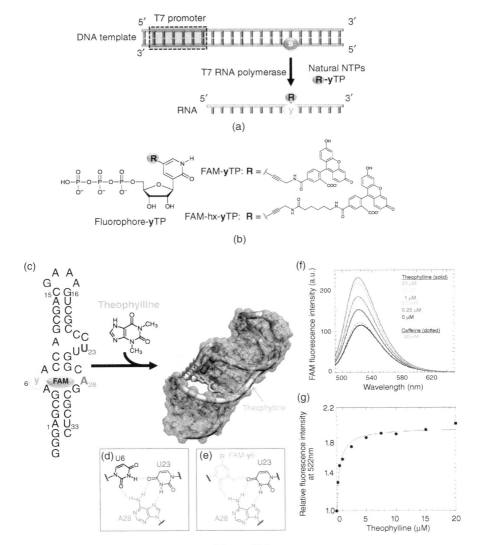

Figure 13.9

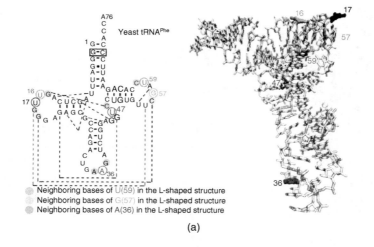

(a)

Tm values of the nonsubstituted, original tRNA

0 mM Mg^{2+} : 56.1°C

2 mM Mg^{2+} : 65.5°C

5 mM Mg^{2+} : 70.1°C

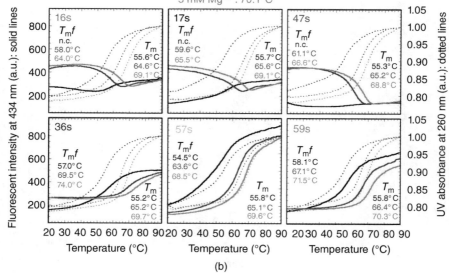

(b)

Figure 13.12

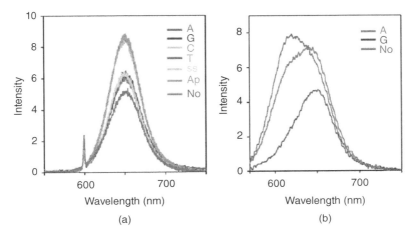

(a)   (b)

Figure 14.4

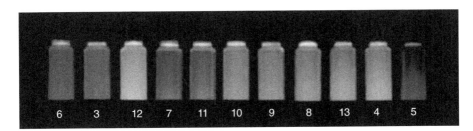

Figure 14.10

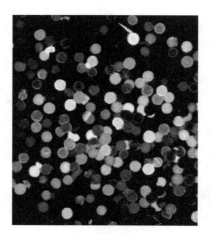

Figure 14.11

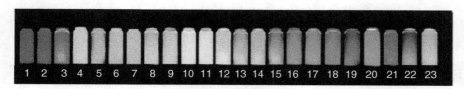

Figure 14.12

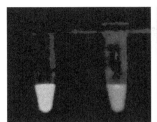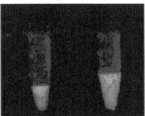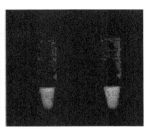

Figure 14.13

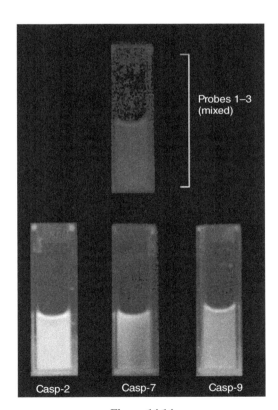

Figure 14.14

(a)   (b)

Figure 14.15

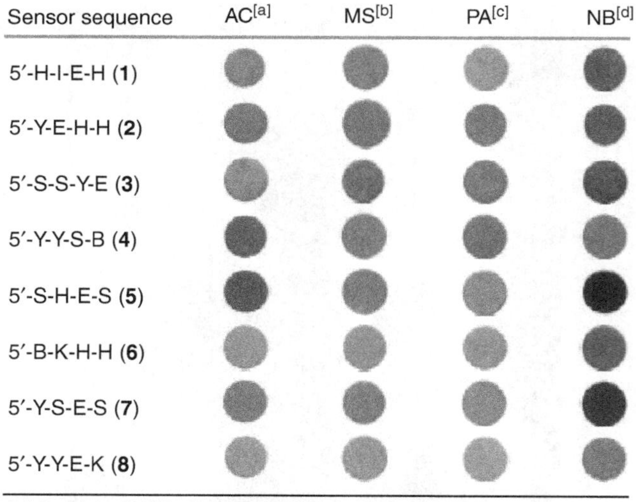

Sensor sequence	AC[a]	MS[b]	PA[c]	NB[d]
5′-H-I-E-H (**1**)	⬤	⬤	⬤	⬤
5′-Y-E-H-H (**2**)	⬤	⬤	⬤	⬤
5′-S-S-Y-E (**3**)	⬤	⬤	⬤	⬤
5′-Y-Y-S-B (**4**)	⬤	⬤	⬤	⬤
5′-S-H-E-S (**5**)	⬤	⬤	⬤	⬤
5′-B-K-H-H (**6**)	⬤	⬤	⬤	⬤
5′-Y-S-E-S (**7**)	⬤	⬤	⬤	⬤
5′-Y-Y-E-K (**8**)	⬤	⬤	⬤	⬤

[a] Acrolein. [b] Mesitylene. [c] Propionic acid. [d] Nitrobenzene.

Figure 14.16

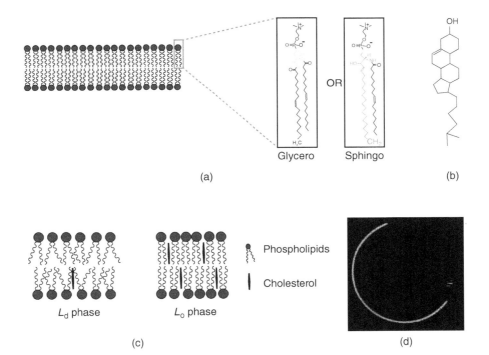

Glycero   Sphingo

(a)   (b)

Phospholipids

Cholesterol

L_d phase L_o phase

(c)   (d)

Figure 16.1

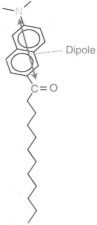

Laurdan

Figure 16.3

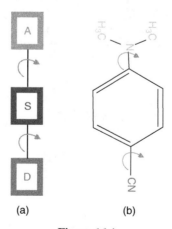

(a)   (b)

Figure 16.4

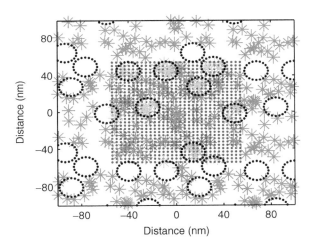

Figure 16.5

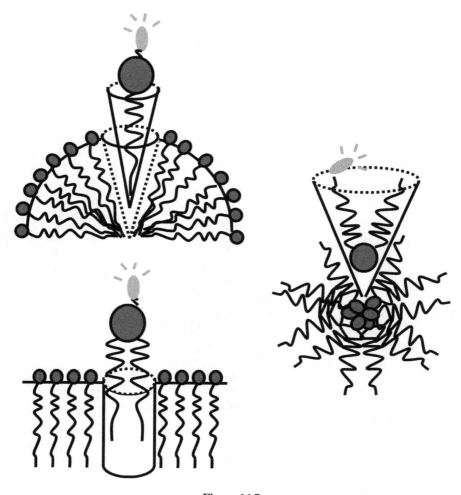

Figure 16.7

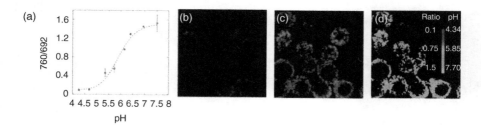

Figure 17.8